GNSS Monitoring of the Terrestrial Environment
Earthquakes, Volcanoes and Climate Change

GNSS Monitoring of the Terrestrial Environment
Earthquakes, Volcanoes and Climate Change

Edited by

Yosuke Aoki
Earthquake Research Institute, The University of Tokyo, Tokyo, Japan

Corné Kreemer
Nevada Geodetic Laboratory, Nevada Bureau of Mines and Geology, University of Nevada, Reno, NV, United States

ELSEVIER

Elsevier
Radarweg 29, PO Box 211, 1000 AE Amsterdam, Netherlands
125 London Wall, London EC2Y 5AS, United Kingdom
50 Hampshire Street, 5th Floor, Cambridge, MA 02139, United States

Copyright © 2024 Elsevier Inc. All rights are reserved, including those for text and data mining, AI training, and similar technologies.

Publisher's note: Elsevier takes a neutral position with respect to territorial disputes or jurisdictional claims in its published content, including in maps and institutional affiliations.

No part of this publication may be reproduced or transmitted in any form or by any means, electronic or mechanical, including photocopying, recording, or any information storage and retrieval system, without permission in writing from the publisher. Details on how to seek permission, further information about the Publisher's permissions policies and our arrangements with organizations such as the Copyright Clearance Center and the Copyright Licensing Agency, can be found at our website: www.elsevier.com/permissions.

This book and the individual contributions contained in it are protected under copyright by the Publisher (other than as may be noted herein).

Notices

Knowledge and best practice in this field are constantly changing. As new research and experience broaden our understanding, changes in research methods, professional practices, or medical treatment may become necessary.

Practitioners and researchers must always rely on their own experience and knowledge in evaluating and using any information, methods, compounds, or experiments described herein. In using such information or methods they should be mindful of their own safety and the safety of others, including parties for whom they have a professional responsibility.

To the fullest extent of the law, neither the Publisher nor the authors, contributors, or editors, assume any liability for any injury and/or damage to persons or property as a matter of products liability, negligence or otherwise, or from any use or operation of any methods, products, instructions, or ideas contained in the material herein.

ISBN: 978-0-323-95507-2

For information on all Elsevier publications
visit our website at https://www.elsevier.com/books-and-journals

Publisher: Candice Janco
Acquisitions Editor: Peter Llewellyn
Editorial Project Manager: Mason Malloy
Production Project Manager: Kumar Anbazhagan
Cover Designer: Christian Bilbow

Typeset by STRAIVE, India

Transferred to Digital Printing in 2024

Contents

Contributors ix
Foreword xi

1. Introduction
Yosuke Aoki and Corné Kreemer

1. History 1
2. Measuring Earth's deformation onshore 2
3. Measuring Earth's deformation offshore 3
4. Unconventional use of GNSS 5
5. GNSS and climate change 5
   5.1. Deformation in response to long- and short-term climate change 5
   5.2. Measuring glacier motions 6
   5.3. Earthquakes and volcanism triggered by climate change 6
6. Future of GNSS 7
References 8

2. Technical aspects of GNSS data processing
Jianghui Geng

1. GNSS measurements 11
   1.1. GNSS observation equations 11
   1.2. GNSS error sources 15
2. GNSS positioning 18
   2.1. Precise point positioning (PPP) 18
   2.2. Carrier-phase-based relative positioning 22
   2.3. Real-time GNSS 24
3. Atmosphere sounding 26
   3.1. Ground-based troposphere sounding 26
   3.2. Ground-based ionosphere sounding 28
   3.3. GNSS radio occultation (GNSS-RO) 31
4. GNSS reflectometry 32
   4.1. GNSS interferometric reflectometry based on single antennas 32
   4.2. GNSS reflectometry based on dual antennas 34
References 35

Part I
Monitoring earthquakes and volcanoes with GNSS

3. On the use of GNSS-inferred crustal strain accumulation in evaluating seismic potential
Corné Kreemer, Ilya Zaliapin, and Dirk Kraaijpoel

1. Introduction 42
2. Estimation of geodetic strain and moment rates 42
   2.1. From velocities to strain rates 42
   2.2. From strain rate to moment rate 43
3. Seismic moment distribution 43
4. Seismic-to-geodetic moment ratio 45
   4.1. Background 45
   4.2. Approach 46
5. Geodetic potency versus earthquake numbers 46
   5.1. Background 46
   5.2. Approach 47
6. Data 47
7. Results 48
8. Discussion 50
Appendix: Approximating cumulative seismic moment distribution 54
   A.1. Existence of two regimes 54
   A.2. Analytic results for each of the regimes 55
   A.3. Approximation equations 56
Acknowledgments 58
References 58

4. GNSS applications for earthquake deformation
Jean-Mathieu Nocquet and Martin Vallée

1. Introduction — 65
2. Observation of the static displacement induced by earthquakes from GNSS — 66
3. Observation of the coseismic static displacement using other techniques — 67
4. Modeling of GNSS static coseismic displacement for imaging earthquake slip distribution — 68
5. Dynamic displacement induced by earthquakes — 72
6. Kinematic inversion: Imaging slip history during earthquake — 73
7. High-rate GNSS as small aperture seismic array — 76
8. Some earthquake properties and perspectives — 77
9. Future issues and opportunities — 78
10. Summary points — 78
Acknowledgment — 78
References — 78

5. GNSS observations of transient deformation in plate boundary zones
Laura M. Wallace and Chris Rollins

1. Introduction — 83
2. Episodic slow slip and creep events — 83
   2.1. Slow slip events at subduction zones — 85
   2.2. Slow slip and creep events in other settings — 91
   2.3. Interplay between slow slip events and earthquakes — 91
   2.4. The ubiquity of slow slip events — 94
3. Postseismic deformation and contributions from GNSS — 94
   3.1. Viscoelastic relaxation — 98
   3.2. Afterslip — 99
   3.3. Poroelastic rebound — 101
   3.4. Toward a holistic understanding of postseismic deformation — 102
4. Conclusions and future directions — 102
References — 103

6. Earthquake and tsunami early warning with GNSS data
Brendan W. Crowell

1. Introduction — 111
2. Real-time GNSS positioning from a historical perspective — 112
3. Peak ground displacement — 114
4. Coseismic-based methods — 118
5. Algorithm development — 122
6. Future directions — 123
7. Final thoughts — 124
References — 124

7. Measuring volcano deformation with GNSS
Yosuke Aoki

1. Introduction — 129
2. Relating observed deformation to subsurface processes — 130
   2.1. Governing equations — 130
   2.2. Analytical modeling — 131
   2.3. Shear on the conduit — 136
   2.4. Adding material complexities — 136
   2.5. Some caveats — 139
3. Observation of volcano deformation — 142
   3.1. Deformation during unrest — 142
   3.2. Coeruptive deformation — 146
   3.3. Posteruptive deformation — 148
4. Observations of volcanic plumes by GNSS — 148
   4.1. Atmospheric disturbance by volcanic plumes — 148
   4.2. GNSS signal decay by volcanic plumes — 150
5. Recommendations — 152
   5.1. Continuing observations — 152
   5.2. Denser observations — 153
   5.3. Modeling — 153
Acknowledgments — 153
References — 153

8. GNSS applications for ionospheric seismology and volcanology
Kosuke Heki

1. Introduction and observation history — 161
2. GNSS-TEC observations — 161
   2.1. Phase difference and TEC — 161
   2.2. From STEC to VTEC — 162
   2.3. Finding signals related to earthquakes and volcanic eruptions — 162
   2.4. Multi-GNSS — 163
3. Ionospheric seismology — 163
   3.1. Three different atmospheric waves — 163
   3.2. Discriminating the three different waves — 164
   3.3. Direct acoustic waves from epicenter — 165

    3.4. Knowing M_w from amplitudes of
         disturbances   166
    3.5. Internal gravity wave signatures   167
  4. Ionospheric volcanology   168
    4.1. Two types of ionospheric
         disturbances by volcanic
         eruptions   168
    4.2. Type 1 disturbance   169
    4.3. Type 2 disturbance   171
  Acknowledgments   174
  References   174

Part II
Monitoring climate change with GNSS

9. GNSS applications for measuring sea level changes

Rüdiger Haas

  1. Introduction   179
  2. GNSS at traditional tide gauges   179
  3. Reflected GNSS signals   180
  4. Coastal GNSS-R with two or more
     antennas   181
  5. Coastal GNSS-IR with single
     antennas   183
  6. Sensing sea level variability with
     GNSS   185
  7. Selected highlights   186
  8. Conclusions and outlook   186
  Acknowledgment   186
  References   186

10. GNSS application for weather and climate change monitoring

Peng Yuan, Mingyuan Zhang, Weiping Jiang, Joseph Awange, Michael Mayer, Harald Schuh, and Hansjörg Kutterer

  1. Introduction   189
  2. Data and methods   190
    2.1. Tropospheric delay   190
    2.2. Water vapor retrieval   191
  3. Extreme weather events   192
  4. Diurnal cycle   194
  5. Annual cycle   195
  6. Interannual variations   197
  7. Long-term trends and climate
     change   197
  8. Summary and outlook   200
  Acknowledgments   201
  References   201

11. Monitoring of extreme weather: GNSS remote sensing of flood inundation and hurricane wind speed

Clara Chew and Chris Ruf

  1. GNSS remote sensing for flood
     inundation mapping   205
    1.1. Amplitude metrics   207
    1.2. Coherency metrics   208
    1.3. Current issues   210
  2. GNSS remote sensing for hurricane
     wind speed retrieval   210
  References   213

12. GNSS and the cryosphere

Tonie van Dam, Pippa Whitehouse, and Lin Liu

  1. Introduction   215
  2. Elastic surface displacements   217
    2.1. Theory   217
    2.2. Half-space loading models   219
    2.3. Glacier dynamics   219
    2.4. Geodynamic processes in
         cryospheric regions   220
  3. The viscoelastic response of the
     Earth to cryospheric change   220
    3.1. GIA: Three pieces to the puzzle   221
    3.2. Using GNSS to measure the
         viscoelastic response to cryospheric
         change   222
    3.3. Horizontal deformation   223
    3.4. Regions of interest   224
    3.5. Polar case studies for GIA   229
  4. GNSS interferometric reflectometry
     for the cryosphere   231
    4.1. Introduction   231
    4.2. Principles and methodology of
         GNSS-IR   231
    4.3. Snow depth   232
    4.4. Ice mass balance   233
    4.5. Freeze and thaw movements
         in permafrost areas   234
    4.6. Summary   235
  Appendix   235
  References   236

13. The role of GNSS monitoring in landslide research

Halldór Geirsson and Þorsteinn Sæmundsson

  1. Introduction   243

    2. Landslide motion and landslide
       types   244
    3. GNSS landslide equipment and data
       processing   245
    4. Case studies   248
       4.1. Åknes, Norway   248
       4.2. Almenningar, Iceland   248
       4.3. El Yunque, Puerto Rico   249
       4.4. Cà Lita, Italy   250
    5. Perspective of GNSS and other
       landslide deformation methods   252
    Acknowledgments   252
    References   252

14. Climate- and weather-driven solid Earth deformation and seismicity

Roland Bürgmann, Kristel Chanard, and Yuning Fu

    1. Introduction   257
    2. Observing and modeling climate-driven
       deformation, stress, and seismicity   258
       2.1. Measuring climate-driven
           deformation   258
       2.2. Modeling climate-driven
           deformation and stress   261
       2.3. Documenting earthquake
           triggering and modulation   263
    3. Deformation and seismicity from
       changing climate and weather   264
       3.1. Ice age climate cycles   264
       3.2. Consequences of recent
           climate change   267
       3.3. Seasonal hydrological and
           atmospheric loads   269
       3.4. "Earthquake weather"   272
    4. Lessons learned from climate-driven
       deformation and seismicity   274
       4.1. Probing the Earth's constitutive
           properties using climate-driven
           deformation   274
       4.2. Insights on frictional fault properties
           and state of stress in the Earth from
           periodic climate forcing   275
    5. Summary and future opportunities   276
    Acknowledgments   277
    References   277

15. Influence of climate change on magmatic processes: What does geodesy and modeling of geodetic data tell us?

Freysteinn Sigmundsson, Michelle Parks, Halldór Geirsson, Fabien Albino, Peter Schmidt, Siqi Li, Finnur Pálsson, Benedikt G. Ófeigsson, Vincent Drouin, Guðfinna Aðalgeirsdóttir, Eyjólfur Magnússon, Andy Hooper, Sigrún Hreinsdóttir, John Maclennan, Erik Sturkell, and Elisa Trasatti

    1. Introduction   287
    2. Ongoing glacier load changes at
       volcanoes   289
    3. Uplift and deformation of volcanoes due
       to climate change and magma
       movements   291
    4. Modeling the effects   292
       4.1. Decomposing observed deformation
           fields to infer GIA and volcano
           processes   292
       4.2. Influence on magma generation   293
       4.3. Influence on magma
           emplacement   295
       4.4. Influence of ice retreat on the stability
           of shallow magma bodies   296
    5. Discussion   297
    6. Conclusions   297
    Acknowledgment   298
    References   298

Index   301

Contributors

Numbers in parenthesis indicate the pages on which the authors' contributions begin.

Guðfinna Aðalgeirsdóttir (287), Institute of Earth Sciences, Science Institute, University of Iceland, Reykjavík, Iceland

Fabien Albino (287), Université Grenoble Alpes, Université Savoie Mont Blanc, CNRS, IRD, Université Gustave Eiffel, ISTerre, Grenoble, France

Yosuke Aoki (1,129), Earthquake Research Institute, The University of Tokyo, Tokyo, Japan

Joseph Awange (189), School of Earth and Planetary Sciences, Curtin University, Perth, Australia

Roland Bürgmann (257), Berkeley Seismological Laboratory; Department of Earth and Planetary Science, University of California Berkeley, Berkeley, CA, United States

Kristel Chanard (257), Institut de Physique du Globe de Paris, Université Paris Cité, CNRS, IGN, Paris, France

Clara Chew (205), Muon Space, Mountain View, CA, United States

Brendan W. Crowell (111), Department of Earth and Space Sciences, University of Washington, Seattle, WA, United States

Vincent Drouin (287), Icelandic Meteorological Office, Reykjavík, Iceland

Yuning Fu (257), School of Earth, Environment and Society, Bowling Green State University, Bowling Green, OH, United States

Halldór Geirsson (243,287), Institute of Earth Sciences, Science Institute, University of Iceland, Reykjavík, Iceland

Jianghui Geng (11), GNSS Research Center, Wuhan University; Innovation Academy for Precision Measurement Science and Technology, Chinese Academy of Sciences, Wuhan, China

Rüdiger Haas (179), Department of Space, Earth and Environment, Chalmers University of Technology, Onsala Space Observatory, Onsala, Sweden

Kosuke Heki (161), Shanghai Astronomical Observatory, Shanghai, China; Department of Earth and Planetary Sciences, Hokkaido University, Sapporo, Japan

Andy Hooper (287), COMET, University of Leeds, Leeds, United Kingdom

Sigrún Hreinsdóttir (287), GNS Science, Lower Hutt, New Zealand

Weiping Jiang (189), GNSS Research Center, Wuhan University, Wuhan, China

Dirk Kraaijpoel (41), TNO, Geological Survey of the Netherlands, Utrecht, the Netherlands

Corné Kreemer (1,41), Nevada Seismological Laboratory, Nevada Bureau of Mines and Geology, University of Nevada, Reno, NV, United States

Hansjörg Kutterer (189), Karlsruhe Institute of Technology, Geodetic Institute, Karlsruhe, Germany

Siqi Li (287), Institute of Earth Sciences, Science Institute, University of Iceland, Reykjavík, Iceland

Lin Liu (215), Earth and Environmental Sciences Programme, Faculty of Science, The Chinese University of Hong Kong, Hong Kong, China

John Maclennan (287), University of Cambridge, Cambridge, United Kingdom

Eyjólfur Magnússon (287), Institute of Earth Sciences, Science Institute, University of Iceland, Reykjavík, Iceland

Michael Mayer (189), Karlsruhe Institute of Technology, Geodetic Institute, Karlsruhe, Germany

Jean-Mathieu Nocquet (65), Université Côte d'Azur, Observatoire de la Côte d'Azur, IRD, CNRS, Geoazur, Valbonne; Université Paris Cité, Institut de Physique du Globe de Paris, Paris, France

Benedikt G. Ófeigsson (287), Icelandic Meteorological Office, Reykjavík, Iceland

Finnur Pálsson (287), Institute of Earth Sciences, Science Institute, University of Iceland, Reykjavík, Iceland

Michelle Parks (287), Icelandic Meteorological Office, Reykjavík, Iceland

Chris Rollins (83), GNS Science, Lower Hutt, New Zealand

Chris Ruf (205), University of Michigan, Ann Arbor, MI, United States

Þorsteinn Sæmundsson (243), Institute of Earth Sciences, Science Institute, University of Iceland, Reykjavík, Iceland

Peter Schmidt (287), University of Uppsala, Uppsala, Sweden

Harald Schuh (189), GFZ German Research Centre for Geosciences, Potsdam; Technische Universität Berlin, Institute of Geodesy and Geoinformation Science, Berlin, Germany

Freysteinn Sigmundsson (287), Institute of Earth Sciences, Science Institute, University of Iceland, Reykjavík, Iceland

Erik Sturkell (287), University of Gothenburg, Gothenburg, Sweden

Elisa Trasatti (287), Istituto Nazionale di Geofisica e Vulcanologia, Rome, Italy

Martin Vallée (65), Université Paris Cité, Institut de Physique du Globe de Paris, Paris, France

Tonie van Dam (215), Department of Geology and Geophysics, University of Utah, Salt Lake City, UT, United States

Laura M. Wallace (83), University of Texas Institute for Geophysics, Austin, TX, United States; GEOMAR Helmholtz Centre for Ocean Research; Kiel University, Institute of Geosciences, Kiel, Germany

Pippa Whitehouse (215), Department of Geography, Durham University, Durham, United Kingdom

Peng Yuan (189), Karlsruhe Institute of Technology, Geodetic Institute, Karlsruhe; GFZ German Research Centre for Geosciences, Potsdam, Germany

Ilya Zaliapin[†] (41), Department of Mathematics and Statistics, University of Nevada, Reno, NV, United States

Mingyuan Zhang (189), GNSS Research Center, Wuhan University, Wuhan, China

---

[†] Deceased

Foreword

It is a great pleasure to be asked to write a foreword to this wonderful compendium of papers about GPS (and now GNSS) applications for Earth sciences. The editors have already written a chapter that outlines the contents of each contribution, so I will instead provide a short perspective on what it was like to have a front-row seat to the early days of GPS in geodesy.

I was first asked to work in GPS in the mid-1980s. I had come to graduate school with the intention to study seismology, so switching to satellite geodesy, and a brand-new satellite system at that, was difficult for me. I used my first AGU meeting in 1986 to "learn about geodesy and GPS." I was told there was going to be a demonstration of GPS equipment outside the AGU conference center. I dutifully set my alarm clock for 4 a.m., so that I could watch the exciting GPS equipment in operation—which anyone who has collected GPS data can tell you is not very exciting at all. Later that month I spent New Year's Eve tracking GPS signals from Mt. Lospe.

My second introduction to GPS came the following summer when I went to JPL to learn how to analyze those data. Now that I look back on those days, what I remember the most are the wonderful friends I made. I also remember the international interest in GPS—as I found myself learning to model refraction, resolve ambiguities, and estimate station coordinates while sitting next to visitors from all around the world. Certainly, some of this interest was driven by a desire to measure things that previously could not be measured. But I think some of it was the growing recognition that GPS was going to change geodesy. One of the first signs I saw that this new community was going to band together to support GPS was the successful proposal to build a global, continuously operating GPS network so that precise orbits would be available to all.

On the application side, the dominant view at that time was that the role of GPS was to measure crustal deformation across plate boundaries. There is no question that this is where GPS measurements made immediate contributions to geophysics. Continental-scale deformation was ostensibly beyond our capabilities and was to be left for the older, established (and more expensive) geodetic techniques. I don't remember thinking that GPS would ever tell us anything useful about the environment. I think most of us were just hopeful that someday we would be able to use the GPS vertical component for geophysics.

After those early campaigns, GPS geodesists almost immediately began to break through these expected limitations. Global GPS studies became quite competitive with, and relatively quickly, surpassed the older geodetic techniques as the global GPS tracking network matured and the reference frame and GPS software improved. The question at some point changed from whether geophysicists believed in the GPS results to whether there was enough money to build more GPS sites. For me, a transformative moment was hearing in the mid-1990s that Japan was building a dense, country-wide GPS network. The discovery of episodic slips by GPS geodesists was another big, big memory for me.

Fairly quickly, geodesists also began to look for ways to push the presumed limitations on how well GPS could measure positions at shorter time scales, i.e., seconds. Measurements from the 2002 Denali earthquake demonstrated that, for large earthquakes, GPS could do this quite well. Subsequent work by others showed that GPS was also especially well-suited for the task of rapidly determining the magnitude of great earthquakes. It makes me happy to see that 20 years after Denali, GPS data can now be streamed in real time and used for both earthquake and tsunami warnings.

When I was first learning about GPS as a student, I felt overwhelmed by the number of error sources that influenced GPS data. And let us be honest—there are a lot of them. What I find particularly satisfying to see in this book is how many of those error sources are now being utilized for Earth science communities far from tectonics. Surface loading studies provide regional maps of water storage in snow and soil. Ionospheric delays are now routinely utilized to provide total electron content measurements and an independent dataset for tsunami warnings. Tropospheric errors are used operationally in many countries to monitor extreme weather events. Finally, after spending the first half of my career thinking that multipath was always a bad thing, I spent the second half of my career using it to measure environmental changes. Several chapters in this book detail this method and provide examples for tides and the cryosphere.

GPS has been a game changer, providing new measurements and insights for an amazing number of Earth science communities. How did this come about? First, I think it is safe to say that I would not be writing this preface today if Earth scientists had had to pay to design, build, launch, and maintain a satellite constellation these last four decades. So the GPS signals being "free" is important. Along with the space segment, the instruments used by geodesists and geophysicists are valuable to communities outside the Earth sciences, which has also helped reduce costs. And once costs came down, it wasn't very long before GPS began to operate 24/7.

Second, geodesists around the world worked together to improve the accuracy and reliability of GPS. Some of these efforts are easy to see, e.g., the global GPS tracking network used to estimate orbits and the international terrestrial reference frame. But, what is less often recognized is how the international GPS community has developed the standards, models, and software that make GPS reliable. By also making the orbits openly available, the international GPS geodesy community has intrinsically inspired the innovation made by geophysicists in this book. Third, most international organizations have open data policies and thus provide the GPS data used in many of the studies in this book; for this, we should be thankful.

Kristine M. Larson
San Diego, CA
June 19, 2024

Chapter 1

Introduction

Yosuke Aoki[a] and Corné Kreemer[b]

[a]Earthquake Research Institute, The University of Tokyo, Tokyo, Japan, [b]Nevada Seismological Laboratory, Nevada Bureau of Mines and Geology, University of Nevada, Reno, NV, United States

1 History

Surveying has long been based on astronomical and terrestrial methods, including leveling, triangulation, and trilateration surveys. However, terrestrial surveying requires human resources, so surveying large areas in a limited time is challenging. While the advent of Electronic Distance Measurement (EDM) in the 1960s advanced terrestrial surveying, EDM surveying is possible only when a site is visible from another place. In other words, Earth's curvature prohibits us from measuring long distances with EDM.

Since the launch of the first satellite, Sputnik 1, some geodesists realized that precise positioning might be possible with fewer human resources. This idea led the US Department of Defense (DoD) to start a project launching Global Positioning System (GPS) satellites in the early 1970s. The DoD launched the first GPS satellite in 1978; 24 satellites became available in 1993 so that at least four satellites, the minimum number of satellites needed for positioning, are visible from everywhere on the Earth.

In parallel with development in the United States, a similar project to launch navigation satellites was ongoing in the former Soviet Union. Their satellite system is called *Global'naya NAvigatsionnaya Sputnikovaya Sistema* (GLONASS). They launched the first satellite in 1982, and 24 satellites became available in 1996. However, due to depression of the Russian economy, the data became globally available only in 2011. In addition, the European Union has been operating Galileo since 2016, while the China National Space Administration started operating Beidou in 2000, and the data has been globally available since 2012. Together with GPS, these systems are collectively called Global Navigation Satellite System (GNSS).

In addition to GNSS, the Indian Space Research Organization has been operating the Navigation Indian Constellation (NavIC) since 2013. Also, the Japan Aerospace Exploration Agency has been operating the Quasi-Zenith Satellite System (QZSS) since 2018. Because NavIC and QZSS are available regionally, not globally, these systems are called Regional Navigation Satellite System (RNSS), although it is often considered a part of GNSS.

The number of GNSS sites, continuous GNSS sites in particular, has rapidly increased since the 1990s. Their spacing is 10–20 km in densely deployed regions, such as Japan, Taiwan, and western North America. Some active volcanoes have even denser network, such as Kīlauea, Piton de la Fournaise, Etna, and Sakurajima. Nowadays, tens of thousands of GNSS sites have been monitoring the Earth (Blewitt et al., 2018). GNSS has been pivotal in space geodetic studies because of its dense deployment and temporal resolution.

Synthetic Aperture Radar (SAR) has been another pillar in space geodetic studies since its emergence in the 1990s. Interferometric SAR (InSAR) detects deformation in unprecedented spatial resolution of ∼10 m. Also, SAR does not require ground-based instruments. Therefore, InSAR is suitable for measuring the deformation of areas where access is difficult or prohibited. However, the measurement by InSAR is not as precise as GNSS because InSAR is more susceptible to atmospheric and ionospheric disturbance. Also, the temporal resolution of InSAR measurements is limited by the recurrence of the InSAR satellites, varying from a few to a few tens of days. The temporal resolution of InSAR measurements is thus worse than that of GNSS, which can be down to seconds. Furthermore, unlike GNSS, which measures three-dimensional displacements, InSAR measures only line-of-sight changes by interferometry. Although InSAR can measure the along-the-satellite-track component of displacement, the precision is much worse than the measurement by interferometry. Since GNSS and InSAR have advantages and disadvantages, as explained so far and summarized in Table 1.1, GNSS and InSAR coexist as dominant players in space geodetic studies, taking advantage of the different characteristics of these measurements.

TABLE 1.1 Comparison between GNSS and InSAR.

	GNSS	InSAR
Measurements	Three components	Line of sight
Precision	2–3 mm (Horizontal) 5–8 mm (Vertical)	10–20 mm
Temporal resolution	Seconds to days	A few to a few tens of days
Spatial resolution	Spacing of GNSS sites (>10 km)	10–100 m

2 Measuring Earth's deformation onshore

GNSS measurements can constrain the antenna's position in millimeter accuracy, as Chapter 2 describes. Therefore, scientists expected that GNSS measurements could measure Earth's slow deformation as small as millimeters per year.

The first measurement of Earth's deformation was by Very Long Baseline Interferometry (VLBI) by Herring et al. (1986) and Satellite Laser Ranging (SLR) by Christodoulidis et al. (1985) instead of by GPS. Both studies measured the relative motion of two plates. The first measurement of relative station motion with GPS was by Larson (1990). VLBI and SLR measurements gave way to GPS to measure Earth's deformation because GPS measurements are much less expensive and more abundant than VLBI and SLR measurements.

The plate motion measured by these space geodetic techniques is consistent with that constrained by geological evidence (e.g., DeMets et al., 1990). Considering that the plate motion based on geological evidence represents the average over hundreds of thousands or millions of years, this consistency indicates that the plate motion has not changed much over hundreds of thousands or millions of years. This insight adds significant constraints on Earth's dynamics. Also, the proliferation of GNSS stations allowed the mapping out of the nonrigid behavior of the Earth's crust within (diffuse) plate boundary zones (Kreemer et al., 2000, 2003, 2014). The strain accumulation obtained in those places allows for a direct comparison with the moment released by earthquakes and the number of earthquakes, as described in Chapter 3.

We have recognized that an earthquake is related to faulting in the late 19th century (Koto, 1893) and measured surface deformation caused by an earthquake in the 20th century (Reid, 1910). Therefore, not long after the dawn of space geodetic observations, we observed surface deformation caused by an earthquake. Since the first observation of coseismic deformation with GPS (Bock et al., 1993), numerous studies have reported coseismic GNSS displacements for earthquakes with magnitude as small as $M=5$ (Blewitt et al., 2008). Chapter 4 provides an overview of these studies. The emergence of GPS allows us to image earthquake deformation in higher spatial resolution than in the era of terrestrial measurements.

The accumulated stress is released not only by earthquakes but also by slow deformation that seismic observations barely detect. Indeed, postseismic deformation has already been observed in the 1960s by terrestrial measurements (Scholz et al., 1969; Smith & Wyss, 1968). Since the first observation of postseismic deformation with GPS (Heki et al., 1997), numerous studies have investigated postseismic deformation of earthquakes larger than about magnitude 6. Because GNSS measurements are not only spatially but also temporally denser than terrestrial measurements, the emergence of GNSS has significantly advanced the understanding of postseismic deformation.

GPS observations found spontaneous slow slips on a fault without being triggered by a large earthquake, for example. Since the first such discovery by Hirose et al. (1999), numerous studies have investigated slow earthquakes that occur globally. In particular, studies on slow earthquakes have accelerated after Rogers and Dragert (2003) discovered that slow slips synchronized with nontectonic tremors that Obara (2002) found. Since then, not only geodesists but also seismologists have shown interest in slow earthquakes. Chapter 5 provides an overview of the state-of-the-art studies of slow earthquakes from a geodetic perspective.

So far, arguments are based on GNSS measurements, which give station coordinates daily. However, because the sampling interval of GNSS measurements is 30 s or shorter, they can provide station coordinates in more temporal resolution in principle. In this regard, using GNSS measurements as a seismometer has shown promise since the early days. Since studies with a shake table (Ge et al., 2000) and an earthquake (Larson et al., 2003) confirmed that GNSS can also be used as a seismometer, many studies have investigated large earthquakes with high-rate GNSS measurements, as described in Chapter 6. The most significant advantage of GNSS measurements over seismic observations is that GNSS measurements lack frequency responses. In other words, GNSS can measure signals from the static displacement to oscillations of Nyquist frequency, half the sampling frequency. In contrast, seismometers can only measure changes greater than at least ~0.1 Hz with short-period sensors and ~0.001 Hz with broadband sensors.

The ability to measure low-frequency oscillations with GNSS measurements allows us to employ them in earthquake early warning, in conjunction with seismic records. Earthquake early warning with GNSS measurements is beneficial for detecting earthquakes with a magnitude greater than 8. While seismic records tend to miss low-frequency signals dominant in seismograms generated by earthquakes of magnitude greater than 8, GNSS measurements do not miss them. Indeed, earthquake early warning with only seismometers initially underestimated the size of the 2011 Tohoku-oki, Japan, earthquake ($M_w = 9.0$). Ohta et al. (2012) and Kawamoto et al. (2017) argued that high-rate GNSS measurements could accurately estimate the size of the earthquake in a few minutes. Similar conclusions were also reached for the 2004 Great Sumatra earthquake (Blewitt et al., 2006).

The Earth deforms not only by earthquakes and plate motion but also by volcanic activity. Chapter 7 discusses the GNSS's contribution to understanding volcanic deformation. Because the migration of magma or hydrothermal fluids deforms the Earth, measuring ground deformation has been essential for understanding the nature of volcanism. The GNSS allows us to monitor ground deformation in active volcanoes with enhanced spatial and temporal resolution. In particular, enhanced temporal resolution is essential because some volcanic activity lasts only briefly. Regardless of the recent dominance of InSAR in monitoring volcano deformation because of its superior spatial resolution, GNSS still plays a significant role in understanding volcanism because of its better temporal resolution than InSAR.

3 Measuring Earth's deformation offshore

Although not covered as an independent chapter, offshore deformation measurements are becoming increasingly important near subduction zones, because the strain rate is high and most major earthquakes occur there. Offshore deformation measurements have been around since the 1980s. The initial offshore deformation measurements were carried out with pressure sensors as a proxy for vertical displacements (Fox, 1993) and direct-path measurements between two transponders (e.g., Spiess, 1980). The measurements conducted by pressure sensors, however, include nontectonic sea-level changes such as those caused by tides. Also, the direct-path measurements are limited to short distances because of the upward refraction of acoustic signals.

Offshore deformation measurements combining GNSS and acoustic measurements were conceived in the 1980s to circumvent these problems (Spiess, 1985) and realized in the late 1990s (Fujimoto et al., 1998; Spiess et al., 1998). The current state-of-the-art system measures the center of transponders at the seafloor by measuring the distance between the transducer at the bottom of the ship and each transponder at the seafloor. The transducer's location is calibrated by locating the position of GNSS antennas on board the vessel and the relative location between the transducer and GNSS antennas (Fig. 1.1).

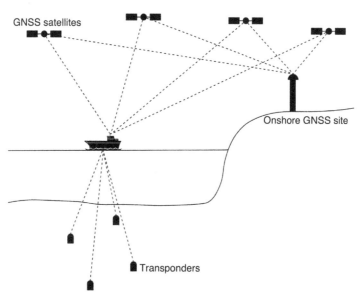

FIG. 1.1 Schematics of the GNSS-acoustics measurement. The gravity center of transponders at the seafloor is measured through the positioning of a ship and the distance measurement between a ship and transponders.

4 GNSS monitoring of the terrestrial environment

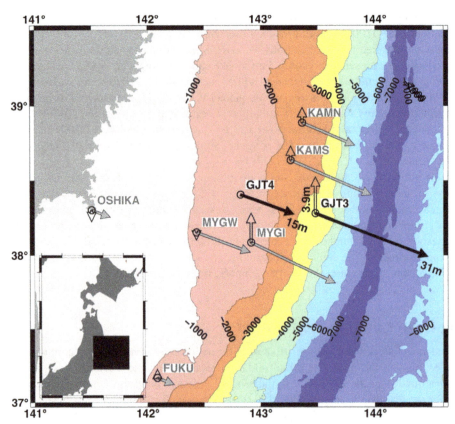

FIG. 1.2 Seafloor deformation by the Tohoku-oki earthquake. *Black arrows* denote the measurements Kido et al. (2011). *Gray arrows* offshore and onshore indicate those by Sato et al. (2011) and Ozawa et al. (2011). *(Figure taken from Kido, M., Osada, Y., Fujimoto, H., Hino, R., Ito, Y. (2011). Trench-normal variation in observed seafloor displacements associated with the 2011 Tohoku-oki earthquake.* Geophysical Research Letters, 38, L24303. https://doi.org/10.1029/2011gl050057.)

Fig. 1.2 depicts coseismic displacements caused by the 2011 Tohoku-oki earthquake ($M_w = 9.0$). As the Tohoku-oki earthquake generated large slips near the trench, the observed displacements offshore are much larger than those onshore (Kido et al., 2011; Sato et al., 2011). While the onshore displacements reach up to 5.5 m, some offshore GNSS sites recorded displacements of more than 30 m. This observation indicates that the slip near the trench is at least 30 m. Indeed, Iinuma et al. (2012) argued that the maximum coseismic slip of the Tohoku-oki earthquake exceeded 50 m near the trench by including offshore measurements, while Ozawa et al. (2011) failed to detect such large slips solely from onshore GNSS measurements.

Offshore GNSS measurements often offer a unique opportunity to gain insights into the mechanics of postseismic deformation. For example, Iinuma et al. (2016) found that postseismic displacements near the trench are in the direction opposite to coseismic displacements, while those far from the trench are in the same direction. This observation is univocal evidence for the significant contribution of viscoelastic relaxation of the asthenosphere, even during the early stage of the postseismic period.

Offshore GNSS measurements are precise enough to measure interseismic deformation. Yokota et al. (2016) measured interseismic displacements near the Nankai trough in southern Japan to demonstrate that the plate interface at shallow depths is strongly locked. Because onshore GNSS measurements far from the trench are not sensitive to the locking status near the trench or shallower depths, offshore GNSS measurements offer a unique opportunity to probe the locking status near the trench. Understanding the locking status near the trench is essential not only from scientific but also practical viewpoints because if the shallow plate interface is locked, it can slip as a large earthquake to generate a tsunami.

So far, we have discussed offshore deformation based on campaign-mode GNSS measurements. However, continuous measurements provide more insights into offshore deformation. Kato et al. (2000) developed a system in which a GPS antenna is mounted on a buoy tethered to an anchor at the seafloor by an iron chain. As the GPS antenna is floated on the seawater, the system measures the sea level rather than the seafloor deformation. This system measures tsunamis generated by the earthquakes occurring off the Kii Peninsula ($M = 7.1$ and 7.4; Kato et al., 2005). Measuring tsunamis offshore significantly contributes to the tsunami early warning. Indeed, the warning system initially underestimated the tsunami caused by the 2011 Tohoku-oki earthquake because it underestimated the size of the earthquake. The tsunami recorded by pressure gauges on the seafloor played a significant role in correctly estimating the size of the earthquake and tsunami.

A GNSS antenna floating on a buoy can measure sea level, but it cannot continuously measure seafloor displacements. Xie et al. (2019) and De Martino et al. (2020) developed seafloor geodesy systems to measure seafloor deformation continuously with a GNSS antenna anchored to the seafloor by a rod. The orientation of the rod is measured with a digital compass to correct the position of the antenna. As this system requires a rod longer than the sea depth, the measurements are possible only where the sea depth is less than ~100 m.

4 Unconventional use of GNSS

GNSS is primarily intended to measure the position of antennas; however, it can also convert noises associated with the positioning into signals for other purposes. For example, fluctuating total electron content in the ionosphere is a noise source for precise positioning because the presence of electrons refracts the radar wave radiated from a GNSS satellite. However, extracting the ionospheric disturbance provides insights into not only space weather, but also the mechanics of interactions among solid Earth, atmosphere, and ionosphere generated by earthquakes, tsunamis, and volcanic explosions. Since the first observation of ionospheric disturbance caused by an earthquake in the mid-1990s (Calais & Minster, 1995), many studies have reported ionospheric disturbances caused by earthquakes, tsunamis, and volcanic explosions. Chapter 8 provides an overview of such observations and insights gained from them.

A GNSS antenna receives not only direct radar waves radiated from a GNSS satellite but also waves reflected by objects such as the Earth's surface and nearby buildings. The reflected component is called multipath. Multipaths from the Earth's surface are particularly unavoidable if a GNSS antenna is attached to the ground. However, multipath allows one to measure the distance between the GNSS antenna and the Earth's surface. Direct and reflected arrivals of the GNSS signal interfere positively or negatively depending on the difference in travel distance, which is a function of the incidence angle and the GNSS signal. Therefore, monitoring temporal changes in the signal strength from a GNSS satellite, whose elevation changes with time, allows us to infer the distance between the GNSS antenna and the reflection surface of the GNSS signal. Since pioneering studies by Larson et al. (2009, 2013), who measured time-varying snow depth and sea levels with GNSS, substantial efforts have been made to monitor sea levels with GNSS signals. Chapter 9 provides an overview of such efforts.

The refraction of radar waves radiated from a GNSS satellite in the atmosphere is another source of positioning errors. Because the amount of water vapor varies over time, it is the most significant contributor to positioning errors by atmospheric disturbance. However, this fact indicates that GNSS can be used to monitor water vapor in the atmosphere. Indeed, since the pioneering research of Bevis et al. (1992), many studies have investigated spatial variations in water vapor from GNSS measurements. Because weather changes come with changes in the distribution of water vapor in the atmosphere, GNSS plays a significant role in monitoring weather changes, including extreme weather. Chapters 10 and 11 describe GNSS's contribution to monitoring weather changes.

5 GNSS and climate change

5.1 Deformation in response to long- and short-term climate change

The Earth is deformed not only by earthquakes and volcanic activity but also in response to climate change. For example, uplifts of up to 10 mm/yr are observed in high latitudes where thick glaciers existed during the last ice age. These observations are due to a viscoelastic response of the Earth due to the unloading of glaciers at the end of the last ice age, ~10,000 years ago. Therefore, these GNSS observations provide insights into the viscosity structure of the Earth as well as the melting history of glaciers.

The Earth also deforms in response to shorter-term climate change. For example, melting glaciers due to recent climate change deform the Earth. This deformation is due to the elastic response of the Earth to unloading caused by glaciers. Therefore, GNSS can provide insights into the melting history of glaciers. Chapter 12 covers the Earth's long- and short-term responses to the changes in the cryosphere.

Short-term climate change, such as heavy rainfall, induces landslides, sometimes occurring in cascades. Earthquakes and volcanic activity can also trigger landslides. GNSS is particularly a powerful tool to monitor landslides because the deformation rate of landslides varies between millimeters per year to hundreds of kilometers per hour. GNSS can cover the whole spectrum of landslide deformation. Chapter 13 discusses insights gained from observing landslides with GNSS. It also discusses the practical aspects of deploying GNSS sites in landslide-prone regions, often in harsh environments.

5.2 Measuring glacier motions

GNSS can also measure displacements of the glaciers themselves if GNSS sites are installed on them, although this topic is not covered as an independent chapter. Because understanding the dynamics of glaciers provides insights into the melting of glaciers, which affects sea levels, it is crucial to monitor glacier motions to understand the evolution of sea levels and climate change.

InSAR is the most popular method to monitor the long-term velocities of glaciers. Several studies (e.g., Andersen et al., 2020; Joughin et al., 2018; Solgaard et al., 2021) have compiled spatial variations of glaciers in Greenland with a spatial resolution of 50–200 m and uncertainty of 10–20 m/yr from the Sentinel-1 SAR satellite. They found that the glacier motions vary from within the measurement errors to faster than 1000 m/yr. InSAR effectively monitors long-term glacier motions because no surface measurements are required. However, the recurrence time of the satellite, which ranges from days to tens of days, limits the temporal resolution of InSAR measurements.

Maintaining GNSS sites in extreme environments, such as polar regions, involves many difficulties. Nonetheless, measuring glacier motions with GNSS offers new insights into the dynamics of glaciers because of higher temporal resolution and smaller measurement errors than InSAR measurements. Hvidberg et al. (2020) deployed 63 GNSS sites with an average spacing of ∼5 km in the middle of the Northeast Greenland Ice Stream to derive spatial variations in the flow velocities of the glacier. They found that the flow velocity at the center of the ice stream is ∼55 m/yr, declining to zero at the edge. They also found that the longitudinal strain rate is 10^{-4} at the center of the ice stream, increasing to 10^{-3} at the edge.

GNSS measurements can take advantage of more precision than InSAR measurements to monitor the acceleration of glacier flow. Khan et al. (2022) observed variable acceleration of glacier flow between 2016 and 2018 in the Northeast Greenland Ice Stream. While the flow velocity is nearly constant between 340 and 390 m/yr, the acceleration is variable, with 5 m/yr^2 near the terminus and 3 m/yr^2 ∼190 km upstream.

The superior temporal resolution of GNSS to InSAR allows us to detect deformation associated with transient dynamics with GNSS. The observed transient deformation mainly originates from the drainage of supraglacial lakes. The drainage occurs because of the flow of the meltwater from the lake to the glacier's base through moulins. Stevens et al. (2015) observed displacements due to the drainage of a supraglacial lake in western Greenland in 2011, 2012, and 2013 with 16 GNSS sites. These transient deformations lasted about a day. They modeled the observed displacements with a shear slip at the glacier bed, opening at the glacier bed, and hydrofracture of the glacier, suggesting that the contribution from slips and opening at the glacier bed is more significant than the hydrofracture.

GNSS can also detect the fluctuations in the glacier flow associated with the drainage of a supraglacial lake. Stevens et al. (2022) observed a transient fluctuation in glacier flow velocities in response to the drainage of a supraglacial lake and lubrication at the glacier bed in southeast Greenland. The fluctuation started with an acceleration of ∼5%, followed by deceleration, lasting for a few days. The excess displacement during the acceleration is up to ∼0.8 m near the lake and ∼0.3 m down-glacier, but no net excess displacements remain after the fluctuation. This fluctuation propagates down-glacier by a velocity of ∼4 km/h. This observation indicates that the drainage of a supraglacial lake has little effect on long-term flow velocities.

Drainage of meltwater from a supraglacial lake leads to not only the fluctuations in glacier flow but also vertical displacements because of the pressure exerted on the glacier bed by meltwater. Lai et al. (2021) observed transient uplift by 0.2–1.5 m associated with the drainage of meltwater, followed by relaxation with a characteristic timescale of 0.1–5 days. This relaxation timescale is related to the subglacial hydraulic transmissivity. Lai et al. (2021) found that the hydraulic transmissivity varies between ∼0.8 and ∼200 mm^2, increasing as a melt season progresses.

GNSS observations measure not only transient but also periodic fluctuation in glacier flow. Stevens et al. (2022) found diurnal and semidiurnal velocity fluctuations in the flow of a marine-terminating glacier in southeast Greenland in the summers of 2007 and 2008. They found that the daily fluctuation in flow velocity peaks ∼6.5 h after the maximum insolation. Also, the fluctuation amplitude positively correlates with the total daily insolation. These observations suggest that lubrication by meltwater at the glacier bed causes daily fluctuation. They also observed semidiurnal changes within ∼10 km from the glacier's terminus. Ocean tide causes this fluctuation; the maximum velocity fluctuation is delayed by 1–2.5 h after low tide.

5.3 Earthquakes and volcanism triggered by climate change

The Earth's deformation is caused not only by the loading of glaciers but also by water. Space geodetic techniques like GNSS and GRACE can directly measure total water storage (TWS) change, while previous hydrologic methods cannot. Because recent global climate change impacts hydrologic cycles, it is becoming more critical to constrain the water budget through direct TWS measurements. GNSS's superior temporal resolution over GRACE plays a significant role in understanding the terrestrial water budget.

Deformation from water loading started to attract attention in the early 2000s. van Dam et al. (2001) demonstrated that the annual variation of continental water loading could cause vertical displacements of up to 30 mm. Blewitt et al. (2001) detected annual deformation signals by observing the annual variation in continental water loading. Heki (2001) showed that the annual variation in snow loading explains most of the annual variation in GNSS vertical displacements in northern Japan.

Terrestrial water does not only deform the Earth through surface loading but also while present in the subsurface. Because the shallow subsurface of the Earth contains pores, the surface water infiltrates and diffuses into the Earth, causing pore pressure changes and poroelastic deformation of the Earth. Chapter 14 discusses terrestrial water's contribution to the Earth's deformation and earthquake generation.

Climate changes affect the stress state of the magma reservoir beneath active volcanoes and the magma pathway from the magma reservoirs to the surface. The melting of glaciers generates elastic deformation of the Earth in the short term and viscoelastic deformation in the long term. Since it decreases the confining pressure of magma reservoirs and pathways, long- and short-term glacier melting increases the eruption potential of volcanoes. Chapter 15 discusses the contribution of climate changes to volcanism, mainly in Iceland, where glacier melting is ongoing.

6 Future of GNSS

As summarized above, recent advancements in the GNSS network and data processing strategies allow us to gain insights into the Earth's deformation associated with various phenomena, including earthquakes, volcanic activity, and hydrological phenomena. Recent studies have revealed that GNSS can measure not only the Earth's deformation but also atmospheric disturbance, ionospheric disturbance, and multipath of microwaves radiated from a GNSS satellite. This section discusses possible avenues for GNSS studies in monitoring the Earth by focusing on (1) dense deployment, (2) offshore measurements, and (3) deploying GNSS antennas on small satellites.

The last few decades have witnessed an explosive expansion of the GNSS network. Blewitt et al. (2018) noted that in 2018 at least 17,000 GNSS sites monitor the Earth and freely deliver data. There may, however, be tens of thousands of GNSS sites, including those not publicly available. Some geologically active regions, such as Japan, Taiwan, and western North America, have dense GNSS networks with an average spacing of 30 km or less. Some areas prone to volcanic activity and landslides have even denser networks. Such dense networks have played significant roles in enhancing our understanding of various geological phenomena.

The recent utilization of GNSS measurements for navigation purposes, such as in autonomous vehicle operation, has increased the private sector's need for GNSS equipment. Since such GNSS equipment requires the positioning precision of geodetic quality, they can also be used to monitor the Earth's deformation. In Japan, such GNSS measurements, with an average spacing of less than 10 km, are already available to measure Earth's deformation associated with geological phenomena (e.g., Ohta & Ohzono, 2022). The coming decades will witness further enhancement in GNSS networks not only to monitor the Earth's deformation but also to probe the atmosphere, ionosphere, and climate change.

Currently, offshore GNSS measurements remain still less dense than onshore measurements because of the strategic difficulties and cost of deployment. Nonetheless, they provide crucial insights because, for example, most of the greatest earthquakes and a significant proportion of volcanic eruptions occur offshore. As described in a previous section, cost-effective strategies for offshore deformation measurements with GNSS have been proposed in the last few years. The next decade may witness more offshore GNSS sites with more temporal resolution or even continuous measurements.

The recent advancement in GNSS reflectometry allows us to monitor the sea surface using small satellites, such as CubeSats, in the low Earth orbit. Various constellations, including Cyclone Global Navigation System (CYGNSS: Ruf et al., 2013; Carreno-Luengo et al., 2021), Spires (e.g., Nguyen et al., 2020; Roessler et al., 2022), and Technology Demonstration Satellite-1 (TechDemoSat-1; Mashburn et al., 2018) are currently in operation to monitor the roughness of the sea surface, as well as sea levels, from microwaves radiated from GNSS satellites that are reflected on the sea surface. With GNSS antennas onboard, these small satellites observe direct and reflected waves from GNSS satellites. These constellations measure ionospheric disturbance as well as sea levels and their roughness. Enhancing the spatial and temporal resolution of the measurements by increasing the number of such small satellites will allow us to detect tsunamis, contributing to tsunami early warning (Colombo & Croteau, 2023).

As discussed so far, GNSS has significantly contributed to various aspects of Earth Science in the last few decades. Its applications include those not related to the positioning, the primary purpose of GNSS observations. The following decades will witness further expansion of GNSS applications, not only of those currently available but also of those unexpected now.

References

Andersen, J. K., Kusk, A., Boncori, J. P. M., Hvidberg, C. S., & Grinsted, A. (2020). Improved ice velocity measurements with sentinel-1 TOPS interferometry. *Remote Sensing, 12*, 2014. https://doi.org/10.3390/rs12122014.

Bevis, M., Businger, S., Herring, T. A., Rocken, C., Anthes, R. A., & Ware, R. H. (1992). GPS meteorology: remote sensing of atmospheric water vapor using the global positioning system. *Journal of Geophysical Research, 97*, 15787. https://doi.org/10.1029/92jd01517.

Blewitt, G., Bell, J., Hammond, W. C., Kreemer, C., Plag, H.-P., & DePolo, C. (2008). GPS and InSAR monitoring of the mogul swarm: Evidence for mainly aseismic fault creep, with implications for seismic hazard. *Eos, Transactions American Geophysical Union, 89*(53). Fall Meet. Suppl., Abstract S53C-03.

Blewitt, G., Hammond, W., & Kreemer, C. (2018). Harnessing the GPS data explosion for interdisciplinary science. *Eos, 99*. https://doi.org/10.1029/2018eo104623.

Blewitt, G., Kreemer, C., Hammond, W. C., Plag, H.-P., Stein, S., & Okal, E. (2006). Rapid determination of earthquake magnitude using GPS for tsunami warning systems. *Geophysical Research Letters, 33*, L11309. https://doi.org/10.1029/2006GL026145.

Blewitt, G., Lavalée, D., Clarke, P., & Nurutdinov, K. (2001). A new global mode of earth deformation: seasonal cycle detected. *Science, 294*, 2342–2345. https://doi.org/10.1126/science.1065328.

Bock, Y., Agnew, D. C., Fang, P., Genrich, J. F., Hager, B. H., Herring, T. A., et al. (1993). Detection of crustal deformation from the landers earthquake sequence using continuous geodetic measurements. *Nature, 361*, 337–340. https://doi.org/10.1038/361337a0.

Calais, E., & Minster, J. B. (1995). GPS detection of ionospheric perturbations following the January 17, 1994, Northridge earthquake. *Geophysical Research Letters, 22*, 1045–1048. https://doi.org/10.1029/95gl00168.

Carreno-Luengo, H., Crespo, J. A., Akbar, R., Bringer, A., Warnock, A., Morris, M., et al. (2021). The CYGNSS mission: On-going science team investigations. *Remote Sensing, 13*, 1814. https://doi.org/10.3390/rs13091814.

Christodoulidis, D. C., Smith, D. E., Kolenkiewicz, R., Klosko, S. M., Torrence, M. H., & Dunn, P. J. (1985). Observing tectonic plate motions and deformations from satellite laser ranging. *Journal of Geophysical Research, 90*, 9249. https://doi.org/10.1029/jb090ib11p09249.

Colombo, O. L., & Croteau, M. J. (2023). *Tsunami early-warning CubeSat design reference mission (DRM)*. NASA Technical Memorandum. TM-20230003987.

De Martino, P., Guardato, S., Donnarumma, G. P., Dolce, M., Trombetti, T., Chierici, F., et al. (2020). Four years of continuous seafloor dis- placement measurements in the Campi Flegrei caldera. *Frontiers in Earth Science, 8*, 615178. https://doi.org/10.3389/feart.2020.615178.

DeMets, C., Gordon, R. G., Argus, D. F., & Stein, S. (1990). Current plate motions. *Geophysical Journal International, 101*, 425–478. https://doi.org/10.1111/j.1365-246x.1990.tb06579.x.

Fox, C. G. (1993). Five years of ground deformation monitoring on axial seamount using a bottom pressure recorder. *Geophysical Research Letters, 20*, 1859–1862. https://doi.org/10.1029/93gl01216.

Fujimoto, H., Ichiro Koizumi, K., Osada, Y., & Kanazawa, T. (1998). Development of instruments for seafloor geodesy. *Earth, Planets and Space, 50*, 905–911. https://doi.org/10.1186/bf03352186.

Ge, L., Han, S., Rizos, C., Ishikawa, Y., Hoshiba, M., Yoshida, Y., et al. (2000). GPS seismometers with up to 20 Hz sampling rate. *Earth, Planets and Space, 52*, 881–884. https://doi.org/10.1186/bf03352300.

Heki, K. (2001). Seasonal modulation of interseismic strain buildup in northeastern Japan driven by snow loads. *Science, 293*, 89–92. https://doi.org/10.1126/science.1061056.

Heki, K., Miyazaki, S., & Tsuji, H. (1997). Silent fault slip following an interplate thrust earthquake at the Japan trench. *Nature, 386*, 595–598. https://doi.org/10.1038/386595a0.

Herring, T. A., Shapiro, I. I., Clark, T. A., Ma, C., Ryan, J. W., Schupler, B. R., et al. (1986). Geodesy by radio interferometry: Evidence for contemporary plate motion. *Journal of Geophysical Research, 91*, 8341. https://doi.org/10.1029/jb091ib08p08341.

Hirose, H., Hirahara, K., Kimata, F., Fujii, N., & Miyazaki, S. (1999). A slow thrust slip event following the two 1996 Hyuganada earthquakes beneath the Bungo channel, Southwest Japan. *Geophysical Research Letters, 26*, 3237–3240. https://doi.org/10.1029/1999gl010999.

Hvidberg, C. S., Grinsted, A., Dahl-Jensen, D., Khan, S. A., Kusk, A., Andersen, J. K., et al. (2020). Surface velocity of the Northeast Greenland ice stream (NEGIS): Assessment of interior velocities derived from satellite data by GPS. *The Cryosphere, 14*, 3487–3502. https://doi.org/10.5194/tc-14-3487-2020.

Iinuma, T., Hino, R., Kido, M., Inazu, D., Osada, Y., Ito, Y., et al. (2012). Coseismic slip distribution of the 2011 off the Pacific coast of Tohoku earthquake (m9.0) refined by means of seafloor geodetic data. *Journal of Geophysical Research: Solid Earth, 117*, B07409. https://doi.org/10.1029/2012jb009186.

Iinuma, T., Hino, R., Uchida, N., Nakamura, W., Kido, M., Osada, Y., et al. (2016). Seafloor observations indicate spatial separation of coseismic and postseismic slips in the 2011 Tohoku earthquake. *Nature Communications, 7*, 13506. https://doi.org/10.1038/ncomms13506.

Joughin, I., Smith, B. E., & Howat, I. (2018). Greenland ice mapping project: Ice flow velocity variation at sub-monthly to decadal timescales. *The Cryosphere, 12*, 2211–2227. https://doi.org/10.5194/tc-12-2211-2018.

Kato, T., Terada, Y., Ito, K., Hattori, R., Abe, T., Miyake, T., et al. (2005). Tsunami due to the 2004 September 5th off the Kii peninsula earthquake, Japan, recorded by a new GPS buoy. *Earth, Planets and Space, 57*, 297–301. https://doi.org/10.1186/bf03352566.

Kato, T., Terada, Y., Kinoshita, M., Kakimoto, H., Isshiki, H., Matsuishi, M., et al. (2000). Real-time observation of tsunami by RTK-GPS. *Earth, Planets and Space, 52*, 841–845. https://doi.org/10.1186/bf03352292.

Kawamoto, S., Ohta, Y., Hiyama, Y., Todoriki, M., Nishimura, T., Furuya, T., et al. (2017). REGARD: A new GNSS-based real-time finite fault modeling system for GEONET. *Journal of Geophysical Research: Solid Earth, 122*, 1324–1349. https://doi.org/10.1002/2016jb013485.

Khan, S. A., Choi, Y., Morlinghem, M., Rignot, E., Helm, V., Humbert, A., ... Bjork, A. A. (2022). Extensive inland thinning and speed-up of Northeast Greenland Ice Stream. *Nature*, *611*, 727–732. https://doi.org/10.1038/s41586-022-05301-z.

Kido, M., Osada, Y., Fujimoto, H., Hino, R., & Ito, Y. (2011). Trench-normal variation in observed seafloor displacements associated with the 2011 Tohoku-oki earthquake. *Geophysical Research Letters*, *38*, L24303. https://doi.org/10.1029/2011gl050057.

Koto, B. (1893). On the cause of the great earthquake in Central Japan, 1891. *The Journal of the College of Science, Imperial University, Japan*, *5*, 295–353. https://doi.org/10.15083/00037587.

Kreemer, C., Blewitt, G., & Klein, E. C. (2014). A geodetic plate motion and global strain rate model. *Geochemistry, Geophysics, Geosystems*, *15*, 3849–3889. https://doi.org/10.1002/2014GC005407.

Kreemer, C., Haines, J., Holt, W. E., Blewitt, G., & Lavallée, D. (2000). On the determination of a global strain rate model. *Earth Planets and Space*, *52*, 765–770. https://doi.org/10.1186/BF03352279.

Kreemer, C., Holt, W. E., & Haines, A. J. (2003). An integrated global model of present-day plate motions and plate boundary deformation. *Geophysical Journal International*, *154*, 8–34. https://doi.org/10.1046/j.1365-246X.2003.01917.x.

Lai, C.-Y., Stevens, L. A., Chase, D. L., Creyts, T. T., Behn, M. D., Das, S. B., & Stone, H. A. (2021). Hydraulic transmissivity inferred from ice-sheet relaxation following Greenland supraglacial lake drainages. *Nature Communications*, *12*, 3955. https://doi.org/10.1038/s41467-021-24186-6.

Larson, K. M. (1990). Evaluation of GPS estimates of relative positions from Central California, 1986-1988. *Geophysical Research Letters*, *17*, 2433–2436. https://doi.org/10.1029/gl017i013p02433.

Larson, K. M., Bodin, P., & Gomberg, J. (2003). Using 1-Hz GPS data to measure deformations caused by the Denali fault earthquake. *Science*, *300*, 1421–1424. https://doi.org/10.1126/science.1084531.

Larson, K. M., Gutmann, E. D., Zavorotny, V. U., Braun, J. J., Williams, M. W., & Nievinski, F. G. (2009). Can we measure snow depth with GPS receivers? *Geophysical Research Letters*, *36*, L17502. https://doi.org/10.1029/2009gl039430.

Larson, K. M., Ray, R. D., Nievinski, F. G., & Freymueller, J. T. (2013). The accidental tide gauge: A GPS reflection case study from Kachemak Bay, Alaska. *IEEE Geoscience and Remote Sensing Letters*, *10*, 1200–1204. https://doi.org/10.1109/lgrs.2012.2236075.

Mashburn, J., Axelrad, P., Lowe, S. T., & Larson, K. M. (2018). Global Ocean altimetry with GNSS reflections from TechDemoSat-1. *IEEE Transactions on Geoscience and Remote Sensing*, *56*, 4088–4097. https://doi.org/10.1109/tgrs.2018.2823316.

Nguyen, V. A., Nogués-Correig, O., Yuasa, T., Masters, D., & Irisov, V. (2020). Initial GNSS phase altimetry measurements from the spire satellite constellation. *Geophysical Research Letters*, *47*, e2020GL088308. https://doi.org/10.1029/2020gl088308.

Obara, K. (2002). Nonvolcanic deep tremor associated with subduction in Southwest Japan. *Science*, *296*, 1679–1681. https://doi.org/10.1126/science.1070378.

Ohta, Y., Kobayashi, T., Tsushima, H., Miura, S., Hino, R., Takasu, T., et al. (2012). Quasi real-time fault model estimation for near-field tsunami forecasting based on RTK-GPS analysis: Application to the 2011 Tohoku-oki earthquake (Mw 9.0). *Journal of Geophysical Research: Solid Earth*, *117*, B02311. https://doi.org/10.1029/2011jb008750.

Ohta, Y., & Ohzono, M. (2022). Potential for crustal deformation monitoring using a dense cell phone carrier global navigation satellite system network. *Earth Planets and Space*, *74*, 25. https://doi.org/10.1186/s40623-22-01585-7.

Ozawa, S., Nishimura, T., Suito, H., Kobayashi, T., Tobita, M., & Imakiire, T. (2011). Coseismic and postseismic slip of the 2011 magnitude-9 tohoku-oki earthquake. *Nature*, *475*, 373–376. https://doi.org/10.1038/nature10227.

Reid, H. F. (1910). *The California earthquake of April 18, 1906: Report of the state earthquake investigation commission, 2. The mechanics of the earthquake*. Washington DC, USA: Carnegie Institute of Washington.

Roessler, C. J., Morton, Y. J., Wang, Y., & Nerem, R. S. (2022). Coherent GNSS-reflection characterization over ocean and sea ice based on spire global CubeSat data. *IEEE Transactions on Geoscience and Remote Sensing*, *60*, 5801918. https://doi.org/10.1109/tgrs.2021.3129999.

Rogers, G., & Dragert, H. (2003). Episodic tremor and slip on the Cascadia subduction zone: The chatter of silent slip. *Science*, *300*, 1942–1943. https://doi.org/10.1126/science.1084783.

Ruf, C., Lyons, A., Unwin, M., Dickinson, J., Rose, R., Rose, D., et al. (2013). CYGNSS: Enabling the future of hurricane prediction. *IEEE Geoscience and Remote Sensing Magazine*, *1*, 52–67. https://doi.org/10.1109/mgrs.2013.2260911.

Sato, M., Ishikawa, T., Ujihara, N., Yoshida, S., Fujita, M., Mochizuki, M., et al. (2011). Displacement above the hypocenter of the 2011 Tohoku-oki earthquake. *Science*, *332*, 1395. https://doi.org/10.1126/science.1207401.

Scholz, C. H., Wyss, M., & Smith, S. W. (1969). Seismic and aseismic slip on the San Andreas fault. *Journal of Geophysical Research*, *74*, 2049–2069. https://doi.org/10.1029/jb074i008p02049.

Smith, S. W., & Wyss, M. (1968). Displacement on the San Andreas fault subsequent to the 1966 Parkfield earthquake. *Bulletin of the Seismological Society of America*, *58*, 1955–1973. https://doi.org/10.1785/bssa0580061955.

Solgaard, A., Kusk, A., Boncori, J. P. M., Dall, J., Mankoff, K. D., Ahlstrøm, A. P., et al. (2021). Greenland ice velocity maps from the PROMICE project. *Earth System Science Data*, *13*, 3491–3512. https://doi.org/10.5194/essd-13-3491-2021.

Spiess, F. (1980). Acoustic techniques for marine geodesy. *Marine Geodesy*, *4*, 13–27. https://doi.org/10.1080/15210608009379369.

Spiess, F. (1985). Suboceanic geodetic measurements. *IEEE Transactions on Geoscience and Remote Sensing*, *GE-23*, 502–510. https://doi.org/10.1109/tgrs.1985.289441.

Spiess, F. N., Chadwell, C., Hildebrand, J. A., Young, L. E., Purcell, G. H., & Dragert, H. (1998). Precise GPS/acoustic positioning of seafloor reference points for tectonic studies. *Physics of the Earth and Planetary Interiors*, *108*, 101–112. https://doi.org/10.1016/s0031-9201(98)00089-2.

Stevens, L. A., Behn, M. D., McGuire, J. J., Das, S. B., Joughin, I., Herring, T., ... King, M. A. (2015). Greenland supraglacial lake drainages triggered by hydrologically induced basal slip. *Nature*, *522*, 73–76. https://doi.org/10.1038/nature14480.

Stevens, L. A., Nettles, M., Davis, J. L., Creyts, T. T., Kingslake, J., Ahlstrom, A. P., & Larsen, T. B. (2022). Helheim glacier diurnal velocity fluctuations driven by surface melt forcing. *Journal of Glaciology*, *68*, 77–89. https://doi.org/10.1017/jog.2021.74.

van Dam, T., Wahr, J., Milly, P. C. D., Shmakin, A. B., Blewitt, G., Lavallée, D., et al. (2001). Crustal displacements due to continental water loading. *Geophysical Research Letters*, *28*, 651–654. https://doi.org/10.1029/2000gl012120.

Xie, S., Law, J., Russell, R., Dixon, T. H., Lembke, C., Malservisi, R., et al. (2019). Seafloor geodesy in shallow water with GPS on an anchored spar buoy. *Journal of Geophysical Research: Solid Earth*, *124*, 12116–12140. https://doi.org/10.1029/2019jb018242.

Yokota, Y., Ishikawa, T., Watanabe, S.-I., Tashiro, T., & Asada, A. (2016). Seafloor geodetic constraints on interplate coupling of the Nankai trough megathrust zone. *Nature*, *534*, 374–377. https://doi.org/10.1038/nature17632.

Chapter 2

Technical aspects of GNSS data processing

Jianghui Geng

GNSS Research Center, Wuhan University, Wuhan, China; Innovation Academy for Precision Measurement Science and Technology, Chinese Academy of Sciences, Wuhan, China

1 GNSS measurements

GNSS receivers produce two fundamental ranging measurements for each satellite: pseudorange and carrier phase. This section introduces GNSS observation equations and error sources. It starts with an introduction of the basic observation equations for pseudorange and carrier-phase measurements, as well as the general mathematical expressions for commonly used linear combination observables. Subsequently, the errors associated with satellites, signal propagation, and receivers are discussed.

1.1 GNSS observation equations

1.1.1 Pseudorange measurements

The pseudorange $P_{r,i}^s$ can be modeled by using the signal reception time $\tilde{t}_r$ measured by the receiver clock (used as the time tag) and the signal emission time $\tilde{t}^s$ measured by the satellite clock as

$$P_{r,i}^s = c \cdot (\tilde{t}_r - \tilde{t}^s) \tag{2.1}$$

where c is the speed of light in vacuum, the superscript s denotes the satellite, while the subscripts r and i denote the receiver and the signal frequency, respectively.

Since the receiver clock is not synchronized with the GPS time (coordinate time, as described by Blewitt, 2007), there is always an offset between the receiver-measured arrival time $\tilde{t}_r$ and the true arrival coordinate time t_r, which can be expressed as the sum of the receiver clock offset from coordinate time dt_r, receiver hardware delay $d_{r,i}$, and unmodeled errors ε_t, including receiver noise and some unmodeled nuisance like clock jumps, that is

$$\tilde{t}_r = t_r + dt_r + d_{r,i} + \varepsilon_t \tag{2.2}$$

Considering the clock offset of the satellite dt^s, the satellite hardware delay d_i^s, and the relativistic effect δt^s, the signal emission time $\tilde{t}^s$ can be decomposed into

$$\tilde{t}^s = t^s + dt^s + d_i^s + \delta t^s \tag{2.3}$$

where t^s is the coordinate time at the satellite. Note that the relativistic effect δt^s here is the delay term caused by different gravitational fields and different speeds, not that the relative time is changed into the absolute time (see Section 1.2.1 for further explanation).

The arrival coordinate time t_r is not just the signal emission coordinate time t^s plus the signal propagation time τ, but is also affected by various propagation delays such as the relativistic signal delay due to space-time curvature δt_r, tropospheric delay T_r^s, ionospheric delay $I_{r,i}^s$, and other propagation terms Δt, as necessary. Consequently, the difference between coordinate time at the receiver and satellite, i.e., the so-called light-time equation, is

$$t_r - t^s = \tau + \delta t_r + T_r^s/c + I_{r,i}^s/c + \Delta t \tag{2.4}$$

Substituting Eqs. (2.2)–(2.4) into Eq. (2.1), the pseudorange is finally derived as

$$P_{r,i}^s = \rho_r^s + c \cdot (dt_r - dt^s + \delta t_r^s) + c \cdot (d_{r,i} - d_i^s) + I_{r,i}^s + T_r^s + \varepsilon_P \tag{2.5}$$

where $\delta t_r^s = \delta t_r - \delta t^s$ is the relativistic correction, which combines the relativistic clock correction δt^s and the relativistic signal delay due to space-time curvature δt_r. ε_P denotes unmodeled errors including multipath and pseudorange noise. Note that $d_{r,i}$ and d_i^s are also called the code biases in the following content.

Note that all the equations in this chapter are solved in an Earth-Centered, Earth-Fixed (ECEF) reference frame at the signal reception time $\tilde{t}_r$ in a specific timescale. However, the ECEF reference frame rotates around its axis from the unknown signal emission time $\tilde{t}^s$ to the recorded signal reception time $\tilde{t}_r$, which adds an extra term $\Delta\rho_r^s$ to the geometric distance:

$$\rho_r^s = \tau/c + \Delta\rho_r^s \tag{2.6}$$

This term $\Delta\rho_r^s$ is called the Earth rotation effect or the Sagnac effect, and it requires iterative steps starting from an initial receiver position to solve this equation. A description of the Earth rotation effect can be found in Section 1.2.1.

1.1.2 Carrier-phase measurements

In addition to pseudorange, another fundamental measurement from GNSS receivers is the carrier phase, which plays a key role in high-precision GNSS. Briefly, the observation equation for carrier-phase measurement $\Phi_{r,i}^s$ in the unit of length can be written as

$$\Phi_{r,i}^s = \rho_r^s + c \cdot (dt_r - dt^s + \delta t_r^s) + c \cdot (\delta_{r,i} - \delta_i^s) - I_{r,i}^s + T_r^s + \lambda_i\left(N_{r,i}^s + \varphi_{pw}\right) + \varepsilon_\Phi \tag{2.7}$$

where $\delta_{r,i}$ and δ_i^s are the phase hardware delays for receiver r and satellite s, respectively, which are also called phase biases throughout. $N_{r,i}^s$ denotes the carrier-phase integer ambiguity and φ_{pw} is the phase wind-up effect (see Section 1.2.1 for further explanation). ε_Φ denotes unmodeled errors including multipath and phase noise. Note that the sign of the ionospheric delay in Eq. (2.7) is opposite to that of the pseudorange in Eq. (2.6). Of particular note, high-precision GNSS often requires identifying the integer quantity of $N_{r,i}^s$ (i.e., ambiguity resolution), which is usually difficult since these ambiguities are highly correlated with clock offset parameters. The integer property of ambiguity will be destroyed due to error propagation.

1.1.3 Linear combination observables

In GNSS data processing, we usually form combination observables using raw pseudorange and carrier phase to facilitate the elimination of nuisance unknowns and integer-cycle resolution of ambiguities. Linear combination observables generally refer to observation combinations between signals from the same satellite and receiver.

For brevity, the geometric terms are usually denoted as $\tilde{\rho}_r^s$, which contains the geometric distance ρ_r^s, the receiver clock dt_r, the satellite clock dt^s, the relativistic term δt_r^s, and the tropospheric delay T_r^s, that is

$$\tilde{\rho}_r^s = \rho_r^s + c \cdot (dt_r - dt^s + \delta t_r^s) + T_r^s \tag{2.8}$$

The first-order ionospheric delay is expressed in terms of the total electron content (TEC) along the signal propagation path from satellite s to receiver r

$$I_{r,i}^s = \frac{40.3\,\text{TEC}}{f_i^2} = \frac{\mu_r^s}{f_i^2} \tag{2.9}$$

where the TEC values are in the unit of TEC units (1 TECU $= 10^{16}$ electrons/m^2).

Substituting Eqs. (2.8), (2.9) into Eqs. (2.5), (2.7), the expressions of the pseudorange and carrier-phase observation equation become

$$\begin{cases} P_{r,i}^s = \tilde{\rho}_r^s + c \cdot d_{r,i}^s + \dfrac{\mu_r^s}{f_i^2} + \varepsilon_P \\ \Phi_{r,i}^s = \tilde{\rho}_r^s + c \cdot \delta_{r,i}^s - \dfrac{\mu_r^s}{f_i^2} + \lambda_i\left(N_{r,i}^s + \varphi_{pw}\right) + \varepsilon_\Phi \end{cases} \tag{2.10}$$

where the satellite and receiver hardware delays are $d_{r,i}^s = d_{r,i} - d_i^s$ and $\delta_{r,i}^s = \delta_{r,i} - \delta_i^s$. Assuming that n frequencies of observations are available for each satellite, the general expression for a linear combination observable consisting of carrier phase and pseudorange can be expressed as (Henkel & Gunther, 2012)

$$Com_r^s = \sum_{i=1}^{n}(\alpha_i \Phi_{r,i}^s + \beta_i P_{r,i}^s)$$

$$= \sum_{i=1}^{n}(\alpha_i + \beta_i)\tilde{\rho}_r^s - \sum_{i=1}^{n}\left(\frac{\alpha_i}{f_i^2} - \frac{\beta_i}{f_i^2}\right)\mu_r^s + c \cdot \sum_{i=1}^{n}(\alpha_i \delta_{r,i}^s - \beta_i d_{r,i}^s) \quad (2.11)$$

$$+ \sum_{i=1}^{n}(\alpha_i \lambda_i N_{r,i}^s) + \sum_{i=1}^{n}(\alpha_i \lambda_i)\varphi_{pw} + \sum_{i=1}^{n}\left(\alpha_i \varepsilon_{\Phi_{r,i}^s} - \beta_i \varepsilon_{P_{r,i}^s}\right)$$

where Com_r^s is the linear combination observable in the unit of length; α_i and β_i are the combination coefficients of the carrier phase $\Phi_{r,i}^s$ and the pseudorange $P_{r,i}^s$, respectively.

Wide-lane- and narrow-lane combinations

Carrier-phase combination observables will result in the combination of raw ambiguities, which is ambiguity $N_{r,c}^s$ with a wavelength of λ_c:

$$\lambda_c N_{r,c}^s = \sum_{i=1}^{n} \alpha_i \lambda_i N_{r,i}^s = \lambda_c \sum_{i=1}^{n} \alpha_i \frac{\lambda_i}{\lambda_c} N_{r,i}^s \quad (2.12)$$

By defining I_i as

$$I_i = \alpha_i \frac{\lambda_i}{\lambda_c} = \alpha_i \frac{f_c}{f_i} \quad (2.13)$$

the combination frequency f_c, wavelength λ_c, and ambiguity $N_{r,c}^s$ can be expressed as

$$\begin{cases} f_c = \sum_{i=1}^{n} I_i f_i \bigg/ \sum_{i=1}^{n} \alpha_i = \sum_{i=1}^{n} I_i k_i f_0 \bigg/ \sum_{i=1}^{n} \alpha_i = k_c f_0 \\ k_c = \sum_{i=1}^{n} I_i k_i \bigg/ \sum_{i=1}^{n} \alpha_i \\ \lambda_c = \frac{c}{f_c} = \frac{\lambda_0}{k_c} \\ N_{r,c}^s = \sum_{i=1}^{n} I_i N_{r,i}^s \end{cases} \quad (2.14)$$

where f_0 is the base frequency, which can be used to recover the frequency of signal i by multiplying an integer k_i which is $f_i = k_i f_0$.

According to Eq. (2.14), the combination wavelength λ_c depends on k_c. By choosing different k_c, combination observables of different wavelengths can be formed, which can be classified as (Cocard et al., 2008)

$$\lambda_c = \begin{cases} \lambda_{WL}, & \lambda_c > \max(\lambda_i) \\ \lambda_{IL}, & \min(\lambda_i) < \lambda_c < \max(\lambda_i), i = 1 \cdots n \\ \lambda_{NL}, & \lambda_c < \min(\lambda_i) \end{cases} \quad (2.15)$$

where λ_{WL} denotes the wavelength of wide-lane combinations, which is even larger than $\max(\lambda_i)$; λ_{NL} denotes the wavelength of narrow-lane combinations, which is shorter than $\min(\lambda_i)$; λ_{IL} is the wavelength of intermediate-lane combinations, which lies between $\max(\lambda_i)$ and $\min(\lambda_i)$.

The wide-lane combinations have longer wavelengths, which are favorable to ambiguity resolution. In particular, the combination coefficients of the pseudorange β_i are all zero, and the sum of the carrier-phase coefficients is 1, i.e., $\sum_{i=1}^{n} \alpha_i = 1$ to ensure that geometry-related parameters (i.e., $\tilde{\rho}_r^s$) can be eliminated. By assigning the integer coefficients as $I_1 = +1$ and $I_2 = -1$, the wide-lane, carrier-phase combination in the unit of cycles can be derived as

$$\varphi_{r,WL}^s = \frac{\Phi_{r,1}^s}{\lambda_1} - \frac{\Phi_{r,2}^s}{\lambda_2} = \varphi_{r,1}^s - \varphi_{r,2}^s \qquad (2.16)$$

The corresponding wide-lane combination in the unit of length can be written as

$$\Phi_{r,WL}^s = \lambda_{WL}\varphi_{r,WL}^s = \frac{f_1}{f_1-f_2}\Phi_{r,1}^s - \frac{f_2}{f_1-f_2}\Phi_{r,2}^s, \lambda_{WL} = \frac{c}{f_1-f_2} \qquad (2.17)$$

When the integer coefficients are assigned as $I_1 = +1$ and $I_2 = +1$, the narrow-lane, carrier-phase combination can be obtained:

$$\begin{cases} \varphi_{r,NL}^s = \varphi_{r,1}^s + \varphi_{r,2}^s \\ \Phi_{r,NL}^s = \lambda_{NL}\varphi_{r,NL}^s = \frac{f_1}{f_1+f_2}\Phi_{r,1}^s + \frac{f_2}{f_1+f_2}\Phi_{r,2}^s, \lambda_{NL} = \frac{c}{f_1+f_2} \end{cases} \qquad (2.18)$$

Modernized GPS, Galileo, BDS (BeiDou System), and QZSS (Quasi-Zenith Satellite System) all support three or even more frequencies of signals, which means that more abundant wide-lane combinations can be formed (e.g., Clara de Lacy et al., 2012; Geng & Bock, 2013; Li et al., 2017; Wang & Rothacher, 2013). It may be noted that wide-lane combination is particularly sensitive to relative rotation (phase wind-up), especially for moving vehicles turning a corner, for instance. On the other hand, the narrow-lane combination is particularly sensitive to measurement errors, such as multipath and receiver noise.

Ionosphere-free combination

If we have $\sum_{i=1}^n \left(\frac{\alpha_i}{f_i^2} - \frac{\beta_i}{f_i^2}\right) = 0$, an ionosphere-free combination observable can be formed. Meanwhile, for a geometry-preserving combination, we also require $\sum_{i=1}^n (\alpha_i + \beta_i) = 1$. Therefore, the dual-frequency, ionosphere-free pseudorange and carrier-phase combination equations are:

$$\begin{cases} P_{r,IF}^s = \frac{f_1^2}{f_1^2-f_2^2}P_{r,1}^s - \frac{f_2^2}{f_1^2-f_2^2}P_{r,2}^s \\ \Phi_{r,IF}^s = \frac{f_1^2}{f_1^2-f_2^2}\Phi_{r,1}^s - \frac{f_2^2}{f_1^2-f_2^2}\Phi_{r,2}^s \end{cases} \qquad (2.19)$$

One of the commonly used ionosphere-free carrier-phase combination equations in the unit of cycle is

$$\varphi_{r,IF}^s = \frac{f_1}{c}\Phi_{r,IF}^s = \frac{f_1}{c}\left(\frac{f_1^2}{f_1^2-f_2^2}\frac{c}{f_1}\varphi_{r,1}^s - \frac{f_2^2}{f_1^2-f_2^2}\frac{c}{f_2}\varphi_{r,2}^s\right)$$
$$= \frac{f_1^2}{f_1^2-f_2^2}\varphi_{r,1}^s - \frac{f_1 f_2}{f_1^2-f_2^2}\varphi_{r,2}^s \qquad (2.20)$$

Geometry-free combination

If $\sum_{i=1}^n (\alpha_i + \beta_i) = 0$, a geometry-free combination observable is formed. Such a combination observable is often used for ionosphere delay extraction, cycle slip detection, multipath error analysis, etc. When the conditions that the sum of ionospheric combination coefficients is 1 and the sum of geometric combination coefficients is 0 are both satisfied, the dual-frequency pseudorange and carrier-phase ionospheric combination observables can be obtained, that is:

$$\begin{cases} P_{r,I}^s = \frac{f_1^2 f_2^2}{f_1^2-f_2^2}(P_{r,2}^s - P_{r,1}^s) = \mu_r^s + \frac{f_1^2 f_2^2}{f_1^2-f_2^2}\left(c \cdot (d_{r,2}^s - d_{r,1}^s) + \varepsilon_{P_2} - \varepsilon_{P_1}\right) \\ \Phi_{r,I}^s = \frac{f_1^2 f_2^2}{f_1^2-f_2^2}(\Phi_{r,1}^s - \Phi_{r,2}^s) \\ \qquad = \mu_r^s + \frac{f_1^2 f_2^2}{f_1^2-f_2^2}\left(c \cdot (\delta_{r,2}^s - \delta_{r,1}^s) + \lambda_1 N_{r,1}^s - \lambda_2 N_{r,2}^s + (\lambda_1 - \lambda_2)\varphi_{pw}\right) + \varepsilon_{\Phi_2} - \varepsilon_{\Phi_1} \end{cases} \qquad (2.21)$$

When the sum of ionospheric combination coefficients and the sum of geometric combination coefficients are both zero, the dual-frequency, geometry-free and ionosphere-free combination, i.e., the Melbourne-Wübbena combination (Melbourne, 1985; Wübbena, 1985), is obtained as

$$\begin{aligned}
C_{r,MW}^s &= \Phi_{r,WL}^s - P_{r,NL}^s = \frac{f_1}{f_1-f_2}\Phi_{r,1}^s - \frac{f_2}{f_1-f_2}\Phi_{r,2}^s - \left(\frac{f_1}{f_1+f_2}P_{r,1}^s + \frac{f_2}{f_1+f_2}P_{r,2}^s\right) \\
&= \lambda_w N_{r,w}^s - c \cdot \left(\frac{f_1}{f_1-f_2}\delta_{r,1}^s + \frac{f_2}{f_1-f_2}\delta_{r,2}^s - \left(\frac{f_1}{f_1+f_2}d_{r,1}^s + \frac{f_2}{f_1+f_2}d_{r,2}^s\right)\right) \\
&\quad + \frac{f_1}{f_1-f_2}\varepsilon_{\Phi_{r,1}^s} + \frac{f_2}{f_1-f_2}\varepsilon_{\Phi_{r,2}^s} - \left(\frac{f_1}{f_1+f_2}\varepsilon_{P_{r,1}^s} + \frac{f_2}{f_1+f_2}\varepsilon_{P_{r,2}^s}\right)
\end{aligned} \quad (2.22)$$

The Melbourne-Wübbena combination includes the wide-lane ambiguity, the pseudorange and carrier-phase biases, and the measurement errors such as observation noise and multipath errors. It eliminates the ionospheric delay as well as geometric terms such as the geometric distance from the satellite to the receiver, the satellite clock offset, the receiver clock offset, and the tropospheric delay.

1.2 GNSS error sources (Fig. 2.1)

1.2.1 Satellite-related errors

Satellite orbit error

The differences between the satellite positions given by the satellite ephemeris and their true position coordinates are called the satellite orbit errors. The magnitude of orbit errors in GNSS positioning mainly depends on the quality of satellite orbit determination and is also directly related to the time interval of ephemeris extrapolation.

Satellite clock offset

The ranging precision of GNSS signals depends on the synchronization precision of the satellite clock and the receiver clock. Satellite clock offset is the difference between the GNSS satellite clock's time system and the standard time system (GPS time), including the physical synchronization error and the mathematical synchronization error.

Satellite antenna PCO/PCV

The GNSS observation is the distance between the phase centers of the receiver antenna and the satellite antenna, both of which have phase-center offsets (PCOs). The physical center of the antenna and the mean phase center (PC) for measurements are generally not the same. Their difference is called PCO, which is usually constant for a specific antenna.

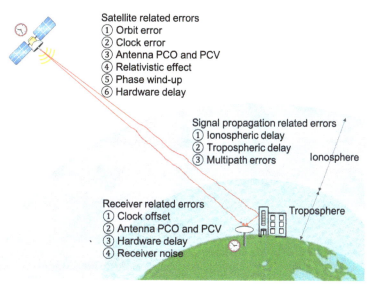

FIG. 2.1 GNSS error sources.

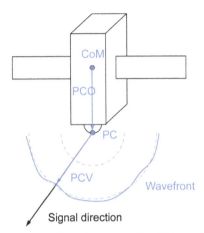

FIG. 2.2 Illustration of the center of mass (CoM), phase center (PC), phase-center offset (PCO), and phase-center variation (PCV) for a satellite antenna.

Specifically, the antenna PCO of a satellite is defined with respect to its center of mass (CoM) (Fig. 2.2). The difference between the instantaneous phase center and the mean phase center is called phase center variations (PCVs), whose values are a function of the satellite azimuth and nadir angles.

Relativistic effect

The relativistic effects are caused by the different velocities and positions of the satellite clock and the receiver clock in the inertial reference frame, which can be divided into the special and the general relativity effect. Their combined effect makes the frequency of the satellite clock become faster, and thus distorts pseudorange and carrier-phase observations. If the GNSS satellite orbit is assumed to be circular, the effect of relativistic effects on the satellite clock frequency is a constant, and GNSS satellites are designed with a reduced clock frequency to correct for this constant. Its periodic term due to eccentricity can be corrected by modeling; for ground receivers, it is generally corrected only for the first-order term, while ignoring the higher order terms.

In addition to the effect on the satellite clock frequency, general relativistic effects include a geometric delay in the signal propagation due to the Earth's gravitational field, often referred to as the gravitational delay, which can also be corrected by modeling. This delay is related to the location of the station and the satellite. The correction formula can be found in the relevant literature (e.g., Wellenhof et al., 2001).

Earth rotation effect (Sagnac effect)

The Earth rotation effect, also known as the Sagnac effect, results from the rotation of the ECEF reference frame during the signal propagation time. This means that the coordinates of a satellite in ECEF at its signal emission time will differ by several meters to several tens of meters from its coordinates at the signal arrival time, although its coordinates in an Earth-Centered Inertial (ECI) reference frame remain unchanged.

Phase wind-up effect

GNSS satellites emit right-handed circularly polarized (RHCP) electromagnetic waves, and thus the carrier phase observed by the receiver is related to the cross-orientation relationship between the satellite and the receiver antennas. The rotation of the receiver antenna or the satellite antenna around the polarization axis would change the phase observation by up to one cycle. Such phase offset caused by the relative rotation between the receiver and satellite antennas is called the phase wind-up effect.

Satellite hardware delay

Satellite hardware delay denotes the signal delay caused in the analog and digital paths of the signal generation unit and antenna. It depends on the satellite, observation type, and frequency. Note that the absolute signal delays are unobservable and only their differences could be calculated. In GNSS data processing, satellite hardware biases are usually absorbed into other parameters such as satellite clocks and ambiguities.

1.2.2 Signal propagation-related errors

Ionospheric delay

The atmosphere between approximately 60 and a few thousand kilometers in terms of altitude is called the ionosphere. As the ionosphere is a dispersive medium, the propagation speed of the satellite signals changes as it passes through the ionosphere, and this change causes errors in the distances measured by a receiver. The ionospheric delay depends on both TEC along the signal propagation path and the frequency of the signal, which vary with time, position, space weather, etc. Note that the ionospheric delays within pseudorange and carrier-phase observations are of the same magnitude but with opposite signs.

Tropospheric delay

The troposphere denotes the atmosphere that is approximately below 50 km in terms of altitude. Tropospheric delay is caused by refraction due to water vapor and dry air contents when GNSS signals pass through the troposphere. Since the troposphere is neutral, the propagation speed of the satellite signals in the troposphere is independent of the signal frequency. The propagation speed of the satellite signals then depends on meteorological factors such as air temperature, pressure, and relative humidity. The magnitudes and signs of tropospheric delay are the same for pseudorange and carrier-phase observations.

Multipath effects

The interference between the line-of-sight signal and the reflected signals around the receiver antenna causes multipath effects. Specifically, the multipath effect has a much greater impact on pseudorange than carrier-phase observations. Pseudorange multipath effect can be up to several meters, while carrier-phase multipath is up to 1/4 cycle (about 5 cm). The multipath error is governed by the surrounding objects of the receiver antenna.

1.2.3 Receiver-related errors

Receiver clock offset

Similar to the satellite clock, the receiver clock offset is defined as the difference between the receiver clock and GPS time, which is caused by the frequency drift of the crystal oscillator inside the receiver. The receiver clock offset depends mainly on the quality of the clock and is also related to the temperature and other environmental effects. Pseudorange and carrier-phase observations share the same receiver clock offset.

Receiver antenna PCO and PCV

Receiver antenna PCO and PCV are defined in the same way as those for satellites, even though the receiver has a physically accessible antenna reference point (ARP) (Fig. 2.3). They depend on the receiver antenna and the signal frequency. Relative and absolute antenna calibration techniques are used to quantify precisely the receiver PCO/PCV effects, and the antenna calibration ANTEX (Antenna Exchange format) files have been provided by the International GNSS Service (IGS).

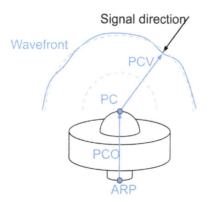

FIG. 2.3 Illustration of the antenna reference point (ARP), phase center (PC), phase center offset (PCO), and phase center variation (PCV) for a receiver antenna.

Receiver hardware delay

Receiver hardware delay arises from the time delay caused by the signal from the antenna to the correlator in the receiver. Similar to satellite hardware delay, it depends on the equipment, signal types, and frequencies. However, for the same signal type, such as pseudorange, the receiver hardware delays on the same frequency are usually presumed the same for all satellite signals from the same constellation. The receiver hardware delays for signals from different GNSSs are distinct, i.e., inter-system bias, and this bias needs to be estimated or corrected in multi-GNSS data processing. For legacy GLONASS signals that rely on the Frequency Division Multiple Access (FDMA) technique, hardware delays are different even among the GLONASS frequencies.

Receiver observation noise

Receiver observation noise is a random error, caused by imperfections of the different electrical components in the signal processing chain and external interference from natural or artificial sources. In general, receiver pseudorange noise might reach about 1 m, with a standard deviation of about one decimeter or less when tracking the modernized signals with high chip rates. Carrier-phase noise is about a few millimeters, and its standard deviation is usually less than 1 mm.

1.2.4 Other errors

Reference frame misalignment

GNSS products (e.g., orbits and station coordinates) are demonstrated in the IGS reference frame, which is aligned to the international terrestrial reference frame (ITRF). To conduct this alignment, a 7-parameter or 14-parameter transformation is often required. In the satellite orbit estimation, ground stations could be fixed or tightly constrained to ITRF solutions, or rather, no solutions are fixed but the no-net-rotation constraints are introduced. Even more, if there is nothing constrained on the station coordinates in estimation, a postalignment of both station and satellite solutions is also required. In this way, the precise orbit products are aligned to ITRF, with which the GNSS solutions are naturally in the same reference frame. Of particular note, the effect of the antenna phase center error could make the scale of the global GNSS solutions deviate from the ITRF scale.

Tide displacement

According to the IERS convention, a ground station bears a series of tide effects, including ocean tide, solid earth tide, pole tide, etc. These tide effects are significant and periodic mainly in the up direction of a station. Ocean tidal loading and solid earth tide are the two major components up to the decimeter level, and they are aroused chiefly by lunisolar attraction. Apart from this, polar motion could also result in small shape changes of the solid earth, and introduce deformation to a station up to the centimeter level. A mass redistribution of the oceanic water could change the gravitational field across the surface of the Earth; on the other hand, this mass redistribution also directly changes the nearby landforms. The tide displacements can be well modeled, as described in the IERS convention (e.g., Petit & Luzum, 2010).

2 GNSS positioning

2.1 Precise point positioning (PPP)

Precise point positioning (PPP) was first proposed by Zumberge et al. (1997). Standard single-point positioning can achieve meter-level positioning precision with pseudorange measurements. For PPP, it can also realize absolute positioning on the globe using a single GNSS receiver, but it further requires carrier-phase measurements and precise orbit/clock corrections to achieve centimeter-level positioning (Fig. 2.4). Note that such satellite orbit/clock corrections are estimated using a global network of continuously operating reference stations (CORS). PPP is suitable for wide-area precise positioning, seismic monitoring, water vapor inversion, ionospheric monitoring, etc.

2.1.1 Basics of precise point positioning

Classic PPP algorithms use the ionosphere-free combination observable, which eliminates the first-order ionospheric delay based on the observed dual-frequency pseudorange and the carrier phase (from the receiver). The ionosphere-free PPP observed minus computed observations (OMC) can be written as

$$\begin{cases} P_{r,IF}^s - l_{r,P_{IF}}^s = -\boldsymbol{\alpha}_r^s \cdot \mathbf{x}_r + c \cdot d\hat{t}_r^S + m_r^s T_r \\ \Phi_{r,IF}^s - l_{r,\Phi_{IF}}^s = -\boldsymbol{\alpha}_r^s \cdot \mathbf{x}_r + c \cdot d\hat{t}_r^S + m_r^s T_r + \lambda_{IF}\hat{N}_{r,IF}^s \end{cases} \quad (2.23)$$

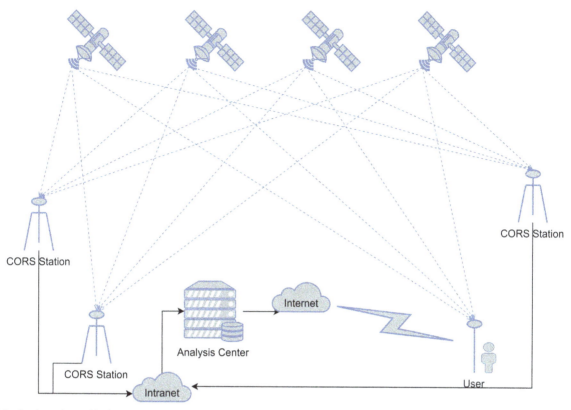

FIG. 2.4 Precise point positioning.

where $P^s_{r,IF} = \frac{g^2}{g^2-1}P^s_{r,1} - \frac{1}{g^2-1}P^s_{r,2}$ and $\Phi^s_{r,IF} = \frac{g^2}{g^2-1}\Phi^s_{r,1} - \frac{1}{g^2-1}\Phi^s_{r,2}$ are the ionosphere-free pseudorange and carrier-phase observations from satellite s to station r, respectively. The subscript IF stands for ionosphere-free. $g = f_1/f_2$ and f_1, f_2 are the two legacy signal frequencies from the constellation S (e.g., GPS, GLONASS, Galileo, and BDS). $l^s_{r,*}$ is the computed observations, i.e., the sum of corrections, which include the geometric distance (calculated using a priori user coordinates), the satellite clock correction, the satellite and receiver antenna PCOs/PCVs, the tropospheric delay, the phase wind-up effect, the relativistic correction, and the observable-specific biases (OSBs) for code and phase, etc. α^s_r is the unit vector from receiver r to satellite . $\mathbf{x}_r$ denotes the correction of the a priori user coordinates. $d\hat{t}^S_r$ is the receiver clock parameter to be estimated. m^s_r is the mapping function and T_r is the residual tropospheric delay. $\hat{N}^s_{r,IF}$ is the ionosphere-free ambiguity. λ_{IF} is the wavelength, which can be presumed to be the wavelength of the L1 signal.

In contrast, the undifferenced, uncombined PPP model is also widely studied due to its easy extensibility to multi-frequency observations. This model uses raw pseudorange and carrier-phase observations, rather than the ionosphere-free combinations. It then needs to estimate slant ionospheric delays and undifferenced raw ambiguities in addition to receiver coordinates, receiver clock error, and zenith tropospheric delay. Although this model has more parameters to be estimated, the ionosphere relevant information can also be obtained, that is,

$$\begin{cases} P^s_{r,i} - l^s_{r,P_i} = -\alpha^s_r \cdot \mathbf{x}_r + c \cdot d\hat{t}^S_r + m^s_r T_r + g_i^2 \hat{I}^s_r \\ \Phi^s_{r,i} - l^s_{r,\Phi_i} = -\alpha^s_r \cdot \mathbf{x}_r + c \cdot d\hat{t}^S_r + m^s_r T_r - g_i^2 \hat{I}^s_r + \lambda_i \hat{N}^s_{r,i} \end{cases} \quad (2.24)$$

where $\hat{I}^s_r$ is the L1 ionospheric delay parameter and $g_i = f_1/f_i$.

Static and kinematic PPP are two major processing modes aiming at geophysical applications. Static PPP assumes that all GNSS measurements are acquired at the same position over a period (e.g., 1 day). Daily static PPP can be used to study inter-seismic crustal movements, etc. Kinematic PPP, on the other hand, estimates positions at each epoch (e.g., 20 samples per second) to identify more rapid ground movements such as those caused by earthquakes and precursory crustal movements before volcano eruptions.

2.1.2 PPP ambiguity resolution (PPP-AR)

PPP-AR is a technique where float ambiguities are fixed to their integer candidates to decrease the correlation between ambiguities and other parameters. Successful PPP-AR can improve the positioning precision significantly and accelerate the convergence speed of PPP. According to the PPP model in Eqs. (2.23), (2.24), the ambiguities are contaminated by the code and phase biases from both the satellite and the receiver. Due to the strong correlation between the biases and other parameters (e.g., clock error), the code and phase biases cannot be estimated due to rank deficiency. In this case, the code and phase biases should be corrected in advance before PPP-AR. An advanced method to prepare the code and phase biases is to compute observable-specific biases (OSBs) for pseudorange and carrier-phase measurements using a reference network (e.g., Geng et al., 2022). One can easily subtract OSBs from raw measurements before PPP-AR (Schaer et al., 2021), that is

$$\begin{cases} \hat{P}^s_{r,i} = P^s_{r,i} - d^s_i \\ \hat{\Phi}^s_{r,i} = \Phi^s_{r,i} - \lambda_i b^s_i \end{cases} \quad (2.25)$$

where d^s_i and b^s_i are the code OSB and phase OSB for frequency i of satellite s, respectively.

PPP-AR is usually based on wide-lane and narrow-lane combination observables (Ge et al., 2008). Wide-lane ambiguities have a relatively long wavelength, and thus they are easily fixed to the correct integers. Wide-lane ambiguities can be first computed using the Melbourne-Wübbena combination $\hat{\Phi}^s_{r,MW}$ such that

$$\hat{\Phi}^s_{r,MW} = \frac{\hat{\Phi}^s_{r,1} - z^s_{r,1}}{\lambda_1} - \frac{\hat{\Phi}^s_{r,2} - z^s_{r,2}}{\lambda_2} - \frac{\mu_2 - 1}{\mu_2 + 1} \cdot \left(\frac{\hat{P}^s_{r,1} - z^s_{r,1}}{\lambda_1} - \frac{\hat{P}^s_{r,2} - z^s_{r,2}}{\lambda_2} \right) \quad (2.26)$$

where $z^s_{r,i}$ denotes the antenna PCO corrections on frequency f_i (Geng et al., 2022).

In contrast, in the case of undifferenced uncombined PPP, the wide-lane ambiguity $\hat{N}^s_{r,WL}$ can also be computed using undifferenced frequency-specific ambiguities, that is

$$\hat{N}^s_{r,WL} = \hat{N}^s_{r,1} - \hat{N}^s_{r,i} \quad (2.27)$$

Note that while the satellite-specific hardware biases can be nominally mitigated by OSBs, the receiver-specific hardware biases should be eliminated by the single-difference operation between a satellite s and a reference satellite k, that is

$$\hat{N}^{sk}_{r,WL} = \hat{N}^s_{r,WL} - \hat{N}^k_{r,WL} \quad (2.28)$$

Once the single-difference, wide-lane ambiguity is fixed, the single-difference, narrow-lane ambiguity $\hat{N}^{sk}_{r,NL}$ can be derived using the ionosphere-free ambiguity $\hat{N}^{sk}_{r,IF}$ and the fixed integer wide-lane ambiguity $\tilde{N}^{sk}_{r,WL}$, that is

$$\begin{cases} \hat{N}^{sk}_{r,IF} = \frac{g_i^2}{g_i^2 - 1} \hat{N}^{sk}_{r,1} - \frac{g_i}{g_i^2 - 1} \hat{N}^{sk}_{r,i} \\ \hat{N}^{sk}_{r,NL} = \frac{g_i + 1}{g_i} \hat{N}^{sk}_{r,IF} - \frac{1}{g_i - 1} \tilde{N}^{sk}_{r,WL} \end{cases} \quad (2.29)$$

The integer-cycle resolution of the wide-lane ambiguity $\tilde{N}^{sk}_{r,WL}$ and the narrow-lane ambiguity $\tilde{N}^{sk}_{r,NL}$ can be achieved using the LAMBDA (Least-squares AMBiguity Decorrelation Adjustment) method (Teunissen et al., 1997). Eqs. (2.27)–(2.29) are also used to convert the raw variance-covariance matrix into that for the wide-lane and narrow-lane ambiguities. By injecting the new variance-covariance matrix into the LAMBDA function, the integer ambiguity candidates can be identified. Once PPP ambiguities are resolved successfully, the precision of other parameters (e.g., positions, troposphere delays, etc.) can potentially be improved.

2.1.3 Multiconstellation and multifrequency GNSS positioning

Apart from PPP-AR, multi-GNSS, and multifrequency signals also pose new opportunities to improve GNSS positioning. An increasing number of GNSS satellites have been launched into space over the past decade. Most of them have multiple frequency bands. Fig. 2.5 shows the frequency bands for different satellite constellations. GPS has three L-bands of signal

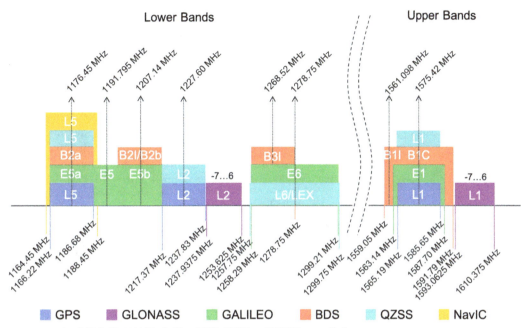

FIG. 2.5 Frequency bands of GPS, GLONASS, Galileo, BDS, QZSS, and IRNSS constellations.

frequencies (i.e., L1, L2, and L5) and GLONASS has the FDMA signals (on G1 and G2). Galileo has five signal frequencies (i.e., E1, E5a, E5b, E5, and E6). BDS satellites nominally emit five signal frequencies (i.e., B1C, B1I, B2a, B2I/B2b, and B3I). In particular, BDS-2 broadcasts B1I, B2I, and B3I signals and BDS-3 emits B1C, B2a, B2b, and B3I signals. QZSS has four signal frequencies where L1, L2, and L5 are the same as those of GPS, and L6/LEX is a new signal. IRNSS (Indian Regional Navigation Satellite System) has only one frequency, L5.

In the case of multi-GNSS PPP by integrating GPS, GLONASS, Galileo, and BDS measurements, one should be aware that the receiver clock parameter absorbs receiver-specific hardware biases. As a result, one receiver clock parameter for each GNSS or inter-system bias parameter should be introduced into the PPP model. In Eq. (2.24), we set a constellation-specific receiver clock parameter for each GNSS. Here, we introduce the inter-system code bias, which can be presumed as a constant. The equation can be reformatted as

$$\begin{cases} P_{r,i}^s - l_{r,P_i}^s = -\boldsymbol{\alpha}_r^s \cdot \mathbf{x}_r + c \cdot d\hat{t}_r^G + h_r^S + m_r^s T_r + g_i^2 \hat{I}_r^s \\ \Phi_{r,i}^s - l_{r,\Phi_i}^s = -\boldsymbol{\alpha}_r^s \cdot \mathbf{x}_r + c \cdot d\hat{t}_r^G + h_r^S + m_r^s T_r - g_i^2 \hat{I}_r^s + \lambda_i \hat{N}_{r,i}^s \end{cases} \quad (2.30)$$

where h_r^S is the inter-system code bias of system S with respect to the GPS. $d\hat{t}_r^G$ is the receiver clock parameter of the GPS.

If the GLONASS constellation is used, the inter-frequency biases (IFBs) for the FDMA signals should also be considered. In PPP-AR, the traditional approach to removing receiver-specific biases is to perform the single-difference operation between satellites. This strategy presumes that the receiver-specific biases with respect to different satellites from the same GNSS are the same. This assumption is roughly acceptable for GNSS-emitting Code Division Multiple Access (CDMA) signals (e.g., GPS, Galileo, BDS, and QZSS), but is invalid for the GLONASS constellation mainly broadcasting FDMA signals with different frequencies on 14 channels. In this case, the classic single-difference operation cannot eliminate the receiver-specific biases. The IFBs are typically stable with respect to the frequency channel number (Wanninger, 2012). The carrier-phase IFBs fitting a linear function of the channel number can be exploited to enable ambiguity resolution for GLONASS. However, the pseudorange IFBs can hardly be mitigated in this manner because they change with respect to receiver types, firmware versions, antenna types, etc. One approach for wide-lane and narrow-lane GLONASS ambiguity resolution is to only use carrier-phase-based, wide-lane combination by precalibrating ionospheric delays.

In addition, the phase IFBs are actually caused by the Differential Code Phase Biases (DCPBs), which are defined as the bias between the pseudorange and carrier-phase observations (Sleewaegen et al., 2012). The phase IFBs can thus be mitigated using the DCPB corrections easily and directly (Geng et al., 2017; Geng & Bock, 2016; Geng & Shi, 2017). After correcting for IFBs, the wide-lane and narrow-lane GLONASS ambiguities can be expressed as

$$\begin{cases} \hat{N}_{i,w}^{jk} = \hat{N}_{i,w}^{j} - \hat{N}_{i,w}^{k} + \left(ichn^j - ichn^k\right)\Delta f_w \Delta \overline{b}_w \\ \hat{N}_{i,n}^{jk} = \frac{\mu_q + 1}{\mu_q}\hat{N}_{i,IF}^{jk} - \frac{1}{\mu_q - 1}\tilde{N}_{i,w}^{jk} + \left(ichn^j - ichn^k\right)\Delta f_n \Delta \overline{b}_n \end{cases} \quad (2.31)$$

where $\Delta \overline{b}_w$ and $\Delta \overline{b}_n$ are the DCPBs of wide-lane and narrow-lane combination ambiguities. Δf_w and Δf_n are the corresponding GLONASS frequency spacing.

When processing multifrequency GNSS data, we need to carefully consider the satellite clock corrections. Since the satellite clock estimation is usually based on dual-frequency ionosphere-free combinations, the satellite clock products will absorb satellite-specific code biases. We usually presume that the code and phase OSBs are both time constant. However, Montenbruck et al. (2012) pointed out that the phase OSBs might vary over time and they cannot be ignored on GPS L5 signals emitted by Block IIF satellites. In practice, the phase OSB can be divided into

$$b_i^s = \overline{b}_i^s - \delta b_i^s \quad (2.32)$$

where $\overline{b}_i^s$ is the time-constant part and δb_i^s is the time-variable part. Normally, δb_i^s must be computed before estimating $\overline{b}_i^s$. Guo and Geng (2018) developed a method that introduces a second satellite clock parameter to accommodate the time-variable OSBs. Thus, the triple-frequency PPP model can be reformulated as

$$\begin{cases} P_{r,1}^s = \rho_r^s + c \cdot d\hat{t}_r^S - c \cdot d\hat{t}_{12}^s + m_r^s T_r + \hat{I}_r^s \\ P_{r,2}^s = \rho_r^s + c \cdot d\hat{t}_r^S - c \cdot d\hat{t}_{12}^s + m_r^s T_r + g_2^2 \hat{I}_r^s \\ P_{r,i}^s = \rho_r^s + c \cdot d\hat{t}_r^S - c \cdot d\hat{t}_i^s + m_r^s T_r + g_i^2 \hat{I}_r^s + \kappa_{r,i}^S \\ \Phi_{r,1}^s = \rho_r^s + c \cdot d\hat{t}_r^S - c \cdot d\hat{t}_{12}^s + m_r^s T_r - \hat{I}_r^s + \lambda_i \hat{N}_{r,1}^s \\ \Phi_{r,2}^s = \rho_r^s + c \cdot d\hat{t}_r^S - c \cdot d\hat{t}_{12}^s + m_r^s T_r - g_2^2 \hat{I}_r^s + \lambda_i \hat{N}_{r,2}^s \\ \Phi_{r,i}^s = \rho_r^s + c \cdot d\hat{t}_r^S - c \cdot d\hat{t}_i^s + m_r^s T_r - g_i^2 \hat{I}_r^s + \lambda_i \hat{N}_{r,i}^s \end{cases} \quad (2.33)$$

where $d\hat{t}_{12}^s$ is the legacy satellite clock and $d\hat{t}_i^s$ is the second satellite clock. $\kappa_{r,i}^S$ is a station-specific, time-constant parameter intended to absorb the satellite inter-frequency code biases. For GPS and BDS, the time-variable phase OSBs should be considered. Since an additional satellite clock product for multifrequency PPP is inconvenient for clock exchange, the time-varying phase OSBs can also be combined with the time-constant OSBs and written into the Bias-SINEX file (Geng et al., 2022).

In the case of multifrequency ambiguity resolution, an additional step is to introduce the extra-wide-lane combination observables, similar to the wide-lane combination shown in Eq. (2.16). The extra-wide-lane, wide-lane, and narrow-lane ambiguities ($\hat{N}_{r,EWL}^{sk}$, $\hat{N}_{r,WL}^{sk}$, and $\hat{N}_{r,NL}^{sk}$) can be generated by mapping the raw undifferenced ambiguities ($\hat{N}_{r,1}^{sk}$, $\hat{N}_{r,2}^{sk}$, and $\hat{N}_{r,i}^{sk}$), that is

$$\begin{bmatrix} \hat{N}_{r,EWL}^{sk} \\ \hat{N}_{r,WL}^{sk} \\ \hat{N}_{r,NL}^{sk} \end{bmatrix} = \begin{bmatrix} 0 & 1 & -1 \\ 1 & -1 & 0 \\ \frac{g_i}{g_i - 1} & \frac{-1}{g_i - 1} & 0 \end{bmatrix} \begin{bmatrix} \hat{N}_{r,1}^{sk} \\ \hat{N}_{r,2}^{sk} \\ \hat{N}_{r,i}^{sk} \end{bmatrix} + \begin{bmatrix} 0 \\ 0 \\ \frac{\tilde{N}_{r,WL}^{sk}}{1 - g_i} \end{bmatrix} \quad (2.34)$$

where i denotes the third frequency f_i. Note that the equation above has carried out the single-difference operation between satellites, and the OSBs have been corrected. The extra-wide-lane and the wide-lane ambiguities should be resolved first. Then, we can use the updated variance–covariance matrix and parameters to resolve narrow-lane ambiguities.

The advantage of multifrequency PPP-AR is to assist in forming an ambiguity-fixed, ionosphere-free, wide-lane combination observable (Geng & Bock, 2013). This combination usually has lower observation noise compared to pseudorange, and thus multifrequency PPP-AR can accelerate the convergence to centimeter-level positioning.

2.2 Carrier-phase-based relative positioning

2.2.1 Basics of relative positioning

Relative positioning technology can be traced back to the 1980s when Counselman III et al. (1983) achieved centimeter-level relative positioning through interferometric measurements based on GPS carrier-phase observations. Subsequently,

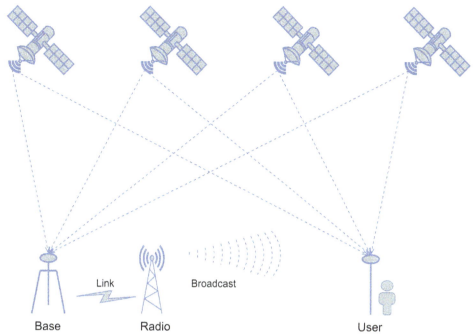

FIG. 2.6 Relative positioning.

the relative positioning technique was attempted for the establishment of a regional geodetic control network (Bock et al., 1985) and satellite orbit determination (Lichten & Border, 1987). In addition to these advancements, Blewitt (1989) focused on the problem of resolving double-difference ambiguity in long baselines. Unlike PPP, relative positioning (also known as differential positioning) needs at least two stations, a rover and a reference forming a baseline. In general, a rover station is the target station with unknown coordinates, and a reference station is a static station with known coordinates.

Differencing operations between the simultaneous observations of two stations are usually applied to GNSS relative positioning. From Fig. 2.6, the simplified observation equation of double-difference, carrier-phase observations can be given as

$$\Phi_{rs,i}^{su} = \rho_{rs,i}^{su} + \lambda_i N_{rs,i}^{su} + \varepsilon_{rs,i}^{su} \qquad (2.35)$$

where s, u represent the two satellites and r, s represent the two receivers, respectively. i and λ_i are the frequency and wavelength of carrier-phase observations, respectively. $\rho_{rs,i}^{su}$ is the double-difference geometric distance between satellites and stations. $N_{rs,i}^{su}$ is the integer double-difference ambiguity in the unit of cycles, and $\varepsilon_{rs,i}^{su}$ contains random noise and unmodeled errors.

Benefiting from the differencing operations, relative positioning has the ability to eliminate, or significantly mitigate the measurement errors such as ephemeris, clock, and atmospheric delays. Therefore, relative positioning, especially over a rather short baseline, generally guarantees a more preferable and stable positioning performance than PPP.

However, depending on the precision requirement, several types of corrections should still be considered when processing baselines of different extents. As the baseline length increases, the spatial correlation of common-mode errors between the two stations, such as atmospheric delays, decreases rapidly. Over longer baselines (e.g., >20 km), the residual atmospheric errors would be significant. In the case of very long baselines (e.g., >1000 km), corrections for Earth curvature and solid Earth tides should also be taken into account (Gupta & Pasquale, 2013).

Differing from PPP ambiguities, double-difference ambiguities naturally preserve their integer nature. To achieve rapid and reliable baseline solutions, it is critical to fix carrier-phase ambiguities to their integer truths. The reliability of ambiguity resolution in relative positioning is the key prerequisite to obtain high-precision baseline solutions. The LAMBDA method can also be used to search for the integer candidates of double-difference ambiguities.

Based on the local or regional GNSS reference stations, relative positioning is capable of maintaining fast and high-precision relative positioning service and providing atmospheric corrections for common users to speed up ambiguity fixing, which is usually called network RTK (real-time kinematic positioning).

2.2.2 Comparison between relative positioning and PPP

Both relative positioning and PPP use carrier-phase observations to achieve centimeter- or even millimeter-level positioning precision. Their difference lies in the products they use and the strategies they adopt to eliminate errors. Relative positioning uses the observations from the reference station or the reference network to assist the rover in eliminating the observation space representation (OSR) errors and achieve high-precision positioning rapidly. Relative positioning has two fundamental requirements for the reference station or network: one is that its coordinates must be precisely known to calculate OSR corrections for the rover; the second is that the reference station should be preferably close enough (e.g., <50 km) to the rover to ensure that the OSR errors (including orbit, atmospheric errors, etc.) for the reference and rover stations are highly spatially correlated. In contrast, PPP is a technique that uses a single receiver to achieve high-precision absolute positioning. Nominally, PPP does not explicitly depend on nearby reference stations, but uses precise ephemeris to mitigate orbit and clock errors, i.e., state space representation (SSR) errors.

Relative or absolute: Due to the different strategies for error mitigation, the results obtained from relative positioning and PPP are different. The positions and atmospheric delays obtained in relative positioning are essentially the difference between the rover and the reference station. For earthquake and volcano monitoring, what we care about is the displacement of the monitoring station. Relative positioning can only compute the displacement difference between the monitoring station and the reference station. If the reference station and the monitoring station are both displaced in an event, the geophysical signals in the displacement time series will be distorted. Similarly, atmosphere sounding in relative positioning suffers from the same problem. PPP has the advantages in this regard because it does not use differential corrections from the reference station, but separates each type of error to form SSR corrections. Due to the release of dependence on the reference station, the absolute positions and atmosphere delays at the rover station can be achieved by PPP.

Initialization efficiency in real time: Since both RTK and PPP use carrier-phase observations, the ambiguity unknowns make both of them suffer from a period of initializations. Over a relatively short baseline, RTK eliminates or weakens the spatially correlated errors, and thus, the underlying GNSS model is relatively strong and can usually be initialized within a few seconds. Conversely, PPP usually has a long initialization time of up to a few tens of minutes in the case of GPS-only constellation. Even worse, long re-initializations might take place repeatedly whenever the communication link for precise satellite products is interrupted.

Service area: The requirement of a dense network by relative positioning makes it impossible to provide wide-area or global positioning services. The interstation distance of a relative positioning reference network is usually less than 50 km, which will be very costly to maintain for wide-area applications. On the opposite, only precise orbit, clock, and code/phase bias products are required by PPP to achieve positions as precise as those from relative positioning. Moreover, the time consumption in the case of GNSS network analysis increases exponentially with the number of stations, while that of PPP is linearly related to the number of stations (Geng & Mao, 2021). Therefore, PPP is more suitable for wide-area, high-precision GNSS services.

2.3 Real-time GNSS

2.3.1 IGS real-time products

Thanks to the ability of providing wide area, absolute positioning service, PPP has been widely used in scientific research, education, and commercial applications over the past decade. However, the latency of postprocessing satellite products is usually about 3–15 days and the predicted satellite clock products cannot satisfy the precision requirement, which limits the application of real-time PPP in geohazard monitoring. In recent years, there has been an increasing demand for very low-latency high-precision satellite products. In this case, IGS initiated real-time service (RTS) in 2007.

IGS is an international academic organization established in 1993 and started in 1994 by the International Association of Geodesy (IAG) to support geodetic and geodynamic research using GNSS. At present, more than 200 organizations and institutions around the world have joined it to share GNSS data, products, and services.

IGS/RTS is a service that provides real-time GNSS observation streams, and broadcasts ephemeris, and high-precision orbit and clock product streams based on a global, real-time reference network supported by dozens of data centers (DCs) and analysis centers (ACs). IGS/RTS broadcasts precise satellite products using the open standard Networked Transport of RTCM via Internet Protocol (NTRIP), which encodes satellite orbit and clock correction streams into Radio Technical Commission for Maritime (RTCM) SSR messages. Currently, each IGS/RTS AC broadcasts multi-GNSS high-precision satellite products at a latency of about 5 s. Table 2.1 lists the real-time GNSS precise products provided by IGS/RTS.

TABLE 2.1 Details of IGS real-time precise satellite products (see https://igs.org/rts/contributors/).

IGS/RTS ACs	Constellation	Description
BKG	GPS/GLONASS/Galileo	Real-time orbits and clocks based on CODE orbits
CAS	GPS/GLONASS/Galileo/BDS	Real-time orbits and clocks based on GFZ orbits. Also includes VTEC (Vertical TEC) from the Global Ionospheric Map
CNES	GPS/GLONASS/Galileo/BDS	Real-time orbits and clocks based on GFZ orbits. Also includes VTEC from the Global Ionospheric Map
DLR/GSOC	GPS/GLONASS/Galileo/BDS/QZSS	Real-time orbits, clocks, and code biases based on IGV/CODE/DLR orbits
ESA/ESOC	GPS	GPS Real-time orbits and clocks using hourly orbits from ESOC
GFZ	GPS/GLONASS/Galileo/BDS	Orbits and clocks based on internal GFZ orbits every 2 h
GMV	GPS/GLONASS/Galileo/BDS	Real-time orbits and clocks based on GMV-generated orbits
NRCan	GPS	Real-time orbits and clocks based on hourly orbits from NRCan software
SHAO	GPS	Real-time orbits and clocks based on IGS Ultras
WUHAN	GPS/GLONASS/Galileo/BDS	Real-time orbits and clocks based on IGU (GPS) and IGS WHU AC orbits. Also transmits VTEC from Global Ionospheric Map

2.3.2 Real-time PPP

After receiving real-time GNSS observation and precise satellite orbit/clock streams, users can perform real-time PPP. Real-time communication in PPP requires encoding data and products. At present, the standard for real-time PPP has not been fully defined. Both RTCM and IGS have carried out part of the standardization work. They have specified a series of coding formats for observation data streams, precise orbits, clock errors, atmospheric delays, and code-/phase bias products. Real-time PPP users can obtain data and product streams in both formats from IGS.

Generally, a real-time PPP service can be mainly divided into two parts: the server and the client ends, both of which rely on NTRIP Caster to access and forward data/product streams. An NTRIP Caster is usually a server, which acts as a bridge among the reference network, the PPP servers, and the clients. It receives the reference network data streams and broadcasts them to the server and the clients, or receives the product streams from servers and broadcasts them to the clients. Fig. 2.7 shows the processing scheme for maintaining a real-time PPP service. Real-time satellite orbits are usually predicted products while the satellite clock and bias products should be estimated in real time. When estimating satellite products on the server, it is necessary to obtain broadcast ephemeris and predicted orbits from the Caster first, and then calculate real-time precise clock errors, phase biases, and other products with a latency as low as possible (e.g., less than a few seconds). The products will be converted into the SSR format using broadcast ephemeris and then broadcast to NTRIP Caster after encoding. The client only needs to receive real-time data, orbit, clock, and bias streams from the Caster to carry out PPP and PPP-AR.

Given that PPP enables the estimation of absolute ionospheric delays, real-time PPP can be used to extract small-scale abnormal signatures of the ionosphere disturbances associated with various geological hazards (e.g., earthquakes, volcanic eruptions, etc.), and therefore holds significant value for geological hazard monitoring.

2.3.3 Convergences of real-time PPP

One of the most significant drawbacks of real-time PPP is its long initialization time (or in other words, convergence time) to achieve high-precision positioning. Due to the high correlations between ambiguity and position parameters in the early PPP convergence phase, it usually takes a few tens of minutes for positions to converge to the centimeter-level precision (Geng et al., 2011). Such a long initialization time seriously undercuts the credibility of applying PPP to geohazard monitoring and other similar time-critical scenarios. If the observation condition is good, the risks of PPP re-initialization can be diminished in the case of continuous or uninterrupted real-time data processing.

Over the past decade, there have been a few strategies for fast PPP initialization: fixing integer ambiguities, integrating multi-GNSS constellations, and introducing atmosphere augmentation. Due to the slow change of satellite geometry and

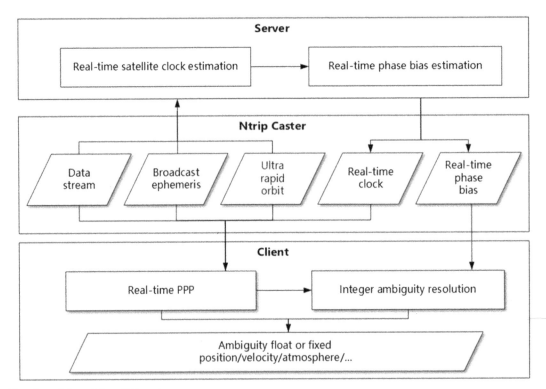

FIG. 2.7 Flowchart of the real-time PPP service.

the poor precision of pseudorange, ambiguities cannot quickly converge to their precise quantities in real-time PPP. Ambiguity resolution is to fix real-valued ambiguities to their correct integers and thus remove their correlation with the position parameters. Ge et al. (2008) proposed a method of using satellite uncalibrated phase delay (UPD) products to achieve dual-frequency PPP-AR. After ambiguity resolution, the convergence time of GPS-only PPP was decreased from 1 h to about 30 min. Moreover, resolving multifrequency ambiguities can further shorten the initialization time. Geng et al. (2019) proposed a triple-frequency PPP-AR method, which shortened the average initialization time from 9 to 6 min.

With the development of GLONASS, Galileo, and BDS constellations, PPP based on multi-GNSS satellites has become an effective approach to shorten initialization times. Compared to GPS, multi-GNSS satellites can provide many more observations, and more importantly, have better satellite geometry and faster geometry variations, which are both beneficial to PPP convergences. It has been reported that the combinations of GPS and other GNSSs contribute significantly to fast PPP-AR and thus efficient PPP initializations (Liu et al., 2020; Min et al., 2017; Pan et al., 2015).

The last effective method is to augment PPP with atmospheric corrections, which is called PPP-RTK conventionally. The key idea is to use precise atmospheric corrections extracted from a regional dense reference network to constrain the atmospheric parameters in PPP and achieve rapid convergences. Geng et al. (2010) realized the fast convergences for GPS PPP with the assistance of ionospheric products generated by a regional network.

3 Atmosphere sounding

As described in Section 2.1, a GNSS signal received by a near-Earth receiver is delayed in the atmosphere, i.e., tropospheric and ionospheric delays. They thus need to be estimated in high-precision GNSS. The estimated atmospheric delays carry valuable information on water vapor distribution and space weather conditions, which drives the application of GNSS atmosphere sounding.

3.1 Ground-based troposphere sounding

3.1.1 GNSS tropospheric delay

The tropospheric delay along the GNSS signal propagation path can be represented by:

$$\text{STD} = M_H(e) \cdot \text{ZHD} + M_W(e) \cdot \text{ZWD} + \Delta L_g \tag{2.36}$$

where STD is the slant total delay of the GNSS signal through the troposphere between a satellite and a receiver. ZHD and ZWD denote the zenith hydrostatic delay and the wet delay, respectively. ZHD is proportional to the surface atmospheric pressure. $M_H(e)$ and $M_W(e)$ are the hydrostatic and wet mapping functions, respectively, which depend on the elevation angle e. ΔL_g is the delay in the GNSS signal due to horizontal troposphere gradients.

Instead of estimating the slant path delay for each satellite, high-precision GNSS uses the signals from all visible GNSS satellites to estimate the atmospheric delay in the zenith direction of the GNSS receiver, i.e., the Zenith Total Delay (ZTD), which is nominally the sum of the ZHD and ZWD.

$$ZTD = ZHD + ZWD \qquad (2.37)$$

Both ZHD and ZWD are first calculated using empirical models, and the residual ZWD has to be estimated as an unknown along with other parameters of interest.

3.1.2 Water vapor content

The ZWD can be converted into the precipitable water vapor (PWV), which is defined in its physical sense as the height of the water column according to the liquid water condensed from the water vapor per unit area (Bevis et al., 1992):

$$PWV = \Pi \cdot ZWD \qquad (2.38)$$

where Π is the dimensionless conversion factor. Its estimation formula can be expressed as:

$$\Pi = \frac{10^6}{\rho_w R_v \left[\dfrac{k_3}{T_m} + k'_2\right]} \qquad (2.39)$$

where ρ_w is the density of liquid water; $R_v = 461.52$ J (kg K)$^{-1}$ is the specific gas constant of water vapor; k'_2 and k_3 are the constants determined from laboratory experiments for the refractivity (Bevis et al., 1994). PWV is sometimes referred to as the integrated water vapor (IWV), which has the unit of kg/m^2, that is, a PWV value of 1 mm is equivalent to an IWV value of 1 kg/m^2. T_m is the weighted mean temperature of water vapor, which can be estimated using the vertical profiles of atmospheric temperature and humidity. In practice, an empirical formula for the surface temperature and T_m for a specific region is established by linear regression using long-term radiosonde data. For example, Bevis et al. (1994) used radiosonde profiles to establish the relationship between the surface temperature and T_m in Northern America using a linear fit (Bevis et al., 1994)

$$T_m = 70.2 + 0.72 T_0 \qquad (2.40)$$

where T_0 is the surface temperature in kelvin around a GNSS antenna. The PWV information derived from the Japan GNSS Earth Observation Network (GEONET) has been successfully assimilated in the numerical weather models (Tsuji et al., 2017).

PWV or IWV plays an important role in global energy transfer and in the formation of clouds via latent heat. However, they only show water vapor information in the zenith direction of the receiver, and cannot reflect the three-dimensional distribution of the water vapor. Slant wet delay (SWD) or slant water vapor (SWV) contains the distribution characteristics of water vapor in the line-of-sight direction and has become an important source for studying the spatial distribution of water vapor. SWV can be calculated from the conversion factor and the SWD, that is,

$$SWV = \Pi \cdot SWD \qquad (2.41)$$

where SWD can be obtained from the ZWD multiplied by the wet delay mapping function (Askne & Nordius, 1987)

$$SWD = M_w(e) \cdot ZWD + \int G(G_E, G_N) de \qquad (2.42)$$

where $G(G_E, G_N)$ is an atmospheric delay gradient vector that encompasses the east and north directions; $\int G(G_E, G_N) de$ is the wet gradient delay, which can be obtained from numerical forecasts such as high-resolution, limited-area modeling (Boehm & Schuh, 2007), or can be obtained by removing or ignoring hydrostatic gradient delay (Bar-Sever et al., 1998).

3.1.3 GNSS troposphere tomography

GNSS troposphere tomography is to reconstruct the detailed 3D tropospheric information under a given mathematical constraint. Most implementations of GNSS troposphere tomography use SWDs or SWVs as observations because they contain the spatial distribution information of water vapor. If a dense GNSS observation network is available in the target area to provide SWDs/SWVs for each epoch, the spatial and temporal distributions of the water vapor in this area can be retrieved by the stratification technique. The GNSS troposphere tomographic results have been assimilated into the numerical weather prediction models and have shown a positive impact on weather forecasting, especially during extreme weather events (Trzcina et al., 2023).

For a theoretical definition, the SWD of the GNSS signal i can be expressed as (Flores et al., 2000):

$$SWD_i = \int N_w^i ds \tag{2.43}$$

where N_w^i denotes the wet refractivity (i.e., the water vapor parameter to be estimated) in the propagation path of signal i. To solve the numerical integration of total water vapor, a discretized method is generally used in GNSS water vapor tomography. For example, using the voxel-based method, the target tropospheric region is discretized into horizontally and vertically oriented self-closing 3D voxels. Assuming that the water vapor parameters of each voxel are represented by n_j, where j stands for the position of voxels, the SWD for signal i can be quantified as

$$SWD_i = \sum_{j=1}^{M} l_{ij} \cdot n_j \tag{2.44}$$

where M denotes the number of voxels; l_{ij} denotes the intercept of the signal i at the jth voxel.

As shown in Eqs. (2.41), (2.42), the difference between SWDs and SWVs is only one water vapor conversion constant. So, the tomographic observation equation corresponding to the SWV can be expressed as

$$SWV_i = \sum_{j=1}^{M} l_{ij} \cdot x_j \tag{2.45}$$

where SWV_i denotes the slant water vapor content for a GNSS signal; x_j denotes the water vapor parameters of the jth voxel.

Once the voxel-based discretization model is established, the water vapor parameters over the target area seem to be obtained by solving a system of linear equations that combine the SWDs or SWVs from a dense, ground GNSS reference network. In practice, due to the limited number of stations, the total number of tropospheric delay signals at a single epoch is not enough to penetrate all voxels. To solve this problem, accumulated observation signals over an interval (i.e., tomography window) are usually used to obtain the average water vapor distribution.

As there are always some voxels that cannot be visited by GNSS signals due to the limited spatial distribution of satellites and ground stations, the troposphere tomographic observation equations are usually ill-posed. This leads to an ill-posed inverse problem characterized by the singular design matrix in the tomography equation, and thus the observation equations cannot be solved directly using a least-squares estimation. In practice, some additional meteorological observations or some mathematical constraints, such as horizontal constraints, vertical constraints, boundary constraints, etc., are usually introduced into the tomography model to obtain a numerically stable solution (Dong & Jin, 2018). However, this makes the tomography observation equation very complex and the direct inversion is, thus, very difficult. Algorithms such as singular value decomposition (Flores et al., 2000; Hirahara, 2000; Yao et al., 2016), wet refractivity Kalman filter (Gradinarsky, 2002; Perler et al., 2011), algebraic reconstruction techniques (Bender et al., 2011), and simultaneous iterative reconstruction techniques (Wang et al., 2014) are used to overcome this challenge.

3.2 Ground-based ionosphere sounding

3.2.1 GNSS ionosphere delay

The ionosphere layer extends from about 50 km to the deep space above the ground. It contains a tremendous quantity of electrons due to the impact of ionizing radiation, which could affect the propagation of GNSS signals. In the GNSS observation equation, as mentioned in Section 1, the first-order ionosphere propagation error is inversely proportional to the signal frequency squared. In the past decades, thousands of GNSS stations have been established around the globe for continuous observations, which facilitates GNSS ionosphere monitoring.

According to Eq. (2.9), the ionosphere delay is proportional to the TEC, which is the integral of the electron density along the ray path. Although the ray-tracing model is mathematically rigorous, it is complex and inefficient. To simplify the model, the electrons within the ionosphere layer are presumed to be totally distributed on a 2D thin shell at a given altitude

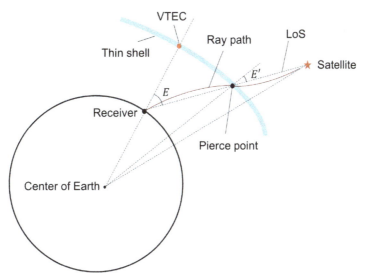

FIG. 2.8 The ray path, the line-of-sight (LoS), the pierce point, and the thin shell of ionosphere modeling.

of 300 km, as shown in Fig. 2.8. To be concrete, all electrons in the ray path are assumed to be concentrated at the intersection point between the thin shell and the signal path, usually referred to as the pierce point. Obviously, the location of the pierce point changes with the satellite position, making the ionosphere TEC estimates vary with the elevation of the satellite. Therefore, a mapping function is usually needed to map the TEC value from different ray paths to the vertical direction, which is normally referred to as the vertical TEC (i.e., VTEC).

3.2.2 Vertical ionosphere delays and global ionosphere map

Generally, the slant TEC (STEC) could be estimated using GPS pseudorange or carrier-phase ionospheric combinations (Eq. (2.21)), and the VTEC could be derived with the mapping function, that is,

$$\text{STEC} = M(E) \cdot \text{VTEC} \qquad (2.46)$$

where the VTEC values are in the unit of TEC units (1 TECU = 10^{16} electrons/m^2); $M(E)$ is the mapping function, which depends on the satellite elevation. Some well-known mapping functions include the SLM (Single Layer Model) (Ya'acob et al., 2008):

$$M(E) = \frac{1}{\sqrt{1 - \sin^2(E')}}, \quad \sin(E') = \frac{R_{earth}}{R_{earth} + H_{ion}} \sin(E) \qquad (2.47)$$

the Klobuchar MF

$$M(E) = 1 + 2\left(\frac{z+6}{90}\right)^3 \qquad (2.48)$$

where z represents the zenith distance, the Q-factor MF

$$\begin{cases} Q(z) = \sum_{i=0}^{3} a_i \left(\frac{z}{90}\right)^{2i} \\ a_0 = 1.0206, a_1 = 0.4663, a_2 = 3.5055, a_3 = -1.8415 \end{cases} \qquad (2.49)$$

as well as the MSLM (modified SLM) (Schaer, 1999),

$$\begin{cases} \sin(E') = \frac{R_{earth}}{R_{earth} + H_{map}} \sin(\alpha E) \\ H_{map} = 506.7\,km, \alpha = 0.9782, R_e = 6371\,km \end{cases} \qquad (2.50)$$

The global ionosphere delays could be expressed with spherical harmonics, fitted by VTEC values derived from globally distributed stations (Schaer, 1999), that is

$$VTEC(\beta, s) = \sum_{n=0}^{n_{\max}} \sum_{m=0}^{n} \left[\tilde{P}_{nm}(\sin \beta) \left(\tilde{C}_{nm} \cos(ms) + \tilde{S}_{nm} \sin(ms) \right) \right] \quad (2.51)$$

where β is the geomagnetic latitude of the anchor point, at a usual interval of 2.5°, while s is the sun-fixed longitude, at a usual interval of 5°; $\tilde{P}_{nm} = P_{nm} N_{nm}$ is normalized Legendre polynomials, and the normalization function N_{nm} yields

$$N_{nm} = \sqrt{\frac{(n-m)!(2n+1)(2-\delta_{0m})}{(n+m)!}}, \delta = \begin{cases} 1, m = 0 \\ 0, m \neq 0 \end{cases} \quad (2.52)$$

where $\tilde{C}_{nm}$ and $\tilde{S}_{nm}$ are the harmonics coefficients.

Once the harmonics coefficients are computed, a global ionosphere map could be generated. The VTEC values are arranged in terms of longitude and latitude, and typically, these values in the unit of TECU are updated every 2 h at the IGS. There exist many online products providing the global VTEC map or harmonics coefficients. With the global ionosphere map, the VTEC above any ground location could be calculated by interpolation with four grid points around. Alternatively, users could also rebuild the spherical harmonics with harmonics coefficient products. The VTEC value from a global map could reach a precision of about several TECUs.

In fact, the global ionosphere map depends on a large-scale modeling of ionosphere delays. A usual piecewise constant model at an interval of 2 h, for example, will average out the variations within this period. A simple linear interpolation on the global map is also under the assumption that the ionosphere delay is stable within a distance of a few hundred kilometers.

Furthermore, the distribution of ionized electrons is hardly isotropic, and thus the mapping functions above are fundamentally incorrect by presuming a symmetrical distribution of TEC around the zenith direction of piece points. Komjathy et al. (2005) showed that the mapping function could cause a 10-m-level error on the ionosphere maps, especially at low elevations. However, a global ionosphere map with a precision of several TECUs is enough for large-scale ionosphere analysis, such as the solar and magnetic activities.

3.2.3 Slant ionosphere delays and their spatiotemporal variations

The STEC, compared with VTEC, could be derived directly from raw GNSS observations. The STEC contains more direct ionospheric information without the nuisance from mapping functions, and the epoch-difference STEC values demonstrate the temporal variations of ionosphere electron density. As shown in Eq. (2.21), the pseudorange and carrier-phase ionospheric combinations are biased by hardware delays (or code/phase biases), while carrier-phase combinations additionally suffer from ambiguities. With an epoch-differencing method, these biases could be favorably eliminated, and the epoch-difference STEC could be derived as

$$\begin{cases} \dfrac{40.3 \Delta STEC}{f_1^2} = -\left(\Delta P_{r,1}^s - \Delta P_{r,2}^s \right) + \zeta \\ \dfrac{40.3 \Delta STEC}{f_1^2} = \left(\Delta \Phi_{r,1}^s - \Delta \Phi_{r,2}^s \right) + \tau \end{cases} \quad (2.53)$$

where Δ is the operation of differencing between adjacent epochs. It is worth mentioning that the ambiguity terms could only be eliminated over a continuous observation arc without cycle slips.

The TEC variations are not only governed by large-scale effects such as the solar and magnetic activities, but also affected by external physical disturbances on the ionosphere. The ionosphere layer could be perturbed by terrestrial-sourced acoustic and gravity waves occurring in phenomena like earthquakes, explosions, tsunamis, volcanic eruptions, and severe tropospheric events (Astafyeva, 2019). When acoustic waves or gravity waves travel to the altitude of the ionosphere, they might result in periodic oscillations on the TEC at the level of several TECUs or sub-TECUs (Hocke & Schlegel, 1996). Such ionosphere disturbances generally take place in a small region spanning hundreds of square kilometers, and last for several hours after a significant event. Actually, the real-time epoch-difference STEC has been applied to monitor small-scale variations of the ionosphere over a regional area, which can thus be used as an indicator of natural hazards such as earthquakes and volcanoes (Calais & Minster, 1995).

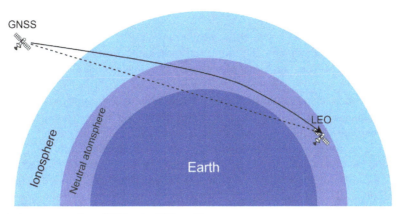

FIG. 2.9 GNSS radio occultation geometry for a GNSS and a LEO satellite.

3.3 GNSS radio occultation (GNSS-RO)

3.3.1 The physical principle of GNSS-RO

GNSS signals are refracted when they travel through the atmosphere to a LEO satellite. The ray path of the GNSS signal is slightly curved and its measurements are similarly affected as a result of the refractive index gradients in the near-Earth space. As shown in Fig. 2.9., this technique is called GNSS-RO and provides atmosphere sounding through the near-Earth space. It reveals the characteristics and the vertical structure of the neutral atmosphere and the ionosphere of the Earth.

GNSS constellations provide reliable and low-cost radio signal sources, as well as precise satellite orbits and clocks. All these features will improve the measurement precision of the GNSS-RO sounding profile. The GNSS-RO receiver onboard an LEO continuously records the GNSS measurements between a GNSS satellite and the LEO satellite. The temperature, pressure, and water vapor can be retrieved from GNSS-RO measurements (Leroy et al., 2009).

3.3.2 GNSS-RO data processing

Each LEO satellite is equipped with four antennas on the front and back sides of the satellite's main frame. Two antennas are mounted on the upper part of the main body for the purpose of precise orbit determination. The other two antennas are mounted on the lower part for atmospheric occultation. The resulting carrier-phase measurements are in synergy with satellite positions and velocities to compute the time delay and excess phase delay. The excess phase delay can be used to derive the bending angles of the GNSS rays and further the refractivity. It is usually presumed that only one ray is observed at each epoch, and then the relationship between the geometrical vector of a typical point on the ray path $\vec{r}$ and the refractive index n can be given by the Eikonal equation (Born et al., 1999) as

$$\frac{d}{ds}\left(n\frac{d\vec{r}}{ds}\right) = \vec{\nabla}n \tag{2.54}$$

where $\vec{\nabla}n$ is the gradient of the refractive index n, and ds is an incremental length along the ray path. Therefore, the accumulation of atmospheric bending along a ray path can be calculated.

The refractive index profile can be derived from a corrected atmospheric bending angle profile. For brevity, refractivity N is often used instead of the refractive index n, that is (Padullés et al., 2018; Thayer, 1974)

$$N = (n-1)\cdot 10^6 = 77.6\frac{P}{T} + 3.73\cdot 10^5\frac{P_w}{T^2} + 40.3\cdot 10^7\frac{n_e}{f^2} + 1.4W \tag{2.55}$$

where P and P_w are the total pressure and water vapor partial pressure, respectively; T is the temperature; n_e is the electron number density per cubic meter; f is the GNSS signal frequency; and W is the liquid water content. Finally, we enable the retrieval of temperature, humidity, and pressure from profiles of refractivity measured by radio occultation based on a 1D variational method.

The inversion of GNSS-RO data is a progressive process. At different processing steps, various approximations and assumptions are applied. Therefore, quality control is an important issue in GNSS-RO data processing.

3.3.3 The advantages of GNSS-RO

Observations of the Earth's atmosphere are essential for weather prediction, mitigation of natural disasters (e.g., meteorological hazards), global climate research, and environmental protection. However, the Earth's atmosphere is a highly dynamic system with global impacts. The monitoring of the near-Earth space needs a large number of timely and globally distributed radiosonde profiles.

The traditional atmosphere monitoring methods including the weather balloon and ground-based stations are limited by geographic locations and vulnerable to extreme weather events. In many circumstances, they cannot meet the rapidly growing demand for atmosphere-sounding profiles in the Earth's atmosphere research. Moreover, the high cost of large-scale deployment of new ground stations and balloons makes them unsustainable.

However, the GNSS-RO missions can be carried out by different satellite platforms coupled with GNSS receivers with low cost and low power consumption. For a better understanding of the atmosphere, the GNSS-RO technique is critical for global atmosphere-sounding profiles. For meteorological applications, the GNSS-RO technique has proven to be a timely and reliable data resource in a series of remarkably successful experiments and missions since the 1960s. The GPS/MET carried out by the University Corporation for Atmospheric Research (UCAR) reveals that the GPS RO measurements are coincident with the vertical structure of temperature. As a worldwide constellation of LEO satellites, the COSMIC project demonstrated the capability of GPS RO in providing global coverage, and timely and sufficient soundings of the near-Earth space.

In summary, in short-, medium-, and long-term weather forecasting, climate research, and space weather monitoring, the GNSS-RO shall operationally provide a dense global observation of fundamental atmospheric variables. An increasing number of countries, regions, and organizations are considering deploying their own next-generation LEO constellation with GNSS-RO receivers, for example, the COSMIC-2 program (Asgarimehr & Hossainali, 2015; Chang & Yang, 2022; Cook et al., 2013). It can be expected that the ever-increasing quality and quantity of GNSS-RO data will have significant benefits to weather forecasting and climate research.

4 GNSS reflectometry

It is usually expected that all error sources should be minimized in high-precision GNSS, but multipath effects are not easy to mitigate completely. The signal reflected by the ground or other surfaces can deteriorate the positioning precision, but these unexpected signals can also reflect the physical properties of the reflecting surface to some extent. In 1993, the concept of using GPS reflection signals for sea surface altimetry was first proposed (Martin-Neira, 1993), and in the following decades, this technology was rapidly developed to form GNSS Reflectometry (GNSS-R). GNSS-R is a new satellite remote-sensing technology, which uses GNSS signals reflected from a particular surface to infer information about the physical parameters of the reflector. The reflected GNSS signal acts as a bistatic remote-sensing radar.

According to the difference of GNSS signal reception, we can generally divide this technique into two branches. The first is GNSS Interferometric Reflectometry (GNSS-IR) based on the signal-to-noise ratio (SNR), and the other is GNSS-R technology based on code/phase delay observations, which needs to distinguish between the direct signal and the reflected signal on the receiving device before performing feature analysis.

4.1 GNSS interferometric reflectometry based on single antennas

Usually, a ground GNSS antenna is in a state of passively receiving signals and cannot screen out or identify reflected signals. Therefore, the received signal is a mixed signal composed of direct and reflected signals (Fig. 2.10). These two signals are not simply superimposed, but also interfere with each other. Since the reflected surface is not an ideal, smooth plane, the antenna receives the scattered signal, which is reflected from a region around the specular point (such as the first Fresnel surface in Fig. 2.10). Therefore, much characteristic information about the state of the surface will be recorded in the SNR observations. An SNR observation is an indicator of evaluating the quality of the received signal. Due to the influence of antenna gain and other effects, the SNR and the signal amplitude have the following relationship

$$SNR = A_c^2 = A_d^2 + A_r^2 + 2A_d A_r \cos\varphi \qquad (2.56)$$

where A_c is the amplitude of the synthesized signal; A_d is the amplitude of the direct signal; A_r is the amplitude of the reflected signal; considering the antenna gain to the direct signal, we can assume that $A_d >> A_r$ and $\cos\varphi$ is the cosine of the phase difference between the direct signal and the reflected signal.

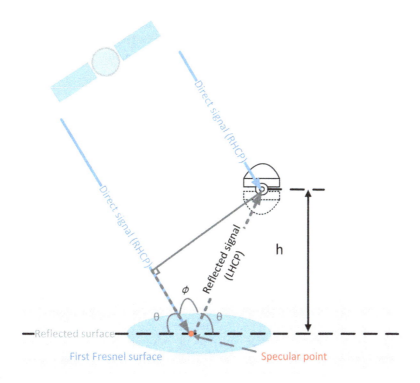

FIG. 2.10 A schematic diagram of GNSS-R.

The SNR is dominated by the direct signal overall and is affected by both the antenna and the signal polarization. An SNR time series shows a parabolic-like characteristic on the whole with elevation changes. In order to see the component of the reflected signal, the overall trend term of the SNR needs to be removed. Larson et al. (2008) suggested that this should be removed using a quadratic polynomial. At both ends of the sequence, after removing the trend term, i.e., at low elevation angles, some significant oscillatory signals appear. This phenomenon is related to the passing path of the reflected signal. In Fig. 2.10, we can see that the reflected signal will take one more path than the direct signal, which indicates that the excess path δ (gray dashed line in Fig. 2.10) will cause a certain phase difference $\Delta\Omega$

$$\Delta\Omega = \frac{2\pi}{\lambda}\delta = \frac{4\pi h}{\lambda}\sin(\theta) \tag{2.57}$$

where θ is the elevation of the GNSS satellite and λ is the carrier wavelength; h is the vertical height from the antenna phase center to the surface.

For a given height h, the interference signal at a low elevation angle can be expressed as

$$SNR_I = A\cos\left(\frac{4\pi h}{\lambda}\sin\theta + \phi\right) \tag{2.58}$$

where SNR_I is the interference signal caused by direct and reflected signals; A is the amplitude of this signal, which is related to the reflecting surface and antenna gain pattern.

If we set $t = \sin\theta$ and $f = 2h/\lambda$, the equation above can be simplified to the standard cosine function

$$SNR_I = A\cos(2\pi f t + \phi) \tag{2.59}$$

When we need to perform GNSS-IR altimetry, we can perform spectral analysis on the signal to obtain its frequency f, and after the conversion, using $h = \lambda \cdot f/2$, we can obtain the vertical distance, h. Although the epochs are evenly sampled and continuous, the SNR residual sequence is not guaranteed to be truncated by the whole cycle because $\sin\theta$ is not evenly sampled. Therefore, to obtain the frequency of the SNR residual sequence, the Lomb-Scargle spectral analysis (Ruf, 1999) can be used in the actual calculation.

As already mentioned before, the physical properties of the reflected surface and the information about its variation affect the reflected signals, which are eventually recorded in the SNR. When we analyze the oscillation frequency, f, we can obtain information about the height, h, of the reflecting surface, which can be used to measure water level changes, such as seas, rivers, reservoirs, and lakes (Larson et al., 2013) and snow depth (Larson et al., 2009). In addition, we can also study the vegetation index, soil moisture, etc., according to the variation of amplitude A and phase ϕ (Larson et al., 2008; Small et al., 2010, 2014).

4.2 GNSS reflectometry based on dual antennas

In most cases, we use a geodetic antenna (one antenna) to acquire GNSS measurements. However, to make better use of the reflected signal, we can improve the signal-receiving device with two different antennas, which can separate the direct signal from the reflected signal. One of the two antennas faces upward to receive the direct signal, and the other faces downward to receive the reflected signal from the reflective surface (the antenna depicted by the dotted line in Fig. 2.11). The signal emitted by the satellite is an RHCP signal, and after reflection, the polarization characteristics of part of the signals will change to Left Hand Circularly Polarized (LHCP). For low satellite elevation angles, the satellite signal is dominated by the RHCP component. However, the amplitude of the LHCP reflection coefficient increases with the increase of the satellite elevation angle, and for the elevation angle over about 8°, the amplitude of the LHCP reflection coefficient is larger than that of the RHCP reflection coefficient (Hannah, 2001).

After obtaining the direct and reflected signals, we need to perform correlation analysis, in which the information corresponding to the direct and reflected signals are synchronized to the signal reception time, t_0. Considering the signal propagation time (Fig. 2.11), we need to invert for the positions of the receiver R and the satellite T at the signal transmission time based on their positions and velocities at the time, t_0. First, according to the position $RX(t_0)$ of the receiver and the position $TX(t_0)$ of the transmitter at the epoch t_0, after obtaining the position $S(t_0)$ of the initial specular point, the propagation times of the direct and reflected signals can be expressed, respectively, as follows

$$\begin{cases} t_d = \dfrac{|TX(t_0) - RX(t_0)|}{c} \\ t_r = \dfrac{|RX(t_0) - SX(t_0)| + |TX(t_0) - SX(t_0)|}{c} \end{cases} \quad (2.60)$$

where c denotes the speed of light in vacuum. Based on the signal propagation time, and the moving speed of the GNSS satellite $TV(t_0)$, the position $TX(t_0 - t_r)$ at the epoch of the reflected signal emission can be computed, and then the updated specular point position $S'(t_0)$ can be calculated. The distance delay of the reflected signal with respect to the direct signal can be expressed as

$$\rho(t_0) = |RX(t_0) - S'(t_0)| + |TX(t_0 - t_r) - S'(t_0)| - |TX(t_0 - t_r) - R(t_0)| \quad (2.61)$$

The GNSS signal travels for a long distance from transmission to reception and is affected by atmosphere delays during the propagation. On shore-based platforms, the direct path is similar to the reflected path, and both are affected approximately.

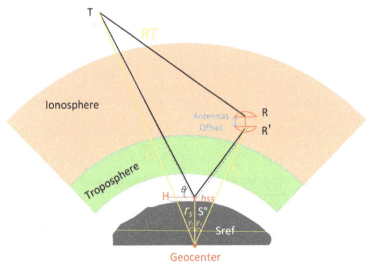

FIG. 2.11 Nonscaled sketch of the space-borne GNSS-R altimetric geometry.

The satellite-based platform is mostly at an altitude of several hundred kilometers, and the errors caused by atmospheric effects are not negligible. Therefore, the path delay model can be expressed as

$$\rho_{mod} = \rho^{geo} + \rho^{ino} + \rho^{tro} \tag{2.62}$$

where ρ^{geo} is the geometric path delay, ρ^{ino} is the ionospheric delay correction, and ρ^{tro} is the tropospheric delay correction.

When calculating the accurate SSH (sea surface height, which is defined as the average of the vertical distances from sea level to the Earth's reference ellipsoid in the area being measured), the method by Wu et al. (1997) can be used. Then, distance delay can be expressed as (Zavorotny et al., 2014)

$$\begin{aligned}\rho_R &= R_{ts} + R_{rs} - R_{rt} \\ &= \sqrt{r_t^2 + (r_s + h_{ss})^2 - 2r_t(r_s + h_{ss})\cos\gamma_t} + \sqrt{r_r^2 + (r_s + h_{ss})^2 - 2r_r(r_s + h_{ss})\cos\gamma_r} - R_{rt}\end{aligned} \tag{2.63}$$

where r_t, r_r, and r_s denote the distances between the geocenter and the GNSS satellite, the altimetry platform receiver, and the specular reflection point, respectively; γ_t denotes the geocentric angle between the GNSS satellite and the specular reflection point; γ_r denotes the geocentric angle between the altimetry platform receiver and the specular reflection point; and h_{ss} denotes the SSH. The partial derivative with respect to h_{ss} for both sides of the above equation is

$$\frac{\sigma_\rho}{\sigma_{ssh}} = \frac{\partial \rho_R}{\partial h_{ss}} = -2\sin\theta \tag{2.64}$$

where θ denotes interaltitude angle. Then, the simplified calculation formula for SSH provided by Helm (2008) can be used

$$\begin{cases} \delta\rho = \rho_{obs} - \rho_{mod} \\ H = -\dfrac{\delta\rho}{2\cos i} \end{cases} \tag{2.65}$$

where i is the incidence angle and ρ_{obs} is the actually observed path delay after retracking. It should be noted that the sea surface height calculated using this formula is relative to the elevation of the specular reflection point used in the computational model. Since the geometric model uses the WGS-84 ellipsoid as the reference surface, the calculated sea surface height is also relative to the WGS-84 ellipsoid.

There are various modes of GNSS-R operations, including ground-based, airborne, shipborne, and space-borne GNSS-R. In different scenarios, GNSS-R can act as ocean altimetry not only for ocean roughness/scattering rate detection and calculation, but also for the detection and identification of earth surface environmental parameters, such as soil moisture and vegetation, snow, and sea ice (Zavorotny et al., 2014).

References

Asgarimehr, M., & Hossainali, M. M. (2015). GPS radio occultation constellation design with the optimal performance in Asia Pacific region. *Journal of Geodesy, 89*(6), 519–536.

Askne, J., & Nordius, H. (1987). Estimation of tropospheric delay for microwaves from surface weather data. *Radio Science, 22*(03), 379–386.

Astafyeva, E. (2019). Ionospheric detection of natural hazards. *Reviews of Geophysics, 57*(4), 1265–1288. https://doi.org/10.1029/2019RG000668.

Bar-Sever, Y. E., Kroger, P. M., & Borjesson, J. A. (1998). Estimating horizontal gradients of tropospheric path delay with a single GPS receiver. *Journal of Geophysical Research-Solid Earth, 103*(B3), 5019–5035.

Bender, M., Dick, G., Ge, M., Deng, Z., Wickert, J., Kahle, H.-G., et al. (2011). Development of a GNSS water vapour tomography system using algebraic reconstruction techniques. *Advances in Space Research, 47*(10), 1704–1720.

Bevis, M., Businger, S., Chiswell, S., Herring, T. A., Anthes, R. A., Rocken, C., et al. (1994). GPS meteorology – mapping zenith wet delays onto precipitable water. *Journal of Applied Meteorology, 33*(3), 379–386.

Bevis, M., Businger, S., Herring, T. A., Rocken, C., Anthes, R. A., & Ware, R. H. (1992). GPS meteorology – remote-sensing of atmospheric water-vapor using the global positioning system. *Journal of Geophysical Research-Atmospheres, 97*(D14), 15787–15801.

Blewitt, G. (1989). Carrier phase ambiguity resolution for the global positioning system applied to geodetic baselines up to 2000 km. *Journal of Geophysical Research: Solid Earth, 94*(B8), 10187–10203.

Blewitt, G. (2007). GPS and space-based geodetic methods. *Geodesy, 3*, 351–390.

Bock, Y., Abbot, R. I., Counselman, C. C., III, Gourevitch, S. A., & King, R. W. (1985). Establishment of three-dimensional geodetic control by interferometry with the global positioning system. *Journal of Geophysical Research: Solid Earth, 90*(B9), 7689–7703.

Boehm, J., & Schuh, H. (2007). Troposphere gradients from the ECMWF in VLBI analysis. *Journal of Geodesy, 81*(6–8), 403–408.

Born, M., Wolf, E., Bhatia, A. B., Clemmow, P. C., Gabor, D., Stokes, A. R., et al. (1999). *Principles of optics: Electromagnetic theory of propagation, interference and diffraction of light* (7th ed.). New York: Cambridge University Press, Cambridge.

Calais, E., & Minster, J. B. (1995). GPS detection of ionospheric perturbations following the January 17, 1994, Northridge earthquake. *Geophysical Research Letters, 22*(9), 1045–1048.

Chang, C. C., & Yang, S.-C. (2022). Impact of assimilating Formosat-7/COSMIC-II GNSS radio occultation data on heavy rainfall prediction in Taiwan: Terrestrial. *Atmospheric and Oceanic Sciences, 33*(1), 7.

Clara de Lacy, M., Reguzzoni, M., & Sanso, F. (2012). Real-time cycle slip detection in triple-frequency GNSS. *GPS Solutions, 16*(3), 353–362.

Cocard, M., Bourgon, S., Kamali, O., & Collins, P. (2008). A systematic investigation of optimal carrier-phase combinations for modernized triple-frequency GPS. *Journal of Geodesy, 82*(9), 555–564.

Cook, K., Fong, C. J., Wenkel, M. J., Wilczynski, P., Yen, N., & Chang, G. S. (2013). FORMOSAT-7/COSMIC-2 GNSS radio occultation constellation mission for global weather monitoring. In *Proceedings 2013 IEEE aerospace conference, March 2013* (pp. 1–8).

Counselman, C. C., III, Abbot, R. I., Gourevitch, S. A., King, R. W., & Paradis, A. R. (1983). Centimeter-level relative positioning with GPS. *Journal of Surveying Engineering, 109*(2), 81–89.

Dong, Z., & Jin, S. (2018). 3-D water vapor tomography in Wuhan from GPS, BDS and GLONASS observations. *Remote Sensing, 10*(1).

Flores, A., Ruffini, G., & Rius, A. (2000). 4D tropospheric tomography using GPS slant wet delays. *Annales Geophysicae-Atmospheres Hydrospheres and Space Sciences, 18*(2), 223–234.

Ge, M., Gendt, G., Rothacher, M., et al. (2008). Resolution of GPS carrier-phase ambiguities in precise point positioning (PPP) with daily observations. *Journal of Geodesy, 82*, 389–399.

Geng, J., & Bock, Y. (2013). Triple-frequency GPS precise point positioning with rapid ambiguity resolution. *Journal of Geodesy, 87*(5), 449–460.

Geng, J., & Bock, Y. (2016). GLONASS fractional-cycle bias estimation across inhomogeneous receivers for PPP ambiguity resolution. *Journal of Geodesy, 90*, 379–396.

Geng, J., Guo, J., Meng, X., et al. (2019). *Speeding up PPP ambiguity resolution using triple-frequency GPS/BeiDou/Galileo/QZSS data.*

Geng, J., & Mao, S. (2021). Massive GNSS network analysis without baselines: Undifferenced ambiguity resolution. *Journal of Geophysical Research: Solid Earth, 126*(10), e2020JB021558.

Geng, J., Meng, X., Dodson, A. H., et al. (2010). Rapid re-convergences to ambiguity-fixed solutions in precise point positioning. *Journal of Geodesy, 84*, 705–714.

Geng, J., & Shi, C. (2017). Rapid initialization of real-time PPP by resolving undifferenced GPS and GLONASS ambiguities simultaneously. *Journal of Geodesy, 91*(4), 361–374.

Geng, J., Teferle, F. N., Meng, X., et al. (2011). Towards PPP-RTK: Ambiguity resolution in real-time precise point Positioning. *Advances in Space Research, 47*, 1664–1673.

Geng, J., Wen, Q., Zhang, Q., Li, G., & Zhang, K. (2022). GNSS observable-specific phase biases for all-frequency PPP ambiguity resolution. *Journal of Geodesy, 96*, 11. https://doi.org/10.1007/s00190-022-01602-3.

Geng, J., Zhao, Q., Shi, C., & Liu, J. (2017). A review on the inter-frequency biases of GLONASS carrier-phase data. *Journal of Geodesy, 91*(3), 329–340.

Gradinarsky, L. P. (2002). *Sensing atmospheric water vapor using radio waves* (pp. 1–25). Göteborg: Department Radio and Space Science, School of Electrical Engineering, Chalmer University of Technology.

Guo, J., & Geng, J. (2018). GPS satellite clock determination in case of inter-frequency clock biases for triple-frequency precise point positioning. *Journal of Geodesy, 92*(10), 1133–1142.

Gupta, H., & Pasquale, V. (2013). *Encyclopedia of solid earth geophysics.* Springer International Publishing.

Hannah, B. M. (2001). *Modelling and simulation of GPS multipath propagation* (Doctoral dissertation). Queensland University of Technology.

Helm, A. (2008). *Ground-based GPS altimetry with the L1 OpenGPS receiver using carrier phase-delay observations of reflected GPS signals* (Doctoral dissertation). Deutsches GeoForschungsZentrum GFZ Potsdam.

Henkel, P., & Gunther, C. (2012). Reliable integer ambiguity resolution: Multi-frequency code carrier linear combinations and statistical a priori knowledge of attitude. *Navigation-Journal of the Institute of Navigation, 59*(1), 61–75.

Hirahara, K. (2000). Local GPS tropospheric tomography. *Earth, Planets and Space, 52*(11), 935–939.

Hocke, K., & Schlegel, K. (1996). A review of atmospheric gravity waves and travelling ionospheric disturbances: 1982–1995. *Annales Geophysicae-Atmospheres Hydrospheres and Space Sciences, 14*(9), 917–940. https://doi.org/10.1007/s00585-996-0917-6.

Komjathy, A., Sparks, L., Mannucci, A. J., & Coster, A. (2005). The ionospheric impact of the October 2003 storm event on wide area augmentation system. *GPS Solutions, 9*(1), 41–50.

Larson, K. M., Gutmann, E. D., Zavorotny, V. U., Braun, J. J., Williams, M. W., & Nievinski, F. G. (2009). Can we measure snow depth with GPS receivers? *Geophysical Research Letters, 36*(17).

Larson, K. M., Löfgren, J. S., & Haas, R. (2013). Coastal sea level measurements using a single geodetic GPS receiver. *Advances in Space Research, 51*(8), 1301–1310.

Larson, K. M., Small, E. E., Gutmann, E., Bilich, A., Axelrad, P., & Braun, J. (2008). Using GPS multipath to measure soil moisture fluctuations: Initial results. *GPS Solutions, 12*(3), 173–177.

Leroy, S. S., Dykema, J. A., Gero, P. J., & Anderson, J. G. (2009). Testing climate models using infrared spectra and GNSS radio occultation. In A. Steiner, B. Pirscher, U. Foelsche, & G. Kirchengast (Eds.), *New horizons in occultation research: Studies in atmosphere and climate* (pp. 195–206). Berlin, Heidelberg: Springer.

Li, J., Yang, Y., He, H., & Guo, H. (2017). An analytical study on the carrier-phase linear combinations for triple-frequency GNSS. *Journal of Geodesy, 91*(2), 151–166.

Lichten, S. M., & Border, J. S. (1987). Strategies for high-precision Global Positioning System orbit determination. *Journal of Geophysical Research: Solid Earth, 92*(B12), 12751–12762.

Liu, T., Jiang, W., Laurichesse, D., et al. (2020). Assessing GPS/Galileo real-time precise point positioning with ambiguity resolution based on phase biases from CNES. *Advances in Space Research, 66*, 810–825.

Martin-Neira, M. (1993). A passive reflectometry and interferometry system (PARIS): Application to ocean altimetry. *ESA Journal, 17*(4), 331–355.

Melbourne, W. (1985). The case for ranging in GPS-based geodetic systems. In *First international symposium on precise positioning with the global positioning system, Rockville* (pp. 373–386).

Min, W., Chai, H., & Yu, L. (2017). Performance analysis of BDS/GPS precise point positioning with undifferenced ambiguity resolution. *Advances in Space Research, 60*, 2581–2595.

Montenbruck, O., Hugentobler, U., Dach, R., Steigenberger, P., & Hauschild, A. (2012). Apparent clock variations of the block IIF-1 (SVN62) GPS satellite. *GPS Solution, 16*(3), 303–313.

Padullés, R., Cardellach, E., Wang, K. N., Ao, C. O., Turk, F. J., & de la Torre-Juárez, M. (2018). Assessment of global navigation satellite system (GNSS) radio occultation refractivity under heavy precipitation. *Atmospheric Chemistry and Physics, 18*(16), 11697–11708.

Pan, Z., Chai, H., Liu, Z., et al. (2015). Integrating BDS and GPS to accelerate convergence and initialization time of precise point Positioning. *Lecture Notes in Electrical Engineering, 342*, 67–80.

Perler, D., Geiger, A., & Hurter, F. (2011). 4D GPS water vapor tomography: New parameterized approaches. *Journal of Geodesy, 85*(8), 539–550.

Petit, G., & Luzum, B. (2010). *IERS Conventions 2010*. Frankfurt am Main: Verlag des Bundesamtes für Kartographie und Geodäsie. IERS Technical Note 36.

Ruf, T. (1999). The Lomb-Scargle periodogram in biological rhythm research: Analysis of incomplete and unequally spaced time-series. *Biological Rhythm Research, 30*(2), 178–201.

Schaer, S. (1999). *Mapping and predicting the Earth's ionosphere using the global positioning system*. Institut für Geodäsie und Photogrammetrie Eidg. Technische Hochschule.

Schaer, S., Villiger, A., Arnold, D., Dach, R., Prange, L., & Jäggi, A. (2021). The CODE ambiguity-fixed clock and phase bias analysis products: Generation, properties, and performance. *Journal of Geodesy, 95*(7), 1–25.

Sleewaegen, J. M., Simsky, A., De Wilde, W., Boon, F., & Willems, T. (2012). Origin and compensation of GLONASS inter-frequency carrier phase biases in GNSS receivers. In *Proceedings of the 25th international technical meeting of the satellite division of the Institute of Navigation (ION GNSS 2012)* (pp. 2995–3001).

Small, E. E., Larson, K. M., & Braun, J. J. (2010). Sensing vegetation growth with reflected GPS signals. *Geophysical Research Letters, 37*(12).

Small, E. E., Larson, K. M., & Smith, W. K. (2014). Normalized microwave reflection index: Validation of vegetation water content estimates from Montana grasslands. *IEEE Journal of Selected Topics in Applied Earth Observations and Remote Sensing, 7*(5), 1512–1521.

Teunissen, P. J. G., De Jonge, P. J., & Tiberius, C. C. J. M. (1997). The least-squares ambiguity decorrelation adjustment: Its performance on short GPS baselines and short observation spans. *Journal of Geodesy, 71*(10), 589–602.

Thayer, G. D. (1974). An improved equation for the radio refractive index of air. *Radio Science, 9*, 803–807.

Trzcina, E., Rohm, W., & Smolak, K. (2023). Parameterisation of the GNSS troposphere tomography domain with optimisation of the nodes' distribution. *Journal of Geodesy, 97*(1), 1–23.

Tsuji, H., Hatanaka, Y., Hiyama, Y., Yamaguchi, K., Furuya, T., Kawamoto, S., et al. (2017). Twenty-year successful operation of Geonet. *Bulletin of the Geospatial Information Authority of Japan, 65*, 20.

Wang, K., & Rothacher, M. (2013). Ambiguity resolution for triple-frequency geometry-free and ionosphere-free combination tested with real data. *Journal of Geodesy, 87*(6), 539–553.

Wang, X., Ziqiang, D., Enhong, Z., Fuyang, K. E., Yunchang, C., & Lianchun, S. (2014). Tropospheric wet refractivity tomography using multiplicative algebraic reconstruction technique. *Advances in Space Research, 53*(1), 156–162.

Wanninger, L. (2012). Carrier-phase inter-frequency biases of GLONASS receivers. *Journal of Geodesy, 86*(2), 139–148.

Wellenhof, B. H., Lichtenegger, H., & Collins, J. (2001). *Global Positioning system: Theory and practice* (4th ed., pp. 181–200). New York: Springer.

Wu, S. C., Meehan, T., & Young, L. (1997). *The potential use of GPS signals as ocean altimetry observables*.

Wübbena, G. (1985). Software developments for geodetic positioning with GPS using TI-4100 code and carrier measurements. In *First international symposium on precise positioning with the global positioning system, Rockville* (pp. 403–412).

Ya'acob, N., Abdullah, M., & Ismail, M. (2008). Determination of GPS total electron content using single layer model (SLM) ionospheric mapping function. *International Journal of Computer Science and Network Security, 8*(9), 154–160.

Yao, Y. B., Zhao, Q. Z., & Zhang, B. (2016). A method to improve the utilization of GNSS observation for water vapor tomography. *Annales Geophysicae, 34*(1), 143–152.

Zavorotny, V. U., Gleason, S., Cardellach, E., & Camps, A. (2014). Tutorial on remote sensing using GNSS bistatic radar of opportunity. *IEEE Geoscience and Remote Sensing Magazine, 2*(4), 8–45.

Zumberge, J. F., Heflin, M. B., Jefferson, D. C., Watkins, M. M., & Webb, F. H. (1997). Precise point positioning for the efficient and robust analysis of GPS data from large networks. *Journal of Geophysical Research: Solid Earth, 102*(B3), 5005–5017.

Part I

Monitoring earthquakes and volcanoes with GNSS

Chapter 3

On the use of GNSS-inferred crustal strain accumulation in evaluating seismic potential

Corné Kreemer[a], Ilya Zaliapin[b,†], and Dirk Kraaijpoel[c]

[a]*Nevada Seismological Laboratory, Nevada Bureau of Mines and Geology, University of Nevada, Reno, NV, United States,* [b]*Department of Mathematics and Statistics, University of Nevada, Reno, NV, United States,* [c]*TNO, Geological Survey of the Netherlands, Utrecht, the Netherlands*

Nomenclature

a	number of events with magnitude $\geq m_T$
A	areal surface area
b	b-value of Gutenberg-Richter relationship
β	Pareto index, equal to $2b/3$
c	seismic coupling coefficient
χ	theoretical ratio of the simulated seismic moment over total expected seismic moment: S_n/E_n
D	seismogenic depth
$E[M]$	expectation value, or the mean M for the tapered Pareto distribution
E_n	expected moment sum of n earthquakes, which equals $nE[M]$.
$\dot{\varepsilon}$	scalar strain rate
$\dot{\varepsilon}_1$	largest principal strain rate axis
$\dot{\varepsilon}_2$	smallest principal strain rate axis
$\Gamma(x,y)$	upper incomplete gamma function
k	parameter that reflects break down of a given space-time; $\max(AT)/AT$
λ	constant, equal to $2\mu D$
m	earthquake magnitude
M	earthquake moment
m_C	corner magnitude
M_C	corner moment
M_G	geodetic moment, equal to $2\mu D\dot{\varepsilon}TA$
M_S	long-term seismic moment
M_Σ	observed seismic moment for a given A and T
m_T	threshold magnitude
M_T	threshold moment
μ	shear modulus
$n(m)$	number of earthquakes larger than or equal to m, based on best-fitting moment-frequency distribution
$n(M)$	number of earthquakes larger than or equal to M, based on best-fitting moment-frequency distribution
$n(M_T)$	number of earthquakes larger than or equal to M_T, based on best-fitting moment-frequency distribution
N	number of observed earthquakes larger than or equal to M_T
ν	earthquake rate density for events with moment larger than or equal to M_T; $n(M_T)/AT$
P_G	geodetic potency, equal to $\dot{\varepsilon}TA$
R	observed ratio of seismic over geodetic moment
S_n	cumulative moment of n earthquakes with moments randomly drawn from tapered Pareto distribution

†. Deceased.

$S(q,n)$ quantile of the maximally skewed stable distribution approximating the distribution of S_n
T time-span, typically of considered earthquake catalog
$Z(q,n)$ quantile of the moment distribution of S_n
z_q quantile of standard normal distribution (zero mean, unit variance)

1 Introduction

Earthquakes are devastating natural phenomena not only because of the damage inflicted by the induced ground shaking but also due to potential side phenomena such as tsunamis and liquefaction. Unfortunately, earthquakes cannot be predicted, i.e., the operational short-term (hours to weeks) forecast of location, timing, and magnitude of large earthquakes is currently not possible due to the complexity of the earthquake process (e.g., Geller et al., 1997; Jordan, 2006; Sykes et al., 1999). However, past earthquake occurrence and/or knowledge of the deformation rate can be used to constrain how many earthquakes are above a certain moment and how much seismic moment release is expected for a certain area and time interval. That is one of the underlying premises of probabilistic seismic hazard assessment (PSHA). Due to the finite length of earthquake catalogs and limitations in knowing the seismic history of each potentially seismogenic fault, strain rates obtained from GNSS-derived velocities are now being considered as additional input to seismic hazard maps (Moschetti et al., 2015; Petersen et al., 2014) or are directly used to create seismicity forecasts (Bird et al., 2015, 2010; Bird & Kreemer, 2015; Rhoades et al., 2017; Rong et al., 2016). These forecasts make use of the observed magnitude-frequency behavior of seismicity. However, given that most seismic moment, and thus hazard, resides in the largest earthquake(s), some studies instead used the strain rates to calculate the repeat time (or "return period") of these "characteristic" earthquakes (Kreemer et al., 2014; Varga, 2011).

Within areas of active tectonic deformation, differences in horizontal crustal velocities obtained from the evolution of GNSS-derived position measurements (when not affected by earthquakes) reveal the strain that is accumulating in the crust. This strain is often assumed to reflect elastic strain accumulation that will convert into permanent deformation on a fault during an earthquake. In particular, the elastic strain rate is expected to localize above (major) faults (i.e., dislocations) that are locked in the seismogenic crust and slip continuously along their down-dip continuation (Savage & Burford, 1973). For example, for strike-slip faults, this results in an elevated shear strain rate in the area around the fault trace corresponding to an "arc-tangent" pattern in the fault-parallel velocity profile (Savage & Burford, 1973; Vernant, 2015). Alternatively, instead of having discrete strike-slip faults, shear might be distributed over a finite zone below the seismogenic crust, which would be expressed as a zone of constant shear strain rate (Prescott & Nur, 1981).

Under the assumption that every fault can be characterized by a locked dislocation, researchers have formulated "block models" in which the geodetic velocity field is assumed to be entirely explained by block rotations and elastic loading on faults. The down-dip slip rate (and sometimes locking depth) is inferred (e.g., Hammond et al., 2011; McCaffrey et al., 2000; Meade & Hager, 2005a). These "geodetic" fault slip rates are often consistent with independently derived fault slip rates (e.g., Evans, 2022; Hammond et al., 2011; Reilinger et al., 2006; Thatcher, 2009) and start to be considered as input to PSHA models. However, it is actually quite uncommon to observe an increased strain signal above most faults (Kreemer et al., 2014; Kreemer & Young, 2022), and earthquakes frequently occur on unidentified faults.

Regardless of where exactly all the faults are located, whether earthquakes occur on them, or whether faults indeed have a continuously slipping down-dip fault continuation, a more generally valid assumption would be that the strain accumulating in a (large) area will be accommodated seismically. If all the accumulating strain is released seismically, records of seismic moment release should be equal to the accumulated geodetic moment over the same time period. However, this could only be tested if the time considered is long enough and/or the area considered large enough. In this chapter, we present an analysis of how the breakdown of those two assumptions translates into an expected variability in the seismic-to-geodetic moment ratio. Additionally, when earthquake populations follow a well-described magnitude-frequency relationship and shear modulus and seismogenic thickness is similar everywhere, there should be a linear relationship between the geodetic moment or potency and the number of earthquakes above a given magnitude [this can be seen, e.g., in similar spatial patterns between earthquake numbers and strain rate (Elliott et al., 2016; Kreemer & Young, 2022; Stevens & Avouac, 2021)]. In this chapter, we show evidence that the observed number of earthquakes in areas with the highest geodetic strain rate is considerably below the expected number derived from the strain rate.

2 Estimation of geodetic strain and moment rates

2.1 From velocities to strain rates

A field of horizontal velocities derived from the GNSS position time-series can be expressed by an underlying velocity gradient tensor field (F). If the GNSS stations are located in an area of deformation, F can be decomposed into a 2D strain rate tensor and rotation rate vector component. It is not the purpose of this chapter to offer a review of the mathematical background. There are multiple proposed methods to derive a continuous strain rate tensor field, varying from using wavelets (Su et al., 2019; Tape et al., 2009; Xu et al., 2020), the least-squares collocation method (Caporali et al., 2003; El-Fiky & Kato, 1998; Kahle et al., 1995; Wu et al., 2011), weighted least-squares regression (Shen et al., 1996, 2007, 2015), spline fitting (Beavan & Haines, 2001; Hackl et al., 2009), elasticity theory (Haines et al., 2015; Noda & Matsu'ura, 2010; Sandwell & Wessel, 2016), Bayesian estimation (Pagani et al., 2021; Xiong et al.; 2021), basis function expansion (Okazaki et al., 2021), and robust imaging (Kreemer et al., 2020, 2018). Some software packages that derive strain rates are publicly available (Cardozo & Allmendinger, 2009; Goudarzi et al., 2015; Materna et al., 2021; Pietrantonio & Riguzzi, 2004; Ramírez-Zelaya et al., 2023; Sandwell & Wessel, 2016; Shen et al., 2015; Teza et al., 2023).

There are multiple proposals on how to convert the strain rate tensor to the scalar $\dot{\varepsilon}$ (e.g., Pancha et al., 2006). For this study, it is not critical which one we use, but a brief discussion is warranted. We follow the definition by Ward (1994): $\max(|\dot{\varepsilon}_1|,|\dot{\varepsilon}_2|)$, where $\dot{\varepsilon}_1$ and $\dot{\varepsilon}_2$ are the horizontal principal values. This definition is equivalent to the one independently proposed by Holt et al. (1995), and this definition gives the lowest moment release based on a strain rate field surrounding arbitrarily oriented faults (Holt et al., 1995). This definition is typically equal to the two definitions proposed by Savage and Simpson (1997) and Bird and Liu (2007) (which are defined differently from each other, but are equivalent), but is lower when both principal values have equal sign (in which case the vertical principal value, which is not considered by the given definitions, is the largest one). Our adopted definition is better suited than the alternatives when using a continuous strain rate field model inferred from horizontal GNSS-derived velocities, as it makes no assumption about deformation in the third dimension. We note that some studies have instead used the shear strain rate (e.g., Zeng et al., 2018) or the second invariant of the tensor (e.g., Chen et al., 2022; Stevens & Avouac, 2021) to define $\dot{\varepsilon}$.

2.2 From strain rate to moment rate

Following Kostrov (1974), the geodetic moment M_G (sometimes called tectonic moment) is defined as

$$M_G = 2\mu D \dot{\varepsilon} T A \tag{3.1}$$

This conversion from the geodetic strain rate (as a scalar $\dot{\varepsilon}$) to moment applies for a given time-span (T) and surface area (A) and requires assumptions of the shear modulus (μ), and the depth (D) up to which elastic strain is built up and released, also called the seismogenic depth. Because strain rates typically vary over A, a more general form of (3.1) is $M_G = 2\mu D T \int \dot{\varepsilon} dA$.

Eq. (3.1) can be written in terms of knowns and unknowns:

$$M_G = \lambda P_G \tag{3.2}$$

where λ is an unknown (assumed) constant ($2\mu D$), and P_G is the measured, purely kinematic, "geodetic potency" ($\dot{\varepsilon} T A$). A similar separation between moment, rigidity, and seismic potency has been advocated for the treatment of seismic moment (Ben-Menahem & Singh, 1981; Ben-Zion, 2001; Heaton & Heaton, 1989; King, 1978).

We assume that the geodetic strain rates are constant in time and represent interseismic strain accumulation (e.g., Hussain et al., 2018; Iezzi et al., 2021). There are, however, epistemic uncertainties related to whether geodetic strain rates, which are inferred over short time-spans, represent the interseismic strain build-up that is accommodated by earthquakes. Strain rates near a fault do vary during the seismic cycle and observed differences between paleoseismic and geodetic rates can be explained by viscoelastic relaxation (e.g., Chuang & Johnson, 2011; Dixon et al., 2003; Hammond et al., 2009). Importantly, if earthquake recurrence times are long, and/or viscosity is low, an observed low geodetic strain rate may underrepresent a fault's seismic potential, given that most of the rapid strain accumulation happened early in the postseismic phase (Wang et al., 2021). The strain rate field we use is based on GNSS data that are minimally affected by rapid postseismic deformation, although postseismic deformation caused by earthquakes that occurred long before the observation period can be present (e.g., Hearn et al., 2013; Young et al., 2023).

3 Seismic moment distribution

In order to understand the relationship between M_G (or P_G) and the seismic moment release or the expected corresponding number of earthquakes, we provide a brief background on seismic moment distribution below (also see Appendix).

The Gutenberg-Richter law (Gutenberg & Richter, 1944; Ishimoto & Iida, 1939) asserts that the number $n(m)$ of earthquakes with magnitude above a given value m in a sufficiently large space-time region decreases exponentially with m:

$$n(m) = a 10^{-b(m-m_T)}, \qquad m \geq m_T \tag{3.3}$$

Here, the constant a controls the total earthquake rate. Specifically, a represents the total number of events with a magnitude above or equal to a threshold m_T, which is assumed to be equal to or above the catalog completeness magnitude. The parameter b, or b-value, varies a bit from region to region and in different tectonic settings (Gulia & Wiemer, 2010; Petruccelli et al., 2019; Schorlemmer et al., 2005), but is universally close to $b=1$ (e.g., Kagan, 1999; Kanamori & Anderson, 1975). The Gutenberg-Richter law (3.3) can be interpreted as a statistical statement that the magnitude of a randomly selected earthquake has an exponential distribution with rate $b\ln(3.10)$ given by its survival function:

$$\text{Prob}(magnitude \geq m) = 10^{-b(m-m_T)}, \quad m \geq m_T \quad (3.4)$$

Assuming that we use the moment magnitude and applying the magnitude-moment conversion (Hanks & Kanamori, 1979) to the seismic moment M in Newton-meters [Nm],

$$m = 2/3(\log_{10} M - 9.05), \quad M = 10^{3/2m+9.05}, \quad (3.5)$$

we find that the exponential magnitude distribution (3.4) corresponds to a power-law, or Pareto, distribution of the seismic moment of a randomly selected event:

$$\text{Prob}(moment \geq M) = \left(\frac{M}{M_T}\right)^{-\beta}, \quad M \geq M_T. \quad (3.6)$$

Here, M_T is the threshold moment that corresponds to m_T via (3.5) and $\beta = \frac{2}{3}b$ is called the Pareto index.

The Pareto distribution (3.6) with index $\beta < 1$ has a "thick tail" with infinite first and second statistical moments (Kagan, 2002a). In reality, however, seismic moment realizations cannot be arbitrarily large, given the ultimately finite dimensions and energy budget of the seismogenic system. Kagan (2002a, 2002b) discussed four alternative models of the seismic moment distribution, each of which imposes some form of a taper on the survival function. That work demonstrated that such a modification provides an appropriate fit to the global earthquake data. Here, we apply the exponentially tapered Pareto distribution, which imposes an exponential taper on the survival function (Vere-Jones et al., 2001) and which results in the closest fit to the data (Kagan, 2002b):

$$Prob(moment \geq M) = \left(\frac{M}{M_T}\right)^{-\beta} \exp\left(\frac{M_T - M}{M_C}\right), M \geq M_T. \quad (3.7)$$

The additional parameter, M_C, called the corner moment (associated with a corner magnitude m_C), controls the transition between the initial power law and the exponential decay of the right tail of the distribution. For the number $n(M)$ of events with a moment above M, we can write:

$$n(M) = n(M_T)\left(\frac{M_T}{M}\right)^{\beta} \exp\left(\frac{M_T - M}{M_C}\right), M \geq M_T. \quad (3.8)$$

where $n(M_T)$ is the total number of events with magnitude above a threshold M_T.

The pth statistical raw moment of the tapered Pareto distribution is given by (Kagan & Schoenberg, 2001):

$$E[M^p] = M_T^p + pM_T^\beta M_C^{p-\beta} \exp\left(\frac{M_T}{M_C}\right) \Gamma\left(p - \beta, \frac{M_T}{M_C}\right), \quad (3.9)$$

where $\Gamma(x,y)$ is the upper incomplete gamma function:

$$\Gamma(x, y) = \int e^{-t} t^{x-1} dt. \quad (3.10)$$

The first moment $E[M]$, i.e., $p=1$, is the expectation value or the mean M for the tapered Pareto distribution. The distribution parameters, β and M_C, are typically estimated with a maximum likelihood estimation using the observed catalog.

Following Molnar (1979), we define the long-term seismic moment (M_S) as follows:

$$M_s = n(M_T)E[M] \quad (3.11)$$

Thus, $n(M_T)$ and M_S are proportional as long as $E[M]$ is a constant, which is true if all seismicity for the entire space-time follows a similar tapered Pareto distribution (i.e., similar β and M_C).

The above derivations are implied for the entire considered space-time. We can, however, extend (3.11) to obtain $n(M_T)$ for any arbitrary subarea, if we consider the ratio of the long-term regional/local moment over the expected mean moment for the entire area. If we use M_G as our best proxy for the long-term regional/local moment, then the number of events is proportional to P_G for areas with similar μ and D. Thus, in areas of similar tectonic setting and earthquake distribution (e.g., subduction zones or continental collision zones), we should expect to see a linear relationship between the number of earthquakes in a given area and P_G, or similarly between earthquake rate density (v, equal to $n(M_T)/AT$) and $\dot{\varepsilon}$:

$$v = \frac{2\mu D \dot{\varepsilon}}{E[M]} \qquad (3.12)$$

4 Seismic-to-geodetic moment ratio

4.1 Background

A common practice in seismotectonic and seismic hazard studies is to compare the rate of seismic moment release with the moment build-up rate for the same area (e.g., D'Agostino, 2014; Graham et al., 2018; Kreemer et al., 2000; Masson et al., 2005; Nishimura, 2022; Ojo et al., 2021; Palano et al., 2018; Pancha et al., 2006; Pondrelli et al., 1995; Rahmadani et al., 2022; Sawires et al., 2021; Sharma et al., 2020; Shen-Tu et al., 1995, 1998; Sparacino et al., 2020; Ward, 1998). If all deformation is accommodated seismically, these two should be equal when averaged over many earthquake cycles. The moment build-up can be inferred from either adjacent plate motions, the long-term rate at which active faults move, or, increasingly more prevalent, crustal strain rates derived from GNSS-inferred velocities. Over a given time T, the constant moment build-up translates to M_G (3.1).

Following multiple other studies, we define the seismic-to-geodetic moment ratio (R) as:

$$R = \frac{M_\Sigma}{M_G} \qquad (3.13)$$

where M_Σ is the sum over the observed seismic moments (i.e., ΣM) of N earthquakes within an area A and over time T. The observed R is typically interpreted in terms of the seismic coupling coefficient (c). A low R, for example, could signify a considerable component of aseismic deformation (i.e., low c) which, if true, would have important implications for understanding seismic hazard, fault mechanics, tectonics, and geodynamics. However, R is highly variable when inferred from an earthquake catalog that is (much) shorter than one or more seismic cycles. Most studies acknowledge this, yet interpretations of R in terms of c are often made regardless.

In some cases, R is a good indicator of c. For example, analyses of mid-oceanic ridges suggest that a low coupling is unambiguous there (e.g., Bird et al., 2002; Frohlich & Apperson, 1992; Kreemer et al., 2002). That conclusion is corroborated by a low seismicity level and low maximum seismic moment. Elsewhere, the relationship between R and c is less conclusive. For subduction zones, for example, it is to be expected that the geodetically observed spatial variation in physical coupling should be reflected in variations in c (Scholz & Campos, 2012). However, the short seismic record for many areas prevents one from confidently reaching such a conclusion (McCaffrey, 1997). The traditional comparison between seismic and geodetic moment for subduction zones is also thwarted by the prevalent occurrence of slow slip events, which, at places, can accommodate a significant portion of the total available moment (e.g., Bekaert et al., 2015; Rolandone et al., 2018; Wallace & Beavan, 2010). We will not consider subduction zones in our study.

For continental plate boundaries, where independent observations of present-day creep (i.e., low c) are rare, findings of low R are often the only proposed direct evidence for aseismic behavior (Chousianitis et al., 2015; Masson et al., 2005; Mazzotti et al., 2011; Palano et al., 2018; Pondrelli et al., 1995; Rahmadani et al., 2022; Sawires et al., 2021; Sparacino et al., 2020). However, because continental boundaries have some of the longest earthquake recurrence times (given the relatively low deformation rates), a low R may simply reflect that the sample time was not long enough to capture the largest earthquake(s) (Brune, 1968). Occasionally, a low R is interpreted to mean that an area is overdue for one or more large events (e.g., Meade & Hager, 2005b). The variability in M_Σ was discussed by Frohlich (2007) and Zaliapin et al. (2005). The resulting variability in R was briefly presented by Kagan (2013), but is more fully developed here.

There are several uncertainties affecting R. Some uncertainties exist in the moment of both preinstrumental and largest instrumental earthquakes. An example of the former is found in the debate of the moment of the 1811–1812 New Madrid earthquake sequence (e.g., Hough et al., 2000; Johnston, 1996) and an example of the latter was given by the discussion of the moment of the Great 2004 Sumatra event (e.g., Ishii et al., 2005; Park et al., 2005; Stein & Okal, 2005; Tsai et al., 2005). While those uncertainties should not be ignored, most of the direct variability in R could be due to uncertainties in the geodetic moment from not knowing μ, and D, but they are also harder to quantify and are arguably much smaller than the variability in the M_Σ resulting from sampling a small A and T.

4.2 Approach

The theoretical variability in R considered in this study follows from the variability in M_Σ. That variability is the result of the sample size, N, which is controlled by A, T, and $\dot{\varepsilon}$, and thus P_G. For any given location, T has to approach the duration of many earthquake cycles in order for M_Σ/T to approach the long-term moment release rate. If that observed moment release is sampled from a long-term earthquake frequency distribution valid for a large area, then the ergodic behavior of seismic moment release allows one to estimate the long-term, time-independent moment release by instead considering a (much) larger A. That is, the "ergodic assumption" implies that you can approximate the time-average moment rate by sampling many places that are in various stages of the earthquake cycle (and thus give different apparent rate estimates) and are not correlated with each other.

Obviously, the longer the T and the larger the A, the more the observed seismic moment release should reliably equate to the long-term average (approximated by the geodetic moment). However, the actual values for what T and A depend on $\dot{\varepsilon}$ (Ward, 1998). The influence of the short earthquake catalog duration on M_Σ or R has already been explored elsewhere (Frohlich, 2007; McCaffrey, 1997; Ward, 1998; Zaliapin et al., 2005). Because of the focus on T, past analyses often framed the problem in terms of how long an earthquake catalog needs to be for the seismic moment to be a reliable estimator of the long-term rate (Mazzotti et al., 2011; Ward, 1998). Pancha et al. (2006) introduced the concept of a "catalog adequacy parameter," which they defined as the product of $\dot{\varepsilon}$, T, and A, and what we call the geodetic potency, P_G.

The goal of this chapter is to provide a more general analysis from which practical uncertainties in the seismic moment, and thus R, can be derived for a given P_G, moment distribution, and number of events with $M \geq M_T$. We use the observed seismicity and geodetic deformation in the Mediterranean-to-Asia continental deformation zone for our analysis. In our exploration of R as a function of N and P_G, we consider different values of A and T, and we fix μ and D such that $M_G = M_S$ for the largest time-window considered. Fixing μ and D as such does not impact any of our following assumptions or conclusions.

The total area is not arbitrarily divided into subareas. Instead, we first sort the grid cells of the strain rate model by decreasing P_G (we sort the potencies and not the strain rates because the grid areas underlying the strain rate model have a latitude dependency). Then we divide the total sorted area into equally sized sequential subareas. For the various areal subdivisions, we then consider different time periods. We define $k = \max(AT)/AT$, where $\max(AT)$ is AT for the largest space-time-window, and we consider $T=100$y, $T=50$y (i.e., the first and second half of the total catalog), and $T=25$y (i.e., four sequential quarter periods of total catalog), and break up the total area into either 1,2,4,8,16, or 32 equal parts. So, k ranges from 1 for the entire space-time to 128 for all 128 cases comprised of 32 areal bins and 4 temporal bins of $T=25$y.

5 Geodetic potency versus earthquake numbers

5.1 Background

Molnar (1979) was the first to explicitly show that there should theoretically be a linear relationship between the tectonic moment and the number of earthquakes in a given space-time (and above a certain magnitude), and the finding was corroborated by Anderson (1979). The first studies that showed that there is a strong correlation between earthquake numbers and tectonic/geodetic moment are those of Kagan (1999) for subduction zones and Kreemer et al. (2002) for subduction zones and areas of continental deformation. These results were later refined for more areas with different deformation regimes (Bird et al., 2009). As a consequence of (3.8), one consequence of there being a strong correlation between earthquake numbers and geodetic moment rate is that there is a greater probability of large earthquakes in areas with higher strain rate (Farolfi et al., 2020). This is, however, opposite to the claim of Riguzzi et al. (2012) for Italy.

There being a linear relationship between earthquake numbers (N) and geodetic moment or potency [or, equivalently, between earthquake rate density (ν) and strain rate (3.12)] is an essential ingredient for recent studies that essentially mapped the spatial variation in strain rate into an earthquake rate forecast (Bayona Viveros et al., 2019; Bird et al., 2015, 2010; Bird & Kreemer, 2015; Nishimura, 2022; Rhoades et al., 2017; Rong et al., 2016). These studies analyze different tectonic regimes separately, and for all combined areas of the same regime calculate $E[M]$ from the best-fitting tapered Pareto distribution. The number of events (above the same M_T) in any subarea is then calculated by dividing the geodetic moment of the subarea by the corresponding regime's $E[M]$ for $M \geq M_T$. For the number (or rate) of events above any other magnitude, $E[M]$ can be estimated for a different M_T (3.9). If one is only concerned with the total number of events above the catalog's completeness moment (as we are), and with the explicit implication that $M_S = M_G$ for the entire space-time (as we do), the number of events for a subregion can be calculated as the total number of events in the entire space-time times the ratio of the geodetic moment (or potency) of that subarea over the entire geodetic moment (or potency): $N = N_{total}(P_G/P_{G\_total})$.

To verify the above-assumed linearity between earthquake numbers and tectonic or geodetic moment, numerous studies have plotted the cumulative M_G or P_G (after sorting the areas following increasing or decreasing values of P_G) and the cumulative number of events (following the same sorted areas) versus cumulative area (Chen et al., 2022; Kreemer et al., 2018; Kreemer & Gordon, 2014; Kreemer & Young, 2022; Shen et al., 2007; Wu et al., 2022; Zeng et al., 2018). These curves typically follow each other rather closely, with a notable exception for intraplate North America (Kreemer et al., 2018). There has also been some evidence that the correlation varies temporarily (Rastin et al., 2022; Wu et al., 2022; Zeng et al., 2018), but despite speculations by these studies, there is no evidence that there is long-term variation in strain rate (except right after an earthquake) (Hussain et al., 2018; Iezzi et al., 2021).

Only occasionally have $\dot{\varepsilon}$ or P_G been shown directly against ν or N, respectively (Kreemer et al., 2018; Kreemer & Young, 2022). Equally rare is when, instead of cumulative plots, discrete plots of P_G versus N (Kreemer & Young, 2022) or $\dot{\varepsilon}$ versus ν (Stevens & Avouac, 2021) are presented. The latter study found relatively more events for the fastest straining areas in the India-Eurasia collision zone; however, they considered bins with unequal areas. Kreemer and Young (2022) found that faster straining areas in the western United States have relatively fewer events. They argued for a different linear relationship between fast and slowly deforming areas. Their inference was misled by them plotting correlated sampling areas simultaneously; however, their general observation of there being fewer events in the highest strain rates is confirmed here.

5.2 Approach

We follow the same approach as the moment analysis and create different sets of equally sized sequential subareas (after sorting the total area for decreasing P_G (see above)). For this analysis, we exclude identified aftershocks. Given that the earthquake rate is much more constant than the moment rate, and to maximize the earthquake count, we only consider the total time period, $T = 100$.

6 Data

We use v.2.1 of the Global Strain Rate Model (Kreemer et al., 2014), which provides the strain rate tensor averaged for grid cells with longitudinal and latitudinal dimensions of 0.25° and 0.2°, respectively. We consider all the deforming areas shown in Fig. 3.1 using their corresponding $\dot{\varepsilon}$. We exclude the Hellenic subduction zone, however, where the relationship between deformation and seismicity may be fundamentally different than for continental areas. Low coupling is found there geodetically (Floyd et al., 2023; Vernant et al., 2014) or through a combination of geodetic strain rates and the seismic moment distribution (Jenny et al., 2004; Sparacino et al., 2022). More generally, for the analysis between earthquake numbers and geodetic potency, subduction zones are typically separated from continental areas, due to significant global differences in average M_C and μ (Bird & Kagan, 2004).

It is worth noting that great effort has been made in the GSRM to minimize the effects of co- and postseismic deformation. As a result, the GSRM is an as good as possible approximation of interseismic strain accumulation, constrained by both GNSS velocity data and imposed plate motions along the edges of the plate boundary zones.

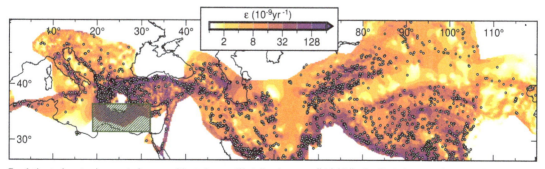

FIG. 3.1 Geodetic strain rates in our study area with strain rate ($\dot{\varepsilon}$) defined as $\max(|\dot{\varepsilon}_1|, |\dot{\varepsilon}_2|)$. *Small circles* are epicenters of earthquakes used in the moment analysis ($m \geq 5.5$ for 1920–2019), with *green circles* being dependent events not used in the number analysis. The *green hatched box* covers the Hellenic subduction zone, which is excluded from our analysis.

For the earthquake analysis, we use the ISC-GEM catalog v.10.0 (Di Giacomo et al., 2018; Storchak et al., 2015, 2013). For the moment analysis, we use $N=2242$ events with $m \geq 5.5$ (i.e., 2×10^{17} Nm) from 1920 to 2019 (inclusive) with depths ≤ 40 km that occurred in the considered deforming areas (minus the Hellenic subduction zone). We use a cut-off depth of 40 km because there is a clear separation at that depth between crustal and deep seismicity in the Pamir-Hindu Kush region, the only area in our study area with subcrustal seismicity (Sippl et al., 2013). That intra-slab deformation has arguably no bearing on the strain rates observed at the surface.

For the analysis of number of events (not the moment analysis), it is important that we isolate mainshocks (i.e., the background seismicity). We identify mainshocks in the catalog used above following the methods of (Zaliapin et al., 2008; Zaliapin & Ben-Zion, 2013, 2016). This provides $N=1682$ mainshocks in our study area.

The total observed seismic moment released by all events with $m \geq 5.5$ over the catalog's 100 years (M_Σ) is 3.15×10^{22} Nm. Fig. 3.2 shows the exceedance probabilities for events with $M \geq 2 \times 10^{17}$ Nm from 1920 to 2019, together with the best-fitting model (3.7) derived from a maximum likelihood estimation (Kagan & Schoenberg, 2001; Vere-Jones et al., 2001). The best-fitting model has $\beta=0.681$ and $M_C=6.38 \times 10^{21}$ ($m_C=8.5$). The model's total predicted seismic moment (M_S) is only 5% more than observed (i.e., $M_S = 3.35 \times 10^{22}$ Nm), which is equivalent to a single $m_C=8.17$ events. We note that, while we observed a temporal variability in the estimated moment release in an earlier global analysis between 1918 and 2014 (Zaliapin & Kreemer, 2017), we do not observe that for the area considered here.

To convert the strain rate to the moment rate, we use $\mu=27.7$ GPa, following the value used by Bird and Kagan (2004) for continental convergent areas. Then, for M_G to be equal to M_S for the entire considered space and time, and under the assumption of complete coupling, we find that it requires (on average) $D=18.8$ km. This value is very reasonable, and is, in fact, close to the 18 km inferred by Bird and Kagan (2004) for continental convergent areas (which most of our area is) when assuming full coupling (as we do). If we had used M_Σ instead of M_S in the above calculations, we would require $D=17.8$ km.

7 Results

Fig. 3.3A summarizes the analysis of R versus N and includes the calculated 2%, 5%, 50%, 95%, and 98% percentiles determined from a simulation of 100k catalogs using the maximum likelihood taper Pareto distribution of Section 6. Fig. 3.3B shows the same for R versus the corresponding P_G. To construct these curves we have, for each P_G, determined

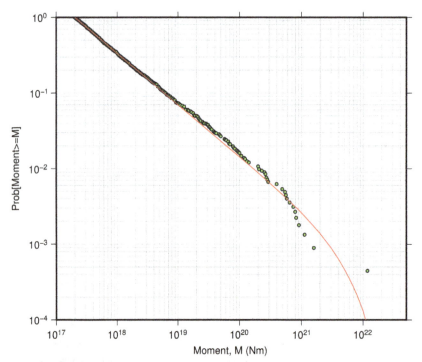

FIG. 3.2 The maximum likelihood tapered Pareto fit of Eq. (3.7) to the observed seismic moment distribution in the examined region. We consider earthquakes during 1920–2019 with magnitude $m \geq 5.5$, which corresponds to $M \geq 2 \times 10^{17}$.

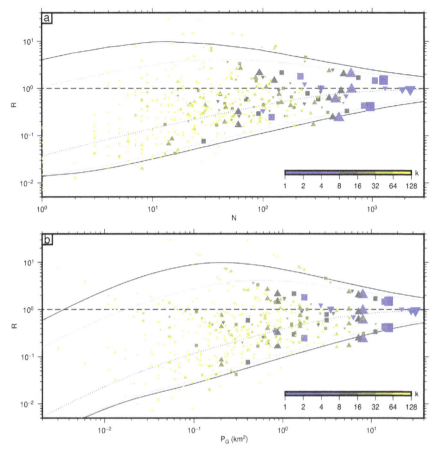

FIG. 3.3 (A) Observed R versus N expressed for different values of k, with the inverted triangle used for cases with $T=100\text{y}$, squares are for $T=50\text{y}$, and triangles are for $T=25\text{y}$. The symbol size is proportional to the area size. *Black lines* are 2% and 98% percentile, *gray lines* are the 5% and 95% percentiles, and the dotted line is the 50% percentiles (B) Same as (A) but for R versus P_G.

the Poisson rate using the fraction (P_G/P_{G_total}) of the total number of events N_{total}. From this Poisson rate, we have determined the probability mass function (pmf) of the discrete Poisson event count distribution. This pmf has subsequently been used to average the percentile curves as a function of the event count of Fig. 3.3A. Table 3.1 lists the approximate percentiles of R for a set of different P_G values. Note that these correspond to our derived values for β and M_C. To derive the percentiles for any other distribution, we are making the source code available (see Appendix). Our observations in Fig. 3.3A closely adhere to the simulations: 89.8% of the cases fall inside the predicted 5%–95% range, while 94.3% of the cases fall inside the 2%–98% range.

TABLE 3.1 Approximate percentile values for R for some selected values of P_G.

P_G (km²)	R (2%)	R (5%)	R (95%)	R (98%)
0.03	0.016	0.018	1.537	5.720
0.1	0.025	0.030	2.600	8.570
0.3	0.046	0.056	3.812	9.346
1	0.090	0.109	3.995	7.287
3	0.161	0.194	3.253	4.730
10	0.293	0.345	2.290	2.982
30	0.463	0.525	1.711	1.971

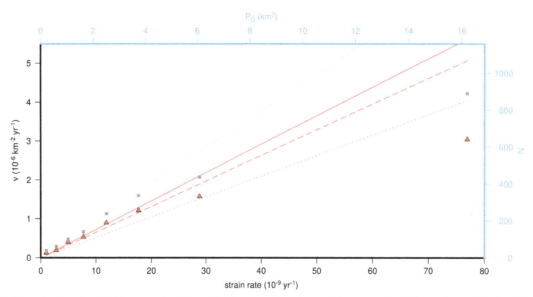

FIG. 3.4 *Red triangles* show for $m \geq 5.5$ v versus $\dot{\varepsilon}$ or N versus P_G for a case where the area is broken up into eight equal areas. The *dotted red line* shows the expected linear relationship and *dashed* and *solid red lines* show a similar relationship when the highest strain rate bin or the two highest bins, respectively, are excluded in the calculation of the prediction. The *gray squares* and *gray lines* are equivalent to the triangles and *solid red lines*, but for a case when we consider all earthquakes (i.e., including aftershocks).

Fig. 3.4 shows the results of the number analysis for $k=8$, expressed either as v versus $\dot{\varepsilon}$, or N versus P_G for $m \geq 5.5$. The dotted red line shows the expected linear relationship. The observations do not follow this linear relationship, particularly with the number of events in the highest strain rate bin being low. We, therefore, also calculated the expected linear relationship when excluding the earthquakes and deformation in the highest strain rate bin (dashed red line). This is closer to the observations (except, of course, the highest strain rate bin), but can be improved if we also calculate the expected linear relationship when we exclude the events and deformation in the two highest strain rate bins (red solid line). We now see that the observed number of events in all bins except the two highest strain rate bins very closely follows the expectation.

8 Discussion

The presentation of R for different-sized areas and time periods highlights some obvious points. Decreasing A and T, and thus P_G, will result in increased variability in R. Going from high to lower P_G, the catalogs become smaller and the proportion of the moment of the largest events relative to the total moment increases. Initially, this leads to stronger variations around the expectation. Going further down, the probability of large events appearing in the catalog becomes lower and lower. These findings are consistent with, for example, the findings of Amemoutou et al. (2023) for the North Anatolian Fault that R (or what they call, or interpret as, coupling) is larger (and closer to parity) when considering a longer seismic catalog. From our analysis, we can conclude that such an observation is entirely a statistical artifact and should not be interpreted in terms of temporal variations in coupling.

Even for the entire space-time interval, for which $P_G = 32 \text{ km}^2$, there is substantial variability, with the expected R ranging from 0.5 to 2, based on 95% confidence intervals. Based on the fact that the 146 years of seismic moment release in the Great Basin of the western United States was about equal to the geodetic moment, Pancha et al. (2006) concluded that a P_G greater or equal to 1.5 km^2 might be sufficient to have a reliable constraint on R. Yet, we find that for such P_G, the 5%–95% quantile range for R is 0.14–3.81. More extremely, Mazzotti et al. (2011) suggested that, based on the similarity between geodetic moment and seismic moment for the Puget Sound in the Pacific Northwest of the United States, a P_G greater or equal to 0.025 km^2 might be sufficient. Clearly, that inference was based on an entirely haphazard parity between seismic and geodetic moments.

From the analysis of a number of events versus P_G, we observe that the number of events for the fastest straining areas is considerably lower (i.e., down to 54% for the highest straining bin) than would be expected from a linear relationship, which can explain the earthquake numbers elsewhere. This finding confirms the observations of Kreemer et al. (2002) for global continental areas and Kreemer and Young (2022) for the western United States (south of the Cascadia subduction zone). For comparison, Fig. 3.4 also shows the results when we consider all earthquakes (i.e., including aftershocks). In that case, a similar deviation for the highest strain rate bins is observed, with the observed number of events in the highest straining bin still being at 58% of the predicted number.

Our result seemingly differs from that of Kreemer and Young (2022) who inferred a different linear relationship for low and high straining areas. However, we think their conclusions were adversely affected by them comparing results for different k's simultaneously. We reanalyze the analysis conducted by Kreemer and Young (2022) using their geodetic strain rate model (which was explicitly corrected for postseismic deformation due to large earthquakes in the last few centuries) and the declustered ANSS/Comcat catalog with $m \geq 2.0$ for the period 2003–2020. Results are shown in Fig. 3.5 in a similar way as for the Mediterranean-Asia area in Fig. 3.4; i.e., we show the case for $k=8$. We find the same result as for the Asia-Mediterranean area; There are a lot fewer events in the fastest straining areas (i.e., down to 22% for the fastest straining bin) than expected from the linear correlation seen for all other areas.

To facilitate the discussion on what could explain the relatively low number of events for the fastest straining areas, we highlight in Fig. 3.6 for the Mediterranean-Asia area and western United States, the extent of the two highest straining bins (when $k=8$) as well as the epicenters of all the earthquakes. For the Mediterranean-Asia area, the highest strain rate bin includes, e.g., the Himalayas, Tien Shan and Zagros mountains, North and East Anatolian and Xianshuihe faults, and the northern Aegean. For the western United States, the highest strain rate bin covers the San Andreas Fault system, and the one-but-highest bin covers the Eastern California Shear Zone and Walker Lane.

Explanations of why the highest straining areas could have relatively few events compared to the linear relationship observed in other areas can be broken down into three different categories: (1) invalid assumption of all strain being accommodated seismically, (2) physical differences, and (3) different moment frequency distributions. We will discuss these below.

Having smaller seismic coupling in high-straining areas or having permanent anelastic strain accommodation would both result in not all geodetic strain being accommodated seismically. Low or zero coupling occurs where faults creep (either continuously or episodically, and not necessarily at full fault slip rate). It is unlikely that low coupling could explain the underestimation of earthquake counts, because there are not many faults (segments) known to creep. However, known examples, such as the creeping segment along the central San Andreas Fault or the Ismetpasa section of the North Anatolian Fault, do occur in high-straining areas. As an apparent paradox from the perspective of the premise our analysis builds on, the creeping segments of the San Andreas Fault have large numbers of microseismicity (Fig. 3.6B). However, recent work (Liu et al., 2022) showed that the number of "nonclustered" events correlates with the creep rate along the central San Andreas Fault creeping segment. A similar positive correlation is found with the b-value (Liu et al., 2022; Tormann et al., 2014; Wyss et al., 2004), which, in itself, will imply more events there.

Concerning anelastic strain, some studies have argued that this 'off-fault' deformation constitutes a substantial portion (~25%–40%) of the overall strain budget (e.g., Herbert et al., 2014; Johnson, 2013; Parsons et al., 2013). Typically, the anelastic strain is hypothesized to explain the discrepancy between the geodetic moment and the moment implied from

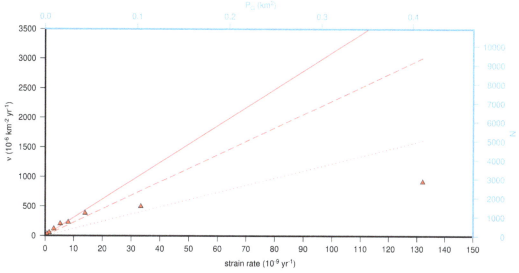

FIG. 3.5 Same as Fig. 3.4, but for the western United States with $m \geq 2$ and using the ANSS/Comcat catalog and the strain rate model of Kreemer and Young (2022).

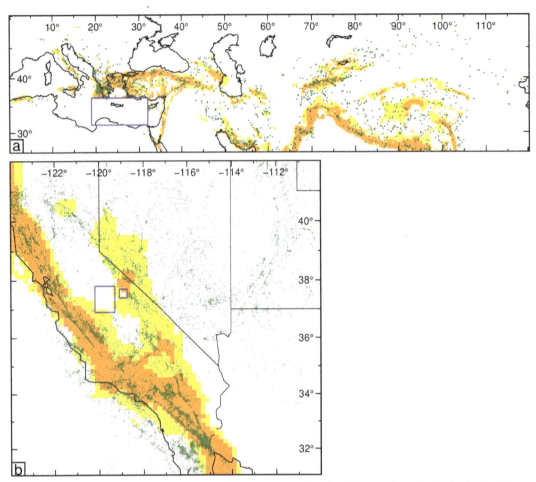

FIG. 3.6 Considered seismicity *(green dots)* and areas with the highest *(orange)* and one-but-highest strain rate *(yellow)*, when breaking up the area into eight equal parts. Areas outlined in *blue* are excluded in analysis (A) Mediterranean-Asia and (B) western United States.

geologic fault slip rate. For a geodetic model alone, it is difficult to distinguish between the elastic strain and any anelastic strain. It is, however, worth noting that one of the strain rate models of Kreemer and Young (2022) that was only based on fault slip rates still yielded considerably fewer events in the fastest straining areas than should be expected there.

If the fastest straining areas have lower seismogenic thickness or lower shear modulus than elsewhere, then fewer events will be predicted. We mostly only know the seismogenic thickness, inferred from either depths of seismicity or geodetically from locking depths, for the faster straining areas. For the southern San Andreas fault system, a few faults indeed have significantly lower locking depths, but most faults have locking depths similar to a similar regional seismogenic depth (Smith-Konter et al., 2011). More generally, while the seismogenic depth does vary across California, it has been found that that variation does not correlate with strain rate (Zuza & Cao, 2020) (but rather heat flow), so we would not expect a systematic bias to have affected our analysis. As for the shear modulus, much less is known about that. It is, however, unlikely that a reduction of 50%–75% in shear modulus for the fastest straining areas relative to the slower deforming areas would not yet be evident.

To assess whether the moment frequency distribution is fundamentally different, we show the moment probabilities separate for events in the two highest strain rate bins and those elsewhere (Fig. 3.7). We note no discernible difference in β. The tails of the distributions are naturally slightly different, but hard to interpret. For the Mediterranean-Asia area (which at least contains some large earthquakes), the difference in the tail between the high (purple) and slow (orange) straining areas, is entirely controlled by the largest earthquake in the catalog (i.e., the 1950 $m=8.68$ Assam earthquake), which occurred in the one-but-highest strain rate bin. Given the ratio of P_G's for the slow- and fast-straining areas, you would not expect the same level of confidence to constrain M_C for the slow-straining areas until the catalog is 229 years long. In any case, if there were a difference in M_C between the fast- and slow-straining areas, it would require a larger M_C for the fast-straining areas to yield fewer expected earthquakes (because the expected mean moment per event would be larger).

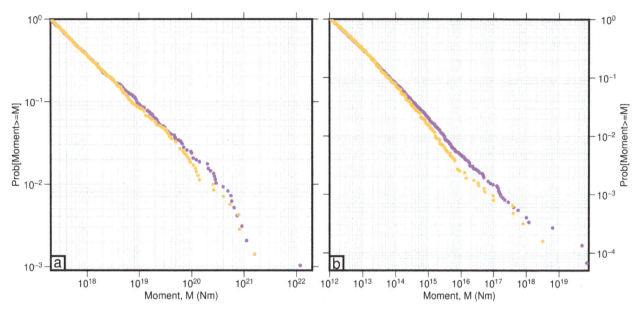

FIG. 3.7 Same as Fig. 3.2, but broken down for events in the two highest strain rate bins *(purple)* and other areas *(orange)* (A) mainshocks Mediterranean-Asia with $m_T \geq 5.5$ and using the ISC-GEM catalog for 1920–2019, and (B) mainshocks western United States with $m_T \geq 2$ and using the ANSS/Comcat catalog for 2003–2020.

Longer faults tend to be found in the areas with the highest strain rate, which would accommodate larger earthquakes. On the other hand, large earthquakes do occur in low-straining areas, such as the 1905 $m = 8.33$ Bolnay (Mongolia) earthquake, which occurred before the start of the catalog that we considered.

One problem in finding a satisfying explanation for there being fewer events in the fastest-straining areas is that those areas can show some large diversity in faulting behavior. The best example is the central part of the San Andreas Fault, which creeps and has many small earthquakes north of Parkfield, California, and is locked and devoid of earthquakes south of there. The latter segment corresponds to the 1857 $m_w = 7.9$ Fort Tejon rupture. There, one could argue that the moment frequency distribution may instead be better characterized by a "characteristic earthquake" behavior; i.e., all or most moment is released by a single event or there is at least a higher rate of large earthquakes than would be expected from a Gutenberg-Richter distribution (Wesnousky, 1994). By mixing areas where seismicity follows a Gutenberg-Richter distribution with areas that follow the characteristic earthquake model, the joint moment frequency behavior may look like a typical Gutenberg-Richter relationship, yet it would yield much fewer events than would be the case if seismicity everywhere follows the same moment frequency behavior, as we assumed. There is increasingly more evidence that, when focusing on single mature faults, seismicity follows characteristic earthquake behavior (Hartleb et al., 2006; Ishibe & Shimazaki, 2012; Parsons et al., 2018; Ran & Wu, 2019; Stirling & Gerstenberger, 2018). Indeed, some recent seismic hazard models put significant weight on faults exhibiting such behavior (Field et al., 2014; Petersen et al., 2014).

The fact that we see a breakdown of the expected relationship between P_G and the actual observed number of events will impact studies that have converted strain rate to seismicity forecasts. Those forecasts would systematically overpredict earthquake rates in faster straining areas, which are also the areas where the hazard is the highest. This discrepancy should have been revealed in prospective tests of earthquake rate predictions based on geodetic strain rate models. This seems to be the case of the global prospective tests of Strader et al. (2018) of Bayona et al. (2021), who found that observed seismicity patterns are broader than those predicted from geodetic strain rate-based forecasts. They concluded that the geodetic models yield too localized strain rates zones and thus too localized seismicity forecasts. We show that instead there is a breakdown in the assumed linear relationship to create the forecasts and this yields an overprediction of the number of events along major tectonic zones such as the North Anatolian Fault, San Andreas Fault, and the Himalayas. Consequently, studies have found seismicity forecasts based on geodetic strain rates to always have less predictive power than those that are based on smoothed past seismicity (Bayona et al., 2021; Bird et al., 2015; Rhoades et al., 2017; Strader et al., 2018). This does not mean that geodesy cannot fulfill an important role in creating forecasts, because the best-performing forecasts are typically hybrid models based on geodetic strain rates and smoothed seismicity (Bayona et al., 2021; Bird et al., 2015; Rhoades et al., 2017; Strader et al., 2018). Ultimately, a closer look is required at the moment frequency distribution for smaller areas (and possibly their mechanical attributes) and those findings need to be built into the conversion of strain rate to seismicity forecasts.

Appendix: Approximating cumulative seismic moment distribution

A.1 Existence of two regimes

We consider the total moment S_n released by n earthquakes with individual moments M_i:

$$S_n = \sum_{i}^{n} M_i. \qquad (3.A1)$$

The moments are assumed to be independent and identically distributed (i.i.d.) random samples from a tapered Pareto distribution (3.7). The tapered Pareto distribution combines a power-law body with an exponentially tapered (right) tail. The taper acts to reduce the probability of large moments, especially those beyond the corner moment, M_C. An example of the survival function of this distribution is shown in Fig. 3.2. While the (nontapered) Pareto distribution features a "thick tail," resulting in infinite expectation value and higher order statistical moments, the taper acts to "thin" the tail and yields finite values for all moments as shown by Kagan and Schoenberg (2001), see (3.9). The separate body and tail descriptions of the tapered Pareto distribution also have a distinguished effect on the behavior of the sum S_n as a function of n. The growth of S_n with increasing n has two principal regimes: an initial power-law increase typical for a (nontapered) Pareto distribution, and, eventually, a linear increase, when the lack of larger moments due to the exponential taper becomes apparent. The crossover between the two regimes is associated with the value of the corner moment, M_C, relative to the lower truncation moment, M_T. This section illustrates these two regimes in synthetic models.

Fig. 3.A1A shows sequences of cumulative seismic moments S_n modeled by random sampling of 10^6 moments from a tapered Pareto distribution (3.7) with a lower truncation moment $M_T = 2 \times 10^{17}$, corresponding to a minimum magnitude $m_T = 5.5$, and a Pareto index $\beta = 2/3$, corresponding to a b-value of 1. For comparison, we also plot the expected moment sum E_n, which equals $nE[M]$.

We examine two values of the corner moment M_C, one higher ($M_C = 2 \times 10^{23}$ or $m_c = 9.5$) and one lower ($M_C = 3.5 \times 10^{19}$ or $m_C = 7.0$), displayed by blue and red lines, respectively. For illustration purposes, we have allowed ourselves two manipulations. First, we have synchronized the random number sampling for the blue and red sequences, meaning that the sequence of sampled moments corresponds to the same quantiles in their respective distributions. Since the "blue distribution" is a bit wider than the "red distribution," the contributing moments of the blue distribution are consistently higher than those in the red distribution, ranging from slightly higher (for low quantiles) to a lot higher (for high quantiles). Second, the random realization has been hand-picked from a small number of tries with various seed values for the random number generator to serve our narrative and illustrate the "typical" behavior that we address here. To provide proper context, we also show a small number of truly random samples for the high corner magnitude case in Fig. 3.A2.

In Fig. 3.A1A, both sequences initially show super-linear (power-law) growth until they converge to a linear trend. For the red curve, that convergence takes place a lot earlier (i.e., lower n) than for the blue curve. In both cases, the convergence

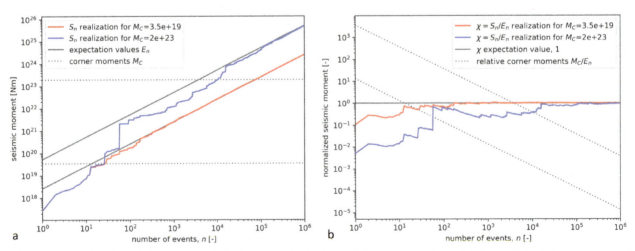

FIG. 3.A1 Illustration of cumulative moment evolution for moments randomly sampled from a tapered Pareto distribution. *Red* and *blue curves* represent distributions with low and high corner moments, respectively. The *gray trend lines* correspond to the theoretical expectation. (A) Total cumulative moment. (B) Normalized moment χ—total cumulative moment S_n normalized by its mathematical expectation, E_n.

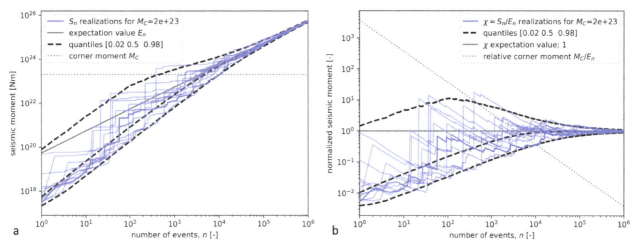

FIG. 3.A2 (A) Provides some more context to Fig. 3.A1, showing multiple random samples of the distribution with the high corner moment, one of which was selected for Fig. 3.A1, here shown with a slightly brighter blue color. Also shown are the 2%, 50%, and 98% percentiles obtained from 10^4 independent draws. (B) The normalized version of panel a, analogous to Fig. 3.A1.

takes place only after the cumulative moment exceeds the respective corner moments (represented by the dashed lines). The linear trends to which the sequences converge are represented by solid gray lines, which equals E_n.

In Fig. 3.A1B, all values are divided by E_n, such that it displays the realization of χ, which is defined as S_n/E_n. This display provides some additional insight into the relative convergence behavior. For the high corner moment case, some more random realizations of cumulative moment sequences are shown in Fig. 3.A2. From the latter figure, it becomes clear that S_n is a random variable with quite a lot of dispersion. In the following section, we study the distribution of S_n and provide approximate expressions for its quantiles.

A.2 Analytic results for each of the regimes

Here, we recall analytic results that exist for the two limiting cases: first, summation of i.i.d. random variables with finite statistical moments, and, second, summation of i.i.d. random variables with infinite first moment (i.e., the mean, or expected value). The former represents the exponential tail of the tapered Pareto distribution, and the latter is power-law body.

Consider a sum S_n of n i.i.d. random variables with a common distribution $F(x)$. In the case of a well-behaved $F(x)$, such that it has a finite expected value and variance, the quantiles of S_n increase approximately linearly with n, for large enough n. In fact, the Central Limit Theorem states that for any distribution with a finite mean m and finite variance σ^2, the sum S_n can, for large enough n, be approximated by a normal distribution with mean nm and variance $n\sigma^2$, that is $S_n \sim \text{Normal}(nm, n\sigma^2)$. Consider an arbitrary quantile $Z(q,n)$ of the sum S_n, which is defined by the equation

$$Prob(S_n \leq Z(q,n)) = q. \tag{3.A2}$$

With $q \in [0,1]$. The quantile $Z(q,n)$ can be approximated by that of the limiting normal distribution:

$$Z(q,n) \approx n \times \left(m + z_q \frac{\sigma}{\sqrt{n}}\right) \sim nm, \text{ as } n \to \infty, \tag{3.A3}$$

where z_q denotes the corresponding quantile of the standard normal distribution Normal(0,1). A simpler, yet less precise, way to derive the linear growth is to recall the Law of Large Numbers (Ross, 2018), stating that the sample average of n independent values X_i from a distribution with mean m converges to m:

$$\frac{1}{n}\sum_i^n X_i \to m, \quad \text{as} \quad n \to \infty. \tag{3.A4}$$

This readily implies that the sum grows linearly as nm.

The heavy-tailed distributions, those with infinite first and second statistical moments, behave differently. The Law of Large Numbers does not apply, indicating that the sample average may show more complex behavior than convergence to a number, and, accordingly, the cumulative sum may exhibit nonlinear growth patterns. This nonlinear growth is particularly simple and well-understood in the case of a power-law Pareto distribution. Specifically, it is known that for Pareto distribution (3.7) with Pareto index $\beta < 1$, the quantiles increase nonlinearly, approximately, as $n^{1/\beta}$ (Huillet & Raynaud, 2001; Pisarenko, 1998; Rodkin & Pisarenko, 2004; Zaliapin et al., 2005). This reflects the fact that the sum of a large number of heavy-tailed random variables is approximated by the maximally skewed stable distribution, according to the Generalized Central Limit Theorem (Nolan, 2020; Samoradnitsky & Taqqu, 2017; Zaliapin et al., 2005). It has been shown in Zaliapin et al. (2005, Eq. 24) that, in this case, the q-quantiles of S_n can be approximated by

$$S(q,n) = n^{1/\beta} s_{q,\beta} [\Gamma(1-\beta)\cos(\pi\beta/2)]^{1/\beta} M_0, \tag{3.A5}$$

where $s_{q,\beta}$ is the q-quantile of the maximally skewed stable distribution with index β, and $\Gamma(x) = \Gamma(x,0)$ is the gamma function.

Consider now an inverse problem of finding the number $n(S_0)$ of summands needed to reach a given value S_0 of the sum S_n (with a given confidence). The above considerations imply that $n(S_0)$ increases linearly with S_0 for the distributions that satisfy the Central Limit Theorem, but may increase nonlinearly for heavy tailed distributions. In particular, for Pareto distribution with power index $\beta < 1$, we have

$$n(S_0) \sim S_0^\beta, \quad \text{as } S_0 \to \infty \tag{3.A6}$$

To examine the cumulative seismic moment release, one needs to combine the above two scenarios. The seismic moments can be approximated by a tapered Pareto distribution with index $\beta \approx 2/3$ (i.e., $\beta < 1$). The behavior of the sum S_n for such summands exhibits two principal regimes: for relatively small n it behaves like a heavy-tailed Pareto distribution, and for large n it behaves as a light-tailed distribution (e.g., Rodkin & Pisarenko, 2004; Zaliapin et al., 2005).

A.3 Approximation equations

Fig. 3.A3 illustrates the double regime effect discussed in the previous section by showing the simulated quantiles of the sum S_n of i.i.d. tapered Pareto random variables (solid black lines) in combination with the two respective approximations $Z(q,n)$ and $S(q,n)$.

The example of Fig. 3.A2 uses $\beta = 2/3$, $M_T = 2 \times 10^{17}$ ($m_T = 5.5$), and $M_C = 2 \times 10^{23}$ ($m_C = 9.5$). In this case, the median of the cumulative moment release closely follows the stable power-law approximation for $n < 2000$ and then switches to the normal approximation. The 98% quantile follows the stable approximation only for $n < 100$, and the 2% quantile follows the stable approximation for a much larger number of summands, up to $n = 10^4$. For larger values of n, both quantiles converge to the respective normal approximations.

Fig. 3.A3B shows the cumulative moment release of Fig. 3.A3A normalized by its expected value χ, which emphasizes the transition from a power-law behavior at small n to normal dynamics at large n. We notice that the variability of the

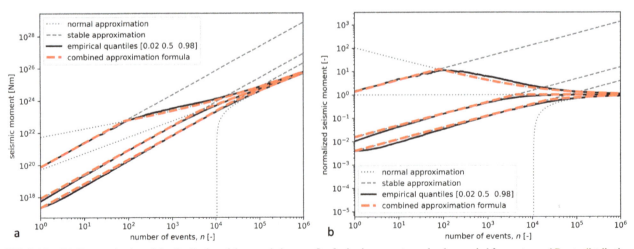

FIG. 3.A3 (A) Characterization of the distribution of the cumulative sum S_n of seismic moments randomly sampled from a tapered Pareto distribution ($\beta = 2/3$, $M_0 = 2 \times 10^{17}$, and $M_C = 2 \times 10^{23}$). The *solid black lines* represent the empirical percentiles shown as *dashed lines* in Fig. 3.A2. The *dashed lines* represent the same percentiles for the approximate maximally skewed stable distribution, while the *dotted lines* represent the same percentiles for the approximate normal distribution. The *red dashed lines* represent the combined approximation formulas (3.A7)–(3.A9). (B) The normalized version of panel (A), analogous to Figs. 3.A1 and 3.A2.

moment release (the range of values between the upper and the lower quantiles) is high and independent of n for the low number of summands ($n < 50$). In addition, in this regime, the total moment is most of the time (high percentiles) substantially lower than its theoretical mean (i.e., $\chi < 1$). For a large number of summands ($n > 5000$), the variability substantially decreases and the total moment is expected to be close to its mean ($\chi \approx 1$). There is also an intermediate regime ($50 < n < 5000$), where the moment variability is large, and one can often see values substantially above the expected mean ($\chi \gg 1$) as well as values substantially below the mean ($\chi \ll 1$).

This qualitative picture with two asymptotic regimes and a transition can be seen for other parameters of the tapered Pareto distribution (with different locations of regimes along the n axis), and for alternative forms of the right-tail taper or cut applied to the Pareto distribution; see Zaliapin et al. (2005) and Kagan (2010).

A closer investigation of Fig. 3.A3 reveals discrepancies between the simulated quantiles of S_n and the closest of their approximations E_n that can reach a factor of 1.5, in particular for small n, before entering the normal asymptotic regime. Nevertheless, this quality seems practically acceptable, taking into account that the range of values of the cumulative moment ranges by a factor of up to 100. In general, it is well known that analytic approximation to the power-law sums for small-to-intermediate n is a hard problem that has no accurate solution (e.g., Zaliapin et al., 2005).

Considering the results of this section, we suggest the following analytic approximation $P(q,n)$ to the q-quantile of the sum S_n of n tapered Pareto i.i.d. random variables with $\beta \approx 2/3$ for $q > 0.5$:

$$P(q,n) = min\,[Z(q,n), S(q,n)], q > 0.5. \qquad (3.A7)$$

For $q < 0.5$ we cannot use the same formula, since the normal distribution breaks down at low n, as revealed by the 2% quantile that shoots down to negative values in Fig. 3.A3. Instead, we suggest using the reciprocal of the upper quantile (perusing the symmetry of the normal approximation):

$$Z_{\text{recip}}(q,n) = \frac{(nE[M])^2}{Z(1-q,n)}. \qquad (3.A8)$$

and thus we get:

$$P(q,n) = min\left[Z_{\text{recip}}(q,n), S(q,n)\right], q < 0.5. \qquad (3.A9)$$

The approximation (3.A7)–(3.A9) is illustrated in Fig. 3.A3. It provides practically useful bounds for the quantiles of the sum S_n of i.i.d. tapered Pareto variables with arbitrary lower moment M_T and corner moment M_C for the power index $\beta \approx 2/3$.

Although we find a good correspondence between the percentiles of the simulation and those predicted by the approximation when we set $\beta \approx 2/3$ and $M_C = 2 \times 10^{23}$ (Fig. 3.A3), the result is a bit less convincing when we consider the same but for the observed $\beta = 0.681$ and $M_C = 6.377 \times 10^{21}$ (Fig. 3.A4).

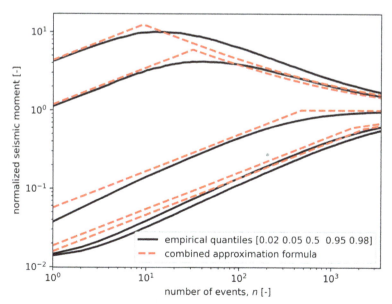

FIG. 3.A4 Quantiles (0.02, 0.05, 0.5, 0.95, and 0.98) of the combined approximation formulas (3.A7)–(3.A9) when $\beta = 0.681$ and $M_C = 6.377 \times 10^{21}$ *(red lines)* versus the empirical quantiles when using the same M_C *(black lines)*.

Therefore, for now, we rely on the empirical results (Fig. 3.3 and Table 3.1). To estimate the percentiles for any distribution, we provide the source code to reproduce all results in this Appendix here: TNO/tapered-pareto: Implementation of the exponentially tapered Pareto distribution (github.com).

Acknowledgments

This work would not have been possible without all the work and insight of our dear friend Ilya Zaliapin. He will be sorely missed in many ways. We appreciate the reviews by Nicola D'Agostino and Yosuke Aoki. Over the years, this work has been supported by USGS NEHRP grants G20AP00080 and G18AP00019, and SCEC grants 17065, 09161, and 08078.

References

Amemoutou, A., Martínez-Garzón, P., Durand, V., Kwiatek, G., Bohnhoff, M., & Dresen, G. (2023). Spatio-temporal variations of seismic coupling along a transform fault: The western north Anatolian fault zone. *Geophysical Journal International*, *235*, 1982–1995. https://doi.org/10.1093/gji/ggad341.

Anderson, J. G. (1979). Estimating the seismicity from geological structure for seismic-risk studies. *Bulletin of the Seismological Society of America*, *69*, 135–158.

Bayona, J. A., Savran, W., Strader, A., Hainzl, S., Cotton, F., & Schorlemmer, D. (2021). Two global ensemble seismicity models obtained from the combination of interseismic strain measurements and earthquake-catalogue information. *Geophysical Journal International*, *224*(3), 1945–1955. https://doi.org/10.1093/gji/ggaa554.

Bayona Viveros, J. A., von Specht, S., Strader, A., Hainzl, S., Cotton, F., & Schorlemmer, D. (2019). A regionalized seismicity model for subduction zones based on geodetic strain rates, geomechanical parameters, and earthquake-catalog data. *Bulletin of the Seismological Society of America*, *109*, 2036–2049. https://doi.org/10.1785/0120190034.

Beavan, J., & Haines, J. (2001). Contemporary horizontal velocity and strain rate fields of the Pacific-Australian plate boundary zone through New Zealand. *Journal of Geophysical Research*, *106*, 741–770. https://doi.org/10.1029/2000JB900302.

Bekaert, D. P. S., Hooper, A., & Wright, T. J. (2015). Reassessing the 2006 Guerrero slow-slip event, Mexico: Implications for large earthquakes in the Guerrero gap. *Journal of Geophysical Research: Solid Earth*, *120*, 1357–1375. https://doi.org/10.1002/2014JB011557.

Ben-Menahem, A., & Singh, S. J. (1981). *Seismic waves and sources*. New York: Springer-Verlag.

Ben-Zion, Y. (2001). On quantification of the earthquake source. *Seismological Research Letters*, *72*, 151–152. https://doi.org/10.1785/gssrl.72.2.151.

Bird, P., Jackson, D. D., Kagan, Y. Y., Kreemer, C., & Stein, R. S. (2015). GEAR1: A global earthquake activity rate model constructed from geodetic strain rates and smoothed seismicity. *Bulletin of the Seismological Society of America*, *105*, 2538–2554. https://doi.org/10.1785/0120150058.

Bird, P., & Kagan, Y. Y. (2004). Plate-tectonic analysis of shallow seismicity: Apparent boundary width, beta, corner magnitude, coupled lithosphere thickness, and coupling in seven tectonic settings. *Bulletin of the Seismological Society of America*, *94*, 2380–2399. https://doi.org/10.1785/0120030107.

Bird, P., Kagan, Y. Y., & Jackson, D. D. (2002). Plate tectonics and earthquake potential of spreading ridges and oceanic transform faults. In S. Stein, & J. T. Freymueller (Eds.), *Plate boundary zones, Geodyn. Ser* (pp. 203–218). American Geophysical Union.

Bird, P., Kagan, Y. Y., Jackson, D. D., Schoenberg, F. P., & Werner, M. J. (2009). Linear and nonlinear relations between relative plate velocity and seismicity. *Bulletin of the Seismological Society of America*, *99*, 3097–3113. https://doi.org/10.1785/0120090082.

Bird, P., & Kreemer, C. (2015). Revised tectonic forecast of global shallow seismicity based on version 2.1 of the global strain rate map. *Bulletin of the Seismological Society of America*, *105*, 152–166. https://doi.org/10.1785/0120140129.

Bird, P., Kreemer, C., & Holt, W. E. (2010). A long-term forecast of shallow seismicity based on the global strain rate map. *Seismological Research Letters*, *81*, 184–194. https://doi.org/10.1785/gssrl.81.2.184.

Bird, P., & Liu, Z. (2007). Seismic hazard inferred from tectonics: California. *Seismological Research Letters*, *78*, 37–48. https://doi.org/10.1785/gssrl.78.1.37.

Brune, J. N. (1968). Seismic moment, seismicity, and rate of slip along major fault zones. *Journal of Geophysical Research*, *73*, 777–784. https://doi.org/10.1029/JB073i002p00777.

Caporali, A., Martin, S., & Massironi, M. (2003). Average strain rate in the Italian crust inferred from a permanent GPS network – II. Strain rate versus seismicity and structural geology. *Geophysical Journal International*, *155*, 254–268. https://doi.org/10.1046/j.1365-246X.2003.02035.x.

Cardozo, N., & Allmendinger, R. W. (2009). SSPX: A program to compute strain from displacement/velocity data. *Computational Geosciences*, *35*, 1343–1357. https://doi.org/10.1016/j.cageo.2008.05.008.

Chen, H., Qu, W., Gao, Y., Zhang, Q., Hao, M., & Wang, Q. (2022). Present-day crustal deformation in the northeastern Tibetan plateau and its correlation with spatiotemporal seismicity characteristics. *Advances in Space Research*, *69*, 2031–2046. https://doi.org/10.1016/j.asr.2021.12.012.

Chousianitis, K., Ganas, A., & Evangelidis, C. P. (2015). Strain and rotation rate patterns of mainland Greece from continuous GPS data and comparison between seismic and geodetic moment release. *Journal of Geophysical Research - Solid Earth*, *120*, 2014JB011762. https://doi.org/10.1002/2014JB011762.

Chuang, R. Y., & Johnson, K. M. (2011). Reconciling geologic and geodetic model fault slip-rate discrepancies in Southern California: Consideration of nonsteady mantle flow and lower crustal fault creep. *Geology*, *39*, 627–630. https://doi.org/10.1130/G32120.1.

D'Agostino, N. (2014). Complete seismic release of tectonic strain and earthquake recurrence in the Apennines (Italy). *Geophysical Research Letters*, *41*, 1155–1162. https://doi.org/10.1002/2014GL059230.

Di Giacomo, D., Engdahl, E. R., & Storchak, D. A. (2018). The ISC-GEM earthquake catalogue (1904–2014): Status after the extension project. *Earth System Science Data, 10*, 1877–1899. https://doi.org/10.5194/essd-10-1877-2018.

Dixon, T. H., Norabuena, E., & Hotaling, L. (2003). Paleoseismology and global positioning system: Earthquake-cycle effects and geodetic versus geologic fault slip rates in the Eastern California shear zone. *Geology, 31*, 55–58.

El-Fiky, G. S., & Kato, T. (1998). Continuous distribution of the horizontal strain in the Tohoku district, Japan, predicted by least-squares collocation. *Journal of Geodynamics, 27*, 213–236. https://doi.org/10.1016/S0264-3707(98)00006-4.

Elliott, J. R., Walters, R. J., & Wright, T. J. (2016). The role of space-based observation in understanding and responding to active tectonics and earthquakes. *Nature Communications, 7*, 13844. https://doi.org/10.1038/ncomms13844.

Evans, E. L. (2022). A dense block model representing Western continental United States deformation for the 2023 update to the national seismic hazard model. *Seismological Research Letters, 93*, 3024–3036. https://doi.org/10.1785/0220220141.

Farolfi, G., Keir, D., Corti, G., & Casagli, N. (2020). Spatial forecasting of seismicity provided from Earth observation by space satellite technology. *Scientific Reports, 10*, 9696. https://doi.org/10.1038/s41598-020-66478-9.

Field, E. H., Arrowsmith, R. J., Biasi, G. P., Bird, P., Dawson, T. E., Felzer, K. R., et al. (2014). Uniform California Earthquake Rupture Forecast, Version 3 (UCERF3)—The time-independent model. *Bulletin of the Seismological Society of America, 104*, 1122–1180. https://doi.org/10.1785/0120130164.

Floyd, M., King, R., Paradissis, D., Karabulut, H., Ergintav, S., Raptakis, K., et al. (2023). Variations in coupling and deformation along the Hellenic Subduction zone. *Turkish Journal of Earth Sciences, 32*, 262–274. https://doi.org/10.55730/1300-0985.1843.

Frohlich, C. (2007). Practical suggestions for assessing rates of seismic-moment release. *Bulletin of the Seismological Society of America, 97*, 1158–1166. https://doi.org/10.1785/0120060193.

Frohlich, C., & Apperson, K. D. (1992). Earthquake focal mechanisms, moment tensors, and the consistency of seismic activity near plate boundaries. *Tectonics, 11*, 279–296. https://doi.org/10.1029/91TC02888.

Geller, R. J., Jackson, D. D., Kagan, Y. Y., & Mulargia, F. (1997). Earthquakes cannot be predicted. *Science, 275*, 1616. https://doi.org/10.1126/science.275.5306.1616.

Goudarzi, M. A., Cocard, M., & Santerre, R. (2015). GeoStrain: An open source software for calculating crustal strain rates. *Computers & Geosciences, 82*, 1–12. https://doi.org/10.1016/j.cageo.2015.05.007.

Graham, S. E., Loveless, J. P., & Meade, B. J. (2018). Global plate motions and earthquake cycle effects. *Geochemistry, Geophysics, Geosystems, 19*, 2032–2048. https://doi.org/10.1029/2017GC007391.

Gulia, L., & Wiemer, S. (2010). The influence of tectonic regimes on the earthquake size distribution: A case study for Italy. *Geophysical Research Letters, 37*. https://doi.org/10.1029/2010GL043066.

Gutenberg, B., & Richter, C. F. (1944). Frequency of earthquakes in California. *Bulletin of the Seismological Society of America, 34*, 185–188.

Hackl, M., Malservisi, R., & Wdowinski, S. (2009). Strain rate patterns from dense GPS networks. *Natural Hazards and Earth System Sciences, 9*, 1177–1187. https://doi.org/10.5194/nhess-9-1177-2009.

Haines, A. J., Dimitrova, L. L., Wallace, L. M., & Williams, C. A. (2015). *Enhanced surface imaging of crustal deformation: Obtaining tectonic force fields using GPS data*. Springer.

Hammond, W. C., Blewitt, G., & Kreemer, C. (2011). Block modeling of crustal deformation of the northern Walker lane and basin and range from GPS velocities. *Journal of Geophysical Research, 116*, B04402. https://doi.org/10.1029/2010JB007817.

Hammond, W. C., Kreemer, C., & Blewitt, G. (2009). Geodetic constraints on contemporary deformation in the northern Walker lane: 3. Central Nevada seismic belt postseismic relaxation. In J. S. Oldow, & P. H. Cashman (Eds.), *Late Cenozoic structure and evolution of the Great Basin-Sierra Nevada transition* (pp. 33–54). Geological Society of America Special Papers.

Hanks, T. C., & Kanamori, H. (1979). A moment magnitude scale. *Journal of Geophysical Research, 84*, 2348–2350. https://doi.org/10.1029/JB084iB05p02348.

Hartleb, R. D., Dolan, J. F., Kozaci, Ö., Akyüz, H. S., & Seitz, G. G. (2006). A 2500-yr-long paleoseismologic record of large, infrequent earthquakes on the North Anatolian fault at Çukurçimen, Turkey. *GSA Bulletin, 118*, 823–840. https://doi.org/10.1130/B25838.1.

Hearn, E. H., Pollitz, F. F., Thatcher, W. R., & Onishi, C. T. (2013). How do "ghost transients" from past earthquakes affect GPS slip rate estimates on southern California faults? *Geochemistry, Geophysics, Geosystems, 14*, 828–838. https://doi.org/10.1002/ggge.20080.

Heaton, T. H., & Heaton, R. E. (1989). Static deformations from point forces and force couples located in welded elastic poissonian half-spaces: Implications for seismic moment tensors. *Bulletin of the Seismological Society of America, 79*, 813–841. https://doi.org/10.1785/BSSA0800041056.

Herbert, J. W., Cooke, M. L., Oskin, M., & Difo, O. (2014). How much can off-fault deformation contribute to the slip rate discrepancy within the eastern California shear zone? *Geology, 42*, 71–75. https://doi.org/10.1130/G34738.1.

Holt, W. E., Li, M., & Haines, A. J. (1995). Earthquake strain rates and instantaneous relative motions within central and eastern Asia. *Geophysical Journal International, 122*, 569–593. https://doi.org/10.1111/j.1365-246X.1995.tb07014.x.

Hough, S. E., Armbruster, J. G., Seeber, L., & Hough, J. F. (2000). On the modified Mercalli intensities and magnitudes of the 1811–1812 New Madrid earthquakes. *Journal of Geophysical Research: Solid Earth, 105*, 23839–23864. https://doi.org/10.1029/2000JB900110.

Huillet, T., & Raynaud, H.-F. (2001). On rare and extreme events. *Chaos, Solitons & Fractals, 12*, 823–844. https://doi.org/10.1016/S0960-0779(00)00046-1.

Hussain, E., Wright, T. J., Walters, R. J., Bekaert, D. P. S., Lloyd, R., & Hooper, A. (2018). Constant strain accumulation rate between major earthquakes on the North Anatolian Fault. *Nature Communications, 9*, 1392. https://doi.org/10.1038/s41467-018-03739-2.

Iezzi, F., Roberts, G., Faure Walker, J., Papanikolaou, I., Ganas, A., Deligiannakis, G., et al. (2021). Temporal and spatial earthquake clustering revealed through comparison of millennial strain-rates from 36Cl cosmogenic exposure dating and decadal GPS strain-rate. *Scientific Reports, 11*, 23320. https://doi.org/10.1038/s41598-021-02131-3.

Ishibe, T., & Shimazaki, K. (2012). Characteristic earthquake model and seismicity around late quaternary active faults in Japan. *Bulletin of the Seismological Society of America, 102*, 1041–1058. https://doi.org/10.1785/0120100250.

Ishii, M., Shearer, P. M., Houston, H., & Vidale, J. E. (2005). Extent, duration and speed of the 2004 Sumatra–Andaman earthquake imaged by the Hi-Net array. *Nature, 435*, 933–936. https://doi.org/10.1038/nature03675.

Ishimoto, M., & Iida, K. (1939). Observations sur les seismes enregistres par le microsismographe construit dernierement. *Bulletin. Earthquake Research Institute, University of Tokyo, 17*, 443–478.

Jenny, S., Goes, S., Giardini, D., & Kahle, H. G. (2004). Earthquake recurrence parameters from seismic and geodetic strain rates in the eastern Mediterranean. *Geophysical Journal International, 157*, 1331–1347.

Johnson, K. M. (2013). Slip rates and off-fault deformation in Southern California inferred from GPS data and models. *Journal of Geophysical Research: Solid Earth, 118*, 5643–5664. https://doi.org/10.1002/jgrb.50365.

Johnston, A. C. (1996). Seismic moment assessment of earthquakes in stable continental regions—III. New Madrid 1811–1812, Charleston 1886 and Lisbon 1755. *Geophysical Journal International, 126*, 314–344. https://doi.org/10.1111/j.1365-246X.1996.tb05294.x.

Jordan, T. H. (2006). Earthquake predictability, brick by brick. *Seismological Research Letters, 77*, 3–6. https://doi.org/10.1785/gssrl.77.1.3.

Kagan, Y. Y. (1999). Universality of the seismic moment-frequency relation. *Pure and Applied Geophysics, 155*, 537–573. https://doi.org/10.1007/s000240050277.

Kagan, Y. Y. (2002a). Seismic moment distribution revisited: I. Statistical results. *Geophysical Journal International, 148*, 520–541. https://doi.org/10.1046/j.1365-246x.2002.01594.x.

Kagan, Y. Y. (2002b). Seismic moment distribution revisited: II. Moment conservation principle. *Geophysical Journal International, 149*, 731–754.

Kagan, Y. Y. (2010). Statistical distributions of earthquake numbers: Consequence of branching process. *Geophysical Journal International, 180*, 1313–1328. https://doi.org/10.1111/j.1365-246X.2009.04487.x.

Kagan, Y. Y. (2013). *Earthquakes: Models, statistics, testable forecasts*. John Wiley & Sons.

Kagan, Y. Y., & Schoenberg, F. (2001). Estimation of the upper cutoff parameter for the tapered Pareto distribution. *Journal of Applied Probability, 38*, 158–175.

Kahle, H.-G., Müller, M. V., Geiger, A., Danuser, G., Mueller, S., Veis, G., et al. (1995). The strain field in northwestern Greece and the Ionian Islands: Results inferred from GPS measurements. *Tectonophysics, 249*, 41–52. https://doi.org/10.1016/0040-1951(95)00042-L.

Kanamori, H., & Anderson, D. L. (1975). Theoretical basis of some empirical relations in seismology. *Bulletin of the Seismological Society of America, 65*, 1073–1095. https://doi.org/10.1785/BSSA0650051073.

King, G. C. P. (1978). Geological faults: Fracture, creep and strain. *Philosophical Transactions of the Royal Society of London A: Mathematical, Physical and Engineering Sciences, 288*, 197–212. https://doi.org/10.1098/rsta.1978.0013.

Kostrov, V. V. (1974). Seismic moment and energy of earthquakes, and seismic flow of rocks. *Izv. Earth Physics, 1*, 23–40.

Kreemer, C., Blewitt, G., & Davis, P. M. (2020). Geodetic evidence for a buoyant mantle plume beneath the Eifel volcanic area, NW Europe. *Geophysical Journal International, 222*, 1316–1332. https://doi.org/10.1093/gji/ggaa227.

Kreemer, C., Blewitt, G., & Klein, E. C. (2014). A geodetic plate motion and global strain rate model. *Geochemistry, Geophysics, Geosystems, 15*, 3849–3889. https://doi.org/10.1002/2014GC005407.

Kreemer, C., & Gordon, R. G. (2014). Pacific plate deformation from horizontal thermal contraction. *Geology, 42*, 847–850.

Kreemer, C., Hammond, W. C., & Blewitt, G. (2018). A robust estimation of the 3-D intraplate deformation of the north American plate from GPS. *Journal of Geophysical Research-Solid Earth, 123*, 4388–4412. https://doi.org/10.1029/2017JB015257.

Kreemer, C., Holt, W. E., Goes, S., & Govers, R. (2000). Active deformation in eastern Indonesia and the Philippines from GPS and seismicity data. *Journal of Geophysical Research - Solid Earth, 105*, 663–680.

Kreemer, C., Holt, W. E., & Haines, A. J. (2002). The global moment rate distribution within plate boundary zones. In S. Stein, & J. T. Freymueller (Eds.), *Plate boundary zones, geodynamics series* (pp. 173–190). Washington DC: American Geophysical Union.

Kreemer, C., & Young, Z. M. (2022). Crustal strain rates in the Western U.S. and their relationship with earthquake rates. *Seismological Research Letters, 93*, 2990–3008.

Liu, Y.-K., Ross, Z. E., Cochran, E. S., & Lapusta, N. (2022). A unified perspective of seismicity and fault coupling along the San Andreas fault. *Science Advances, 8*, eabk1167. https://doi.org/10.1126/sciadv.abk1167.

Masson, F., Chéry, J., Hatzfeld, D., Martinod, J., Vernant, P., Tavakoli, F., et al. (2005). Seismic versus aseismic deformation in Iran inferred from earthquakes and geodetic data. *Geophysical Journal International, 160*, 217–226. https://doi.org/10.1111/j.1365-246X.2004.02465.x.

Materna, K., Maurer, J., & Sandoe, L. (2021). *kmaterna/Strain_2D: First release*. https://doi.org/10.5281/zenodo.5240908.

Mazzotti, S., Leonard, L. J., Cassidy, J. F., Rogers, G. C., & Halchuk, S. (2011). Seismic hazard in western Canada from GPS strain rates versus earthquake catalog. *Journal of Geophysical Research, 116*, B12310. https://doi.org/10.1029/2011JB008213.

McCaffrey, R. (1997). Statistical significance of the seismic coupling coefficient. *Bulletin of the Seismological Society of America, 87*, 1069–1073.

McCaffrey, R., Long, M. D., Goldfinger, C., Zwick, P. C., Nabelek, J. L., Johnson, C. K., et al. (2000). Rotation and plate locking at the southern Cascadia subduction zone. *Geophysical Research Letters, 27*, 3117–3120.

Meade, B. J., & Hager, B. H. (2005a). Block models of crustal motion in southern California constrained by GPS measurements. *Journal of Geophysical Research: Solid Earth, 110*, B03403. https://doi.org/10.1029/2004JB003209.

Meade, B. J., & Hager, B. H. (2005b). Spatial localization of moment deficits in southern California. *Journal of Geophysical Research: Solid Earth, 110*. https://doi.org/10.1029/2004JB003331.

Molnar, P. (1979). Earthquake recurrence intervals and plate tectonics. *Bulletin of the Seismological Society of America, 69*, 115–133.

Moschetti, M. P., Powers, P. M., Petersen, M. D., Boyd, O. S., Chen, R., Field, E. H., et al. (2015). Seismic source characterization for the 2014 update of the U.S. National seismic hazard model. *Earthquake Spectra, 31*, S31–S57. https://doi.org/10.1193/110514EQS183M.

Nishimura, T. (2022). Time-independent forecast model for large crustal earthquakes in Southwest Japan using GNSS data. *Earth, Planets and Space, 74*, 58. https://doi.org/10.1186/s40623-022-01622-5.

Noda, A., & Matsu'ura, M. (2010). Physics-based GPS data inversion to estimate three-dimensional elastic and inelastic strain fields. *Geophysical Journal International, 182*, 513–530. https://doi.org/10.1111/j.1365-246X.2010.04611.x.

Nolan, J. P. (2020). *Univariate stable distributions: Models for heavy tailed data* (1st ed. 2020 ed.). Cham, Switzerland: Springer.

Ojo, A. O., Kao, H., Jiang, Y., Craymer, M., & Henton, J. (2021). Strain accumulation and release rate in Canada: Implications for Long-term crustal deformation and earthquake hazards. *Journal of Geophysical Research: Solid Earth, 126*, e2020JB020529. https://doi.org/10.1029/2020JB020529.

Okazaki, T., Fukahata, Y., & Nishimura, T. (2021). Consistent estimation of strain-rate fields from GNSS velocity data using basis function expansion with ABIC. *Earth, Planets and Space, 73*, 153. https://doi.org/10.1186/s40623-021-01474-5.

Pagani, C., Bodin, T., Metois, M., & Lasserre, C. (2021). Bayesian estimation of surface strain rates from global navigation satellite system measurements: Application to the southwestern United States. *Journal of Geophysical Research: Solid Earth, 126*, e2021JB021905. https://doi.org/10.1029/2021JB021905.

Palano, M., Imprescia, P., Agnon, A., & Gresta, S. (2018). An improved evaluation of the seismic/geodetic deformation-rate ratio for the Zagros fold-and-thrust collisional belt. *Geophysical Journal International, 213*, 194–209. https://doi.org/10.1093/gji/ggx524.

Pancha, A., Anderson, J. G., & Kreemer, C. (2006). Comparison of seismic and geodetic scalar moment rates across the basin and range province. *Bulletin of the Seismological Society of America, 96*, 11–32.

Park, J., Song, T.-R. A., Tromp, J., Okal, E., Stein, S., Roult, G., et al. (2005). Earth's free oscillations excited by the 26 December 2004 Sumatra-Andaman earthquake. *Science, 308*, 1139–1144. https://doi.org/10.1126/science.1112305.

Parsons, T., Geist, E. L., Console, R., & Carluccio, R. (2018). Characteristic earthquake magnitude frequency distributions on faults calculated from consensus data in California. *Journal of Geophysical Research: Solid Earth, 123*, 10761–10784. https://doi.org/10.1029/2018JB016539.

Parsons, T., Johnson, K., Bird, P., Bormann, J., Dawson, T., Field, E., et al. (2013). *Appendix C—Deformation models for UCERF3*. USGS Open-File Report.

Petersen, M. D., Moschetti, M. P., Powers, P. M., Mueller, C. S., Haller, K. M., Frankel, A. D., et al. (2014). *Documentation for the 2014 update of the United States National Seismic Hazard Maps (No. Open-File Report 2014–1091)*. U.S. Geological Survey.

Petruccelli, A., Schorlemmer, D., Tormann, T., Rinaldi, A. P., Wiemer, S., Gasperini, P., et al. (2019). The influence of faulting style on the size-distribution of global earthquakes. *Earth and Planetary Science Letters, 527*, 115791. https://doi.org/10.1016/j.epsl.2019.115791.

Pietrantonio, G., & Riguzzi, F. (2004). Three-dimensional strain tensor estimation by GPS observations: Methodological aspects and geophysical applications. *Journal of Geodynamics, 38*, 1–18. https://doi.org/10.1016/j.jog.2004.02.021.

Pisarenko, V. F. (1998). Non-linear growth of cumulative flood losses with time. *Hydrological Processes, 12*, 461–470. https://doi.org/10.1002/(SICI)1099-1085(19980315)12:3<461::AID-HYP584>3.0.CO;2-L.

Pondrelli, S., Morelli, A., & Boschi, E. (1995). Seismic deformation in the Mediterranean area estimated by moment tensor summation. *Geophysical Journal International, 122*, 938–952. https://doi.org/10.1111/j.1365-246X.1995.tb06847.x.

Prescott, W. H., & Nur, A. (1981). The accommodation of relative motion at depth on the San Andreas fault system in California. *Journal of Geophysical Research: Solid Earth, 86*, 999–1004. https://doi.org/10.1029/JB086iB02p00999.

Rahmadani, S., Meilano, I., Susilo, S., Sarsito, D. A., Abidin, H. Z., & Supendi, P. (2022). Geodetic observation of strain accumulation in the Banda arc region. *Geomatics, Natural Hazards and Risk, 13*, 2579–2596. https://doi.org/10.1080/19475705.2022.2126799.

Ramírez-Zelaya, J., Peci, L. M., Fernández-Ros, A., Rosado, B., Pérez-Peña, A., Gárate, J., et al. (2023). Q-Str2–Models: A software in PyQGIS to obtain Stress–Strain models from GNSS geodynamic velocities. *Computers & Geosciences, 172*, 105308. https://doi.org/10.1016/j.cageo.2023.105308.

Ran, H., & Wu, G. (2019). Seismicity around late quaternary active faults in China. *Bulletin of the Seismological Society of America, 109*, 1498–1523. https://doi.org/10.1785/0120180340.

Rastin, S. J., Rhoades, D. A., Rollins, C., & Gerstenberger, M. C. (2022). How useful are strain rates for estimating the long-term spatial distribution of earthquakes? *Applied Sciences, 12*, 6804. https://doi.org/10.3390/app12136804.

Reilinger, R., McClusky, S., Vernant, P., Lawrence, S., Ergintav, S., Cakmak, R., et al. (2006). GPS constraints on continental deformation in the Africa-Arabia-Eurasia continental collision zone and implications for the dynamics of plate interactions. *Journal of Geophysical Research, 111*, B05411. https://doi.org/10.1029/2005JB004051.

Rhoades, D. A., Christophersen, A., & Gerstenberger, M. C. (2017). Multiplicative earthquake likelihood models incorporating strain rates. *Geophysical Journal International, 208*, 1764–1774. https://doi.org/10.1093/gji/ggw486.

Riguzzi, F., Crespi, M., Devoti, R., Doglioni, C., Pietrantonio, G., & Pisani, A. R. (2012). Geodetic strain rate and earthquake size: New clues for seismic hazard studies. *Physics of the Earth and Planetary Interiors, 206–207*, 67–75. https://doi.org/10.1016/j.pepi.2012.07.005.

Rodkin, M. V., & Pisarenko, V. F. (2004). Earthquake losses and casualties: a statistical analysis. In *Selected papers from Volume 31 of Vychislitel'naya Seysmologiya* (pp. 85–102). American Geophysical Union (AGU). https://doi.org/10.1002/9781118669853.ch11.

Rolandone, F., Nocquet, J.-M., Mothes, P. A., Jarrin, P., Vallée, M., Cubas, N., et al. (2018). Areas prone to slow slip events impede earthquake rupture propagation and promote afterslip. *Science Advances, 4*, eaao6596. https://doi.org/10.1126/sciadv.aao6596.

Rong, Y., Bird, P., & Jackson, D. D. (2016). Earthquake potential and magnitude limits inferred from a geodetic strain-rate model for southern Europe. *Geophysical Journal International, 205*, 509–522. https://doi.org/10.1093/gji/ggw018.

Ross, S. (2018). *First course in probability, a* (10th ed.). Harlow, UK: Pearson.

Samoradnitsky, G., & Taqqu, M. S. (2017). *Stable non-Gaussian random processes: Stochastic models with infinite variance* (1st ed.). Routledge.

Sandwell, D. T., & Wessel, P. (2016). Interpolation of 2-D vector data using constraints from elasticity. *Geophysical Research Letters, 43*, 10703–10709. https://doi.org/10.1002/2016GL070340.

Savage, J. C., & Burford, R. O. (1973). Geodetic determination of relative plate motion in Central California. *Journal of Geophysical Research, 78*, 832–845.

Savage, J. C., & Simpson, R. W. (1997). Surface strain accumulation and the seismic moment tensor. *Bulletin of the Seismological Society of America, 87*, 1345–1353.

Sawires, R., Peláez, J. A., Sparacino, F., Radwan, A. M., Rashwan, M., & Palano, M. (2021). Seismic and geodetic crustal moment-rates comparison: New insights on the seismic hazard of Egypt. *Applied Sciences, 11*, 7836. https://doi.org/10.3390/app11177836.

Scholz, C. H., & Campos, J. (2012). The seismic coupling of subduction zones revisited. *Journal of Geophysical Research, 117*, B05310. https://doi.org/10.1029/2011JB009003.

Schorlemmer, D., Wiemer, S., & Wyss, M. (2005). Variations in earthquake-size distribution across different stress regimes. *Nature, 437*, 539–542. https://doi.org/10.1038/nature04094.

Sharma, Y., Pasari, S., Ching, K.-E., Dikshit, O., Kato, T., Malik, J. N., et al. (2020). Spatial distribution of earthquake potential along the Himalayan arc. *Tectonophysics, 791*, 228556. https://doi.org/10.1016/j.tecto.2020.228556.

Shen, Z.-K., Jackson, D. D., & Ge, B. X. (1996). Crustal deformation across and beyond the Los Angeles basin from geodetic measurements. *Journal of Geophysical Research, 101*, 27957–27980. https://doi.org/10.1029/96JB02544.

Shen, Z.-K., Jackson, D. D., & Kagan, Y. Y. (2007). Implications of geodetic strain rate for future earthquakes, with a five-year forecast of M5 earthquakes in southern California. *Seismological Research Letters, 78*, 116–120. https://doi.org/10.1785/gssrl.78.1.116.

Shen, Z.-K., Wang, M., Zeng, Y., & Wang, F. (2015). Optimal interpolation of spatially discretized geodetic data. *Bulletin of the Seismological Society of America, 105*, 2117–2127. https://doi.org/10.1785/0120140247.

Shen-Tu, B., Holt, W. E., & Haines, A. J. (1995). Intraplate deformation in the Japanese Islands: A kinematic study of intraplate deformation at a convergent plate margin. *Journal of Geophysical Research, 100*, 24275–24293. https://doi.org/10.1029/95JB02842.

Shen-Tu, B., Holt, W. E., & Haines, A. J. (1998). Contemporary kinematics of the western United States determined from earthquake moment tensors, very long baseline interferometry, and GPS observations. *Journal of Geophysical Research, 103*, 18087–18117.

Sippl, C., Schurr, B., Yuan, X., Mechie, J., Schneider, F. M., Gadoev, M., et al. (2013). Geometry of the Pamir-Hindu Kush intermediate-depth earthquake zone from local seismic data. *Journal of Geophysical Research - Solid Earth, 118*, 1438–1457. https://doi.org/10.1002/jgrb.50128.

Smith-Konter, B. R., Sandwell, D. T., & Shearer, P. (2011). Locking depths estimated from geodesy and seismology along the San Andreas fault system: Implications for seismic moment release. *Journal of Geophysical Research: Solid Earth, 116*. https://doi.org/10.1029/2010JB008117.

Sparacino, F., Galuzzi, B. G., Palano, M., Segou, M., & Chiarabba, C. (2022). Seismic coupling for the Aegean – Anatolian region. *Earth-Science Reviews, 228*, 103993. https://doi.org/10.1016/j.earscirev.2022.103993.

Sparacino, F., Palano, M., Peláez, J. A., & Fernández, J. (2020). Geodetic deformation versus seismic crustal moment-rates: Insights from the Ibero-Maghrebian region. *Remote Sensing, 12*, 952. https://doi.org/10.3390/rs12060952.

Stein, S., & Okal, E. A. (2005). Seismology: Speed and size of the Sumatra earthquake. *Nature, 434*, 581–582. https://doi.org/10.1038/434581a.

Stevens, V. L., & Avouac, J.-P. (2021). On the relationship between strain rate and seismicity in the India–Asia collision zone: implications for probabilistic seismic hazard. *Geophysical Journal International, 226*, 220–245. https://doi.org/10.1093/gji/ggab098.

Stirling, M., & Gerstenberger, M. (2018). Applicability of the Gutenberg–Richter relation for major active faults in New Zealand. *Bulletin of the Seismological Society of America, 108*, 718–728. https://doi.org/10.1785/0120160257.

Storchak, D. A., Di Giacomo, D., Engdahl, E. R., Harris, J., Bondár, I., Lee, W. H. K., et al. (2015). The ISC-GEM global instrumental earthquake catalogue (1900–2009): Introduction. *Physics of the Earth and Planetary Interiors, ISC-GEM Catalogue, 239*, 48–63. https://doi.org/10.1016/j.pepi.2014.06.009.

Storchak, D. A., Giacomo, D. D., Bondár, I., Engdahl, E. R., Harris, J., Lee, W. H. K., et al. (2013). Public release of the ISC–GEM global instrumental earthquake catalogue (1900–2009). *Seismological Research Letters, 84*, 810–815. https://doi.org/10.1785/0220130034.

Strader, A., Werner, M., Bayona, J., Maechling, P., Silva, F., Liukis, M., et al. (2018). Prospective evaluation of global earthquake forecast models: 2 yrs of observations provide preliminary support for merging smoothed seismicity with geodetic strain rates. *Seismological Research Letters, 89*, 1262–1271. https://doi.org/10.1785/0220180051.

Su, X., Yao, L., Wu, W., Meng, G., Su, L., Xiong, R., et al. (2019). Crustal deformation on the Northeastern margin of the Tibetan plateau from continuous GPS observations. *Remote Sensing, 11*, 34. https://doi.org/10.3390/rs11010034.

Sykes, L. R., Shaw, B. E., & Scholz, C. H. (1999). Rethinking earthquake prediction. *Pure and Applied Geophysics, 155*, 207–232. https://doi.org/10.1007/s000240050263.

Tape, C., Musé, P., Simons, M., Dong, D., & Webb, F. (2009). Multiscale estimation of GPS velocity fields. *Geophysical Journal International, 179*, 945–971. https://doi.org/10.1111/j.1365-246X.2009.04337.x.

Teza, G., Pesci, A., & Meschis, M. (2023). A MATLAB toolbox for computation of velocity and strain rate field from GNSS coordinate time series. *Annals of Geophysics, 66*. https://doi.org/10.4401/ag-8933.

Thatcher, W. (2009). How the continents deform: The evidence from tectonic geodesy. *Annual Review of Earth and Planetary Sciences, 37*, 237–262. https://doi.org/10.1146/annurev.earth.031208.100035.

Tormann, T., Wiemer, S., & Mignan, A. (2014). Systematic survey of high-resolution b value imaging along Californian faults: Inference on asperities. *Journal of Geophysical Research: Solid Earth, 119*, 2029–2054. https://doi.org/10.1002/2013JB010867.

Tsai, V. C., Nettles, M., Ekström, G., & Dziewonski, A. M. (2005). Multiple CMT source analysis of the 2004 Sumatra earthquake. *Geophysical Research Letters, 32*. https://doi.org/10.1029/2005GL023813.

Varga, P. (2011). Geodetic strain observations and return period of the strongest earthquakes of a given seismic source zone. *Pure and Applied Geophysics, 168*, 289–296. https://doi.org/10.1007/s00024-010-0112-2.

Vere-Jones, D., Robinson, R., & Yang, W. (2001). Remarks on the accelerated moment release model: Problems of model formulation, simulation and estimation. *Geophysical Journal International, 144*, 517–531. https://doi.org/10.1046/j.1365-246x.2001.01348.x.

Vernant, P. (2015). What can we learn from 20 years of interseismic GPS measurements across strike-slip faults? *Tectonophysics, 644–645*, 22–39. https://doi.org/10.1016/j.tecto.2015.01.013.

Vernant, P., Reilinger, R., & McClusky, S. (2014). Geodetic evidence for low coupling on the Hellenic subduction plate interface. *Earth and Planetary Science Letters, 385*, 122–129. https://doi.org/10.1016/j.epsl.2013.10.018.

Wallace, L. M., & Beavan, J. (2010). Diverse slow slip behavior at the Hikurangi subduction margin, New Zealand. *Journal of Geophysical Research: Solid Earth, 115*. https://doi.org/10.1029/2010JB007717.

Wang, K., Zhu, Y., Nissen, E., & Shen, Z.-K. (2021). On the relevance of geodetic deformation rates to earthquake potential. *Geophysical Research Letters, 48*, e2021GL093231. https://doi.org/10.1029/2021GL093231.

Ward, S. N. (1994). A multidisciplinary approach to seismic hazard in southern California. *Bulletin of the Seismological Society of America, 84*, 1293–1309.

Ward, S. N. (1998). On the consistency of earthquake moment rates, geological fault data, and space geodetic strain: The United States. *Geophysical Journal International, 134*, 172–186.

Wesnousky, S. G. (1994). The Gutenberg-Richter or characteristic earthquake distribution, which is it? *Bulletin of the Seismological Society of America, 84*, 1940–1959. https://doi.org/10.1785/BSSA0840061940.

Wu, Y., Jiang, Z., Pang, Y., & Chen, C. (2022). Statistical correlation of seismicity and geodetic strain rate in the Chinese mainland. *Seismological Research Letters, 93*, 268–276. https://doi.org/10.1785/0220200048.

Wu, Y., Jiang, Z., Yang, G., Wei, W., & Liu, X. (2011). Comparison of GPS strain rate computing methods and their reliability. *Geophysical Journal International, 185*, 703–717. https://doi.org/10.1111/j.1365-246X.2011.04976.x.

Wyss, M., Sammis, C. G., Nadeau, R. M., & Wiemer, S. (2004). Fractal dimension and b-value on creeping and locked patches of the San Andreas fault near Parkfield, California. *Bulletin of the Seismological Society of America, 94*, 410–421. https://doi.org/10.1785/0120030054.

Xiong, Z., Zhuang, J., Zhou, S., Matsuura, M., Hao, M., & Wang, Q. (2021). Crustal strain-rate fields estimated from GNSS data with a Bayesian approach and its correlation to seismic activity in Mainland China. *Tectonophysics, 815*, 229003. https://doi.org/10.1016/j.tecto.2021.229003.

Xu, K., Liu, J., Liu, X., Liu, J., & Zhao, F. (2020). Multiscale crustal deformation around the southeastern margin of the Tibetan Plateau from GNSS observations. *Geophysical Journal International, 223*, 1188–1209. https://doi.org/10.1093/gji/ggaa289.

Young, Z. M., Kreemer, C., Hammond, W. C., & Blewitt, G. (2023). Interseismic strain accumulation between the Colorado plateau and the eastern California shear zone: Implications for the seismic Hazard near Las Vegas, Nevada. *Bulletin of the Seismological Society of America, 113*, 856–876.

Zaliapin, I., & Ben-Zion, Y. (2013). Earthquake clusters in southern California I: Identification and stability. *Journal of Geophysical Research - Solid Earth, 118*, 2847–2864. https://doi.org/10.1002/jgrb.50179.

Zaliapin, I., & Ben-Zion, Y. (2016). A global classification and characterization of earthquake clusters. *Geophysical Journal International, 207*, 608–634. https://doi.org/10.1093/gji/ggw300.

Zaliapin, I., Gabrielov, A., Keilis-Borok, V., & Wong, H. (2008). Clustering analysis of seismicity and aftershock identification. *Physical Review Letters, 101*, 018501. https://doi.org/10.1103/PhysRevLett.101.018501.

Zaliapin, I. V., Kagan, Y. Y., & Schoenberg, F. P. (2005). Approximating the distribution of Pareto sums. *Pure and Applied Geophysics, 162*, 1187–1228. https://doi.org/10.1007/s00024-004-2666-3.

Zaliapin, I., & Kreemer, C. (2017). Systematic fluctuations in the global seismic moment release. *Geophysical Research Letters, 44*, 4820–4828. https://doi.org/10.1002/2017GL073504.

Zeng, Y., Petersen, M. D., & Shen, Z.-K. (2018). Earthquake potential in California-Nevada implied by correlation of strain rate and seismicity. *Geophysical Research Letters, 45*, 1778–1785. https://doi.org/10.1002/2017GL075967.

Zuza, A. V., & Cao, W. (2020). Seismogenic thickness of California: Implications for thermal structure and seismic hazard. *Tectonophysics, 782–783*, 228426. https://doi.org/10.1016/j.tecto.2020.228426.

Chapter 4

GNSS applications for earthquake deformation

Jean-Mathieu Nocquet[a,b] and Martin Vallée[b]

[a]Université Côte d'Azur, Observatoire de la Côte d'Azur, IRD, CNRS, Geoazur, Valbonne, France, [b]Université Paris Cité, Institut de Physique du Globe de Paris, Paris, France

1 Introduction

Large earthquakes are among the most sudden and energetic events in our terrestrial environment. Most of the earthquakes—and among them the largest—occur at plate boundary zones. With rapid urbanization, about one-third of the world's largest cities are located in earthquake-prone areas (Bilham, 2009) and the number of people exposed to seismic hazard has increased by 93% in 40 years, from 1.4 billion in 1975 to 2.7 billion in 2015 (Pesaresi et al., 2017). In addition to risk reduction policies, there is a need for a better knowledge of large earthquakes.

GNSS technology has been instrumental in earthquake-related research. Specifically, GNSS provides key constraints on the deformation before, during, and after large earthquakes, paving the way to understand how earthquakes are related to the slow tectonic loading preceding them.

This chapter focuses on the description of the coseismic phase using GNSS. An earthquake results from a rapid slip along a fault, here defined as the surface of contact between two tectonic blocks. This rapid slip generates a seismic wavefield, resulting in dynamic displacements at the Earth's surface. The dynamic displacement finally results in a permanent displacement, referred to as static offset or coseismic displacement.

Earthquakes have been—and still are—first studied using the seismic wavefield recorded by seismometers. For any earthquake with a magnitude larger than 5.5, the seismic waves are clearly observed by broadband seismometers worldwide. Global seismological networks (Ringler et al., 2022) allow to estimate routinely the location, magnitude, and focal mechanisms (Dziewonski et al., 1981; Ekström et al., 2012), as well as the time evolution of moment rate release (Tanioka & Ruff, 1997; Vallée & Douet, 2016). Some generic properties of earthquakes can be derived through these approaches (Meier et al., 2017; Renou et al., 2019).

More details about the rupture itself are obtained through kinematic slip inversion, which provides images of the evolution of slip in time and space during the seismic rupture along the fault surface. Although kinematic inversions are also routinely performed using teleseismic data for earthquakes with magnitude larger than 7 (Hayes, 2017; Ye et al., 2016), the fine description of where and how the seismic slip developed requires data in the near field of large earthquakes. In particular, static displacements measured at the Earth's surface are sensitive to the slip distribution along the fault at depth. Unlike seismic waves that can travel over long distances, the magnitude of coseismic displacements decreases as the inverse of the distance from the earthquake. As a consequence, relatively dense measurements close to potentially seismogenic faults are required to capture the pattern of the coseismic displacement field. GNSS, because of its ease of use and relatively low cost has enabled scientists to develop networks of benchmarks in seismically active areas. These benchmarks are remeasured after the earthquake (GNSS campaign mode), allowing us to quantify the static displacement for many earthquakes.

The development of continuously recording GNSS networks in seismically active areas, together with the increasing memory storage capabilities of GNSS receivers and telemetry have enabled to record GNSS observables at a high rate, typically one sample per second. Through kinematic processing, high-rate GNSS allows to measure dynamic displacement during large earthquakes (Larson, 2009).

In the following, we review the static and dynamic observations of displacement during earthquakes captured by GNSS measurements. We also briefly described other observables of the displacements induced by earthquakes, complementing GNSS measurements. We then introduce how these observations are used to image the slip occurring during large earthquakes. Finally, we show an original use of high-rate GNSS data to track the rupture propagation using array analysis.

66 PART | I Monitoring earthquakes and volcanoes with GNSS

2 Observation of the static displacement induced by earthquakes from GNSS

Measuring static displacements from GNSS time series is in principle straightforward by making the difference from a position immediately after the earthquake minus the position before the earthquake. Most often, GNSS time series are processed using 24-h long sessions to estimate a single position for the entire day. As a result, the position estimated for the day of the earthquake might be a mix of pre- and post-earthquake observation, requiring some care on the selection of data to estimate the coseismic displacement. Dedicated processing, splitting data before and after the time of the earthquake, is preferable. In addition, some earthquakes are followed by rapid deformation, mostly due to rapid aseismic slip following the seismic rupture, referred to as afterslip (see Chapter 5). Early afterslip has been shown to contribute to a few millimeters for the 2004 Parkfield earthquake (Langbein et al., 2006), but rates of a few centimeters per hour were observed after several large megathrust earthquakes (Liu et al., 2022; Munekane, 2012; Twardzik et al., 2019). Fig. 4.1 provides an illustration of

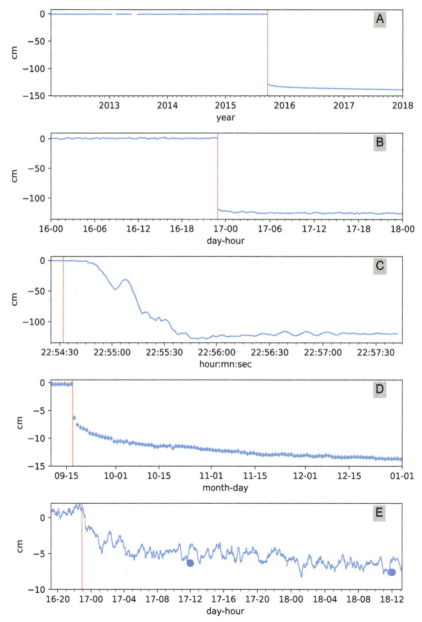

FIG. 4.1 East component time series showing the coseismic displacement at GNSS site CNBA for the Illapel 2015 Mw 8.3 earthquake. The *vertical red line* is the origin time of the earthquake. (A) Daily time series from Klein et al. (2022). The time series has been detrended from its pre-earthquake velocity. (B) 30s kinematic processing from Twardzik et al. (2019). (C) 1s seismogram from Ruhl et al. (2019). It shows a coseismic offset of −120cm. (D) Daily solution corrected from a coseismic offset of −120cm estimated from (C). The position of the first day (September 17) exceeds the coseismic offset by 6cm. (E) 30s kinematic processing corrected from the −120-cm coseismic offset. It shows the rapid displacement due to afterslip during the hours following the earthquake. The two *blue dots* show the position from the daily solution of (D).

this issue observed at a coastal site in Chile for the 2015 Mw 8.3 megathrust earthquake. Neglecting early afterslip can lead to an overestimation of coseismic displacements of tens of percent (Hill et al., 2012a), or even more for some specific cases (Twardzik et al., 2019). In complement of continuous GNSS networks, remeasurements of GNSS benchmarks after an earthquake are often essential to improve the spatial sampling of the coseismic displacement field, although afterslip contribution might contribute to a significant fraction of the GNSS-derived offsets.

3 Observation of the coseismic static displacement using other techniques

GNSS provides three-dimensional (3D) coseismic static offsets at discrete points located on emerged lands. Other geodetic techniques or observations complement GNSS data to improve the coverage of the static displacement field during earthquakes: interferometric synthetic aperture radar (InSAR) measures the ground displacement in the line-of-sight of an emitting radar source, most often carried on board a low-orbiting satellite. InSAR offers the advantage of allowing measurement in remote areas or with difficult access. Another important advantage of this technique is that InSAR provides a continuous displacement field. A continuous field is particularly powerful for finding the location of the fault rupture if it reaches the surface. It is also efficient to obtain good estimates of the fault parameters at depth (Elliott et al., 2016). Compared to GNSS, InSAR shows a few limitations for measuring coseismic displacements. Radar phase decorrelation sometimes prevents measurements in highly vegetated areas. The return time required for the satellite to fly again over the earthquake area can be several days, making measurements potentially subject to postseismic motion. The high angle (typically 30° from zenith) of the satellite line-of-sight together with near-polar orbiting satellites makes InSAR measurements very sensitive to vertical displacement and poorly sensitive to displacement along the north-south component (Wright et al., 2004), although several approaches have been developed to extract the full 3D displacements (Hu et al., 2014). Subpixel correlation of aerial/satellite optical images is another technique that provides submetric continuous horizontal displacements (Puymbroeck et al., 2000). Differential air-bone Lidar topographic data allow to measure coseismic displacements in the near-fault zone (Nissen et al., 2014). Optical correlation and differential Lidar are particularly powerful at imaging fault surface rupture and near-fault deformation, but less suitable for buried faults and subduction earthquakes.

In coastal areas, the sea level provides a vertical reference that can be used to estimate earthquake-induced vertical displacements from the observation of marker height changes with respect to the sea. This approach allows some estimates of vertical displacements induced by past earthquakes (Shaw et al., 2008). Markers are as diverse as coral reefs (Sieh et al., 2008), dead algae raised above the inter tidal zone (Farías et al., 2010), and man-made infrastructure like jetties (Hubert et al., 1996).

Many earthquakes, and among them the largest occurring at subduction zones, are located at sea where the geodetic techniques mentioned above cannot be used. Techniques of seafloor geodesy have been developed to overcome this limitation. A first technique referred to as GNSS-A uses a combination of acoustic measurements between fixed transponders at the seafloor and a moving acoustic source at the sea surface, precisely located using a kinematic GNSS and inertial navigation system (Bürgmann & Chadwell, 2014; Yokota et al., 2018). Long measurement sessions from 24h to a few days are used to take benefit from numerous surface-to-seafloor beacons travel-time measurements allowing to average the unknown variability of the sound speed in the water column. GNSS-A reaches a centimeter accuracy in the global frame (Chadwell & Spiess, 2008; Fujita et al., 2006; Gagnon et al., 2005). Using this technique, impressive coseismic displacements of 5–31 m were measured for the Tohoku 2011 Mw 9.0 Japan earthquakes (Kido et al., 2011; Sato et al., 2011). For GNSS-A, high equipment costs together with the heavy logistics involved in implementing them have limited their use to a few subduction segments so far, mainly located in Japan (Yokota et al., 2016). In order to reduce the costs, new approaches using unmanned surface vehicle (USV) or water landing unmanned aerial vehicle (UAV) instead of ships are currently being developed (Iinuma et al., 2021; Yokota et al., 2023). Large megathrust earthquakes induce seafloor vertical motion, changing the height of the water column. Ocean bottom pressure gauges (OBPG) can record this change. For instance, an OBPG located 20km from the trench, recorded a negative pressure change of ~500 hPa, equivalent to 5m uplift for the Tohoku 2011 earthquake (Ito et al., 2011). Seafloor vertical displacement uplifts the sea level above its equilibrium level, triggering a tsunami wave. Tsunami waves can be observed at the coast using tide gauges. Since the 2000s, a network of buoys Deep-ocean Assessment and Reporting of Tsunamis (DART) is operated by National Oceanic and Atmospheric Administration (NOAA). DART buoys are mooring systems deployed in the open sea that transmit in real-time pressure measurements every minute or less from a sensor at the seafloor. DART records the tsunami wave, whose timing, amplitude, and waveform can be related to the spatial distribution of vertical motion induced by the earthquake. DART data are often essential to get a reliable estimate of the slip amount close to the trench (Bletery et al., 2014). Finally, for great earthquakes, gravity changes are large enough over a great distance so that they can be detected from satellite gravimetry (Matsuo & Heki, 2011).

4 Modeling of GNSS static coseismic displacement for imaging earthquake slip distribution

For time scales of the order of seconds to minutes that are typical durations of earthquakes, the Earth behaves elastically outside the fault zone. Coseismic offsets observed at the Earth's surface are thus the response of the elastic medium to the slip that occurred along the fault at depth.

Analytical solutions—often referred to as static Green's functions—relating unit slip-along dip and along strike of a finite screw dislocation to surface displacements have been proposed for infinitely long faults, for finite rectangular faults (Okada, 1992), or triangular dislocations elements (Meade, 2007; Nikkhoo & Walter, 2015) in a homogeneous linear isotropic elastic half-space. Such formulas are derived from the stress field solution verifying the Laplace equation within the elastic medium, a free surface condition at the surface and along the dislocation surface, and a prescribed constant slip along the dislocation. Calculations for the more complex elastic structures of the Earth require numerical approaches. The case of a horizontally layered elastic medium is still fast either for Cartesian coordinates (Wang et al., 2003) or spherically layered medium (Pollitz, 1996). Complex geometry and elastic structure like in subduction zones require finite-element approaches (Williams & Wallace, 2015). However, unlike in seismology where knowledge of the elastic medium is essential to reproduce the observation of seismograms, there have been only very few studies spotting the deficiency of the homogeneous elastic half-space (Hori et al., 2021; Pollitz et al., 2011; Williams & Wallace, 2015) which therefore seems a fairly good approximation, widely used. Because of the linearity of elasticity, surface displacements are proportional to slip and the displacement produced by two faults is the sum of displacement induced by slip at each fault separately.

Imaging the fault coseismic slip consists of three steps: (1) defining a parametrization of the fault. (2) Relating the fault slip parameters to observed surface displacements. These two steps are referred to as the forward problem. (3) Finding the slip parameters satisfying the observations, referred to as the inverse problem.

The simplest parametrization of a fault is a single rectangular dislocation. This parametrization is adequate when only scarce data are available, when there is no data close to the slip area, or as a preliminary search when the fault location is unknown. In that latter case, the fault is often parametrized by the location of its top left corner, width, length, strike, and dip, while slip parameters are the slip magnitude and the slip direction (rake). Including the fault geometry parameters in the inversion breaks the linearity of the forward problem and some optimized scheme of iterative search must be used (Bagnardi & Hooper, 2018; Fukahata & Wright, 2008; Johnson et al., 2001).

Fig. 4.2 shows examples of the 3D coseismic displacements observed by GNSS for different types of earthquakes in various tectonic settings together with the prediction from a single rectangular dislocation. Fig. 4.2 illustrates that even a simple model already captures the first-order characteristics of the coseismic displacements.

We then focus here on the case when data are dense enough and the fault geometry is known. In that case, spatially variable slip along a discretized fault plane can be searched (Fig. 4.3). The static Green's function $g_{i,j,k,l}$ provides the contribution of the slip component $s_{k,l}$ ($l=1,2$ for strike-slip and along-dip components, respectively) at subfault k to the component of displacement $d_{i,j}$ ($j=1,2,3$ for East, North, and Up) observed at GNSS site i. This contribution, noted $\tilde{d}_{i,j,k,l}$, writes:

$$\tilde{d}_{i,j,k,l} = g_{i,j,k,l}\, s_{k,l}$$

The observed displacement component $d_{i,j}$ is thus the sum over all subfaults k and for the along-dip and strike-slip components of slip:

$$d_{i,j} = \sum_{l=1}^{2} \sum_{k=1}^{N_k} g_{i,j,k,l}\, s_{k,l}$$

Such equation can be expressed in matrix form:

$$G\, s = d$$

An optimal model minimizes the misfit to the data given by the likelihood function

$$L(s) = (G\, s - d)^T C_d^{-1} (G\, s - d)$$

where C_d^{-1} is the variance-covariance matrix describing uncertainties in the observed displacements. Despite its linearity, determining s from surface measurement remains an underdetermined ill-posed problem, and there is usually not a single vector s minimizing $L(s)$. The illness of the inverse problem originates from a combination of the partial spatial sampling of the deformation field and a rapidly decreasing sensitivity of displacement to slip when the distance from the sources increases (Nocquet, 2018).

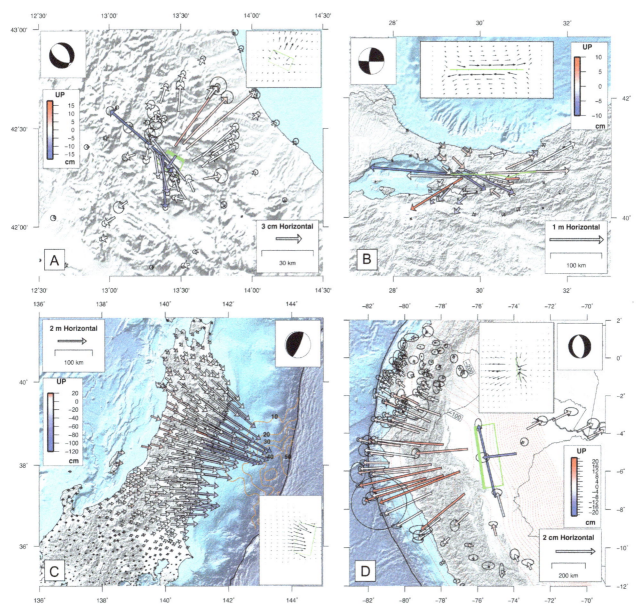

FIG. 4.2 Examples of static coseismic displacements recorded by GNSS for various earthquakes in different tectonic contexts. Insets indicate the focal mechanism for each earthquake and the prediction for horizontal displacements from a single rectangular dislocation based on the focal mechanism. Arrows indicate horizontal displacements and are color-coded according to vertical displacement. The green rectangle highlights the rupture area, with the upper edge indicated by the bold green line. (A) Crustal normal-fault earthquake, L'Aquila, Mw 6.3, 2009. Coseismic displacements show a diverging pattern. Displacement magnitude rapidly decreases because of the shallow rupture. Uplift and subsidence are observed for the footwall and hanging wall, respectively. (B) Crustal right-lateral strike-slip earthquake, Izmit, Mw 7.6, 1999. (C) Shallow subduction megathrust earthquake, Tohoku, Mw 9.0, 2011. The Coseismic slip model from Bletery et al. (2014) is shown every 10 m *(orange contours)*. Horizontal displacements point toward the maximum slip area. Large subsidence is observed along the coast, while more distant sites show uplift. (D) Intermediate depth (130 km) intraslab normal earthquake, northern Peru, Mw 8.0, 2019. Because of the depth of the earthquake, displacement spreads over a wide area (∼1000 km) and shows a relatively smooth decay. *(A: Data from Serpelloni, E., Anderlini, L., & Belardinelli, M. E. (2012). Fault geometry, coseismic-slip distribution and Coulomb stress change associated with the 2009 April 6, Mw 6.3, L'Aquila earthquake from inversion of GPS displacements. Geophysical Journal International, 188(2), 473–489. https://doi.org/10.1111/j.1365-246X.2011.05279.x. B: Data from Reilinger, R. E., Ergintav, S., Bu, R., Mcclusky, S., Lenk, O., Barka, A., Gurkan, O., Hearn, L., Feigl, K. L., Cakmak, R., Aktug, B., Ozener, H., & To, M. N. (2000). Coseismic and postseismic fault slip for the 17 August 1999, M= 7.5, Izmit, Turkey earthquake. Science, 289(5484), 1519–1524. https://doi.org/10.1126/science.289.5484.1519. C: Data from ARIA_coseismic_offsets.v0.3.table. D: Data from Vallée, M., Xie, Y., Grandin, R., Villegas-Lanza, J. C., Nocquet, J. M., Vaca, S., Meng, L., Ampuero, J. P., Mothes, P., Jarrin, P., Sierra Farfán, C., & Rolandone, F. (2023). Self-reactivated rupture during the 2019 Mw=8 northern Peru intraslab earthquake. Earth and Planetary Science Letters, 601, 117886. https://doi.org/10.1016/j.epsl.2022.117886.)*

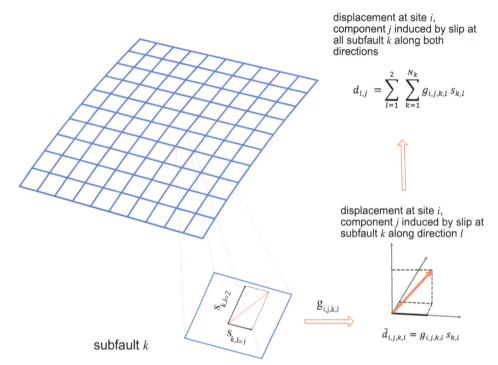

FIG. 4.3 Sketch illustrating the forward model for slip inversion from GNSS coseismic displacements. See text for details.

A widely used approach to solve the inverse problem is to add regularization constraints. Regularization constraints are penalty functions avoiding to depart too much from a priori properties of the model. Regularization constraints are also usually tuned so that they exclude too oscillatory unphysical solutions and so that they avoid overfitting the data. There has been a large diversity of proposed regularization constraints. The simplest constraint is called "damping" or "minimum length" (Menke, 2012). Such constraint penalizes slip that departs too much from an a priori value, usually taken as 0. The cost function is defined:

$$C(s) = (G\,s - d)^T C_d^{-1} (G\,s - d) + \lambda\, s^T s$$

where λ is a hyperparameter controlling the weight given to the regularization constraint. Equivalently, this condition is written in matrix form:

$$\begin{bmatrix} G \\ \sqrt{\lambda}\, I \end{bmatrix} s = \begin{bmatrix} d \\ 0 \end{bmatrix}$$

where I is the identity matrix. Smoothing constraints penalize sharp spatial variations of slip. The most common smoothing approach is to minimize the Laplacian of slip. This is technically achieved in the form of a discrete Laplacian operator, equivalent to write that slip at a given subfault equals—in the least squares sense—the average of slip at its neighbor subfaults. The discrete Laplacian operator D writes:

$D_{ij} = -1$ if subfaults i and j are neighbors.
$D_{ij} = 0$ if $i \neq j$ and subfaults i and j are not neighbors.
$D_{ii} = +\, n_i$ where n_i is the number of neighbors of subfault i.

When added in the cost function:

$$C(s) = (G\,s - d)^T C_d^{-1} (G\,s - d) + \lambda\, s^T (D^T D) s$$

Or in the linear system:

$$\begin{bmatrix} G \\ \sqrt{\lambda}\, D \end{bmatrix} s = \begin{bmatrix} d \\ 0 \end{bmatrix}$$

Regularization can also be made through a model covariance matrix C_m, implicitly mixing damping and smoothing constraints. Such an approach, detailed in Tarantola and Valette (1982) and Tarantola (2005) is grounded in a stochastic view of the inverse problems, where C_m reflects the prior information. Radiguet et al. (2011) used C_m with a form of an exponentially decreasing correlation function:

$$C_m(i,j) = \left(\sigma_m \frac{\rho_0}{\rho}\right)^2 exp\left(-\frac{\rho_{i,j}}{\rho}\right)$$

Here, ρ is a critical length controlling the distance over which the slip at two subfaults i,j remains correlated, $\rho_{i,j}$ is the distance between the centroids of the two subfaults i and j, and σ_m is the a priori standard deviation on slip. ρ_0 is a reference length corresponding to the smallest length scale expected to be resolved.

In that case, the cost function becomes

$$C(s) = (G\,s - d)^T C_d^{-1}(G\,s - d) + (s - s_0)^T C_m^{-1}(s - s_0)$$

Alternative choices for C_m have been proposed (Benavente et al., 2019; Tago et al., 2021). Because all regularization schemes described above are linear, they ensure both the existence and unicity of the solution.

As an additional constraint, it is a reasonable and a physically grounded assumption to search for slip to be positive in an a priori prescribed direction, limiting the range of possible models (Fukuda & Johnson, 2008; Minson et al., 2013). For instance, a slip for megathrust earthquakes is expected to accommodate the relative plate motion. Thus, slip parameters are constrained to be positive in the direction of plate convergence. Such constraint is achieved by minimizing the cost function using non-negative least-squares algorithms (Lawson & Hanson, 1974; Stark et al., 1995).

Within the approaches described above, one question is the choice of weight given to regularization, controlled by one or several hyperparameters. Hyperparameters are chosen to provide an optimal trade-off between the overall smoothness of the model and the misfit to the data. The so-called L-curve (Hansen, 1992) plots the likelihood function versus the hyperparameter. The best model is chosen as the turning point of the curve, above which rougher models result in only marginal improvement of the misfit. Alternative, more objective and statistics-grounded choices are the Akaike criterion and the Akaike Bayesian information criterion (ABIC) (Yabuki & Matsu'ura, 1992). Cross-validation is another approach where data are iteratively removed to compute prediction errors as a function of the hyperparameter (Árnadóttir & Segall, 1994).

A drawback of the above approaches is that the weight given to regularization is homogeneous throughout the fault plane. In particular, it does not account for the fact that regularization weight could be weaker where data provide good resolution of model parameters and regularization could be stronger where data have little sensitivity. Spatially variable regularization schemes have been proposed by changing the size of the fault elements in order to obtain a quasi-uniform resolving power, requiring little additional regularization (Barnhart & Lohman, 2010). Alternatively, Wang et al. (2019) propose a method to define spatially variable weighting of the discrete Laplacian constraint based on slip estimates. Finally, it has been argued that regardless of its formulation, l2 norm minimization (least-squares) might over-smooth the slip spatial distribution. Approaches using regularization constraints based on the l1 norm (the sum of absolute values) promote sparsity of the estimated parameters, suitable for detecting sharp changes in the slip spatial distribution (Evans & Meade, 2012; Nakata et al., 2017).

Bayesian approaches are alternative to optimization approaches. Rather than providing a single "best" model, Bayesian approaches aim at exploring the range of models allowed by the data. More precisely, the solution to the inverse problem is the posterior joint probability density function (pdf) of model parameters, which describes the likelihood of model parameters constrained by the combination of prior information and observations. Fukuda and Johnson (2008) and Minson et al. (2013) use a Monte Carlo Markov chain (MCMC) sampler based on a cascading Metropolis algorithm to approximate the joint posterior pdf. The true result of a Bayesian inversion is a large set of models allowing statistics to be computed. Within this approach, any type of prior can be handled. The less informative prior is a uniform pdf, where all values within an interval are equally probable. The interval is often chosen to be bounded by 0 to impose a nonnegativity constraint on slip and some very large values. Model prediction errors due to imperfect physical models can also be handled (Duputel et al., 2014). The main advantage of Bayesian approaches is that they overcome the issue of subjective regularization and therefore provide reliable uncertainty of the slip estimates in the form of the marginal pdf. Nonetheless, this approach has a few drawbacks. First, the posterior marginal pdfs are directly related to the discretization size of the fault, which needs to be chosen a priori. This problem can be overcome through iterative approaches by setting the number of subfaults as an unknown in the inversion and letting the data determine an optimal parametrization (Dettmer et al., 2014; Tomita et al., 2021), at the price however of additional computing cost. Second, posterior pdfs are approximated using a necessary large number of samples, which also requires large computing power. Finally, although the actual results are a large set of models sampling the posterior covariance, it is practically impossible to look at all models. The results are often shown as the mean or median model, and marginal pdf are only projections of the sampled range of models. On the parts of the fault where data have no or little resolution, the value in the mean or median model is equal to the mean of the prior model, which can be misleading for nonspecialists.

5 Dynamic displacement induced by earthquakes

In addition to static displacement, GNSS data allow us to quantify dynamics motion during earthquakes. In that case, a position is estimated at every epoch of measurements, as opposed to static positioning where satellite observations are accumulated over long sessions to estimate a single fixed position. As they do not benefit from many observations, dynamic positions are more sensitive to the geometry of the satellites seen by the antenna, to any mismodeling of the carrier wave delay through the troposphere, and to multipath induced by reflecting surfaces near the antenna. Some mitigation of these effects is obtained using a so-called sidereal filtering. Sidereal filtering uses the property that a given satellite is seen at the same location in the sky every 23 h 56 min 4 s and that some of the errors are related to the receiver-satellite vector position (Choi et al., 2004; Genrich & Bock, 2006). As a consequence, patterns of apparent displacement tend to repeat every sidereal day, or more exactly the mean of the repeat time for the seen satellites (Choi et al., 2004). Applying such a filter improves the time series accuracy by a few millimeters and removes even more efficiently spurious long periods of apparent motion. Fig. 4.4 shows an example of the 1 sample per second time series observed for the Pedernales (Ecuador) 2016 Mw 7.8 megathrust earthquake. A notable feature is that the dynamic displacement exceeded 1 m, which is significantly greater than the final static displacement. The same feature was also observed for the 2011 Tohoku earthquake, and Fukahata et al. (2012) attributed it to the intermediate field term of S-wave.

Kinematic GNSS has a precision of the order of 1 cm, about 5 times less accurate than the precision typically achieved for static positioning using 24 h long positions. Such a sensitivity limitation prevents GNSS for instance from detecting the onset of the P-wave even for large earthquakes. But GNSS-seismograms remain nonetheless a powerful tool for large dynamic displacements complementing seismological records. Modern broadband seismometers are inertial sensors where ground velocity is obtained from the electromagnetic force required to counteract the motion of a small mass. In their usual frequency range of operation (down to about 0.001 Hz), these seismometers are about four orders of magnitude more sensitive than GNSS, making them particularly suitable for the study of moderate-size earthquakes. However, at close distances from a large earthquake, ground velocity will exceed the range allowed by broadband seismometers. In that case, the

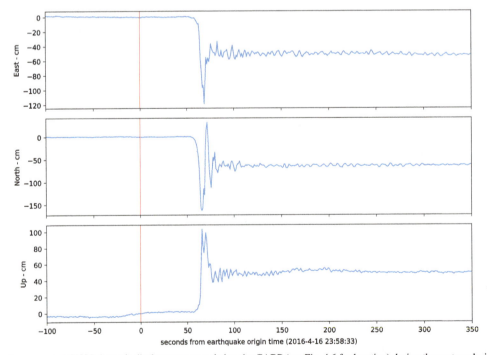

FIG. 4.4 1 sample-per-second GNSS dynamic displacement recorded at site CABP (see Fig. 4.6 for location) during the rupture during the Pedernales (Ecuador) Mw 7.82016 megathrust earthquake.

resulting seismograms are saturated or clipped, making most of the records unsuitable for the earthquake study. Accelerometers on the contrary have a low gain and do not saturate during large ground motion. However, dynamic rotation component of the seismic wavefield and possibly mechanical issues contaminate the measured translational acceleration. This results in a poor recording of the long periods of the dynamic motion and in particular an unphysical doubly integrated dynamic displacement, although some approaches have been proposed to mitigate this bias (Melgar et al., 2013).

GNSS, despite a lower sensitivity, directly measures the dynamic displacement in a geometric external frame. GNSS does not need integration, does not show loss of sensitivity at long periods, and does not saturate during large and rapid ground motion. This makes GNSS particularly suitable to record the dynamic displacement in the vicinity of the seismic rupture. Although GNSS receivers can commonly record data up to 30 Hz, storage and/or transmission constraints limit most networks to record and to archive data at one sample per second. The best description of the dynamic wavefield in the near field of the rupture therefore comes from the joint use of GNSS and accelerometric data. Both data can be included as separated input data for kinematic source inversion (see below). Combination at the observation level to retrieve the full wavefield over a broad frequency range from 100 to 0.001 Hz is also a topic of research through the optimal combination of GNSS and accelerometer records (Melgar et al., 2013).

The first dynamic displacements from single-epoch measurements were obtained for the 1999, Mw 7.1 Hector Mine earthquake in California, but using 30 s samples data preventing to see the details of the waveform (Nikolaidis et al., 2001). Using 1-Hz data, Larson et al. (2003) could demonstrate the similarity of the seismic waves between filtered accelerometer records and GPS dynamic displacement for a GPS site located ~140 km away from the epicenter of the 2002 Mw 7.9 Denali earthquake in Alaska. In addition, Larson et al. (2003) showed that 1-Hz GPS data could detect seismic surface waves up to ~3400 km from the earthquake. The 2003 Mw 8.0 Tokachi-oki earthquake was the first earthquake studied using high-rate GNSS (Miyazaki et al., 2004). High-rate GNSS data have been instrumental in studying the great 2010 Mw 8.8 Maule megathrust earthquake in the context of very few near-field accelerometric records available (Delouis et al., 2010; Vigny et al., 2011). An impressive data set of dynamic motion has been produced for the densely instrumented Japan region during the 2011 Mw 9.0 Tohoku earthquake. During the last decade, many earthquakes have been observed using high-rate GNSS. A public repository of high-rate GNSS displacement seismograms in the standard seismological mini-seed format has been made available by Ruhl et al. (2019).

Displacement with a rate higher than 1 Hz was also recorded for a few earthquakes. Avallone et al. (2011) recorded 10 Hz data during the L'Aquila Mw 6.3 (Italy) 2009 and showed that 10 Hz data are useful for the study of moderate-size earthquakes. We show later in this chapter an original use of 5-Hz GNSS seismograms derived for the 2008 Wenchuan Mw 7.9 earthquake in China.

6 Kinematic inversion: Imaging slip history during earthquake

In order to image the evolution of the seismic slip along the fault, slip is now parametrized as a function of time and space along the fault plane, and observations are time series of the dynamic displacement.

Fig. 4.5 illustrates the principle of the so-called multi-time window approach (Olson & Apsel, 1982), which leads to a linear system similar to the static case. First, the displacement response $g_{i,j,k,l}(t)$ at sensor locations i and component j is calculated for a synthetic shape $b(t)$ of the local slip velocity, for each subfault k and component of slip l. b is chosen as a unit integral boxcar or isosceles triangle function, with a duration $\Delta\tau$ equal to the shortest period expected to be resolved. $g_{i,j,k,l}(t)$ is obtained from Green's functions for horizontally layered elastic medium, which are commonly computed with the discrete wave number approach (Bouchon, 1981). For a given subfault k, the slip velocity $\dot{s}_{k,l}$ is parametrized as a weighted sum of the time-shifted functions b, overlapping by half of their duration $\Delta\tau$. Thus,

$$\underline{\dot{s}}_{k,l}(t) = \sum_{m=1}^{N_m} \widetilde{s}_{k,l,m} b(t - (m-1)\Delta\tau/2)$$

where N_m is the chosen number of time windows, and $\widetilde{s}_{k,l,m}$ is the slip amplitude of the subfault k in the direction l and for the time window m. The contribution of $\dot{s}_{k,l}(t)$ to the displacement $d_{i,j}$ at the GNSS position i and component l, noted $\widetilde{d}_{i,j,k,l}$ as in the static case, writes

$$\widetilde{d}_{i,j,k,l}(t) = \sum_{m=1}^{N_m} \widetilde{s}_{k,l,m}\, g_{i,j,k,l}(t - (m-1)\Delta\tau/2)$$

leading to the observation equation by summation over all subfaults and the two components of slip:

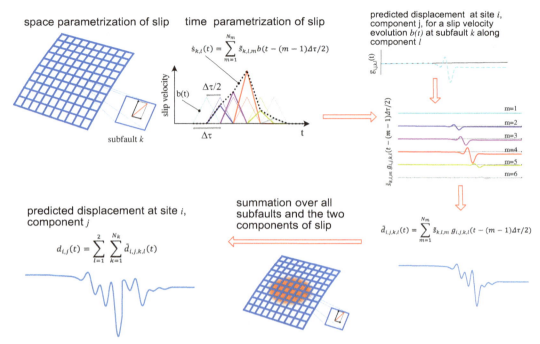

FIG. 4.5 Sketch illustrating the multi-time window approach for earthquake kinematic slip inversion. See text for details.

$$d_{i,j}(t) = \sum_{l=1}^{2} \sum_{k=1}^{N_k} \tilde{d}_{i,j,k,l}(t) = \sum_{l=1}^{2} \sum_{k=1}^{N_k} \sum_{m=1}^{N_m} \tilde{s}_{k,l,m}\, g_{i,j,k,l}(t - (m-1)\Delta\tau/2)$$

This formulation results in a linear system, similar to the static case but now involving the ($N_k \times 2 \times N_m$) unknown $\tilde{s}_{k,l,m}$ parameters. After the resolution of the inverse problem, snapshots of the distribution of the cumulative slip or during any time window can be obtained by summation of the $\tilde{s}_{k,l,m}$ for m corresponding to the chosen time windows (Fig. 4.6). In particular, the final static slip $s_{k,l}$ of the subfault k in the direction l is

$$s_{k,l} = \sum_{m=1}^{N_m} \tilde{s}_{k,l,m}$$

Some approaches prefer to work in the frequency domain and therefore express the displacement time series as coefficients of the Fourier or wavelet time series, allowing to give more weight on user-chosen frequencies. Because such transformations can be written as matrix operators, they still preserve the linearity. For both the time and frequency approaches, the large number of unknowns can be reduced using two assumptions. The first and most physically grounded one is that slip at subfault k can only start after a time greater than D_k/V_r where D_k is the distance of subfault k from the hypocenter and V_r is the rupture velocity, always smaller than the P-wave speed. The second one is that slip at a given subfault is not active during the whole earthquake duration T, but only during a shorter time following the rupture front (Heaton, 1990). A common approach is therefore to impose a maximum rupture front velocity and to consider only several time windows, rather than the value $N_m = (2T/\Delta\tau - 1)$ which has to be chosen in the most general case.

Alternative approaches, while retaining a similar scheme, treat the onset time of slip or the local rupture propagation velocity as unknown. The latter option enforces causality in the sense that slip at a given patch starts only if one of its neighbors already started. Considering onset times or rupture propagation velocity as unknown breaks the linearity of the problem, with the consequence of the possible existence of secondary minima of the cost function, requiring a specific optimization scheme like a simulated annealing algorithm (Delouis et al., 2010).

Observation data are usually low pass filtered because high frequencies are more sensitive to usually unknown small-scale variations of the Earth's elastic structure. Together with the range of the chosen frequency range, using either velocity or displacement impacts the result, having a different sensitivity to moment or slip, and radiating areas of the fault (Ide, 2007). As a result of the numerous choices offered to the analyst and the availability of data at the time of the study, some diversity is observed in the different models that are published after a large earthquake, as shown in Fig. 4.7 for the Maule

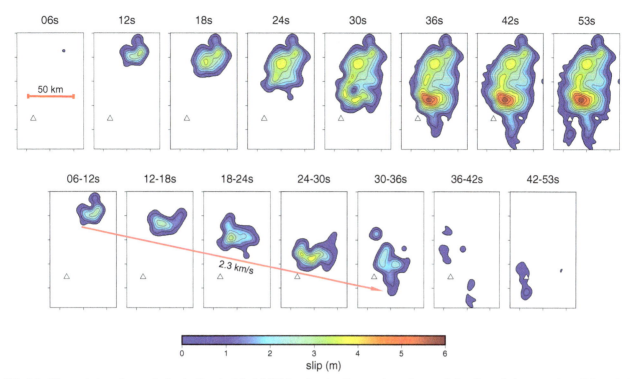

FIG. 4.6 Kinematic inversion results for the Ecuador Mw 7.8 2016 earthquake. *Top row* shows the cumulative slip evolution every 6 s until the end of the rupture (53 s). It highlights the rupture of two main patches, one close to the epicenter during the first 24 s and then a second one where slip develops between 24 and 36 s, reaching 6 m. *Bottom row* shows snapshots of slip occurring during successive 6 s long time windows. The rupture propagates as a slip pulse moving southward at 2.3 km/s. The triangle indicates the location of GNSS site CABP whose dynamic displacement is shown in Fig. 4.4. *(Data and analysis from Nocquet, J.-M., Jarrin, P., Vallée, M., Mothes, P. A., Grandin, R., Rolandone, F., Delouis, B., Yepes, H., Font, Y., Fuentes, D., Montes, L., & Charvis, P. (2017). Supercycle at the Ecuadorian subduction zone revealed after the 2016 Pedernales earthquake. Nature Geoscience, 10(2), 145–149. https://doi.org/10.1038/ngeo2864.)*

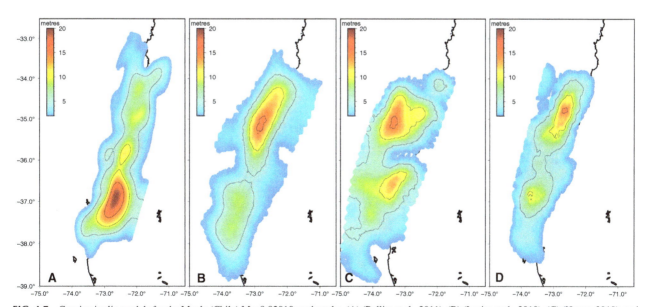

FIG. 4.7 Coseismic slip models for the Maule (Chile) Mw 8.8 2010 earthquake. (A) (Pollitz et al., 2011), (B) (Lorito et al., 2010), (C) (Hayes, 2010), and (D) (Delouis et al., 2010). Model parameters from http://equake-rc.info/srcmod (Mai & Thingbaijam, 2014).

Mw 8.8, 2010 earthquake. The srcmod project (Mai & Thingbaijam, 2014) is a community effort in collecting finite-fault rupture models for intercomparison. Patterns common to several studies can usually be considered as reliable. Fig. 4.6 shows results from a slip kinematic inversion for the Ecuador Mw 7.8 2016 earthquake and highlights several properties observed from many earthquakes, that we discuss at the end of this chapter.

7 High-rate GNSS as small aperture seismic array

To further illustrate GNSS's potential contribution to earthquake study, we show that GNSS-derived seismograms have the accuracy required to successfully apply array analysis. Seismic array analysis methods provide an alternative approach to finite source inversions. Rather than inverting for the slip evolution, seismic array analysis aims at identifying highly energetic coherent seismic pulses during the rupture, providing a direct observation of the rupture propagation. Here, we use body and surface waves generated by the Wenchuan (China) 2008 Mw = 7.9 earthquake, recorded by an array of seven high-rate (5 Hz) GNSS stations located about 950 km northwest of the epicenter (Fig. 4.8A). We apply seismic array analysis techniques, widely used for networks of seismometers, in the frequency range of 0.02–0.12 Hz.

Because each site is located at a different distance from the emitting seismic source, a given waveform reaches each site at different times. Owing to the small aperture (~100 km) of the GNSS network and the selected frequency range, the characteristics of a given wave (waveform and propagation velocity) are very similar within the network. As a consequence, the time shift $\Delta t_{i,j}$ at which a given waveform is observed at two sites i and j only depends on the location x of the emitting seismic source and the apparent travel phase velocity across the network, V_Φ. Using an arbitrary reference station i, for each realistic value of x and V_Φ, we calculate the predicted $\Delta t_{i,j}$ at each station j, and shift the windowed seismograms from this time amount. Then, all time-shifted normalized seismograms are stacked, squared, and integrated over the considered time window. The obtained quantity $E_c(x,V_\Phi)$ reflects the amount of seismic energy coherent across the network for the tested value (x,V_Φ). Alternatively, windowed-shifted seismograms are individually squared, then stacked and integrated over the same time window. The obtained quantity E_t reflects the energy observed across the whole network. We then define the semblance value as $E_c(x,V_\Phi)/(N.E_t)$, which indicates the fraction of coherent energy with respect to the total observed

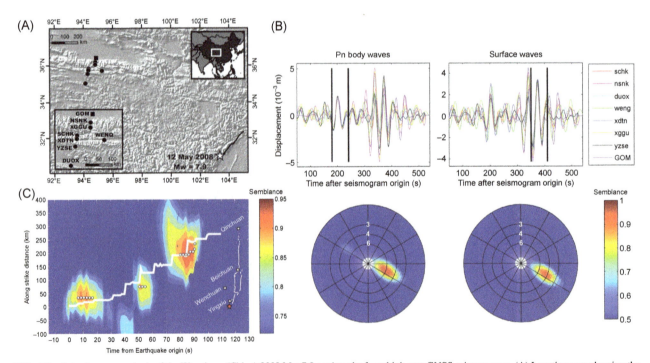

FIG. 4.8 Seismic array analysis of the Wenchuan (China) 2008 Mw 7.9 earthquake from high-rate GNSS seismograms. (A) Location map showing the epicenter *(white star)* and surface rupture of the Wenchuan earthquake together with the local GNSS network *(solid circles)* and broadband station GOM *(solid square)*. (B) Result of the back projection for the single emitting source. *Top row* shows the waveforms (filtered in the (0.02–0.04 Hz frequency range) time shifted for the optimal values obtained from the semblance maximum displayed in the *bottom row*. *Left column* displays the results for Pn body wave time window and the *right column* displays the surface wave time window, both shown by the vertical black lines. (C) Semblance as a function of the source location in time and along the fault (computed from waveforms in the 0.07–0.12 Hz frequency range). The *white lines* are back-projection results from Xu et al. (2009) from regional arrays of broadband seismometers for comparison.

energy over the N stations. If the emitted seismic energy originates from a given point, the semblance increases close to 1 for a well-defined value of (x,V_ϕ). To be successful, this method requires the waveforms to be accurately determined across the whole network with the right timing.

We first assess the suitability of 5-Hz GNSS time series for array analysis, by considering the earthquake as a single emitting source. The input data are the horizontal 5-Hz GNSS time series augmented by horizontal seismograms of a broadband seismic station GOM (Fig. 4.8A), all projected onto the radial component and band-pass filtered between 0.02 and 0.04 Hz. The source location x is defined by its polar coordinates (θ,R), where θ is the back-azimuth and R is the distance to the emitting source. Because the array dimension is much smaller than the source-array distance, very little resolution is expected on parameter R and we only solve for θ and V_ϕ. We use a 60-s long-time window, larger than the longest period present in the filtered GNSS seismograms. Fig. 4.8B shows the results obtained separately for the body and surface wave domains. High values of semblance (>0.93) for both types of wave show that all high-rate GNSS time series are internally coherent and are also coherent with the broadband seismogram. The optimal back-azimuths detected in the body wave window (110°) and in the surface wave window (112°) are consistent with the theoretical back-azimuth of the rupture zone (105–120°).

The location of the GNSS network, with an azimuth roughly perpendicular to the Wenchuan fault, is favorable for tracking the along-strike propagation of the rupture. We now consider a higher-frequency range (0.07–0.12 Hz) and a shorter time window (12 s) of the surface waves domain of the 5-Hz GNSS time series. The location of the source emission x is now constrained to move along the Wenchuan fault trace only. The analysis identifies three zones of coherent, highly energetic emission of surface waves occurring at 0–20 s, around 55 s, and during 75–90 s (Fig. 4.8C). Their distance from the epicenter, increasing with time, demonstrates a unilateral northeastward propagation of the rupture. The latest source indicates that high seismic energy was radiated 250 km northeast of the epicenter and 90 s after earthquake initiation, providing lower bound values for the rupture length and earthquake duration. As a further validation, we compare our results with an independent result from Xu et al. (2009), who performed a back-projection of P-wave energy analysis using seismograms from three regional arrays of broadband seismometers located in Europe, Australia, and Alaska. In Fig. 4.8C, we report their estimate of the rupture location as a function of time for the combined regional arrays at 0.2 Hz, close to the frequency range used in our analysis. Their results correlate, within the uncertainty, with the location and time of the three bursts identified by our analysis. Furthermore, the 5-Hz GNSS array analysis shows that in a slow initial stage, 80 km of the fault ruptured during ~50 s. During the second half of the earthquake, the rupture propagated over ~200 km in 50 s. The back-projection results from Xu et al. (2009) suggest a similar pattern with a rupture acceleration occurring after 40 s. This example shows that high-rate dynamic displacements are accurate enough to apply seismic array analysis.

8 Some earthquake properties and perspectives

There are many aspects of the rupture of large earthquakes that can be studied. We briefly mention them and provide more details on the relation between large earthquakes and the interseismic strain, for which GNSS has provided key observations.

As shown in Fig. 4.6, many earthquakes show a pulse-like propagation of the rupture (Heaton, 1990). The pulse is relatively small compared to the final extent of the rupture and implies that slip occurs over short durations compared to the total duration of the earthquake. Propagation usually occurs at a velocity of 2–3 km/s, but a few earthquakes show rupture velocity exceeding the S-wave velocity. These supershear ruptures are well highlighted through seismic array analysis. Most large earthquakes show heterogeneous slip distribution, corresponding to the successive rupture of several patches, as highlighted in Fig. 4.6 for the Ecuador 2016 earthquake. The hypocenter is usually located outside the maximum slip areas (Mai, 2005). For crustal earthquakes, there is a tendency for the hypocenter to be located in the deeper part of the fault, with high slip occurring at shallower depths. Another pattern often observed is that aftershocks are scarce at high coseismic slip areas and more abundant are their periphery (Wetzler et al., 2018).

With improved resolution of coseismic slip models, one important direction of research has been to correlate the location of the seismic slip with interseismic coupling maps derived from GNSS velocity measured prior to the earthquake, with the ultimate goal to know whether the extent of future seismic ruptures could be anticipated. At first order, most well-documented large earthquakes ruptured fault segments found to be highly locked during the years prior to the rupture (Chlieh et al., 2008; Madariaga et al., 2010; Moreno et al., 2010; Protti et al., 2013). A finer correlation between the extent of the seismic rupture area and interseismic coupling has been found for several large earthquakes (Chlieh et al., 2008; Moreno et al., 2010; Protti et al., 2013) and in a few cases, patches of high coseismic slip could even be correlated to spatial variations of heterogenous interseismic coupling (Nocquet et al., 2017). The Tohoku (Japan) Mw 9.0 2011 earthquake however showed extremely large slip at shallow depth close to the trench where interseismic models published before the earthquake had not evidenced significant coupling. The Tohoku case raised questions about the limited resolution of models based on onshore-only measurements (Almeida et al., 2018; Avouac, 2011; Lindsey et al., 2021; Loveless & Meade, 2011) and the mechanism for shallow slip (Cubas et al., 2015; Kubota et al., 2022; Noda & Lapusta, 2013). Finally, some studies noticed that slow slip events (SSEs, see Chapter 5) occur at the periphery of future coseismic ruptures (Dixon et al., 2014; Nishikawa et al., 2019; Rolandone et al., 2018). These results suggest that SSE information could be used in complement to the steady interseismic coupling maps to better infer the location of persisting asperities. Tsunami earthquakes, where slow rupture produces little high-frequency seismic waves but significant displacement (Hill et al., 2012b) are a specific case of earthquakes occurring in areas of low interseismic coupling.

These observations fuel physics-based models, trying to simultaneously explain patterns observed before and during earthquakes through dynamical simulation with spatially variable friction laws along faults (Barbot et al., 2012; Kaneko et al., 2010). More broadly, understanding how slip occurs along faults throughout the seismic cycle remains a significant challenge in earthquake research, crucial for improving our ability to anticipate these events.

9 Future issues and opportunities

1. GNSS networks in actively deforming areas need to be maintained in the long term to capture the signature of large earthquakes
2. Many earthquakes have their source at offshore faults. The development of seafloor geodesy is crucial to gain resolution and better document the dynamic at the shallow portion of subduction faults
3. Comparison of coseismic ruptures with pre-earthquake interseismic deformation, slow slip events, and postseismic slip highlights the heterogeneity of slip modes at faults. These observations are key for developing physics-based models of faults and earthquakes.

10 Summary points

1. GNSS and other geodetic techniques complement seismological observation to better characterize earthquake-induced displacements.
2. GNSS measurements located in the near field of active faults have become a standard tool to image coseismic slip distribution and seismic rupture.
3. Systematic imaging of earthquakes allows to derive generic properties of earthquakes to decipher the underlying physics.
4. Precise model of coseismic ruptures allow to investigate how earthquakes are related to the slow tectonic loading at faults.

Acknowledgment

We thank P. Jarrin for creating Fig. 2. The data used for the array analysis were collected and analyzed in the frame of the LACUNES (SEISMIC GAPS) project (grant ANR-05-CATT-0006) with contributions from C. Guihua, X. Xiwei, C. Romieu, M. Ueno, and Y. Klinger.

References

Almeida, R., Lindsey, E. O., Bradley, K., Hubbard, J., Mallick, R., & Hill, E. M. (2018). Can the updip limit of frictional locking on megathrusts be detected geodetically? Quantifying the effect of stress shadows on near-trench coupling. *Geophysical Research Letters, 45*, 4754–4763. https://doi.org/10.1029/2018GL077785.

Árnadóttir, T., & Segall, P. (1994). The 1989 Loma Prieta earthquake imaged from inversion of geodetic data. *Journal of Geophysical Research: Solid Earth, 99*(B11), 21835–21855. https://doi.org/10.1029/94JB01256.

Avallone, A., Marzario, M., Cirella, A., Piatanesi, A., Rovelli, A., Di Alessandro, C., et al. (2011). Very high rate (10 Hz) GPS seismology for moderate-magnitude earthquakes: The case of the M w 6.3 L'Aquila (Central Italy) event. *Journal of Geophysical Research, 116*(B2), B02305. https://doi.org/10.1029/2010JB007834.

Avouac, J.-P. (2011). The lessons of Tohoku-Oki. *Nature, 475*(7356), 300. https://doi.org/10.1038/nature10265.

Bagnardi, M., & Hooper, A. (2018). Inversion of surface deformation data for rapid estimates of source parameters and uncertainties: A Bayesian approach. *Geochemistry, Geophysics, Geosystems, 19*(7), 2194–2211. https://doi.org/10.1029/2018GC007585.

Barbot, S., Lapusta, N., & Avouac, J.-P. (2012). Under the hood of the earthquake machine: Toward predictive modeling of the seismic cycle. *Science, 336*(6082), 707–710. https://doi.org/10.1126/science.1218796.

Barnhart, W. D., & Lohman, R. B. (2010). Automated fault model discretization for inversions for coseismic slip distributions. *Journal of Geophysical Research: Solid Earth, 115*(10), 1–17. https://doi.org/10.1029/2010JB007545.

Benavente, R., Dettmer, J., Cummins, P. R., & Sambridge, M. (2019). Efficient Bayesian uncertainty estimation in linear finite fault inversion with positivity constraints by employing a log-normal prior. *Geophysical Journal International, 217*(1), 469–484. https://doi.org/10.1093/gji/ggz044.

Bilham, R. (2009). *The seismic future of cities*. https://doi.org/10.1007/s10518-009-9147-0.

Bletery, Q., Sladen, A., Delouis, B., Vallée, M., Nocquet, J.-M. J., Rolland, L., et al. (2014). A detailed source model for the M w 9.0 Tohoku-Oki earthquake reconciling geodesy, seismology, and tsunami records. *Journal of Geophysical Research: Solid Earth, 119*(10), 7636–7653. https://doi.org/10.1002/2014JB011261.Received.

Bouchon, M. (1981). A simple method to calculate Green's functions for elastic layered media. *Bulletin of the Seismological Society of America, 71*(4), 959–971. https://doi.org/10.1785/BSSA0710040959.

Bürgmann, R., & Chadwell, D. (2014). Seafloor Geodesy. *Annual Review of Earth and Planetary Sciences, 42*(1), 509–534. https://doi.org/10.1146/annurev-earth-060313-054953.

Chadwell, C. D., & Spiess, F. N. (2008). Plate motion at the ridge-transform boundary of the south cleft segment of the Juan de Fuca ridge from GPS-acoustic data. *Journal of Geophysical Research, 113*(B4), B04415. https://doi.org/10.1029/2007JB004936.

Chlieh, M., Avouac, J. P., Sieh, K., Natawidjaja, D. H., & Galetzka, J. (2008). Heterogeneous coupling of the Sumatran megathrust constrained by geodetic and paleogeodetic measurements. *Journal of Geophysical Research: Solid Earth, 113*(5), 1–31. https://doi.org/10.1029/2007JB004981.

Choi, K., Bilich, A., Larson, K. M., & Axelrad, P. (2004). Modified sidereal filtering: Implications for high-rate GPS positioning. *Geophysical Research Letters, 31*(22), 1–4. https://doi.org/10.1029/2004GL021621.

Cubas, N., Lapusta, N., Avouac, J. P., & Perfettini, H. (2015). Numerical modeling of long-term earthquake sequences on the NE Japan megathrust: Comparison with observations and implications for fault friction. *Earth and Planetary Science Letters, 419*, 187–198. https://doi.org/10.1016/j.epsl.2015.03.002.

Delouis, B., Nocquet, J.-M., & Vallée, M. (2010). Slip distribution of the February 27, 2010 Mw = 8.8 Maule earthquake, Central Chile, from static and high-rate GPS, InSAR, and broadband teleseismic data. *Geophysical Research Letters, 37*(17), L17305. https://doi.org/10.1029/2010GL043899.

Dettmer, J., Benavente, R., Cummins, P. R., & Sambridge, M. (2014). Trans-dimensional finite-fault inversion. *Geophysical Journal International, 199*(2), 735–751. https://doi.org/10.1093/gji/ggu280.

Dixon, T. H., Jiang, Y., Malservisi, R., McCaffrey, R., Voss, N., Protti, M., et al. (2014). Earthquake and tsunami forecasts: Relation of slow slip events to subsequent earthquake rupture. *Proceedings of the National Academy of Sciences, 111*(48), 17039–17044. https://doi.org/10.1073/pnas.1412299111.

Duputel, Z., Agram, P. S., Simons, M., Minson, S. E., & Beck, J. L. (2014). Accounting for prediction uncertainty when inferring subsurface fault slip. *Geophysical Journal International, 197*(1), 464–482. https://doi.org/10.1093/gji/ggt517.

Dziewonski, A. M., Chou, T.-A., & Woodhouse, J. H. (1981). Determination of earthquake source parameters from waveform data for studies of global and regional seismicity. *Journal of Geophysical Research: Solid Earth, 86*(B4), 2825–2852. https://doi.org/10.1029/JB086iB04p02825.

Ekström, G., Nettles, M., & Dziewoński, A. M. (2012). The global CMT project 2004–2010: Centroid-moment tensors for 13,017 earthquakes. *Physics of the Earth and Planetary Interiors, 200–201*, 1–9. https://doi.org/10.1016/j.pepi.2012.04.002.

Elliott, J. R., Jolivet, R., González, P. J., Avouac, J.-P., Hollingsworth, J., Searle, M. P., et al. (2016). Himalayan megathrust geometry and relation to topography revealed by the Gorkha earthquake. *Nature Geoscience, 9*(2), 174–180. https://doi.org/10.1038/ngeo2623.

Evans, E. L., & Meade, B. J. (2012). Geodetic imaging of coseismic slip and postseismic afterslip: Sparsity promoting methods applied to the great Tohoku earthquake. *Geophysical Research Letters, 39*(11). https://doi.org/10.1029/2012GL051990.

Farías, M., Vargas, G., Tassara, A., Carretier, S., Baize, S., Melnick, D., et al. (2010). Land-level changes produced by the M w 8.8 2010 Chilean earthquake. *Science, 329*(5994), 916.

Fujita, M., Ishikawa, T., Mochizuki, M., Sato, M., Toyama, S., Katayama, M., et al. (2006). GPS/acoustic seafloor geodetic observation: Method of data analysis and its application. *Earth, Planets and Space, 58*(3), 265–275. https://doi.org/10.1186/BF03351923.

Fukahata, Y., & Wright, T. J. (2008). A non-linear geodetic data inversion using ABIC for slip distribution on a fault with an unknown dip angle. *Geophysical Journal International, 173*(2), 353–364. https://doi.org/10.1111/j.1365-246X.2007.03713.x.

Fukahata, Y., Yagi, Y., & Miyazaki, S. (2012). Constraints on the early-stage rupture process of the 2011 Tohoku-oki earthquake from 1-Hz GPS data. *Earth, Planets and Space, 64*, 1093–1099. https://doi.org/10.5047/eps.2012.09.007.

Fukuda, J., & Johnson, K. M. (2008). A fully Bayesian inversion for spatial distribution of fault slip with objective smoothing. *Bulletin of the Seismological Society of America, 98*(3), 1128–1146. https://doi.org/10.1785/0120070194.

Gagnon, K., Chadwell, C. D., & Norabuena, E. (2005). Measuring the onset of locking in the Peru-Chile trench with GPS and acoustic measurements. *Nature, 520*(2004), 516–520.

Genrich, J. F., & Bock, Y. (2006). Instantaneous geodetic positioning with 10-50 Hz GPS measurements: Noise characteristics and implications for monitoring networks. *Journal of geophysical research: Solid Earth, 111*(B3). https://doi.org/10.1029/2005JB003617.

Hansen, P. C. (1992). Analysis of discrete ill-posed problems by means of the L-curve. *SIAM Review, 34*(4), 561–580.

Hayes, G. (2010). *Updated result of the Feb 27, 2010 Mw 8.8 Maule, Chile Earthquake*.

Hayes, G. P. (2017). The finite, kinematic rupture properties of great-sized earthquakes since 1990. *Earth and Planetary Science Letters, 468*, 94–100. https://doi.org/10.1016/j.epsl.2017.04.003.

Heaton, T. H. (1990). Evidence for and implications of self-healing pulses of slip in earthquake rupture. *Physics of the Earth and Planetary Interiors, 64*(1), 1–20. https://doi.org/10.1016/0031-9201(90)90002-F.

Hill, E. M., Borrero, J. C., Huang, Z., Qiu, Q., Banerjee, P., Natawidjaja, D. H., et al. (2012a). The 2010 Mw 7.8 Mentawai earthquake: Very shallow source of a rare tsunami earthquake determined from tsunami field survey and near-field GPS data. *Journal of geophysical research: Solid Earth, 117*(6), 1–21. https://doi.org/10.1029/2012JB009159.

Hill, E. M., Borrero, J. C., Huang, Z., Qiu, Q., Banerjee, P., Natawidjaja, D. H., et al. (2012b). The 2010 Mw 7.8 Mentawai earthquake: Very shallow source of a rare tsunami earthquake determined from tsunami field survey and near-field GPS data. *Journal of geophysical research: Solid Earth, 117* (B6). https://doi.org/10.1029/2012JB009159.

Hori, T., Agata, R., Ichimura, T., Fujita, K., Yamaguchi, T., & Iinuma, T. (2021). High-fidelity elastic Green's functions for subduction zone models consistent with the global standard geodetic reference system. *Earth, Planets Space, 73*(41), 1–12. https://doi.org/10.1186/s40623-021-01370-y.

Hu, J., Li, Z. W., Ding, X. L., Zhu, J. J., Zhang, L., & Sun, Q. (2014). Resolving three-dimensional surface displacements from InSAR measurements: A review. *Earth-Science Reviews, 133*, 1–17. https://doi.org/10.1016/j.earscirev.2014.02.005.

Hubert, A., King, G., Armijo, R., Meyer, B., & Papanastasiou, D. (1996). Fault re-activation, stress interaction and rupture propagation of the 1981 Corinth earthquake sequence. *Earth and Planetary Science Letters, 142*(3–4), 573–585. https://doi.org/10.1016/0012-821X(96)00108-2.

Ide, S. (2007). Slip inversion. In *Treatise on geophysics* (pp. 193–224). Amsterdam:Elsevier.

Iinuma, T., Kido, M., Ohta, Y., Fukuda, T., Tomita, F., & Ueki, I. (2021). GNSS-acoustic observations of seafloor crustal deformation using a wave glider. *Frontiers in Earth Science, 9*(600946). https://doi.org/10.3389/feart.2021.600946.

Ito, Y., Tsuji, T., Osada, Y., Kido, M., Inazu, D., Hayashi, Y., et al. (2011). Frontal wedge deformation near the source region of the 2011 Tohoku-Oki earthquake. *Geophysical Research Letters, 38*(7). https://doi.org/10.1029/2011GL048355.

Johnson, K. M., Hsu, Y. J., Segall, P., & Yu, S. B. (2001). Fault geometry and slip distribution of the 1999 chi-chi, Taiwan earthquake imaged from inversion of gps data. *Geophysical Research Letters, 28*(11), 2285–2288.

Kaneko, Y., Avouac, J.-P., & Lapusta, N. (2010). Towards inferring earthquake patterns from geodetic observations of interseismic coupling. *Nature Geoscience, 3*(5), 363–369. https://doi.org/10.1038/ngeo843.

Kido, M., Osada, Y., Fujimoto, H., Hino, R., & Ito, Y. (2011). Trench-normal variation in observed seafloor displacements associated with the 2011 Tohoku-Oki earthquake. *Geophysical Research Letters, 38*(L24303). https://doi.org/10.1029/2011GL050057.

Klein, E., Vigny, C., Nocquet, J. M., & Boulze, H. (2022). A 20 year-long GNSS solution across South-America with focus in Chile. *Bulletin de la Société Géologique de France, 193*, 1–32. https://doi.org/10.1051/bsgf/2022005.

Kubota, T., Saito, T., & Hino, R. (2022). A new mechanical perspective on a shallow megathrust near-trench slip from the high-resolution fault model of the 2011 Tohoku-Oki earthquake. *Progresses in Earth and Planetary Science, 9*(68). https://doi.org/10.1186/s40645-022-00524-0.

Langbein, J., Murray, J. R., & Snyder, H. A. (2006). Coseismic and initial postseismic deformation from the 2004 Parkfield, California, earthquake, observed by global positioning system, electronic distance meter, creepmeters, and borehole strainmeters. *Bulletin of the Seismological Society of America, 96*(4 B), 304–320. https://doi.org/10.1785/0120050823.

Larson, K. M. (2009). GPS seismology. *Journal of Geodesy, 83*(3–4), 227–233. https://doi.org/10.1007/s00190-008-0233-x.

Larson, K. M., Bodin, P., & Gomberg, J. (2003). Using 1-Hz GPS data to measure deformations caused by the Denali fault earthquake. *Science, 300*(5624), 1421–1424. https://doi.org/10.1126/science.1084531.

Lawson, C. W., & Hanson, R. J. (1974). *Solving least squares problems*. New York: John Wiley and Sons, Inc.

Lindsey, E. O., Mallick, R., Hubbard, J. A., et al. (2021). Slip rate deficit and earthquake potential on shallow megathrusts. *Nature Geoscience, 14*, 321–326. https://doi.org/10.1038/s41561-021-00736-x.

Liu, K., Geng, J., Wen, Y., Ortega-Culaciati, F., & Comte, D. (2022). Very early Postseismic deformation following the 2015 Mw 8.3 Illapel earthquake, Chile revealed from kinematic GPS. *Geophysical Research Letters, 49*(11), 1–11. https://doi.org/10.1029/2022GL098526.

Lorito, S., Romano, F., Atzori, S., Tong, X., Avallone, A., Mccloskey, J., et al. (2010). Limited overlap between the seismic gap and coseismic slip of the great 2010 Chile earthquake. *Nature Geoscience, 4*(3), 173–177. https://doi.org/10.1038/NGEO1073.

Loveless, J. P., & Meade, B. J. (2011). Spatial correlation of interseismic coupling and coseismic rupture extent of the 2011 Mw= 9.0 Tohoku-oki earthquake. *Geophysical Research Letters, 38*(17). https://doi.org/10.1029/2011GL048561.

Madariaga, R., Métois, M., Vigny, C., & Campos, J. (2010). Central Chile finally breaks. *Science, 328*(5975), 181–182. https://doi.org/10.1126/science.1189197.

Mai, P. M. (2005). Hypocenter locations in finite-source rupture models. *Bulletin of the Seismological Society of America, 95*(3), 965–980. https://doi.org/10.1785/0120040111.

Mai, P. M., & Thingbaijam, K. K. S. (2014). SRCMOD: An online database of finite-fault rupture models. *Seismological Research Letters, 85*(6), 1348–1357. https://doi.org/10.1785/0220140077.

Matsuo, K., & Heki, K. (2011). Coseismic gravity changes of the 2011 Tohoku-Oki earthquake from satellite gravimetry. *Geophysical Research Letters, 38*(7). https://doi.org/10.1029/2011GL049018.

Meade, B. J. (2007). Algorithms for the calculation of exact displacements, strains, and stresses for triangular dislocation elements in a uniform elastic half space. *Computers and Geosciences, 33*(8), 1064–1075. https://doi.org/10.1016/j.cageo.2006.12.003.

Meier, M. A., Ampuero, J. P., & Heaton, T. H. (2017). The hidden simplicity of subduction megathrust earthquakes. *Science, 357*(6357), 1277–1281.

Melgar, D., Bock, Y., Sanchez, D., & Crowell, B. W. (2013). On robust and reliable automated baseline corrections for strong motion seismology. *Journal of Geophysical Research: Solid Earth, 118*(3), 1177–1187. https://doi.org/10.1002/jgrb.50135.

Menke, W. (2012). Chapter 3 - solution of the linear, Gaussian inverse problem, viewpoint 1: The length method. In W. Menke (Ed.), *Geophysical data analysis: Discrete inverse theory* (3rd Edition, pp. 39–68). Academic Press. https://doi.org/10.1016/B978-0-12-397160-9.00003-5.

Minson, S. E., Simons, M., & Beck, J. L. (2013). Bayesian inversion for finite fault earthquake source models I-theory and algorithm. *Geophysical Journal International, 194*(3), 1701–1726. https://doi.org/10.1093/gji/ggt180.

Miyazaki, S., Larson, K. M., Choi, K., Hikima, K., Koketsu, K., Bodin, P., et al. (2004). Modeling the rupture process of the 2003 September 25 Tokachi-Oki (Hokkaido) earthquake using 1-Hz GPS data. *Geophysical Research Letters, 31*(21). https://doi.org/10.1029/2004GL021457.

Moreno, M., Rosenau, M., & Oncken, O. (2010). 2010 Maule earthquake slip correlates with pre-seismic locking of Andean subduction zone. *Nature, 467*(7312), 198–202. https://doi.org/10.1038/nature09349.

Munekane, H. (2012). Coseismic and early postseismic slips associated with the 2011 off the Pacific coast of Tohoku earthquake sequence: EOF analysis of GPS kinematic time series. *Earth, Planets and Space, 64*, 1077–1091. https://doi.org/10.5047/eps.2012.07.009I.

Nakata, R., Hino, H., Kuwatani, T., Yoshioka, S., Okada, M., & Hori, T. (2017). Discontinuous boundaries of slow slip events beneath the Bungo Channel, southwest Japan. *Scientific Reports, 7*(6129). https://doi.org/10.1038/s41598-017-06185-0.

Nikkhoo, M., & Walter, T. R. (2015). Triangular dislocation: An analytical, artefact-free solution. *Geophysical Journal International, 201*(2), 1119–1141. https://doi.org/10.1093/gji/ggv035.

Nikolaidis, R. M., Bock, Y., De Jonge, P. J., Shearer, P., Agnew, D. C., & Van Domselaar, M. (2001). Seismic wave observations with the global positioning system. *Journal of Geophysical Research: Solid Earth, 106*(B10), 21897–21916. https://doi.org/10.1029/2001JB000329.

Nishikawa, T., Matsuzawa, T., Ohta, K., Uchida, N., Nishimura, T., & Ide, S. (2019). The slow earthquake spectrum in the Japan trench illuminated by the S-net seafloor observatories. *Science, 365*(6455), 808–813. https://doi.org/10.1126/science.aax5618.

Nissen, E., Maruyama, T., Ramon Arrowsmith, J., Elliott, J. R., Krishnan, A. K., Oskin, M. E., et al. (2014). Coseismic fault zone deformation revealed with differential lidar: Examples from Japanese Mw ~7 intraplate earthquakes. *Earth and Planetary Science Letters, 405*, 244–256. https://doi.org/10.1016/j.epsl.2014.08.031.

Nocquet, J.-M. (2018). Stochastic static fault slip inversion from geodetic data with non-negativity and bound constraints. *Geophysical Journal International, 214*(1), 366–385. https://doi.org/10.1093/gji/ggy146.

Nocquet, J.-M., Jarrin, P., Vallée, M., Mothes, P. A., Grandin, R., Rolandone, F., et al. (2017). Supercycle at the Ecuadorian subduction zone revealed after the 2016 Pedernales earthquake. *Nature Geoscience, 10*(2), 145–149. https://doi.org/10.1038/ngeo2864.

Noda, H., & Lapusta, N. (2013). Stable creeping fault segments can become destructive as a result of dynamic weakening. *Nature, 493*(7433), 518–521. https://doi.org/10.1038/nature11703.

Okada, Y. (1992). Internal deformation due to shear and tensile faults in a half-space. *Bulletin of the Seismological Society of America, 82*(2), 1018–1040. https://doi.org/10.1785/BSSA0820021018.

Olson, A. H., & Apsel, R. J. (1982). Finite faults and inverse theory with applications to the 1979 Imperial Valley earthquake. *Bulletin of the Seismological Society of America, 72*(6A), 1969–2001. https://doi.org/10.1785/BSSA07206A1969.

Pesaresi, M., Ehrlich, D., Kemper, T., Siragusa, A., Florczyk, A., Freire, S., et al. (2017). Atlas of the human planet 2017: Global exposure to natural hazards. *Publications Office of the European Union*. https://doi.org/10.2760/19837.

Pollitz, F. (1996). Coseismic deformation from earthquake faulting on a layered spherical earth. *Geophysical Journal International, 125*(1), 1–14. https://doi.org/10.1111/j.1365-246X.1996.tb06530.x.

Pollitz, F., Brooks, B., Tong, X., Bevis, M. G., Foster, J. H., Bürgmann, R., et al. (2011). Coseismic slip distribution of the February 27, 2010 Mw 8.8 Maule, Chile earthquake. *Geophysical Research Letters, 38*(9). https://doi.org/10.1029/2011GL047065. 2011GL047065.

Protti, M., González, V., Newman, A. V., Dixon, T. H., Schwartz, S. Y., Marshall, J. S., et al. (2013). Nicoya earthquake rupture anticipated by geodetic measurement of the locked plate interface. *Nature Geoscience, 7*(2), 117–121. https://doi.org/10.1038/ngeo2038.

Puymbroeck, V., Binet, R., & Avouac, J. (2000). Measuring earthquakes from optical satellite images. *Applied Optics, 39*(20), 3486–3494. https://doi.org/10.1364/AO.39.003486.

Radiguet, M., Cotton, F., Vergnolle, M., Campillo, M., Valette, B., Kostoglodov, V., et al. (2011). Spatial and temporal evolution of a long term slow slip event: The 2006 Guerrero slow slip event. *Geophysical Journal International, 184*(2), 816–828. https://doi.org/10.1111/j.1365-246X.2010.04866.x.

Renou, J., Vallée, M., & Dublanchet, P. (2019). How does seismic rupture accelerate? Observational insights from earthquake source time functions. *Journal of Geophysical Research: Solid Earth, 124*(8), 8942–8952. https://doi.org/10.1029/2019JB018045.

Ringler, A. T., Anthony, R. E., Aster, R. C., Ammon, C. J., Arrowsmith, S., Benz, H., et al. (2022). Achievements and prospects of global broadband seismographic networks after 30 years of continuous geophysical observations. *Reviews of Geophysics, 60*(3). https://doi.org/10.1029/2021RG000749.

Rolandone, F., Nocquet, J. M., Mothes, P. A., Jarrin, P., Vallée, M., Cubas, N., et al. (2018). Areas prone to slow slip events impede earthquake rupture propagation and promote afterslip. *Science Advances, 4*(1). https://doi.org/10.1126/sciadv.aao6596.

Ruhl, C. J., Melgar, D., Allen, R. M., Geng, J., Goldberg, D. E., Bock, Y., et al. (2019). A global database of strong-motion displacement GNSS recordings and an example application to PGD scaling. *Seismological Research Letters, 90*(1). https://doi.org/10.1785/0220180177.

Sato, M., Ishikawa, T., Ujihara, N., Yoshida, S., Fujita, M., Mochizuki, M., et al. (2011). Displacement above the hypocenter of the 2011 Tohoku-Oki earthquake. *Science, 332*(6036), 1395. https://doi.org/10.1126/science.1207401.

Shaw, B., Ambraseys, N. N., England, P. C., Floyd, M. A., Gorman, G. J., Higham, T. F. G., et al. (2008). Eastern Mediterranean tectonics and tsunami hazard inferred from the AD 365 earthquake. *Nature Geoscience, 1*(4). https://doi.org/10.1038/ngeo151.

Sieh, K., Natawidjaja, D. H., Meltzner, A. J., Shen, C. C., Cheng, H., Li, K. S., et al. (2008). Earthquake supercycles inferred from sea-level changes recorded in the corals of West Sumatra. *Science, 322*(5908), 1674–1678. https://doi.org/10.1126/science.1163589.

Stark, P. B., Ca, B., & Parker, R. L. (1995). Bounded-variable least-squares: An algorithm and applications. *Computational Statistics, 10*, 129–141.

Tago, J., Cruz-Atienza, V. M., Villafuerte, C., Nishimura, T., Kostoglodov, V., Real, J., et al. (2021). Adjoint slip inversion under a constrained optimization framework: Revisiting the 2006 Guerrero slow slip event. *Geophysical Journal International, 226*(2), 1187–1205. https://doi.org/10.1093/gji/ggab165.

Tanioka, Y., & Ruff, L. J. (1997). Source time functions. *Seismological Research Letters*, *68*(3), 386–400. https://doi.org/10.1785/gssrl.68.3.386.

Tarantola, A. (2005). Inverse problem theory. *Siam*, *2005*. https://doi.org/10.1137/1.9780898717921.

Tarantola, A., & Valette, B. (1982). Generalized nonlinear inverse problems solved using the least squares criterion. *Reviews of Geophysics*, *20*(2), 219–232. https://doi.org/10.1029/RG020i002p00219.

Tomita, F., Iinuma, T., Agata, R., & Hori, T. (2021). Development of a trans-dimensional fault slip inversion for geodetic data. *Journal of Geophysical Research: Solid Earth*, *126*(5). https://doi.org/10.1029/2020JB020991.

Twardzik, C., Vergnolle, M., Sladen, A., & Avallone, A. (2019). Unravelling the contribution of early postseismic deformation using sub-daily GNSS positioning. *Scientific Reports*, *9*(1), 1775. https://doi.org/10.1038/s41598-019-39038-z.

Vallée, M., & Douet, V. (2016). A new database of source time functions (STFs) extracted from the SCARDEC method. *Physics of the Earth and Planetary Interiors*, *257*, 149–157. https://doi.org/10.1016/j.pepi.2016.05.012.

Vigny, C., Socquet, A., Peyrat, S., Ruegg, J.-C., Metois, M., Madariaga, R., et al. (2011). The 2010 Mw 8.8 Maule megathrust earthquake of Central Chile, monitored by GPS. *Science*, *332*(6036), 1417–1421. https://doi.org/10.1126/science.1204132.

Wang, R., Martin, F. L., & Roth, F. (2003). Computation of deformation induced by earthquakes in a multi-layered elastic crust - FORTRAN programs EDGRN/EDCMP. *Computers and Geosciences*, *29*(2), 195–207. https://doi.org/10.1016/S0098-3004(02)00111-5.

Wang, L., Zhao, X., Xu, W., Xie, L., & Fang, N. (2019). Coseismic slip distribution inversion with unequal weighted Laplacian smoothness constraints. *Geophysical Journal International*, *218*(1), 145–162. https://doi.org/10.1093/gji/ggz125.

Wetzler, N., Lay, T., Brodsky, E. E., & Kanamori, H. (2018). Systematic deficiency of aftershocks in areas of high coseismic slip for large subduction zone earthquakes. *Science Advances*, *4*(2). https://doi.org/10.1126/sciadv.aao3225.

Williams, C. A., & Wallace, L. M. (2015). Effects of material property variations on slip estimates for subduction interface slow-slip events. *Geophysical Research Letters*, *42*(4), 1113–1121. https://doi.org/10.1002/2014GL062505.

Wright, T. J., Parsons, B. E., & Lu, Z. (2004). Toward mapping surface deformation in three dimensions using InSAR. *Geophysical Research Letters*, *31*(1). https://doi.org/10.1029/2003GL018827.

Xu, Y., Koper, K. D., Sufri, O., Zhu, L., & Hutko, A. R. (2009). Rupture imaging of the Mw 7.9 12 May 2008 Wenchuan earthquake from back projection of teleseismic P waves. *Geochemistry, Geophysics, Geosystems*, *10*(4). https://doi.org/10.1029/2008GC002335.

Yabuki, T., & Matsu'ura, M. (1992). Geodetic data inversion using a Bayesian information criterion for spatial distribution of fault slip. *Geophysical Journal International*, *109*(2), 363–375. https://doi.org/10.1111/j.1365-246X.1992.tb00102.x.

Ye, L., Lay, T., Kanamori, H., & Rivera, L. (2016). Rupture characteristics of major and great (Mw $\geq$ 7.0) megathrust earthquakes from 1990 to 2015: 1. Source parameter scaling relationships. *Journal of Geophysical Research: Solid Earth*, *121*(2), 826–844. https://doi.org/10.1002/2015JB012426.

Yokota, Y., Ishikawa, T., & Watanabe, S. (2018). Seafloor crustal deformation data along the subduction zones around Japan obtained by GNSS-A observations. *Scientific Data*, *5*(1), 180182. https://doi.org/10.1038/sdata.2018.182.

Yokota, Y., Ishikawa, T., Watanabe, S., Tashiro, T., & Asada, A. (2016). Seafloor geodetic constraints on interplate coupling of the Nankai trough megathrust zone. *Nature*, *534*(7607), 374–377. https://doi.org/10.1038/nature17632.

Yokota, Y., Kaneda, M., Hashimoto, T., Yamaura, S., Kouno, K., & Hirakawa, Y. (2023). Experimental verification of seafloor crustal deformation observations by UAV-based GNSS-A. *Scientific Reports*, *13*(1). https://doi.org/10.1038/s41598-023-31214-6.

Chapter 5

GNSS observations of transient deformation in plate boundary zones

Laura M. Wallace[a,b,c] and Chris Rollins[d]

[a]University of Texas Institute for Geophysics, Austin, TX, United States, [b]GEOMAR Helmholtz Centre for Ocean Research, Kiel, Germany, [c]Kiel University, Institute of Geosciences, Kiel, Germany, [d]GNS Science, Lower Hutt, New Zealand

1 Introduction

The advent of GNSS geodesy over the last 30 years has played a pivotal role in revealing a rich array of previously unknown transient deformation processes at plate boundaries. This includes the recognition that deformation rates vary substantially throughout the earthquake cycle, particularly in the years to decades following major earthquakes (e.g., Freed & Bürgmann, 2004; Freed, Bürgmann, Calais, & Freymueller, 2006; Freed, Bürgmann, Calais, Freymueller, & Hreinsdóttir, 2006; Heki et al., 1997; Pollitz et al., 2001; Suito & Freymueller, 2009; Wang et al., 2012). Moreover, continuously operating GNSS networks at many plate boundaries worldwide have enabled the discovery of episodic slow slip events (SSEs) lasting days to years (Dragert et al., 2001; Schwartz & Rokosky, 2007), which has revolutionized our understanding of the broad spectrum of fault slip and earthquake processes (Avouac, 2015; Bürgmann, 2018; Ide, Beroza, et al., 2007; Obara & Kato, 2016; Peng & Gomberg, 2010).

SSEs (sometimes called "silent earthquakes") are episodic fault slip events that typically involve centimeters to tens of centimeters of slip over days to years, and recur on a relatively frequent basis (months to several years). They are expressed at the Earth's surface as millimeter- to centimeter-level horizontal and/or vertical surface displacement and are most commonly observed by continuously operating GNSS networks (Fig. 5.1). The energy released during an SSE can be equivalent to that which would occur in an Mw 6–7 earthquake, albeit more slowly, and with a lower stress drop (a few to tens of kPa; Brodsky & Mori, 2007) compared to a typical seismic event (a few to tens of MPa; Allmann & Shearer, 2009). SSEs have been observed on most types of faults and are now known to play a major role in the accommodation of plate motion.

Likewise, modern geodesy has shown that earthquakes are followed by a phase of accelerated deformation often lasting years to decades and decaying with time (Bürgmann & Thatcher, 2013, and references therein). This postseismic deformation is driven by aseismic lithospheric processes—viscoelastic relaxation, afterslip and poroelastic rebound. and dilatancy recovery—that are kicked into a higher gear by the stress changes imposed by an earthquake, and gradually relax those stress changes away. This phenomenon has been observed following earthquakes in all corners of the world (Fig. 5.2). Although geodetic data only measure deformation at the surface, the spatiotemporal character of this deformation can be used (in conjunction with modeling) to determine the specific subsurface processes that drive it, shedding light on the rheology of the lithosphere.

Prior to widespread GNSS geodetic observations, plate boundary deformation and fault slip were largely viewed in terms of two end-members: (a) seismic slip in an earthquake (centimeters to meters per second) and (b) steady creep at plate motion rates (centimeters per year). The realization that a rich spectrum of deformation processes occur throughout the earthquake cycle at plate boundaries has raised numerous questions about the rheology of the Earth's crust and mantle, the physical processes and fault mechanical properties that control fault slip, and the relationship between transient deformation events and earthquake occurrence. Recognition of this spatiotemporal diversity of crustal deformation processes in plate boundary zones is arguably one of the most important scientific contributions of GNSS geodetic techniques.

2 Episodic slow slip and creep events

Although a few studies since the 1960s indicated that faults underwent creep events based on creepmeter and strainmeter data (Bilham, 1989; Langbein et al., 1999; Linde et al., 1996; Smith & Wyss, 1968), the concept did not gain widespread

84 PART | I Monitoring earthquakes and volcanoes with GNSS

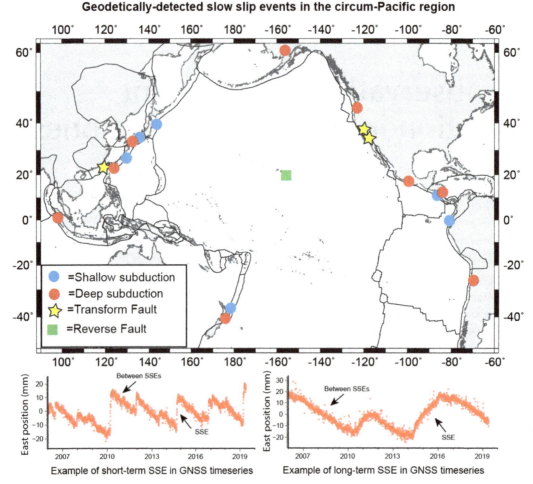

FIG. 5.1 Map of geodetically detected slow slip/creep events in the circum-Pacific region. See Table 5.1 for a summary of these events and their sources. *Lower panels* show examples of short-term *(left)* and long-term *(right)* slow slip events in GNSS timeseries (examples shown are from continuous GNSS sites GISB and KAPT in New Zealand; www.geonet.org.nz). Plate boundaries are shown with heavy *gray lines*. *(Data for dark grey lines on the figure are from Bird, P. (2003), An updated digital model of plate boundaries. Geochemistry, Geophysics, Geosystems, 4, 1027, https://doi.org/10.1029/2001GC000252.)*

recognition until slow slip events were discovered by scientists at the Cascadia subduction zone in western North America and the Nankai subduction zone in southwest Japan, using data from continuously operating GNSS sites (Dragert et al., 2001; Hirose et al., 1999). Scientists working in these regions recognized that temporal changes in the direction of movement of continuous GNSS sites overlying these subduction zones represented episodic creep events occurring on the subduction plate boundary fault (e.g., the subduction interface). Around the same time, seismologists working in Cascadia and Nankai noticed high-frequency seismic tremor signals (similar to volcanic tremors occurring at active volcanoes), now called tectonic tremor (Obara, 2002; Rogers & Dragert, 2003). For example, SSEs in Cascadia observed on GNSS stations at remarkably regular intervals (~14 months) also coincided with episodes of tectonic tremor (Rogers & Dragert, 2003). These and other observations have since revealed a strong association between many SSEs (observed geodetically) and a diverse array of seismological phenomena, including tremor and low-frequency earthquakes (Obara et al., 2004; Shelly et al., 2006; Wech & Creager, 2012). In the last 20 years, transient creep episodes have been observed on subduction zones (Bartlow et al., 2011; Dixon et al., 2014; Radiguet et al., 2012; Wallace & Beavan, 2010), transform faults (Murray & Segall, 2005; Wei et al., 2013, 2015), and reverse faults (Brooks et al., 2006; Montgomery-Brown et al., 2009). The discovery of SSEs has raised many questions about the physical mechanisms that produce them (Avouac, 2015; Bürgmann, 2018; Saffer & Wallace, 2015), as well as their role in the earthquake cycle (Obara & Kato, 2016).

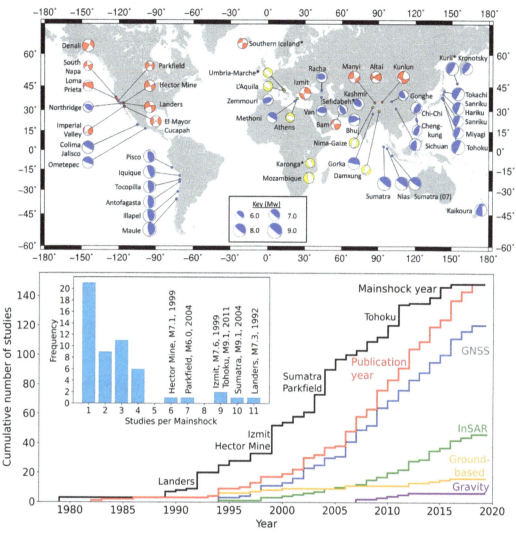

FIG. 5.2 *Top*: Recent earthquakes worldwide that exhibit geodetic evidence for afterslip (with published models). *Bottom*: Cumulative number of studies of afterslip through time *(red line)*, subdivided by method *(lines of other colors)*. Note the strong contribution of GNSS *(dark blue)*. (The *black line* shows when the mainshocks featured in the afterslip studies in the *colored lines* occurred.) Inset: Afterslip studies per event. *(Reproduced from Churchill, R. M., Werner, M. J., Biggs, J., & Fagereng, Å. (2022). Afterslip moment scaling and variability from a global compilation of estimates.* Journal of Geophysical Research: Solid Earth, *127(4), e2021JB023897.)*

2.1 Slow slip events at subduction zones

SSEs have been identified at most subduction zones where sufficient observational capability to detect them exists. Continuous GNSS (cGNSS) networks in Japan, Cascadia, New Zealand, Costa Rica, Mexico, Ecuador, Chile, and Alaska have produced the most notable SSE observations thus far (Figs. 5.3 and 5.4). These networks have revealed a vast diversity of SSE duration (days to years), recurrence (months to decades), depth (0–50 km), and magnitude characteristics (equivalent Mw 6–7.3). Typical signatures of SSE cycles in continuous GNSS timeseries involve long periods of landward movement during the time between SSEs (the inter-SSE period), with intermittent periods where the direction of motion is reversed (typically in a trenchward direction) during the SSE (Fig. 5.1). Landward movement during the inter-SSE period is largely driven by elastic strain accumulation in the Earth's crust due to locking on the plate boundary within the SSE source area. This elastic strain is then released when slip in the SSE source occurs, thus reversing the sense of horizontal motion of cGNSS stations in the area. These cycles of inter-SSE locking and slow slip produce the characteristic "sawtooth" pattern of many GNSS timeseries impacted by SSEs (Fig. 5.1). Depending on where the GNSS stations are located relative to the SSE source, vertical deformation can also be observed. If the GNSS station is located directly above or just updip of the SSE, uplift is typically recorded; in contrast, if the GNSS station is downdip of the SSE then subsidence can result. Up to

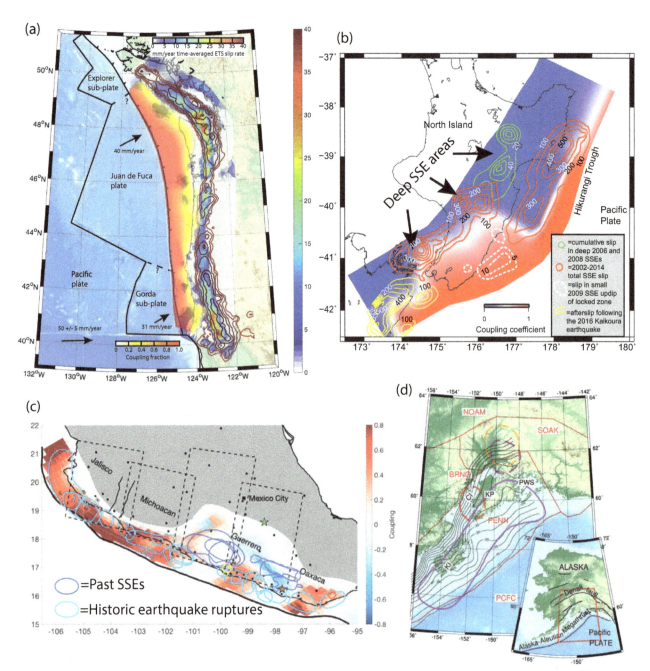

FIG. 5.3 Examples of deep slow slip events and their relationship to interseismic coupling and/or historical earthquakes. (A) Slow slip *(rainbow colors)* and interseismic coupling *(hot colors)* at the Cascadia subduction zone, (B) interseismic coupling *(red to blue)* and slow slip events *(colored contours)* at the Hikurangi subduction zone in New Zealand, (C) interseismic coupling *(red to blue colors)* and slow slip events and historic earthquakes at the Middle America subduction zone in Mexico, and (D) the location of the 1964 Alaska Mw 9.2 rupture *(purple)* and outlines of previously identified slow slip events *(red and yellow/orange dashed lines)* at the Alaska subduction zone . *(A: Reproduced from Bartlow, N. M. (2020). A long-term view of episodic tremor and slip in Cascadia.* Geophysical Research Letters, *47, e2019GL08530; B: Reproduced from Wallace, L.M. (2020), Slow slip events in New Zealand.* Annual Reviews of Earth and Planetary Sciences, *48, 175–203; C: Reproduced from Maubant, L., Radiguet, M., Pathier, E., Doin, M. P., Cotte, N., Kazachkina, E., & Kostoglodov, V. (2022). Interseismic coupling along the Mexican subduction zone seen by InSAR and GNSS.* Earth and Planetary Science Letters, *586, 117534; D: Reproduced from Li, S., Freymueller, J., & McCaffrey, R. (2016). Slow slip events and time-dependent variations in locking beneath lower cook inlet of the Alaska-Aleutian subduction zone.* Journal of Geophysical Research - Solid Earth, *121, 1060–1079, https://doi.org/10.1002/2015JB012491.)*

GNSS observations of transient deformation in plate boundary zones **Chapter | 5** **87**

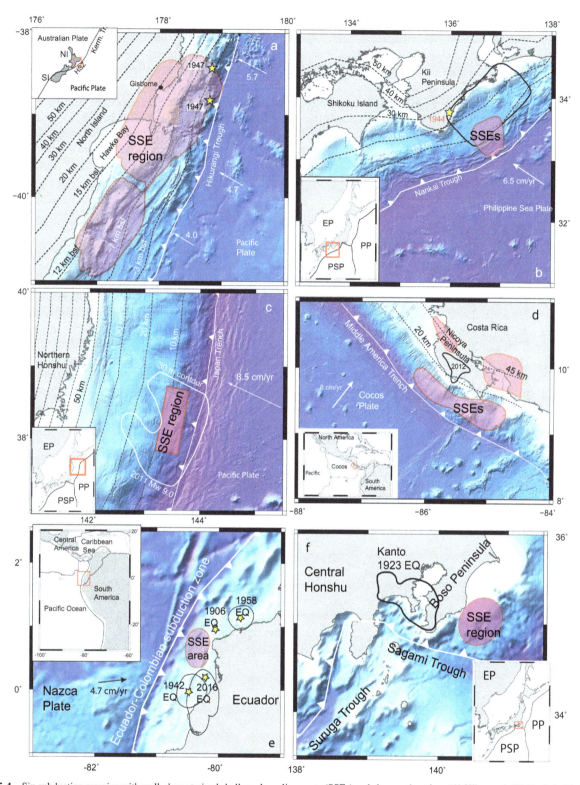

FIG. 5.4 Six subduction margins with well-characterized shallow slow slip events (SSEs) and slow earthquakes: (A) Hikurangi, (B) Nankai, (C) Japan Trench, (D) Costa Rica, (E) Ecuador, and (F) central Japan (Boso Peninsula). *Black dashed contours* are depths to the interface (in kilometers below the Earth's surface); *pink shaded areas* show the locations of SSEs (Araki et al., 2017; Dixon et al., 2014; Ito et al., 2013; Vaca et al., 2018; Wallace et al., 2012); *black/white solid contours* outline slip in previous large plate interface earthquakes at Nankai, Costa Rica, Northern Japan, Ecuador, and Boso Peninsula. *Yellow stars* show epicenters of historical large subduction interface earthquakes at Hikurangi and Nankai.

5 cm of horizontal and vertical displacements have been recorded at GNSS stations overlying the largest recorded SSEs in New Zealand and Mexico (Fig. 5.3; Radiguet et al., 2012; Wallace, 2020). Episodic tremor and slip (ETS) episodes recorded in Cascadia and Nankai (southwest Japan) produce smaller surface displacements—typically a few to several millimeters (Bartlow et al., 2011; Nishimura et al., 2013; Schmidt & Gao, 2010). Although continuous GNSS networks are conventionally used to investigate subduction SSEs, large subduction SSEs in Mexico and New Zealand have also been captured with Interferometric Synthetic Aperture Radar (InSAR; Bekaert et al., 2015; Hamling & Wallace, 2015). Crustal deformation during deep ETS episodes at Nankai was initially recognized on highly sensitive borehole tiltmeters (Obara et al., 2004); it was only later more careful analysis of the GNSS data revealed that Nankai ETS episodes were also visible in GNSS timeseries (Nishimura et al., 2013). Shallow, offshore SSEs have also been recorded using seafloor geodetic methods (Araki et al., 2017; Davis et al., 2015; Wallace et al., 2016; Yokota & Ishikawa, 2020). In the following sections, we discuss the characteristics of subduction zone SSEs, in terms of deep (>15 km depth) SSEs and shallow (<15 km depth) SSEs.

2.1.1 Deep SSEs at subduction zones

SSEs are widely observed along the downdip transition from the locked seismogenic zone to aseismic creep, typically at depths between 20 and 50 km (Figs. 5.3 and 5.5) (Bartlow, 2020; Obara & Kato, 2016; Wallace & Beavan, 2010). In many cases, this transition coincides with a thermally controlled transition from brittle to ductile behavior (Gao & Wang, 2017). Arguably, some of the best-studied examples of deep SSEs occurs at the Cascadia subduction zone (Bartlow, 2020; Bartlow et al., 2011; Dragert et al., 2001; Schmidt & Gao, 2010) and the Nankai subduction zone (Hirose & Obara, 2006, 2010; Nishimura et al., 2013), where tectonic tremor is also observed during the transient slip events (ETS events; Rogers & Dragert, 2003; Wech & Creager, 2012; Obara et al., 2004). Deep SSEs have also been identified in Mexico (Kostoglodov et al., 2003; Larson et al., 2004; Radiguet et al., 2012), New Zealand (Wallace & Beavan, 2006, 2010), southwest Japan (Hirose et al., 1999; Miyazaki et al., 2006; Ozawa, Suito, Imakiire, & Murakmi, 2007), Costa Rica (Dixon et al., 2014; Outerbridge et al., 2010), Alaska (Fu & Freymueller, 2013; Ohta et al., 2006; Wei et al., 2012), Chile (Klein et al., 2018), and the southwestern Ryukyu Trench (Chen et al., 2018). See Table 5.1 for a summary of deep subduction SSEs observed thus far.

Cascadia and Nankai ETS events typically involve 1–5 cm of slip on the plate boundary over days to weeks (equivalent Mw 6.0–7.0), and the surface displacements observed at GNSS sites located above the SSEs are on the order of a few to several millimeters (the displacements, slip, and equivalent Mw for Nankai ETS events are on the lower end of the ranges suggested above). Deep SSEs in Mexico, New Zealand, Alaska, and southwest Japan (Bungo Channel) tend to be much larger (15–30 cm slip on the plate interface, with up to 5 cm surface displacements at GNSS sites and equivalent

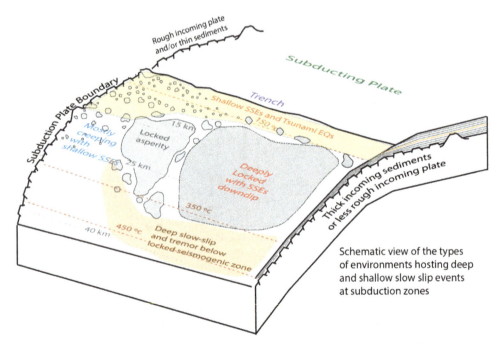

FIG. 5.5 Generalized schematic showing the types of environments that host slow slip events at subduction zones.

TABLE 5.1 Summary of locations of slow slip events (SSEs) shown in Fig. 5.1.

Fault type	Location	Depth (km)	Duration	References
Subduction zone (shallow)	Nankai Trough, SW Japan	<10	Days to several weeks	Araki et al. (2017)
Subduction zone (shallow)	Boso Peninsula, central Japan	10–20	1–3 weeks	Ozawa, Suito, and Tobita (2007)
Subduction zone (shallow)	Japan Trench, northern Japan	10–15	A few to several weeks	Ito et al. (2013)
Subduction zone (shallow)	Hikurangi SZ, New Zealand	0–15	a few to several weeks	Wallace and Beavan (2010) and Wallace (2020)
Subduction zone (shallow)	Costa Rica	0–20	A few weeks	Dixon et al. (2014), Davis et al. (2015), and Jiang et al. (2012)
Subduction zone (shallow)	Ryukyu Trench	10–20	A few to several weeks	Nishimura (2014)
Subduction zone (shallow)	Ecuador	10	A week or more	Vallee et al. (2013)
Subduction zone (deep)	Nankai Trough, SW Japan	20–50	Days (Nankai ETS) to years (Bungo Channel and Tokai long-term events)	Obara and Hirose (2006), Nishimura et al. (2013), Hirose et al. (1999), and Miyazaki et al. (2006)
Subduction zone (deep)	Cascadia Subduction Zone	20–40	Weeks to a few months	Dragert et al. (2001), Schmidt and Gao (2010), and Bartlow (2020)
Subduction zone (deep)	Hikurangi SZ, New Zealand	20–50	Months to years	Wallace and Beavan (2010) and Wallace (2020)
Subduction zone (deep)	Alaska	30–70	A few years	Ohta et al. (2006), Wei et al. (2012), and Fu and Freymueller (2013)
Subduction zone (deep)	Mexico	20–50	1–2 years	Radiguet et al. (2012)
Subduction zone (deep)	Costa Rica	30–50	A few to several months	Outerbridge et al. (2010), Dixon et al. (2014), and Jiang et al. (2012)
Subduction zone (deep)	Chile	40–60	Several months to more than a year	Klein et al. (2018)
Subduction zone (deep)	Sumatra	30–60	15 years	Tsang et al. (2015)
Subduction zone (deep)	Southwest Ryukyu Trench	30–40	A few months to a year or more	Chen et al. (2018) and Nishimura (2014)
Transform fault	San Andreas, Fault	0–10	Days to years	Linde et al. (1996), Langbein et al. (1999), and Murray and Segall (2005)
Transform fault	Superstition Hills Fault, California	0–4	Weeks?	Bilham (1989) and Wei et al. (2009)
Transform fault	Longitudinal Valley Fault, Taiwan	2–4	Days	Canitano et al. (2019)
Transform fault	North Anatolian Fault	0–3	1 month	Rousset et al. (2016)
Reverse fault	Kilauea south Flank	~8	Days	Montgomery-Brown et al. (2009)

Mw of 7.0–7.5) and are longer-lived (1–3 years), and less frequent (5–7 year recurrence) than their Cascadia and Nankai deep SSE counterparts. The longer duration of the largest SSEs, compared to shorter, smaller deep events in Cascadia and Nankai, is consistent with the notion that SSE magnitude scales with duration, similar to what is observed for earthquakes, though whether that relationship is linear (Mo $\sim$ t, Ide, Beroza, et al., 2007) or cubic (Mo $\sim$ t^3, as is the case for earthquakes) is currently debated (Frank & Brodsky, 2019; Michel et al., 2019). Gomberg et al. (2016) suggest that SSEs may exhibit both linear and cubic moment-duration scaling depending on whether the growth is unbounded (usually the case for smaller SSEs) or bounded (larger SSEs) by the updip and downdip boundaries of the region of the fault that can undergo slow slip.

Although GNSS observations are currently the primary means of detecting deep subduction SSEs, great advances in understanding the spatiotemporal evolution of these events have been made using seismological observations of tectonic tremor and low-frequency earthquakes during the SSEs (e.g., Frank et al., 2014; Wech & Creager, 2012). Shelly et al. (2006) suggest that tremor episodes are actually composed of numerous low-frequency earthquakes (LFEs), and earthquake focal mechanisms (Ide, Shelly, & Beroza, 2007) obtained from subduction zone LFEs are consistent with shear slip on the plate interface. In locations where tremor and/or low-frequency earthquakes are closely linked to SSE slip (such as Cascadia and Nankai), these seismic phenomena can be used to probe the details of SSE evolution and migration. Frank et al. (2018) suggest that LFE and tremor bursts accompanying slow slip in Mexico indicate that long-lived slow slip events might actually be composed of many shorter-lived slip events (with rapid plate locking between the LFE and/or tremor bursts). If correct, this has important implications for establishing magnitude-duration scaling relationships for SSEs (Frank & Brodsky, 2019). Tremor occurrence can also be highly sensitive to small changes in stress during tidal cycles, suggesting that megathrusts hosting tremor and slow slip are mechanically very weak (Hawthorne & Rubin, 2010; Houston, 2015).

2.1.2 Shallow SSEs at subduction zones

Although early studies of SSEs focused largely on deep SSEs (Section 2.1.1) that were easily detected by land-based continuous GNSS networks overlying the SSE source regions, it has become increasingly clear that shallow SSEs (<15 km depth) on the offshore portion of subduction plate boundaries may be just as common (Table 5.1). Onshore continuous GNSS networks in New Zealand (Douglas et al., 2005; Wallace & Beavan, 2010), central Japan (Hirose et al., 2012; Ozawa, Suito, & Tobita, 2007), Costa Rica (Dixon et al., 2014), and Ecuador (Vallee et al., 2013) revealed some of the first evidence for shallow SSEs (Fig. 5.4). At these subduction zones, the shallow (<15 km deep) portion of the plate interface is located close to shore, such that coastal geodetic stations are well positioned to record crustal deformation during these shallow events. This contrasts with many other subduction plate boundaries such as Cascadia, Nankai, Alaska, the Japan Trench, and the Peru-Chile Trench, where the shallow plate interface (<15 km) is located >50–100 km offshore, and if shallow SSEs occur in those locations they may be too far offshore to be detected by onshore geodetic networks. More recently, seafloor geodetic investigations and subseafloor International Ocean Discovery Program (IODP) observatories have revealed that shallow SSEs propagate close to the trench (Davis et al., 2015; Wallace et al., 2016). Although most shallow SSEs are observed at subduction plate boundaries that appear to be dominated by aseismic creep and/or heterogeneous interseismic coupling (Saffer & Wallace, 2015), observations of SSEs at the Nankai Trough using seafloor and subseafloor observatories indicate that they can also occur updip of locked seismogenic zones (Araki et al., 2017; Yokota & Ishikawa, 2020).

The vast majority of shallow SSEs observed thus far are short-lived (weeks to months; Ozawa, Suito, & Tobita, 2007; Hirose et al., 2012; Dixon et al., 2014; Vallee et al., 2013; Wallace et al., 2012; Araki et al., 2017; Fig. 5.4) as compared to long-lived SSEs (>1 year) observed in some deep SSE regions (Fu & Freymueller, 2013; Hirose et al., 1999; Li et al., 2016; Radiguet et al., 2012; Wallace & Beavan, 2010) (Table 5.1). The amount of slip on the plate boundary during the largest recorded shallow SSEs can reach 10–20 cm (Wallace et al., 2016), comparable to the slip magnitudes of the longer-lived deep SSEs (Fig. 5.4). The reasons behind the shorter duration of most shallow SSEs (compared to some deep events) are unclear, although this may be related to differences in effective normal stresses in the two environments (e.g., Matsuzawa et al., 2010; Shibazaki et al., 2019), and/or differences in the types of deformation mechanisms at play in deep vs. shallow environments (e.g., Behr & Burgmann, 2021; Fagereng & den Hartog, 2017).

Another commonality between most shallow SSE regions is the tendency for shallow SSEs to trigger microseismicity and earthquake swarms. In New Zealand, shallow SSEs are typically accompanied by elevated levels of Mw 1–5 earthquakes in the region of the SSE (Delahaye et al., 2009; Wallace et al., 2012; Yarce et al., 2019), and a 2016 SSE at central Hikurangi was notably accompanied by a Mw $\sim$6.0 earthquake on the subduction interface (Wallace et al., 2017). A large SSE detected with seafloor geodetic data offshore northern Japan in the weeks leading up to the 2011 Mw 9.0 Tohoku-Oki earthquake (Ito et al., 2013) was accompanied by a swarm of earthquakes up to Mw 7.3 (Kato et al., 2012). Boso Peninsula (central Japan) and Ecuador regularly experience swarms of small to moderate earthquakes during shallow SSEs (Hirose et al., 2012; Vallee et al., 2013). Shallow SSEs at the Nankai trough, Costa Rica, and offshore Kyushu are typically accompanied by tremors and low-frequency earthquakes (Araki et al., 2017; Walter et al., 2013; Yamashita et al., 2015), while some shallow Hikurangi SSEs (Todd et al., 2018) also coincide with tremor. One emerging pattern is that shallow SSEs appear to be accompanied by seismic swarms more commonly than deep SSE regions (which are more typically associated with tremor). If this pattern holds true in other regions, then it may be telling us something about the scale of frictional heterogeneities in shallow SSE regions vs. tremor-dominated deep SSE regions. Indeed, many shallow SSE regions are the site of seamount subduction and/or a geometrically rough incoming plate (Figs. 5.4 and 5.5; Saffer & Wallace, 2015; Barnes et al., 2020). Such geometric and lithological diversity may play an important role in the coexistence of shallow SSEs and seismicity.

2.2 Slow slip and creep events in other settings

Transform fault systems have been the site of the earliest robust observations of transient creep events since the 1960s in the San Juan Bautista and Parkfield segments of the San Andreas faults, largely through the use of creepmeters and borehole strainmeters (Gao et al., 2000; Langbein et al., 1999; Linde et al., 1996; Smith & Wyss, 1968). Such events have proven challenging to detect with GNSS data due to their comparatively smaller size, though there are some exceptions. A 1992 transient slip event was well documented near San Juan Bautista, California with two borehole strainmeters and two creepmeters installed near the San Andreas fault; these data indicate that the transient event lasted ~5 days, and released moment equivalent to an Mw 4.8 earthquake (Linde et al., 1996). A transient deformation episode on the Parkfield segment of the San Andreas fault lasted from 1993 to 1996 and was accompanied by repeating earthquakes (Langbein et al., 1999; Nadeau & McEvilly, 1999). More detailed spatiotemporal modeling of the Parkfield event using an Electronic Distance Meter network, strainmeters, and a creepmeter reveals significant time variability during the 1993–1996 transient event, with increases in creep rate coinciding with Mw 4.3–4.7 earthquakes (Murray & Segall, 2005). Multiple creep events have been identified on the Superstition Hills Fault (part of the San Jacinto Fault system in southern California, Bilham, 1989), including one in 2006 (equivalent to an Mw 4.7 earthquake) that was also observed using InSAR data (Wei et al., 2009, and references therein; Fig. 5.6). InSAR data has also been used to identify transient creep events on the North Anatolian fault in Turkey, involving a few centimeters of slip-releasing moment equivalent to Mw 5.2–5.5 (Rousset et al., 2016). Creepmeters installed across a creeping section of the North Anatolian fault indicate that the creep in that section is highly episodic, with many small creep events occurring at roughly 8-month intervals (Bilham et al., 2016). Changes in seismicity rate have also been used to discern transient slip events on transform faults, with the most notable examples on a highly active oceanic transform fault, the GOFAR transform (Liu et al., 2020, and references therein).

Shallow-dipping reverse faults below the south flank of Kilauea volcano host episodic SSEs at ~5 km depth, based on continuous GNSS data and seismological observations (Cervelli et al., 2002; Brooks et al., 2006; Montgomery-Brown et al., 2009, 2013; Fig. 5.6). These events are relatively short-lived (lasting a few days), involve slip of several centimeters or more with equivalent moment release to an Mw 6.0, and recur on approximately yearly intervals. Similar to shallow subduction SSEs, these events are accompanied by an increase in rates of microseismicity (Brooks et al., 2006; Montgomery-Brown et al., 2009, 2013). Transient creep events on offshore faults flanking Mount Etna volcano in Italy have also been detected using seafloor acoustic direct-path ranging techniques (Urlaub et al., 2018).

2.3 Interplay between slow slip events and earthquakes

SSEs play a major role in the accommodation of plate motion on faults aseismically, thus reducing the number of earthquakes and overall seismic hazards. However, during the time that SSEs are occurring, they can increase stressing rates on nearby faults, with the potential for triggering larger seismic events if those faults are close to failure (Obara & Kato, 2016). SSEs were observed in the weeks to months prior to the Mw 9.0 Tohoku-Oki earthquake in northern Japan (Ito et al., 2013; Kato et al., 2012), the 2014 Mw 8.1 Iquique earthquake in Chile (Ruiz et al., 2014; Socquet et al., 2017; Fig. 5.7), and the 2014 Mw 7.3 Papanoa earthquake in Mexico (Radiguet et al., 2016). It is suggested that these earthquakes were triggered by the nearby SSEs. As discussed earlier in the chapter, some SSEs are also associated with increased rates of microseismicity (Delahaye et al., 2009; Vallee et al., 2013) and even moderate earthquakes (Mw ~6; Wallace et al., 2017). Such observations raise the potential to incorporate information about ongoing SSEs into improved near-term (days to months) earthquake forecasting. However, an improved understanding of the influence of SSEs on earthquake rates is required to utilize SSEs in operational earthquake forecasts.

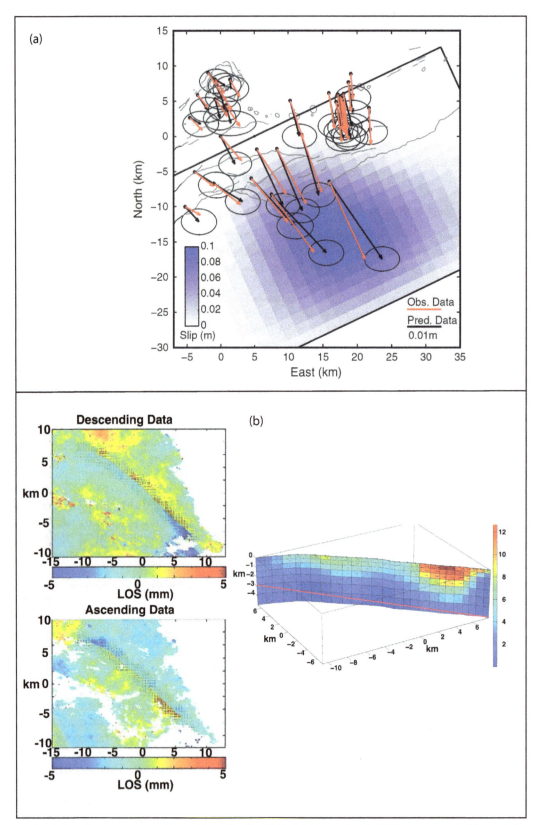

FIG. 5.6 Examples of slow slip events on crustal faults: (A) an Mw 5.9 SSE in 2010 detected with GNSS on a reverse fault beneath the south flank of Kilauea volcano and (B) an Mw 4.7 slow slip event in 2006 on the Superstition Hills Fault in southern California observed using InSAR data. *(A: Reproduced from Montgomery-Brown, E. K., Thurber, C. H., Wolfe, C. J. & Okubo, P. (2013). Slow slip and tremor search at Kilauea volcano, Hawaii.* Geochemistry, Geophysics, Geosystems, 14, *367–384, doi: 10.1002/ggge.20044.; B: From Wei, M., Sandwell, D., & Fialko, Y. (2009). A silent Mw 4.7 slip event of October 2006 on the Superstition Hills fault, southern California.* Journal of Geophysical Research, 114, *B07402, https://doi.org/10.1029/2008JB006135.)*

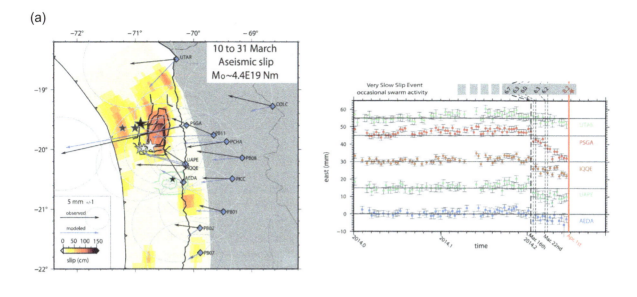

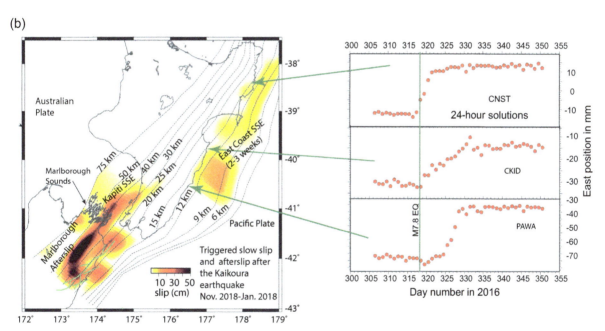

FIG. 5.7 Examples of (A) slow slip events preceding a large earthquake and (B) a large slow slip event being triggered by a large earthquake. (A) The slow slip event and foreshock sequence preceding the 2014 M8.1 Iquique, Chile earthquake. The *left panel* shows the slip distribution *(hot colors)* of the SSE determined from onshore GNSS displacements (the east component of GNSS timeseries on the right panel). *Blue arrows* on the left show observed horizontal displacements at GNSS, and *black arrows* are displacements based on the best-fitting model. (B) *Left panel* shows triggered slow slip and afterslip following the 2016 M7.8 Kaikoura earthquake *(hot colors)*, with timeseries of GNSS sites along the east coast of the North Island revealing immediate triggering of slow slip, up to 600 km away from the earthquake. *(A: Reproduced from Ruiz, S., Metois, M., Fuenzalida, A., Ruiz, J., Leyton, F., Grandin, R., Vigny, C., Madariaga, R., & Campos, J. (2014). Intense foreshocks and a slow slip event preceded the 2014 Iquique Mw 8.1 earthquake. Science, 345, 1165–1169; B: Modified from Wallace, L. M., Hreinsdóttir, S., Ellis, S., Hamling, I., D'Anastasio, E., & Denys, P. (2018). Triggered slow slip and afterslip on the southern Hikurangi subduction zone following the Kaikōura earthquake. Geophysical Research Letters, 45, https://doi.org/10.1002/2018GL077385.)*

Large earthquakes have also been observed to trigger SSEs, from tens to thousands of kilometers away. One of the most notable examples of this was the widespread slow slip triggered over most of the Hikurangi subduction zone's SSE regions by the 2016 Mw 7.8 Kaikoura earthquake (Wallace et al., 2017, 2018; Fig. 5.7). SSEs triggered more proximally to the Kaikoura earthquake (Kapiti SSE and Marlborough afterslip) were plausibly due to large static stress changes induced by the Kaikoura earthquake. The more distant triggering of SSEs offshore the North Island's northeast coast (up to 600 km away from the Kaikoura earthquake) is attributed to dynamic stress changes induced by passing seismic waves of the earthquake, which were enhanced by the presence of an ultralow velocity sedimentary wedge overlying the northern Hikurangi subduction plate boundary (Kaneko et al., 2019; Wallace et al., 2017). We note that stress-driven aseismic afterslip typically occurs following major earthquakes on portions of a fault that are proximal to the rupture and in the region of largest coseismic stress changes (see Section 3.2). Locally triggered slow slip events near the rupture area (such as that at the southern Hikurangi subduction zone following the Kaikoura earthquake; Fig. 5.7) can sometimes be difficult to distinguish from afterslip, unlike distant and regional triggered SSEs that are not proximal to the earthquake rupture, which can be more clearly categorized as triggered slow slip events. The southern Hikurangi deep triggered slow slip in the Kapiti area (Fig. 5.7A) was classified as a triggered SSE in part because it ruptured an area of previously identified recurring slow slip events (Wallace et al., 2018).

Triggered shallow slow slip events were also observed on the offshore Nankai Trough subduction zone following the 2011 Mw 9.0 Tohoku-Oki earthquake and the 2016 Mw 7.1 Kumamoto earthquake, using formation pressure sensing (as a proxy for volumetric strain) in offshore borehole observatories (Araki et al., 2017). Creep events have been triggered by local to regional earthquakes on faults in California, such as the Superstition Hills Fault (Wei et al., 2015) and the Parkfield section of the San Andreas fault (Murray & Segall, 2005). Triggering of SSEs, tremors, and low-frequency earthquakes (Gomberg et al., 2008; Kundu et al., 2016; Shelly et al., 2011) by distant earthquakes (thousands of kilometers away) is particularly remarkable, such as SSEs observed with InSAR and creepmeters on the southern San Andreas fault immediately following the 2017 Mw 8.2 earthquake in Chiapas Mexico (Tymofyeyeva et al., 2019). Together, these observations indicate that SSE source regions are mechanically weak and highly susceptible to small perturbations in stress.

2.4 The ubiquity of slow slip events

The wide range of pressure and temperature conditions, fault types, and rock types that host SSEs globally (Figs. 5.1 and 5.5) demonstrate that the physical conditions and fault properties that promote transient slow slip event behavior may in fact be relatively common. At some subduction zones, slow slip events accommodate nearly half of the overall plate boundary moment budget (Wallace & Beavan, 2010), and in many locales, they accommodate the vast majority of the plate motion on specific portions of the fault that undergoes slow slip. This makes episodic slow slip a fundamentally important mode of tectonic strain release. The observations that regional and distant earthquakes commonly trigger SSEs, and the sensitivity of tremor and slow slip episodes to tidal cycles (Hawthorne & Rubin, 2010; Houston, 2015) suggest that the faults that host SSEs are mechanically weak. This extreme mechanical weakness is thought to be due to elevated fluid pressures (and concomitant low effective normal stress) within slow slip zones (e.g., Audet et al., 2009; Kodaira et al., 2004; Saffer & Tobin, 2011) and/or the frictional properties of the rocks within the fault zone (Beeler et al., 2018; Ikari et al., 2013).

Although SSEs are most commonly observed at subduction zones, this does not necessarily mean that they are less likely to occur on other types of faults. The apparent prevalence of SSEs at subduction zones relative to other types of faults may simply be an observational bias. The gentle dip of subduction plate boundaries beneath the overriding plate (Fig. 5.5) provides an optimal situation to easily detect SSEs with surface geodetic networks overlying subduction zones. GNSS networks have to be much denser to effectively capture SSEs on more steeply dipping crustal faults. Moreover, many crustal faults have slower long-term slip rates and dip more steeply, with a smaller surface area (compared to subduction zones) available to undergo slip in SSEs. This may yield smaller and/or less-frequent SSEs in crustal faulting environments compared to subduction environments, further exacerbating the observational challenges. More widespread use of highly sensitive instruments (such as strainmeters, creepmeters, and tiltmeters), improvements in InSAR techniques, and denser continuous GNSS networks capable of sensing smaller, more localized transients on crustal faults may help to overcome these challenges and observational gaps in the future.

3 Postseismic deformation and contributions from GNSS

Postseismic deformation is accelerated aseismic deformation in the crust and mantle (Fig. 5.8) that is driven by the sudden stress changes during an earthquake (and by the evolving stress state as those coseismic stress changes are relaxed). It can include viscoelastic relaxation in ductile portions of the crust and mantle, afterslip (accelerated fault creep) around and/or within the coseismic rupture area, and poroelastic rebound and dilatancy recovery around the rupture (e.g., Bürgmann & Thatcher, 2013). Each of these processes deforms the lithosphere, producing accelerated deformation at the Earth's surface, at a rate that steadily decays over the months, years, and decades following an earthquake. This transient surface deformation can be captured in geodetic data and can be combined with modeling to resolve (albeit with some nonuniqueness) which deformation processes are operating at depth.

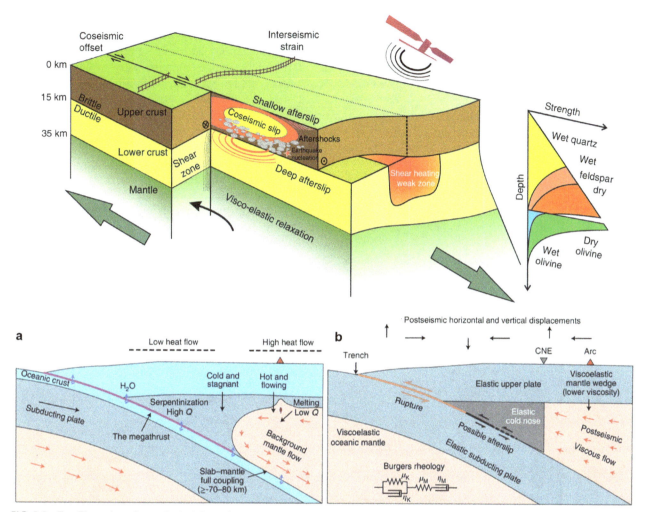

FIG. 5.8 *Top*: The variety of postseismic deformation processes that might follow a strike-slip earthquake, and alternate rheological conceptions of the effective strength of the crust and mantle modulated by composition, water content, and temperature. *Bottom*: Thermal/compositional (A) and rheological (B) elements that might control postseismic deformation in a subduction setting (Luo & Wang, 2021, Fig. 5.1). (Note that the Burgers viscoelastic rheology shown at right is frequently but not universally invoked.) *(Top: From Elliott, J. R., Walters, R. J., & Wright, T. J. (2016). The role of space-based observation in understanding and responding to active tectonics and earthquakes.* Nature Communications, *7(1), 1–16.)*

This has several uses in both fundamental science and seismic hazard assessment. First, it provides a window into the rheology of the solid earth as a function of depth, location, and other variables. Rheology dictates how plate tectonics mechanically operates, how deeply and shallowly earthquakes can rupture, and many other fundamental features of the solid earth. Some key rheology-related questions that postseismic deformation can help answer are (1) whether the strength of plates is contained more in the crust or the mantle (Fig. 5.8, top right), (2) how this varies laterally in the vicinity of faults (Bürgmann & Dresen, 2008), (3) whether faults extend into the upper mantle (as shear zones) or sole out into distributed viscous flow (Fig. 5.8, top; Thatcher, 1983), and (4) how large a role postseismic and time-dependent deformation plays in long-term strain accommodation (e.g., Hager et al., 1999; Hearn, 2022; Hussain et al., 2018). Postseismic deformation may also play a role in triggering subsequent earthquakes (e.g., Freed & Lin, 2001; Piombo et al., 2005; Tsang et al., 2016) and aftershocks (e.g., Gualandi et al., 2020; He & Peltzer, 2008; Hsu et al., 2006; Huang et al., 2017; Peng & Zhao, 2009; Perfettini & Avouac, 2004; Ross et al., 2017), and it can drive lasting coastal uplift and subsidence (e.g., Goto et al., 2021; Sawai et al., 2004; Suito, 2018). These impacts make effective monitoring of postseismic deformation all the more essential.

GNSS geodetic data help provide this monitoring, and their advent in the 1990s led to a flowering in the study of postseismic deformation (Fig. 5.2, bottom). This success is in part because daily GNSS position data are frequent and precise enough to track postseismic deformation throughout its evolution, in three dimensions. Continuous GNSS timeseries have revealed that, in both space and time, postseismic deformation often consists of both localized deformation lasting months to years (near the earthquake), and long-wavelength (hundreds of kilometers), long-lived deformation (e.g., Freed, Bürgmann, Calais, Freymueller, & Hreinsdóttir, 2006). This spatiotemporal complexity is often difficult to explain with a single deformation process or timescale (e.g., Freed et al., 2010, 2012) and instead is often attributed to one or more processes (Fig. 5.8, top) with multiple timescales (e.g., Pollitz, 2003). Such hypotheses can be tested by comparing predicted displacement timeseries from physics-driven models to the data (e.g., Barbot et al., 2009; Barbot & Fialko, 2010; Freed et al., 2007; Freed & Bürgmann, 2004; Freed, Bürgmann, Calais, & Freymueller, 2006; Freed, Bürgmann, Calais, Freymueller, & Hreinsdóttir, 2006; Masuti et al., 2016) or by inverting the data timeseries or cumulative displacements for kinematic and rheological parameters (e.g., Hines & Hetland, 2016a, 2016b; Moore et al., 2017; Pollitz, 2015; Pollitz et al., 2012; Qiu et al., 2018, 2019; Weiss et al., 2020). Importantly, for these tests, daily GNSS timeseries also show how postseismic deformation is superimposed on the signals of interseismic, seasonal, coseismic, and anthropogenic surface deformation (Fig. 5.9). These signals affect all high-precision geodetic data, but in temporally sparse data their effects can be difficult to estimate. Given sufficiently long continuous GNSS timeseries (e.g., Blewitt & Lavallée, 2002), the postseismic signals can be relatively cleanly extracted from the timeseries via least-squares curve fitting or other methods (e.g., Bevis & Brown, 2014; Gualandi et al., 2017; Liu-Zeng et al., 2020; Fig. 5.9), which reduces biases when using them to evaluate models.

Another research-friendly attribute of GNSS (and other space geodetic) data is the open availability of datasets. This is fostered by the many organizations that operate networks and process and publish GNSS data (Nevada Geodetic Laboratory, SOPAC, NASA MeaSUREs, UNAVCO/Earthscope Consortium, New Zealand's GeoNet). This availability means that a single postseismic transient (if monitored by publicly available continuous GNSS data) can be studied by a great

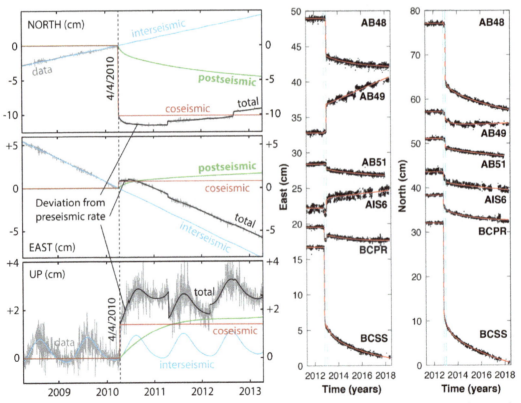

FIG. 5.9 Signals of postseismic deformation within daily GNSS timeseries. *Left panel*: Three-component position timeseries at station P497 in the Imperial Valley, California, spanning the 2010 El Mayor-Cucapah earthquake *(vertical dashed black lines)*. *Gray*: total timeseries (obtained from SOPAC), *black*: fit to total timeseries, *blue*: background interseismic and seasonal deformation, *red*: coseismic deformation in the El Mayor-Cucapah earthquake, and *green*: postseismic deformation signal. *Right panel*: Timeseries of east and north components of position at continuous GNSS stations in the vicinity of the 2012 Haida Gwaii and 2013 Craig earthquakes *(vertical dashed cyan lines)*. *Black dots*: data and *red*: cumulative fit. Note that the timeseries at the right are detrended (background interseismic and seasonal terms are removed) in the plots. *(Left panel: Adapted from Rollins, C., Barbot, S., & Avouac, J. P. (2015). Postseismic deformation following the 2010 M= 7.2 El Mayor-Cucapah earthquake: Observations, kinematic inversions, and dynamic models.* Pure and Applied Geophysics, 172(5), 1305–1358.; *Right panel: Adapted from Fig. 3 of Guns, K. A., Pollitz, F. F., Lay, T., & Yue, H. (2021). Exploring GPS observations of postseismic deformation following the 2012 MW7.8 Haida Gwaii and 2013 MW7.5 Craig, Alaska earthquakes: Implications for viscoelastic earth structure.* Journal of Geophysical Research: Solid Earth, 126(7), e2021JB021891.)

number of researchers with varying approaches and perspectives. It also improves reproducibility and comparability, which is crucial as the interpretation of deformation processes at depth can often be nonunique and model-dependent (e.g., Montési, 2004; Savage, 1990).

A third benefit of GNSS monitoring is that it can robustly detect millimeter-level postseismic deformation signals hundreds to thousands of kilometers from a mainshock (Fig. 5.10, top), which often provide crucial constraints on viscoelastic relaxation at depth (e.g., Freed et al., 2007; Hines & Hetland, 2016a, 2016b; Park et al., 2023; Trubienko et al., 2013, 2014;

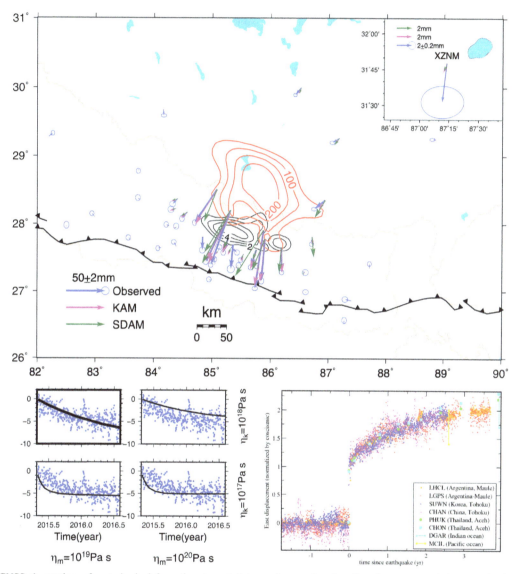

FIG. 5.10 GNSS observations of postseismic deformation at great distances from earthquakes. *Top*: first-year cumulative postseismic displacements *(blue arrows)* at GNSS stations surrounding the 2015 Gorkha (Nepal) earthquake. Inset shows cumulative displacement at station XZNM, located >500 km NNE of the earthquake (off the main map). *Pink* and *green* vectors show predictions from a kinematic afterslip model (KAM) and a stress-driven afterslip model (SDAM), neither of which reproduces the observed displacement at XZNM. *Bottom left*: the extracted postseismic displacement timeseries at XZNM *(blue dots)* can be reproduced by models of viscoelastic relaxation *(black lines)*, used to constrain the ductile rheology of the Tibetan lithosphere (frame of preferred model is bolded). *Bottom right*: postseismic deformation at far-field stations following the 2010 Maule, 2011 Tohoku-oki, and 2004 Sumatra-Andaman earthquakes. All timeseries are normalized by the coseismic displacement at each station. *(Top: Fig. 8a from Zhao, B., Bürgmann, R., Wang, D., Tan, K., Du, R., & Zhang, R. (2017). Dominant controls of downdip afterslip and viscous relaxation on the postseismic displacements following the Mw7.9 Gorkha, Nepal, earthquake. Journal of Geophysical Research: Solid Earth, 122(10), 8376–8401; Bottom left: Extract from Fig. 10 of Zhao, B., Bürgmann, R., Wang, D., Tan, K., Du, R., & Zhang, R. (2017). Dominant controls of downdip afterslip and viscous relaxation on the postseismic displacements following the Mw7.9 Gorkha, Nepal, earthquake. Journal of Geophysical Research: Solid Earth, 122(10), 8376–8401; Bottom right: Fig. 5 from Trubienko, O., Garaud, J. D., & Fleitout, L. (2014). Models of postseismic deformation after megaearthquakes: The role of various rheological and geometrical parameters of the subduction zone. Solid Earth Discussions, 6(1), 427–466.)*

Zhao et al., 2017). Because GNSS is a point measurement, it can detect these signals regardless of the terrain between the mainshock and the station (Fig. 5.10, top). By contrast, long-wavelength signals can be difficult to identify using Interferometric Synthetic Aperture Radar (InSAR) because InSAR interferograms often contain bilinear and long-wavelength artifacts introduced in processing, and because water, mountains, vegetation, or other decorrelating terrain features can compromise large sections of interferograms. GNSS' unique ability to robustly capture these far-field signals (and accurately measure displacements in three dimensions) helps to secure its important role in postseismic (and other tectonic) deformation research.

In recent years, GNSS data have found a new utility in quantitatively orienting and ground-truthing InSAR data as the latter have become increasingly high-rate, reliable, and accessible. The two make a natural pairing due to their complementary strengths. InSAR provides high-resolution, (nominally) continuous spatial coverage of the Earth's surface, but generally detects only two components of deformation (two satellite look directions), has lower temporal resolution (one to multiple weeks), and is a relative measurement (deformation is relative to a reference SAR image and a chosen reference point within each frame). GNSS data are typically much sparser spatially but temporally denser (seconds to days), and unambiguously three-dimensional positions can be derived within a known terrestrial reference frame (e.g., Altamimi et al., 2016). Therefore, GNSS and InSAR can be combined to resolve aspects of surface deformation that would be challenging for either individual dataset to resolve, and this has fostered many postseismic studies that have combined the two (e.g., Bruhat et al., 2011; Guns et al., 2021; Hong & Liu, 2021; Huang et al., 2014, 2016; Wang & Fialko, 2018).

Excellent overviews of postseismic deformation research are provided by Bürgmann and Thatcher (2013) (who also cover the contributions of GNSS to tectonic geodesy), Bürgmann and Dresen (2008), Freymueller (2017), Bock and Melgar (2016), Blewitt (2015), Segall and Davis (1997), Scholz (2019), Thatcher and Pollitz (2008), Handy et al. (2007), Watts et al. (2013), Wright et al. (2013), and Feigl and Thatcher (2006), among others. Pollitz (2019) and Churchill et al. (2022) also provide global overviews of studies of viscoelastic relaxation and afterslip, respectively, and Segall (2010) provides excellent quantitative coverage of poroelastic effects (and postseismic deformation in general). Here, we briefly review the commonly recognized postseismic processes (Fig. 5.8) and discuss how GNSS and other data can help distinguish between them.

3.1 Viscoelastic relaxation

Postseismic viscoelastic relaxation is solid-state bulk creep in crustal and mantle rocks (movement of molecules and grains), which acts to gradually relax the stress changes imparted by an earthquake (e.g., Freed, Bürgmann, Calais, & Freymueller, 2006; Freed, Bürgmann, Calais, Freymueller, & Hreinsdóttir, 2006; Nur & Mavko, 1974; Pollitz, 2019; Pollitz et al., 2001). It can take several micromechanical forms. One is diffusion creep, which has a linear (Newtonian) rheology in which the viscosity is independent of the stress state and constant through time. Another is dislocation creep, which has a nonlinear (non-Newtonian) rheology in which the strain rate is proportional to the stress raised to a power (often 3–3.5, e.g., Freed & Bürgmann, 2004; Watts et al., 2013). With this and other nonlinear rheologies, the effective viscosity will be lowest just after the earthquake and progressively increase over the postseismic period. This increase will be substantial if the coseismic stress change dwarfs the background stress, resulting in viscoelastic relaxation that is initially rapid but quickly decays to a lower rate (e.g., Freed & Bürgmann, 2004). If the coseismic stress change is much smaller than the background stress, the viscosity will not change as much and the response will be quasi-Newtonian. These processes may also be progressively modulated by strain hardening, which may result in postseismic deformation with a time evolution like that produced by a Burgers rheology (e.g., Freed et al., 2012; Masuti et al., 2016, among others). They are also strongly influenced by temperature, grain size, water content, and other variables.

An aspect of viscoelastic relaxation that can help identify it in postseismic data is that it tends to produce the longest-wavelength surface deformation of all of the commonly invoked postseismic mechanisms. This is for two reasons. First, a deeper deformation source produces a proportionally broader surface deformation pattern (e.g., Segall, 2010), and viscoelastic relaxation is generally the deepest postseismic process invoked, especially if the mantle is considered (e.g., Pollitz, 2003; Pollitz et al., 2001). Viscous processes become important at depth, as increasing temperature with depth leads to a downward transition from brittle to ductile deformation mechanisms in silicate rocks (Fig. 5.8, top right; e.g., Blanpied et al., 1991). Brittle deformation is most plausible *above* the brittle-ductile transition, whereas viscoelastic relaxation dominates *below* this transition, and is effectively unbounded as to how deep it can operate. Second, viscoelastic relaxation is distributed rather than localized, which also broadens its surface displacement pattern. Because of this broad nature, viscoelastic relaxation is usually the best candidate to explain any long-wavelength postseismic deformation observed in data (e.g., Freed et al., 2007; Zhao et al., 2017, etc.). As such, GNSS geodetic data's unique ability to capture this long-wavelength deformation with high precision means that it is an important source of information to constrain viscoelastic relaxation and the viscosity structure at depth (e.g., Park et al., 2023).

Viscoelastic relaxation is also considered to be the longest duration postseismic deformation mechanism. This is because the effective viscosities that commonly emerge from studies of glacial isostatic adjustment or plate flexure imply characteristic relaxation timescales of hundreds to thousands of years (e.g., Watts et al., 2013), implying that postseismic viscoelastic relaxation should (to some extent) continue for similarly long timescales. This is in contrast to afterslip and poroelastic rebound, which typically span shorter timescales (months to years). Therefore, viscoelastic relaxation is also the natural candidate to explain any observed long-lived postseismic deformation (e.g., decades or longer). Evidence for viscoelastic relaxation lasting several decades has been found in Chile (Wang et al., 2012), Alaska (Huang et al., 2020), Tibet (Ryder et al., 2014), southwest Japan (Johnson & Tebo, 2018), and California (Hearn, 2022), and so it may be important to consider it at all stages of the earthquake cycle (e.g., Li et al., 2015, 2018; Pollitz & Evans, 2017).

Viscoelastic relaxation can also be distinguished from other postseismic deformation processes in ways that depend on the tectonic setting and earthquake source characteristics. In a strike-slip setting, mantle viscoelastic relaxation and afterslip should produce opposite patterns of vertical displacement around the mainshock: quadrants that undergo uplift by afterslip undergo subsidence by deep viscoelastic relaxation, and vice versa (e.g., Pollitz et al., 2001). This has helped to distinguish between the two mechanisms as the vertical component of GNSS geodetic measurements has become better constrained in recent years (e.g., Rollins et al., 2015). A similar opposing polarity in co- and postseismic vertical displacements is expected in a normal-faulting setting (e.g., Pousse-Beltran et al., 2020). In a subduction setting, afterslip on the subduction interface and viscoelastic relaxation in the overriding mantle wedge should move the surface (i.e., the GNSS stations) seaward, while viscoelastic relaxation in the mantle of the downgoing plate should move the surface landward. Therefore, landward postseismic motion observed in data is typically interpreted as an indicator of lower-plate viscoelastic relaxation (e.g., Sun et al., 2014; Fig. 5.11). Patterns of vertical displacements can also be used to distinguish between viscoelastic relaxation and afterslip and shed light on the viscosity structure at subduction zones (e.g., Hu et al., 2016; Luo & Wang, 2021; Fig. 5.8, bottom).

3.2 Afterslip

Afterslip is accelerated postseismic fault creep that occurs around and/or within the rupture zone of an earthquake on sections of fault that are capable of slipping stably and aseismically. Like viscoelastic relaxation, afterslip is (initially) driven by the shear stress changes in an earthquake and acts to relax them. The key difference (formally) is that viscoelastic relaxation is a bulk process and is described by solid-state creep rheologies, whereas afterslip occurs on discrete faults or shear zones and is commonly described by rate and state friction laws (e.g., Dieterich, 1979; Ruina, 1983). Under this framework, large stress perturbations induced by unstable seismic slip yield a time evolution of afterslip in nearby stable (velocity strengthening) regions that approximately follows the logarithm of time since the mainshock (Marone et al., 1991; Perfettini & Avouac, 2004).

Laboratory experiments on granite suggest that stable sliding is favored at temperatures of ~ 350–$600°C$, while unstable (seismic) slip is favored at ~ 100–$350°C$ (Blanpied et al., 1991). This would imply a simple stratification in which afterslip (and interseismic creep) predominantly occur below the seismogenic zone (e.g., Thatcher, 1983), a model that has been borne out following earthquakes such as the 2015 Gorkha (Nepal) event (e.g., Gualandi et al., 2017). However, shallow afterslip has also been observed or inferred following numerous earthquakes (e.g., Bucknam et al., 1978; Scholz et al., 1969; Smith & Wyss, 1968; Wang & Bürgmann, 2020), showing that the shallow portions of many faults are also capable of stable slip, which is likely related to frictional properties of some rock types at shallow depths and/or the presence of high fluid pressures in some settings (such as subduction zones). Moreover, episodic slow slip events are also commonly observed on the shallow portions of faults (see Section 2) often coinciding with regions of high fluid pressure (Saffer & Wallace, 2015). At the same time, events such as the 1999 Mw 7.6 Chi-Chi and 2011 Mw 9 Tohoku-oki earthquakes demonstrate that the shallow sections of faults can also host very large slip in earthquakes (Fujiwara et al., 2011; Johnson et al., 2001). Observations of overlapping coseismic slip and afterslip inferred for several earthquake sequences (e.g., Perfettini et al., 2010) may indicate large regions of conditional frictional stability that are capable of seismic slip when they are hit by large stress changes during dynamic rupture (e.g., Noda & Lapusta, 2013), and aseismic afterslip during the more slowly evolving postseismic phase. Given some of the potential similarities between frictional properties on portions of faults undergoing afterslip and episodic slow slip events (SSEs are often thought to occur in regions of conditional frictional stability), slow slip event source areas (see Section 2) may also be prime candidates for hosting afterslip. We note that while conditional frictional stability (slightly velocity-weakening) is typically invoked for faults hosting slow slip events (e.g., Liu & Rice, 2007), and velocity-strengthening friction is commonly invoked for afterslip (e.g., Marone et al., 1991),

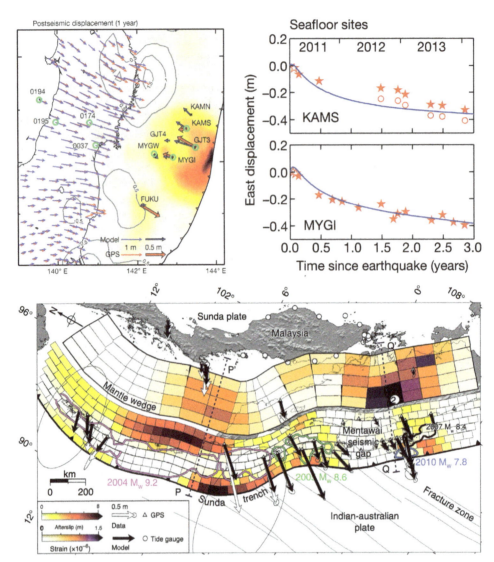

FIG. 5.11 GNSS observations of postseismic displacement and benefits to the understanding of regional rheology. *Top*: postseismic deformation *(left)* and seafloor geodetic timeseries *(right)* following the 2011 Tohoku-oki earthquake, including observations at offshore GPS-Acoustic stations *(arrows on the right)*, which imaged landward motion that can be explained by viscoelastic relaxation in the mantle beneath the slab. *Bottom*: postseismic deformation (afterslip and viscoelastic deformation) of recent subduction earthquakes along the Sunda trench, which has been in part constrained by GNSS stations on barrier islands in the Sumatran Global Positioning System (GPS) Array (SuGAr). *White* and *black arrows* are observed and modeled postseismic displacements; *red/yellow colored* patches are preferred afterslip model; and *orange/purple patches* are preferred viscoelastic model, colored by total strain in subsurface octahedral volumes. *(Top: Fig. 1b from Sun, T., Wang, K., Iinuma, T., Hino, R., He, J., Fujimoto, H., ... & Hu, Y. (2014). Prevalence of viscoelastic relaxation after the 2011 Tohoku-oki earthquake. Nature, 514(7520), 84–87; Bottom: Fig. 3a from Qiu, Q., Moore, J. D., Barbot, S., Feng, L., & Hill, E. M. (2018). Transient rheology of the Sumatran mantle wedge revealed by a decade of great earthquakes. Nature Communications, 9(1), 1–13.)*

unstable slip (including SSEs) can be reproduced numerically with rate-strengthening friction when subjected to external stress perturbations (Perfettini & Ampuero, 2008), such as temporal changes in normal stress related to fluid cycling and pore pressure changes (Perez-Silva et al., 2023).

Afterslip is usually limited to depths shallower than viscoelastic relaxation and so produces comparatively shorter-wavelength surface deformation. Physically, this is because afterslip is initially and primarily driven by coseismic stress changes, which decay quickly with distance from coseismic slip. The surface deformation from afterslip is also shorter-wavelength because it is a localized rather than distributed process. Longer-wavelength deformation can sometimes be fit kinematically to afterslip models, but often only by invoking slip at great depths, which in some cases may be difficult to explain frictionally (e.g., Huang et al., 2014; Liu-Zeng et al., 2020; Rollins et al., 2015).

Surface-breaching or shallow afterslip can often be unambiguously diagnosed in geodetic data as a sharp shear deformational feature (e.g., Jiang et al., 2021; Zhou et al., 2018). However, mid-to-lower-crustal afterslip has more overlap with other mechanisms and is more difficult to unambiguously identify and diagnose. One person's interpretation of lower-crustal afterslip can be another person's interpretation of lower-crustal viscoelastic relaxation, particularly if a power-law rheology is invoked (e.g., Montési, 2004). The stress concentrations at the edges of coseismic rupture models can also pose challenges in stress-driven afterslip modeling. Therefore, individual afterslip models come with many caveats.

Some of these trade-offs can be better understood by synoptically examining the collective body of afterslip models that have been produced for earthquakes around the world in the past three decades. A recent survey (Churchill et al., 2022) found a strong positive correlation between the total moment release in the afterslip and that in the corresponding mainshock, which makes sense as larger mainshocks impart large stress changes over a larger region. The ratio of afterslip to coseismic moment is modestly negatively correlated with rupture aspect ratio: longer ruptures appear to have proportionally less afterslip compared to coseismic moment. This ratio is modestly positively correlated with the local fault slip rate, as faster-slipping and more mature faults may more effectively dominate the local strain accumulation pattern and stress-shadow other faults. Superimposed upon these correlations are large variations in afterslip characteristics among individual earthquakes. For example, the ratio of afterslip moment to coseismic moment ranges from $<1\%$ for the 2008 Mw 7.2 Altai and Mw 8.0 Sichuan earthquakes to $\sim300\%$ for the 2004 Parkfield earthquake. A considerable number of the models studied included overlapping afterslip and coseismic slip, which supports the potential for conditionally stable fault rheologies.

That afterslip can occur at the surface introduces uncertainty into paleoseismological interpretations of past slip events on faults, as it is likely that single-event displacements resolved by paleoearthquake studies are in some cases an aggregate of coseismic and postseismic displacements. A potential research direction going forward is to examine how often surface-breaching afterslip occurs, whether the typical amount of surface afterslip is significant compared to coseismic surface slip, and whether approaches can be developed that can distinguish seismic slip vs. aseismic afterslip in studies of paleoearthquakes (e.g., Cashman et al., 2007; Toké et al., 2011).

3.3 Poroelastic rebound

Coseismic deformation during an earthquake causes some areas of the crust to expand, and other areas to contract, depending on the slip distribution in the earthquake. This crustal strain can lead to poroelastic deformation (or "rebound") in the weeks following the earthquake, produced by the migration of fluid away from regions that underwent contraction in the earthquake (where fluid pressure increased) toward areas that underwent dilatation (where fluid pressure decreased) (Peltzer et al., 1996, 1998). Typically, poroelastic rebound produces subsidence and radial displacement away from the region that underwent coseismic contraction (e.g., Hu et al., 2014; Peltzer et al., 1996). Only a handful of studies have attempted to resolve poroelastic deformation following major earthquakes (Fielding et al., 2009; Hu et al., 2014; Jónsson et al., 2003; McCormack et al., 2020; Peltzer et al., 1996, 1998; Peña et al., 2022; Yang et al., 2022), in part due to the smaller (and shorter-duration) surface deformation response compared to afterslip and viscoelastic deformation. Most published studies estimate the contribution of poroelastic rebound by taking the difference between the predicted coseismic surface deformation using a Poisson's ratio consistent with an undrained state, and one using a Poisson's ratio that reflects a return to hydrostatic equilibrium (e.g., a drained state; Peltzer et al., 1998; Jónsson et al., 2003), but a few studies explore the time dependence of deformation using poroelastic modeling (e.g., McCormack et al., 2020). Similar to afterslip and viscoelastic deformation, poroelastic rebound evolves over time; however, the timescales for poroelastic rebound are typically shorter—lasting a few months or less.

The 1992 Landers earthquake provided one of the first opportunities to document poroelastic rebound with geodesy (in this case InSAR; Peltzer et al., 1996, 1998). Following the Landers earthquake, the authors invoked poroelastic rebound to explain postseismic subsidence observed within a restraining bend (that experienced coseismic uplift) of the fault system that ruptured, and postseismic uplift within a releasing bend (that subsided coseismically). Since then, the poroelastic rebound has been inferred following strike-slip (Fielding et al., 2009; Jónsson et al., 2003), normal fault (He & Peltzer, 2008), and subduction megathrust earthquakes (Hu et al., 2014; McCormack et al., 2020; Peña et al., 2022; Yang et al., 2022). In addition to a conventional poroelastic rebound response, Fielding et al. (2009) used InSAR data from the 2003 Bam earthquake to demonstrate that shallow fault zone dilatancy recovery also strongly influenced the near-fault postseismic deformation signal. Modeling of postseismic deformation following the 2011 Mw 9.0 Tohoku earthquake suggests that poroelastic rebound may have produced a few to tens of centimeters of horizontal and vertical surface displacement offshore and onshore (Hu et al., 2014). Following the 2014 Mw 8.3 Illapel earthquake and the 2010 Mw

8.8 Maule earthquake in Chile, Yang et al. (2022) and Peña et al. (2022) used models fitting GNSS data to demonstrate that accounting for poroelastic rebound strongly influences the resulting afterslip distributions at subduction zones, suggesting that poroelastic rebound should be accounted for to accurately resolve the source of other forms of postseismic deformation.

3.4 Toward a holistic understanding of postseismic deformation

Recent work points the way toward models of postseismic deformation that more fully account for the uncertainties and trade-offs in discriminating various deformation processes at depth using surface data. One key step in this progress has been the derivation of closed-form Green's functions describing the surface deformation due to viscoelastic relaxation in subsurface 2D and 3D volumes (Barbot et al., 2017; Hines & Hetland, 2016a, 2016b; Lambert & Barbot, 2016). These Green's functions can be used in conjunction with Green's functions corresponding to fault afterslip (e.g., Okada, 1985; Wang et al., 2003), making it possible to jointly estimate the distribution of afterslip and effective viscosity from geodetic data in an inversion framework (e.g., Moore et al., 2017; Qiu et al., 2018, 2019; Fig. 5.11B). This is complemented by methods that enable searching of the parameter space within complex stress-driven forward models in a computationally efficient way (e.g., Agata et al., 2019; Masuti et al., 2016). These and other advances will allow us to compare and distinguish between candidate deformation mechanisms in a systematic and holistic way. This will ultimately help to resolve some of the questions posed at the beginning of this section, particularly those that relate to whether deformation below the seismogenic zone is localized (e.g., afterslip), distributed (e.g., viscoelastic relaxation), or somewhere in between (e.g., shear zones), and whether the relevant viscoelastic rheology is linear, bilinear, power law, or a combination of all of these.

Another frontier is to extend the "visible spectrum" of postseismic deformation, in both space and time, spanning from meters to hundreds of kilometers from the fault rupture, and from seconds (e.g., Premus et al., 2022; Twardzik et al., 2019) to many decades (e.g., Hearn, 2022; Hearn et al., 2013; Hearn & Thatcher, 2015; Pollitz et al., 2008; Wang et al., 2012) following an earthquake. One relevant question is whether deformation rates really ever "return" to a "background interseismic rate" (Hussain et al., 2018) or, conversely, how much of what we infer as interseismic is actually "spun-up" postseismic deformation from previous earthquakes and earthquake cycles (e.g., Liu et al., 2021). This suggests that interseismic and postseismic deformation should be considered jointly to infer earthquake cycle deformation parameters (Hearn & Thatcher, 2015). Sustained observations (for decades to centuries) of crustal deformation throughout multiple earthquake cycles are needed to address this. One potential simplifying factor is that long-lived postseismic deformation should probably mostly be viscoelastic relaxation, rather than afterslip or poroelastic rebound (which are shorter-lived), so it can be accounted for with less complexity and nonuniqueness than shorter-term (e.g., first-year) postseismic deformation. Indeed, long-lasting viscoelastic deformation models have already been incorporated into estimates of interseismic deformation (e.g., Hearn, 2022; Hearn et al., 2013; Hearn & Thatcher, 2015; Li et al., 2015; Pollitz et al., 2008; Pollitz & Evans, 2017; Wang et al., 2012).

If, where, and when lithospheric and fault rheology can be adequately constrained—with appropriate uncertainties and covariances conveyed—it should then be possible to incorporate postseismic deformation into earthquake forecasts and time-dependent hazard models, both as a potential consideration when interpreting geodetic strain rates and as a potential driver of seismicity in its own right.

4 Conclusions and future directions

GNSS geodetic data have been instrumental in revealing the diverse array of crustal deformation processes that occur in plate boundary zones, spanning from episodic slow slip events during the interseismic period, to large changes in deformation rates that occur following major earthquakes. These discoveries have revealed the diverse ways in which plate boundary strain is accommodated along faults, and throughout the solid Earth. Although GNSS techniques have precipitated a revolution in our understanding of transient crustal deformation throughout the earthquake cycle, there is still much work to be done to fully reveal and understand these processes and their implications for seismic and tsunami hazards.

Major observational gaps still exist, particularly in terms of our ability to detect episodic slow slip events in crustal faulting environments where SSEs may be smaller and/or less frequent, and in offshore regions where technological challenges must be overcome. More widespread installation of dense continuously operating GNSS networks, in a range of onshore plate boundary environments will yield a clearer understanding of the diversity of tectonic environments that host SSEs; such networks will also be well-positioned to accurately capture postseismic deformation processes of future earthquakes. Observational gaps must be filled to enable the detection of smaller, more frequent transient deformation signals that are difficult to observe using more conventional surface geodetic methods (GNSS, InSAR). Sensitive borehole instrumentation (tiltmeters, strainmeters), both onshore and offshore, holds great promise for bridging the gap between geodetic and seismological observations of transient deformation, particularly in the case of episodic slow slip events. Emerging seafloor geodetic techniques (including GNSS-Acoustic techniques) are already enabling the detection of vertical and horizontal crustal deformation during slow slip events at offshore plate boundaries (Araki et al., 2017; Davis et al., 2015; Wallace et al., 2016; Yokota & Ishikawa, 2020), and postseismic deformation in the years following major earthquakes (Brooks et al., 2023; Iinuma et al., 2016; Sun et al., 2014; Watanabe et al., 2021), although such methods have only been applied in a small handful of locations thus far. Developing more cost-effective, widely usable, and lower-noise approaches to measure offshore crustal deformation will ultimately offer greater scope to expand our understanding of transient plate boundary deformation beyond the shoreline.

Slow slip events and postseismic deformation induce changes in stress on nearby faults, and in some cases may influence the likelihood of future large earthquakes. The potential societal value of GNSS geodetic techniques in detecting and quantifying transient deformation and post-earthquake deformation processes will ultimately be realized through the incorporation of time-varying deformation processes into operational earthquake forecasts and probabilistic seismic and tsunami hazard models. However, this requires an improved understanding of the spatiotemporal relationship between slow slip events, postseismic deformation, stressing rates, and seismicity rates. This relationship can be revealed through sustained and expanded investigation of transient deformation and related seismicity at plate boundaries worldwide.

In the same way that GNSS geodetic techniques have spurred on a new age in our ability to detect transient deformation of the Earth's crust, InSAR remote-sensing techniques are making immense advances in increasing the spatial resolution of crustal deformation processes. Although InSAR is providing significant advancements in spatial resolution, it currently lacks the dense (seconds to days) temporal coverage and accurate three-dimensional positioning (relative to a known terrestrial reference frame) provided by continuous GNSS. These aspects of GNSS are instrumental in detecting and characterizing the full spatiotemporal evolution of SSEs and postseismic deformation processes. Moreover, unlike GNSS, InSAR has greater difficulty resolving longer wavelength deformation signals (>100km), which can occur as a consequence of viscoelastic deformation. Ideally, InSAR and GNSS techniques should be used in tandem to maximize our understanding of transient deformation processes (e.g., Guns et al., 2021; Hong & Liu, 2021; Huang et al., 2014, 2016; Wang & Fialko, 2018). Alongside and in conjunction with InSAR and other geodetic techniques, GNSS will continue to play a major role in the investigation of transient deformation processes at plate boundary zones going well into the future.

References

Agata, R., Barbot, S. D., Fujita, K., Hyodo, M., Iinuma, T., Nakata, R., ... Hori, T. (2019). Rapid mantle flow with power-law creep explains deformation after the 2011 Tohoku mega-quake. *Nature Communications, 10*(1), 1385.

Allmann, B. P., & Shearer, P. M. (2009). Global variations of stress drop for moderate to large earthquakes. *Journal of Geophysical Research, 114*, B01310. https://doi.org/10.1029/2008JB005821.

Altamimi, Z., Rebischung, P., Métivier, L., & Collilieux, X. (2016). ITRF2014: A new release of the International Terrestrial Reference Frame modeling nonlinear station motions. *Journal of Geophysical Research, 121*(8), 6109–6131.

Araki, E., Saffer, D. M., Kopf, A., Wallace, L. M., Kimura, T., Machida, Y., & Ide, S. (2017). Recurring and triggered slow slip events near the trench at the Nankai trough subduction megathrust. *Science, 356*, 1157–1160. https://doi.org/10.1126/science.aan3120.

Audet, P., Bostock, M. G., Christensen, N. I., & Peacock, S. M. (2009). Seismic evidence for overpressured subducted oceanic crust and megathrust fault sealing. *Nature, 457*, 76–78.

Avouac, J.-P. (2015). From geodetic imaging of the seismic and aseismic fault slip to dynamic modeling of the seismic cycle. *Annual Reviews of Earth and Planetary Sciences, 43*, 233–271.

Barbot, S., & Fialko, Y. (2010). A unified continuum representation of post-seismic relaxation mechanisms: Semi-analytic models of afterslip, poroelastic rebound and viscoelastic flow. *Geophysical Journal International, 182*(3), 1124–1140.

Barbot, S., Fialko, Y., & Bock, Y. (2009). Postseismic deformation due to the Mw 6.0 2004 Parkfield earthquake: Stress-driven creep on a fault with spatially variable rate-and-state friction parameters. *Journal of Geophysical Research: Solid Earth, 114*(B7).

Barbot, S., Moore, J. D., & Lambert, V. (2017). Displacement and stress associated with distributed anelastic deformation in a half-space. *Bulletin of the Seismological Society of America, 107*(2), 821–855.

Barnes, P. M., et al. (2020). Slow slip source characterized by lithological and geometric heterogeneity, science. *Advances, 6*, eaay3314.

Bartlow, N. M. (2020). A long-term view of episodic tremor and slip in Cascadia. *Geophysical Research Letters, 47*, e2019GL08530.

Bartlow, N. M., Miyazaki, S. I., Bradley, A. M., & Segall, P. (2011). Space–time correlation of slip and tremor during the 2009 Cascadia slow slip event. *Geophysical Research Letters*. https://doi.org/10.1029/2011GL048714.

Beeler, N. M., Thomas, A., & Bürgmann, R., & Shelly, D. (2018). Constraints on friction, dilatancy, diffusivity, and effective stress from low-frequency earthquake rates on the deep San Andreas fault. *Journal of Geophysical Research: Solid Earth, 123*. https://doi.org/10.1002/2017JB015052.

Behr, W., & Burgmann, R. (2021). What's down there? The structures, materials and environment of deep-seated slow slip and tremor. *Philosophical Transactions of the Royal Society A, 379*. https://doi.org/10.1098/rsta.2020.0218.

Bekaert, D. P. S., Hooper, A., & Wright, T. J. (2015). Reassessing the 2006 Guerrero slow-slip event, Mexico: Implications for large earthquakes in the Guerrero gap. *Journal of Geophysical Research - Solid Earth, 120*, 1357–1375. https://doi.org/10.1002/2014JB011557.

Bevis, M., & Brown, A. (2014). Trajectory models and reference frames for crustal motion geodesy. *Journal of Geodesy, 88*(3), 283–311.

Bilham, R. (1989). Surface slip subsequent to the 24 November 1987 Superstition Hills, California, earthquake monitored by digital creepmeters. *Bulletin of the Seismological Society of America, 79*(2), 424–450.

Bilham, R., et al. (2016). Surface creep on the north Anatolian fault at Ismetpasa, Turkey, 1944–2016. *Journal of Geophysical Research - Solid Earth, 121*, 7409–7431. https://doi.org/10.1002/2016JB013394.

Blanpied, M. L., Lockner, D. A., & Byerlee, J. D. (1991). Fault stability inferred from granite sliding experiments at hydrothermal conditions. *Geophysical Research Letters, 18*(4), 609–612.

Blewitt, G. (2015). GPS and space-based geodetic methods. In *Treatise on geophysics* (2nd Edition). Elsevier. https://doi.org/10.1016/B978-0-444-53802-4.00060-9.

Blewitt, G., & Lavallée, D. (2002). Effect of annual signals on geodetic velocity. *Journal of Geophysical Research: Solid Earth, 107*(B7), ETG-9.

Bock, Y., & Melgar, D. (2016). Physical applications of GPS geodesy: A review. *Reports on Progress in Physics, 79*(10), 106801.

Brodsky, E. E., & Mori, J. (2007). Creep events slip less than ordinary earthquakes. *Geophysical Research Letters, 34*, L16309. https://doi.org/10.1029/2007GL030917.

Brooks, B. A., Foster, J. A., Bevis, M., Frazer, L. N., Wolfe, C. J., & Behn, M. (2006). Periodic slow earthquakes on the flank of Kilauea volcano. *Hawai'i, 246*(3–4), 207–216.

Brooks, B. A., Goldberg, D., DeSanto, J., Ericksen, T. L., Webb, S. C., Nooner, S. L., … Haeussler, P. (2023). Rapid shallow megathrust afterslip from the 2021 M8. 2 Chignik, Alaska earthquake revealed by seafloor geodesy. *Science Advances, 9*(17), eadf9299.

Bruhat, L., Barbot, S., & Avouac, J. P. (2011). Evidence for postseismic deformation of the lower crust following the 2004 Mw6. 0 Parkfield earthquake. *Journal of Geophysical Research: Solid Earth, 116*(B8).

Bucknam, R. C., Plafker, G., & Sharp, R. V. (1978). Fault movement (afterslip) following the Guatemala earthquake of February 4, 1976. *Geology, 6*(3), 170–173.

Bürgmann, R. (2018). The geophysics, geology, and mechanics of slow fault slip. *Earth Planet Science Letters, 495*, 112–134.

Bürgmann, R., & Dresen, G. (2008). Rheology of the lower crust and upper mantle: Evidence from rock mechanics, geodesy, and field observations. *Annual Review of Earth and Planetary Sciences, 36*, 531–567.

Bürgmann, R., & Thatcher, W. (2013). Space geodesy: A revolution in crustal deformation measurements of tectonic processes. *Geological Society of America Special Papers, 500*, 397–430.

Canitano, A., Gonzalez-Huizar, H., Hsu, Y. J., Lee, H. M., Linde, A. T., & Sacks, S. (2019). Testing the influence of static and dynamic stress perturbations on the occurrence of a shallow, slow slip event in eastern Taiwan. *Journal of Geophysical Research: Solid Earth, 124*(3), 3073–3087.

Cashman, S., Baldwin, J. N., Cashman, K. V., Swanson, K., & Crawford, R. (2007). Microstructures developed by coseismic and aseismic faulting in near-surface sediments, San Andreas fault, California. *Geology, 35*(7), 611–614. https://doi.org/10.1130/G23545A.1.

Cervelli, P., Segall, P., Johnson, K., et al. (2002). Sudden aseismic fault slip on the south flank of Kilauea volcano. *Nature, 415*, 1014–1018. https://doi.org/10.1038/4151014a.

Chen, S. K., Wu, Y.-M., & Chan, Y.-C. (2018). Episodic slow slip events and overlying plate seismicity at the southernmost Ryukyu trench. *Geophysical Research Letters, 45*, 10369–10377. https://doi.org/10.1029/2018GL079740.

Churchill, R. M., Werner, M. J., Biggs, J., & Fagereng, Å. (2022). Afterslip moment scaling and variability from a global compilation of estimates. *Journal of Geophysical Research: Solid Earth, 127*(4), e2021JB023897.

Davis, E. E., Villinger, H., & Sun, T. (2015). Slow and delayed deformation and uplift of the outermost subduction prism following ETS and seismogenic slip events beneath Nicoya peninsula, Costa Rica. *Earth and Planetary Science Letters, 410*, 117–127.

Delahaye, E. J., Townend, J., Reyners, M. E., & Rogers, G. (2009). Microseismicity but no tremor accompanying slow slip in the Hikurangi subduction zone, New Zealand. *Earth and Planetary Science Letters, 277*, 21–28. https://doi.org/10.1016/j.epsl.2008.09.038.

Dieterich, J. H. (1979). Modeling of rock friction: 1. Experimental results and constitutive equations. *Journal of Geophysical Research: Solid Earth, 84*(B5), 2161–2168.

Dixon, T. H., et al. (2014). Earthquake and tsunami forecasts: Relation of slow slip events to subsequent earthquake rupture. *Proceedings of the National Academy of Sciences of the United States of America, 111*, 17039–17044.

Douglas, A., Beavan, J., Wallace, L., & Townend, J. (2005). Slow slip on the northern Hikurangi subduction interface, New Zealand. *Geophysical Research Letters, 32*. https://doi.org/10.1029/2005GL023607.

Dragert, H., Kelin, W., & James, T. S. (2001). A silent slip event on the deeper Cascadia subduction interface. *Science, 292*, 1525–1528.

Fagereng, Å., & den Hartog, S. (2017). Subduction megathrust creep governed by pressure solution and frictional–viscous flow. *Nature Geoscience, 10*, 51–57.

Feigl, K. L., & Thatcher, W. (2006). Geodetic observations of post-seismic transients in the context of the earthquake deformation cycle. *Comptes Rendus Geoscience, 338*(14–15), 1012–1028.

Fielding, E. J., Lundgren, P. R., Bürgmann, R., & Funning, G. J. (2009). Shallow fault-zone dilatancy recovery after the 2003 Bam earthquake in Iran. *Nature, 458*(7234), 64–68.

Frank, W. B., & Brodsky, E. (2019). Daily measurement of slow slip from low-frequency earthquakes is consistent with ordinary earthquake scaling. *Science Advances, 5*(10). https://doi.org/10.1126/sciadv.aaw9386.

Frank, W. B., Rousset, B., Lasserre, C., & Campillo, M. (2018). Revealing the cluster of slow transients behind a large low slip event. *Science Advances, 4* (5). https://doi.org/10.1126/sciadv.aat0661.

Frank, W. B., Shapiro, N. M., Husker, A. L., Kostoglodov, V., Romanenko, A., & Campillo, M. (2014). Using systematically characterized low-frequency earthquakes as a fault probe in Guerrero, Mexico. *The Journal of Geophysical Research, 119*, 7686–7700. https://doi.org/10.1002/2014JB011457.

Freed, A. M., & Bürgmann, R. (2004). Evidence of power-law flow in the Mojave desert mantle. *Nature, 430*(6999), 548–551.

Freed, A. M., Bürgmann, R., Calais, E., & Freymueller, J. (2006). Stress-dependent power-law flow in the upper mantle following the 2002 Denali, Alaska, earthquake. *Earth and Planetary Science Letters, 252*(3–4), 481–489.

Freed, A. M., Bürgmann, R., Calais, E., Freymueller, J., & Hreinsdóttir, S. (2006). Implications of deformation following the 2002 Denali, Alaska, earthquake for postseismic relaxation processes and lithospheric rheology. *Journal of Geophysical Research: Solid Earth, 111*(B1).

Freed, A. M., Bürgmann, R., & Herring, T. (2007). Far-reaching transient motions after Mojave earthquakes require broad mantle flow beneath a strong crust. *Geophysical Research Letters, 34*(19).

Freed, A. M., Herring, T., & Bürgmann, R. (2010). Steady-state laboratory flow laws alone fail to explain postseismic observations. *Earth and Planetary Science Letters, 300*(1–2), 1–10.

Freed, A. M., Hirth, G., & Behn, M. D. (2012). Using short-term postseismic displacements to infer the ambient deformation conditions of the upper mantle. *Journal of Geophysical Research: Solid Earth, 117*(B1).

Freed, A. M., & Lin, J. (2001). Delayed triggering of the 1999 Hector mine earthquake by viscoelastic stress transfer. *Nature, 411*(6834), 180–183.

Freymueller, J. (2017). Geodynamics. In *Springer handbook of global navigation satellite systems* (pp. 1063–1106). Cham: Springer.

Fu, Y., & Freymueller, J. (2013). Repeated large slow slip events at the southcentral Alaska subduction zone. *Earth and Planetary Science Letters, 375*, 303–311.

Fujiwara, T., Kodaira, S., No, T., Kaiho, Y., Takahashi, N., & Kaneda, Y. (2011). The 2011Tohoku-Oki earthquake: Displacement reaching the trench axis. *Science, 334*, 1240.

Gao, S. S., Silver, P. G., & Linde, A. T. (2000). Analysis of deformation data at Parkfield, California: Detection of a long-term strain transient. *Journal of Geophysical Research, 105*(B2), 2955–2967. https://doi.org/10.1029/1999JB900383.

Gao, X., & Wang, K. (2017). Rheological separation of the megathrust seismogenic zone and episodic tremor and slip. *Nature, 543*, 416–419. https://doi.org/10.1038/nature21389.

Gomberg, J., Rubinstein, J. L., Peng, Z., Creager, K. C., Vidale, J. E., & Bodin, P. (2008). Widespread triggering of nonvolcanic tremor in California. *Science, 319*(5860), 173.

Gomberg, J., Wech, A., Creager, K., Obara, K., & Agnew, D. (2016). Reconsidering earthquake scaling. *Geophysical Research Letters, 43*, 6243–6251.

Goto, K., Ishizawa, T., Ebina, Y., Imamura, F., Sato, S., & Udo, K. (2021). Ten years after the 2011 Tohoku-oki earthquake and tsunami: Geological and environmental effects and implications for disaster policy changes. *Earth-Science Reviews, 212*, 103417.

Gualandi, A., Avouac, J. P., Galetzka, J., Genrich, J. F., Blewitt, G., Adhikari, L. B., ... Liu-Zeng, J. (2017). Pre-and post-seismic deformation related to the 2015, Mw7. 8 Gorkha earthquake, Nepal. *Tectonophysics, 714*, 90–106.

Gualandi, A., Liu, Z., & Rollins, C. (2020). Post-large earthquake seismic activities mediated by aseismic deformation processes. *Earth and Planetary Science Letters, 530*, 115870.

Guns, K. A., Pollitz, F. F., Lay, T., & Yue, H. (2021). Exploring GPS observations of postseismic deformation following the 2012 MW7. 8 Haida Gwaii and 2013 MW7. 5 Craig, Alaska earthquakes: Implications for viscoelastic earth structure. *Journal of Geophysical Research: Solid Earth, 126*(7), e2021JB021891.

Hager, B. H., Lyzenga, G. A., Donnellan, A., & Dong, D. (1999). Reconciling rapid strain accumulation with deep seismogenic fault planes in the Ventura Basin, California. *Journal of Geophysical Research: Solid Earth, 104*(B11), 25207–25219.

Hamling, I. J., & Wallace, L. M. (2015). Silent triggering: Aseismic crustal faulting induced by a subduction slow slip event. *Earth and Planetary Science Letters, 421*, 13–19.

Handy, M. R., Hirth, G., & Burgmann, R. (2007). *Continental fault structure and rheology from the frictional-to-viscous transition downward*.

Hawthorne, J. C., & Rubin, A. M. (2010). Tidal modulation of slow slip in Cascadia. *Journal of Geophysical Research: Solid Earth, 115*(B9).

He, J., & Peltzer, G. (2008). Poroelastic triggering in the 9–22 January 2008 Nima-Gaize (Tibet) earthquake sequence. *Geology, 38*(10), 907–910.

Hearn, E. (2022). "Ghost transient" corrections to the southern California GPS velocity field from San Andreas fault seismic cycle models. *Seismological Society of America, 93*(6), 2973–2989.

Hearn, E. H., Pollitz, F. F., Thatcher, W. R., & Onishi, C. T. (2013). How do "ghost transients" from past earthquakes affect GPS slip rate estimates on southern California faults? *Geochemistry, Geophysics, Geosystems, 14*(4), 828–838.

Hearn, E. H., & Thatcher, W. R. (2015). Reconciling viscoelastic models of postseismic and interseismic deformation: Effects of viscous shear zones and finite length ruptures. *Journal of Geophysical Research: Solid Earth, 120*(4), 2794–2819.

Heki, K., Miyazaki, S. I., & Tsuji, H. (1997). Silent fault slip following an interplate thrust earthquake at the Japan trench. *Nature, 386*(6625), 595–598.

Hines, T. T., & Hetland, E. A. (2016a). Rheologic constraints on the upper mantle from 5 years of postseismic deformation following the El Mayor-Cucapah earthquake. *Journal of Geophysical Research: Solid Earth, 121*(9), 6809–6827.

Hines, T. T., & Hetland, E. A. (2016b). Rapid and simultaneous estimation of fault slip and heterogeneous lithospheric viscosity from post-seismic deformation. *Geophysical Journal International, 204*(1), 569–582.

Hirose, J., Hirahara, K., Kimata, F., Fjii, N., & Miyazaki, S. (1999). A slow thrust slip event following the two 1996 Hyuganada earthquakes beneath the Bongo Channel, Southwest Japan. *Geophysical Research Letters, 26*, 3237–3240.

Hirose, H., Kimura, H., Enescu, B., & Aoi, S. (2012). Recurrent slow slip event likely hastened by the 2011 Tohoku earthquake. *Proceedings of the National Academy of Sciences*, *109*(39), 15157–15161.

Hirose, H., & Obara, K. (2006). Short-term slow slip and correlated tremor episodes in the Tokai region, Central Japan. *Geophysical Research Letters*, *33*, L17311. https://doi.org/10.1029/2006GL026579.

Hirose, H., & Obara, K. (2010). Recurrence behavior of short-term slow slip and correlated nonvolcanic tremor episodes in western Shikoku, Southwest Japan. *Journal of Geophysical Research*, *115*, B00A21. https://doi.org/10.1029/2008JB006050.

Hong, S., & Liu, M. (2021). Postseismic deformation and afterslip evolution of the 2015 Gorkha earthquake constrained by InSAR and GPS observations. *Journal of Geophysical Research: Solid Earth*, *126*(7), e2020JB020230.

Houston, H. (2015). Low friction and fault weakening revealed by rising sensitivity of tremor to tidal stress. *Nature Geoscience*, *8*, 409–415.

Hsu, Y. J., Simons, M., Avouac, J. P., Galetzka, J., Sieh, K., Chlieh, M., & Bock, Y. (2006). Frictional afterslip following the 2005 Nias-Simeulue earthquake, Sumatra. *Science*, *312*(5782), 1921–1926.

Hu, Y., Bürgmann, R., Freymueller, J. T., Banerjee, P., & Wang, K. (2014). Contributions of poroelastic rebound and a weak volcanic arc to the postseismic deformation of the 2011 Tohoku earthquake. *Earth, Planets and Space*, *66*(1), 106. https://doi.org/10.1186/1880-5981-66-106.

Hu, Y., Bürgmann, R., Uchida, N., Banerjee, P., & Freymueller, J. T. (2016). Stress-driven relaxation of heterogeneous upper mantle and time-dependent afterslip following the 2011 Tohoku earthquake. *Journal of Geophysical Research: Solid Earth*, *121*(1), 385–411.

Huang, M. H., Bürgmann, R., & Freed, A. M. (2014). Probing the lithospheric rheology across the eastern margin of the Tibetan plateau. *Earth and Planetary Science Letters*, *396*, 88–96.

Huang, M. H., Bürgmann, R., & Pollitz, F. (2016). Lithospheric rheology constrained from twenty-five years of postseismic deformation following the 1989 Mw 6.9 Loma Prieta earthquake. *Earth and Planetary Science Letters*, *435*, 147–158.

Huang, K., Hu, Y., & Freymueller, J. T. (2020). Decadal viscoelastic postseismic deformation of the 1964 Mw9.2 Alaska earthquake. *Journal of Geophysical Research: Solid Earth*, *125*(9), e2020JB019649.

Huang, H., Xu, W., Meng, L., Bürgmann, R., & Baez, J. C. (2017). Early aftershocks and afterslip surrounding the 2015 Mw 8.4 Illapel rupture. *Earth and Planetary Science Letters*, *457*, 282–291.

Hussain, E., Wright, T. J., Walters, R. J., Bekaert, D. P., Lloyd, R., & Hooper, A. (2018). Constant strain accumulation rate between major earthquakes on the north Anatolian fault. *Nature Communications*, *9*(1), 1–9.

Ide, S., Beroza, G. C., Shelly, D. R., & Uchide, T. (2007). A scaling law for slow earthquakes. *Nature*, *447*, 76–79. https://doi.org/10.1038/nature05780.

Ide, S., Shelly, D. R., & Beroza, G. C. (2007). Mechanism of deep low frequency earthquakes: Further evidence that deep non-volcanic tremor is generated by shear slip on the plate interface. *Geophysical Research Letters*, *34*, L03308. https://doi.org/10.1029/2006GL028890.

Iinuma, T., Hino, R., Uchida, N., et al. (2016). Seafloor observations indicate spatial separation of coseismic and postseismic slips in the 2011 Tohoku earthquake. *Nature Communications*, *7*, 13506. https://doi.org/10.1038/ncomms13506.

Ikari, M. J., Marone, C., Saffer, D. M., & Kopf, A. J. (2013). Slip weakening as a mechanism for slow earthquakes. *Nature Geoscience*, *6*, 468–472.

Ito, Y., Hino, R., Kido, M., Fujimoto, H., Osada, Y., Inazu, D., et al. (2013). Episodic slow slip events in the Japan subduction zone before the 2011 Tohoku-Oki earthquake. *Tectonophysics*, *600*, 14–26.

Jiang, J., Bock, Y., & Klein, E. (2021). Coevolving early afterslip and aftershock signatures of a San Andreas fault rupture. *Science Advances*, *7*(15), eabc1606.

Jiang, Y., Wdowinski, S., Dixon, T. H., Hackl, M., Protti, M., & Gonzalez, V. (2012). Slow slip events in Costa Rica detected by continuous GPS observations, 2002–2011. *Geochemistry, Geophysics, Geosystems*, *13*(4).

Johnson, K. M., Hsu, Y.-J., Segall, P., & Yu, S.-B. (2001). Fault geometry and slip distribution of the 1999 Chi-Chi, Taiwan earthquake imaged from inversion of GPS data. *Geophysical Research Letters*, *28*, 2285–2288.

Johnson, K. M., & Tebo, D. (2018). Capturing 50 years of postseismic mantle flow at Nankai subduction zone. *Journal of Geophysical Research: Solid Earth*, *123*(11), 10–091.

Jónsson, S., Segall, P., Pedersen, R., & Björnsson, G. (2003). Post-earthquake ground movements correlated to pore-pressure transients. *Nature*, *424*, 179–183. https://doi.org/10.1038/nature01776.

Kaneko, Y., Ito, Y., Chow, B., Wallace, L. M., Tape, C., Grapenthin, R., et al. (2019). Ultralong duration of seismic ground motion arising from a thick, low-velocity sedimentary wedge. *Journal of Geophysical Research: Solid Earth*. https://doi.org/10.1029/2019JB017795.

Kato, A., Obara, K., Igarashi, T., Tsuruoka, H., Nakagawa, S., & Hirata, N. (2012). Propagation of slow slip leading up to the 2011 M9 Tohoku-Oki earthquake. *Science*, *335*, 705–708.

Klein, E., Duputel, Z., Zigone, D., Vigny, C., Boy, J.-P., Doubre, C., & Meneses, G. (2018). Deep transientslow slip detected by survey GPS in the region of Atacama, Chile. *Geophysical Research Letters*, *45*, 12263–12273.

Kodaira, S., et al. (2004). High pore fluid pressure may cause silent slip in the Nankai trough. *Science*, *304*, 1295–1298.

Kostoglodov, V., Singh, S. K., Santiago, J. A., Franco, S. I., Larson, K. M., Lowry, A. R., & Bilham, R. (2003). A large silent earthquake in the Guerrero seismic gap, Mexico. *Geophysical Research Letters*, *30*, 1807. https://doi.org/10.1029/2003GL017219.

Kundu, B., Ghosh, A., Mendoza, M., Bürgmann, R., Gahalaut, V. K., & Saikia, D. (2016). Tectonic tremor on Vancouver Island, Cascadia, modulated by the body and surface waves of the Mw 8.6 and 8.2, 2012 East Indian Ocean earthquakes. *Geophysical Research Letters*, *43*(17), 9009–9017.

Lambert, V., & Barbot, S. (2016). Contribution of viscoelastic flow in earthquake cycles within the lithosphere-asthenosphere system. *Geophysical Research Letters*, *43*(19), 10–142.

Langbein, J., Gwyther, R. L., Hart, R. H. G., & Gladwin, M. T. (1999). Slip-rate increase at Parkfield in 1993 detected by high-precision EDM and borehole tensor strainmeters. *Geophysical Research Letters*, *26*, 2529–2532.

Larson, K. M., Kostoglodov, V., Lowry, A., Hutton, W., Sanchez, O., Hudnut, K., & Suarez, G. (2004). Crustal deformation measurements in Guerrero, Mexico. *Journal of Geophysical Research*, *109*, B04409. https://doi.org/10.1029/2003JB002843.

Li, S., Freymueller, J., & McCaffrey, R. (2016). Slow slip events and time-dependent variations in locking beneath lower cook inlet of the Alaska-Aleutian subduction zone. *Journal of Geophysical Research - Solid Earth, 121*, 1060–1079. https://doi.org/10.1002/2015JB012491.

Li, S., Moreno, M., Bedford, J., Rosenau, M., & Oncken, O. (2015). Revisiting viscoelastic effects on interseismic deformation and locking degree: A case study of the Peru-North Chile subduction zone. *Journal of Geophysical Research - Solid Earth, 120*, 4522–4538. https://doi.org/10.1002/2015JB011903.

Li, S., Wang, K., Wang, Y., Jiang, Y., & Dosso, S. E. (2018). Geodetically inferred locking state of the Cascadia megathrust based on a viscoelastic Earth model. *Journal of Geophysical Research: Solid Earth, 123*(9), 8056–8072.

Linde, A. T., Gladwin, M. T., Johnston, M. J. S., Gwyther, R. L., & Bilham, R. G. (1996). A slow earthquake sequence on the San Andreas fault. *Nature, 383*, 65–68.

Liu, Y., McGuire, J., & Behn, M. (2020). Aseismic transient slip on the Gofar transform fault, East Pacific rise. *Proceedings of the National Academy of Science, 117*(19), 10188–10194.

Liu, Y., & Rice, J. R. (2007). Spontaneous and triggered aseismic deformation transients in a subduction fault model. *Journal of Geophysical Research, 112*, B09404. https://doi.org/10.1029/2007JB004930.

Liu, S., Shen, Z. K., Bürgmann, R., & Jónsson, S. (2021). Thin crème brûlée rheological structure for the Eastern California shear zone. *Geology, 49*(2), 216–221.

Liu-Zeng, J., Zhang, Z., Rollins, C., Gualandi, A., Avouac, J. P., Shi, H., ... Li, Z. (2020). Postseismic deformation following the 2015 Mw7.8 Gorkha (Nepal) earthquake: New GPS data, kinematic and dynamic models, and the roles of afterslip and viscoelastic relaxation. *Journal of Geophysical Research: Solid Earth, 125*(9), e2020JB019852.

Luo, H., & Wang, K. (2021). Postseismic geodetic signature of cold forearc mantle in subduction zones. *Nature Geoscience, 14*(2), 104–109.

Marone, C. J., Scholtz, C. H., & Bilham, R. (1991). On the mechanics of earthquake afterslip. *Journal of Geophysical Research: Solid Earth, 96*(B5), 8441–8452.

Masuti, S., Barbot, S. D., Karato, S. I., Feng, L., & Banerjee, P. (2016). Upper-mantle water stratification inferred from observations of the 2012 Indian Ocean earthquake. *Nature, 538*(7625), 373–377.

Matsuzawa, T., Hirose, H., Shibazaki, B., & Obara, K. (2010). Modeling short- and long-term slow slip events in the seismic cycles of large subduction earthquakes. *Journal of Geophysical Research, 115*, B12301. https://doi.org/10.1029/2010JB007566.

McCormack, K., Hesse, M. A., Dixon, T., & Malservisi, R. (2020). Modeling the contribution of poroelastic deformation to postseismic geodetic signals. *Geophysical Research Letters, 47*(8), e2020GL086945.

Michel, S., Gualandi, A., & Avouac, J.-P. (2019). Similar scaling laws for earthquakes and Cascadia slow-slip events. *Nature, 574*(7779), 522–526.

Miyazaki, S., Segall, P., McGuire, J. J., Kato, T., & Hatanaka, Y. (2006). Spatial and temporal evolution of stress and slip rate during the 2000 Tokai slow earthquake. *Journal of Geophysical Research, 111*, B03409. https://doi.org/10.1029/2004JB003426.

Montési, L. G. (2004). Controls of shear zone rheology and tectonic loading on postseismic creep. *Journal of Geophysical Research: Solid Earth, 109*(B10).

Montgomery-Brown, E. K., Segall, P., & Miklius, A. (2009). Kilauea slow slip events: Identification, source inversions, and relation to seismicity. *Journal of Geophysical Research, 114*, B00A03. https://doi.org/10.1029/2008JB006074.

Montgomery-Brown, E. K., Thurber, C. H., Wolfe, C. J., & Okubo, P. (2013). Slow slip and tremor search at Kilauea volcano, Hawaii. *Geochemistry, Geophysics, Geosystems, 14*, 367–384. https://doi.org/10.1002/ggge.20044.

Moore, J. D., Yu, H., Tang, C. H., Wang, T., Barbot, S., Peng, D., & Shibazaki, B. (2017). Imaging the distribution of transient viscosity after the 2016 M w 7.1 Kumamoto earthquake. *Science, 356*(6334), 163–167.

Murray, J. R., & Segall, P. (2005). Spatiotemporal evolution of a transient slip event on the San Andreas fault near Parkfield, California. *Journal of Geophysical Research, 110*, B09407. https://doi.org/10.1029/2005JB003651.

Nadeau, R. M., & McEvilly, T. V. (1999). Fault slip rates at depth from recurrence intervals of repeating microearthquakes. *Science, 285*(5428), 718–721.

Nishimura, T. (2014). Short-term slow slip events along the Ryukyu trench, southwestern Japan, observed by continuous GNSS. *Earth, Planets and Space, 1*(1), 22. https://doi.org/10.1186/s40645-014-0022-5.

Nishimura, T., Matsuzawa, T., & Obara, K. (2013). Detection of short-term slow slip events along the NankaiTrough, Southwest Japan, using GNSS data. *Journal of Geophysical Research - Solid Earth, 118*, 3112–3125. https://doi.org/10.1002/jgrb.50222.

Noda, H., & Lapusta, N. (2013). Stable creeping fault segments can become destructive as a result of dynamic weakening. *Nature, 493*(7433), 518–521.

Nur, A., & Mavko, G. (1974). Postseismic viscoelastic rebound. *Science, 183*(4121), 204–206.

Obara, K. (2002). Nonvolcanic deep tremor associated with subduction in Southwest Japan. *Science, 296*, 1679–1681.

Obara, K., & Hirose, H. (2006). Non-volcanic deep low-frequency tremors accompanying slow slips in the Southwest Japan subduction zone. *Tectonophysics, 417*(1–2), 33–51.

Obara, K., Hirose, H., Yamamizu, F., & Kasahara, K. (2004). Episodic slow slip events accompanied by non-volcanic tremors in Southwest Japan subduction zone. *Geophysical Research Letters, 31*, L23602. https://doi.org/10.1029/2004GL020848.

Obara, K., & Kato, A. (2016). Connecting slow earthquakes to huge earthquakes. *Science, 353*, 253–257. https://doi.org/10.1126/science.aaf1512.

Ohta, Y., Freymueller, J., Hreinsdottir, S., & Suito, H. (2006). A large slow slip event and the depth of the seismogenic zone in the south Central Alaska subduction zone. *Earth and Planetary Science Letters, 247*, 108–116.

Okada, Y. (1985). Surface deformation due to shear and tensile faults in a half-space. *Bulletin of the Seismological Society of America, 75*(4), 1135–1154.

Outerbridge, K. C., et al. (2010). A tremor and slip event on the Cocos-Caribbean subduction zone as measured by a global positioning system (GPS) and seismic network on the Nicoya peninsula, Costa Rica. *The Journal of Geophysical Research, 115*, B10408.

Ozawa, S., Suito, H., Imakiire, T., & Murakmi, M. (2007). Spatiotemporal evolution of aseismic interplate slip between 1996 and 1998 and between 2002 and 2004, in Bungo channel, Southwest Japan. *Journal of Geophysical Research, 112*, B05409. https://doi.org/10.1029/2006JB004643.

Ozawa, S., Suito, H., & Tobita, M. (2007). Occurrence of quasi-periodic slow slip off the east coast of the Boso peninsula, Central Japan. *Earth, Planets and Space, 59*, 1241–1245.

Park, S., Avouac, J. P., Zhan, Z., & Gualandi, A. (2023). Weak upper-mantle base revealed by postseismic deformation of a deep earthquake. *Nature, 615*(7952), 455–460.

Peltzer, G., Rosen, P., Rogez, F., & Hudnut, K. (1996). Postseismic rebound in fault step-overs caused by pore fluid flow. *Science, 273*, 1202–1204.

Peltzer, G., Rosen, P., Rogez, F., & Hudnut, K. (1998). Poroelastic rebound along the landers 1992 earthquake surface rupture. *Journal of Geophysical Research, 103*(98), 30131–30145. https://doi.org/10.1029/98JB02302.

Peña, C., Metzger, S., Heidbach, O., Bedford, J., Bookhagen, B., Moreno, M., et al. (2022). Role of poroelasticity during the early postseismic deformation of the 2010 Maule megathrust earthquake. *Geophysical Research Letters, 49*, e2022GL098144.

Peng, Z., & Gomberg, J. (2010). An integrated perspective of the continuum between earthquake and slow slip phenomena. *Nature Geoscience, 3*, 599–607. https://doi.org/10.1038/ngeo940.

Peng, Z., & Zhao, P. (2009). Migration of early aftershocks following the 2004 Parkfield earthquake. *Nature Geoscience, 2*(12), 877–881.

Perez-Silva, A., Kaneko, Y., Savage, M., Wallace, L., & Warren-Smith, E. (2023). Characteristics of slow slip events explained by rate-strengthening faults subject to periodic pore fluid pressure changes. *Journal of Geophysical Research: Solid Earth*, e2022JB026332.

Perfettini, H., & Ampuero, J. P. (2008). Dynamics of a velocity strengthening fault region: Implications for slow earthquakes and postseismic slip. *Journal of Geophysical Research: Solid Earth, 113*(B9).

Perfettini, H., & Avouac, J. P. (2004). Postseismic relaxation driven by brittle creep: A possible mechanism to reconcile geodetic measurements and the decay rate of aftershocks, application to the Chi-Chi earthquake, Taiwan. *Journal of Geophysical Research: Solid Earth, 109*(B2).

Perfettini, H., Avouac, J. P., Tavera, H., Kositsky, A., Nocquet, J. M., Bondoux, F., ... Soler, P. (2010). Seismic and aseismic slip on the central Peru megathrust. *Nature, 465*(7294), 78–81.

Piombo, A., Martinelli, G., & Dragoni, M. (2005). Post-seismic fluid flow and coulomb stress changes in a poroelastic medium. *Geophysical Journal International, 162*(2), 507–515.

Pollitz, F., Banerjee, P., Grijalva, K., Nagarajan, B., & Bürgmann, R. (2008). Effect of 3-D viscoelastic structure on post-seismic relaxation from the 2004 M= 9.2 Sumatra earthquake. *Geophysical Journal International, 173*(1), 189–204.

Pollitz, F. F. (2003). Transient rheology of the uppermost mantle beneath the Mojave Desert, California. *Earth and Planetary Science Letters, 215*(1–2), 89–104.

Pollitz, F. F. (2015). Postearthquake relaxation evidence for laterally variable viscoelastic structure and water content in the Southern California mantle. *Journal of Geophysical Research: Solid Earth, 120*(4), 2672–2696.

Pollitz, F. F. (2019). Lithosphere and shallow asthenosphere rheology from observations of post-earthquake relaxation. *Physics of the Earth and Planetary Interiors, 293*, 106271.

Pollitz, F. F., Bürgmann, R., & Thatcher, W. (2012). Illumination of rheological mantle heterogeneity by the M7. 2 2010 El Mayor-Cucapah earthquake. *Geochemistry, Geophysics, Geosystems, 13*(6).

Pollitz, F. F., & Evans, E. L. (2017). Implications of the earthquake cycle for inferring fault locking on the Cascadia megathrust. *Geophysical Journal International, 209*(1), 167–185.

Pollitz, F. F., Wicks, C., & Thatcher, W. (2001). Mantle flow beneath a continental strike-slip fault: Postseismic deformation after the 1999 Hector mine earthquake. *Science, 293*(5536), 1814–1818.

Premus, J., Gallovič, F., & Ampuero, J. P. (2022). Bridging time scales of faulting: From coseismic to postseismic slip of the M w 6.0 2014 South Napa, California earthquake. *Science Advances, 8*(38), eabq2536.

Qiu, Q., Feng, L., Hermawan, I., & Hill, E. M. (2019). Coseismic and postseismic slip of the 2005 Mw 8.6 Nias-Simeulue earthquake: Spatial overlap and localized viscoelastic flow. *Journal of Geophysical Research: Solid Earth, 124*(7), 7445–7460.

Qiu, Q., Moore, J. D., Barbot, S., Feng, L., & Hill, E. M. (2018). Transient rheology of the Sumatran mantle wedge revealed by a decade of great earthquakes. *Nature Communications, 9*(1), 1–13.

Radiguet, M., Cotton, F., Vergnolle, M., Campillo, M., Walpersdorf, A., Cotte, N., & Kostoglodov, V. (2012). Slow slip events and strain accumulation in the Guerrero gap, Mexico. *Journal of Geophysical Research: Solid Earth, 117*. https://doi.org/10.1029/2011JB008801.

Radiguet, M., Perfettini, H., Cotte, N., et al. (2016). Triggering of the 2014 M_w7.3 Papanoa earthquake by a slow slip event in Guerrero, Mexico. *Nature Geoscience, 9*, 829–833. https://doi.org/10.1038/ngeo2817.

Rogers, G., & Dragert, H. (2003). Episodic tremor and slip on the Cascadia subduction zone: The chatter of silent slip. *Science, 300*, 1942–1943.

Rollins, C., Barbot, S., & Avouac, J. P. (2015). Postseismic deformation following the 2010 M= 7.2 El Mayor-Cucapah earthquake: Observations, kinematic inversions, and dynamic models. *Pure and Applied Geophysics, 172*(5), 1305–1358.

Ross, Z. E., Rollins, C., Cochran, E. S., Hauksson, E., Avouac, J. P., & Ben-Zion, Y. (2017). Aftershocks driven by afterslip and fluid pressure sweeping through a fault-fracture mesh. *Geophysical Research Letters, 44*(16), 8260–8267.

Rousset, B., Jolivet, R., Simons, M., Lasserre, C., Riel, B., Milillo, P., ... Renard, F. (2016). An aseismic slip transient on the north Anatolian fault. *Geophysical Research Letters, 43*, 3254–3262. https://doi.org/10.1002/2016GL068250.

Ruina, A. (1983). Slip instability and state variable friction laws. *Journal of Geophysical Research: Solid Earth, 88*(B12), 10359–10370.

Ruiz, S., Metois, M., Fuenzalida, A., Ruiz, J., Leyton, F., Grandin, R., ... Campos, J. (2014). Intense foreshocks and a slow slip event preceded the 2014 Iquique Mw 8.1 earthquake. *Science, 345*, 1165–1169.

Ryder, I., Wang, H., Bie, L., & Rietbrock, A. (2014). Geodetic imaging of late postseismic lower crustal flow in Tibet. *Earth and Planetary Science Letters, 404*, 136–143.

Saffer, D. M., & Tobin, H. (2011). Hydrogeology and mechanics of subduction zone forearcs: Fluid flow and pore pressure. *Annual Reviews of Earth and Planetary Sciences, 39*, 157–186.

Saffer, D. M., & Wallace, L. M. (2015). The frictional, hydrologic, metamorphic, and thermal habitat of shallow slow earthquakes. *Nature Geoscience*, *8*, 594–600. https://doi.org/10.1038/ngeo2490.

Savage, J. C. (1990). Equivalent strike-slip earthquake cycles in half-space and lithosphere-asthenosphere earth models. *Journal of Geophysical Research: Solid Earth*, *95*(B4), 4873–4879.

Sawai, Y., Satake, K., Kamataki, T., Nasu, H., Shishikura, M., Atwater, B. F., … Yamaguchi, M. (2004). Transient uplift after a 17th-century earthquake along the Kuril subduction zone. *Science*, *306*(5703), 1918–1920.

Schmidt, D. A., & Gao, H. (2010). Source parameters and time-dependent slip distributions of slow slip events on the Cascadia subduction zone from 1998 to 2008. *Journal of Geophysical Research*, *115*, B00A18. https://doi.org/10.1029/2008JB006045.

Scholz, C. H. (2019). *The mechanics of earthquakes and faulting*. Cambridge University Press.

Scholz, C. H., Wyss, M., & Smith, S. W. (1969). Seismic and aseismic slip on the San Andreas fault. *Journal of Geophysical Research*, *74*(8), 2049–2069.

Schwartz, S. Y., & Rokosky, J. M. (2007). Slow slip events and seismic tremor at circum-Pacific subduction zones. *Reviews of Geophysics*, *45*, RG3004. https://doi.org/10.1029/2006RG000208.

Segall, P. (2010). Earthquake and volcano deformation. In *Earthquake and volcano deformation* Princeton University Press.

Segall, P., & Davis, J. L. (1997). GPS applications for geodynamics and earthquake studies. *Annual Review of Earth and Planetary Sciences*, *25*(1), 301–336.

Shelly, D. R., Beroza, G. C., Ideo, S., & Nakamula, S. (2006). Low-frequency earthquakes in Shikoku, Japan, and their relationship to episodic tremor and slip. *Nature*, *442*, 188–191.

Shelly, D. R., Peng, Z., Hill, D. P., & Aiken, C. (2011). Triggered creep as a possible mechanism for delayed dynamic triggering of tremor and earthquakes. *Nature Geoscience*, *4*(6), 384–388.

Shibazaki, B., Wallace, L. M., Kaneko, Y., Hamling, I., Ito, Y., & Matsuzawa, T. (2019). Three-dimensional modeling of spontaneous and triggered slow-slip events at the Hikurangi subduction zone, New Zealand. *Journal of Geophysical Research: Solid Earth*, *124*. https://doi.org/10.1029/2019JB018190.

Smith, S. W., & Wyss, M. (1968). Displacement on the San Andreas fault subsequent to the 1966 Parkfield earthquake. *Bulletin of the Seismological Society of America*, *58*(6), 1955–1973.

Pousse-Beltran, L., Socquet, A., Benedetti, L., Doin, M. P., Rizza, M., & d'Agostino, N. (2020). Localized afterslip at geometrical complexities revealed by InSAR after the 2016 Central Italy seismic sequence. *Journal of Geophysical Research: Solid Earth*, *125*(11), e2019JB019065.

Socquet, A., Valdes, J. P., Jara, J., Cotton, F., Walpersdorf, A., Cotte, N., … Norabuena, E. (2017). An 8 month slow slip event triggers progressive nucleation of the 2014 Chile megathrust. *Geophysical Research Letters*, *44*, 4046–4053. https://doi.org/10.1002/2017GL073023.

Suito, H. (2018). Current status of postseismic deformation following the 2011 Tohoku-Oki earthquake. *Journal of Disaster Research*, *13*(3), 503–510.

Suito, H., & Freymueller, J. T. (2009). A viscoelastic and afterslip postseismic deformation model for the 1964 Alaska earthquake. *Journal of Geophysical Research: Solid Earth*, *114*(B11).

Sun, T., Wang, K., Iinuma, T., Hino, R., He, J., Fujimoto, H., & Hu, Y. (2014). Prevalence of viscoelastic relaxation after the 2011 Tohoku-oki earthquake. *Nature*, *514*(7520), 84–87.

Thatcher, W. (1983). Nonlinear strain buildup and the earthquake cycle on the San Andreas fault. *Journal of Geophysical Research: Solid Earth*, *88*(B7), 5893–5902.

Thatcher, W., & Pollitz, F. F. (2008). Temporal evolution of continental lithospheric strength in actively deforming regions. *GSA Today*, *18*(4/5), 4.

Todd, E. K., Schwartz, S. Y., Mochizuki, K., Wallace, L. M., Sheehan, A. F., Webb, S. C., et al. (2018). Earthquakes and tremor linked to seamount subduction durineg shallow slow slip at the Hikurangi margin, New Zealand. *Journal of Geophysical Research: Solid Earth*, *123*, 6769–6783.

Toké, N. A., Arrowsmith, J. R., Rymer, M. J., Landgraf, A., Haddad, D. E., Busch, M., … Hannah, A. (2011). Late Holocene slip rate of the San Andreas fault and its accommodation by creep and moderate-magnitude earthquakes at Parkfield, California. *Geology*, *39*(3), 243–246. https://doi.org/10.1130/G31498.1.

Trubienko, O., Fleitout, L., Garaud, J. D., & Vigny, C. (2013). Interpretation of interseismic deformations and the seismic cycle associated with large subduction earthquakes. *Tectonophysics*, *589*, 126–141.

Trubienko, O., Garaud, J. D., & Fleitout, L. (2014). Models of postseismic deformation after megaearthquakes: The role of various rheological and geometrical parameters of the subduction zone. *Solid Earth Discussions*, *6*(1), 427–466.

Tsang, L. L., Hill, E. M., Barbot, S., Qiu, Q., Feng, L., Hermawan, I., … Natawidjaja, D. H. (2016). Afterslip following the 2007 Mw 8.4 Bengkulu earthquake in Sumatra loaded the 2010 Mw 7.8 Mentawai tsunami earthquake rupture zone. *Journal of Geophysical Research: Solid Earth*, *121*(12), 9034–9049.

Tsang, L. L., Meltzner, A. J., Philibosian, B., Hill, E. M., Freymueller, J. T., & Sieh, K. (2015). A 15 year slow-slip event on the Sunda megathrust offshore Sumatra. *Geophysical Research Letters*, *42*(16), 6630–6638.

Twardzik, C., Vergnolle, M., Sladen, A., & Avallone, A. (2019). Unravelling the contribution of early postseismic deformation using sub-daily GNSS positioning. *Scientific Reports*, *9*(1), 1–12.

Tymofyeyeva, E., Fialko, Y., Jiang, J., Xu, X., Sandwell, D., Bilham, R., et al. (2019). Slow slip event on the southern San Andreas fault triggered by the 2017 M_w8.2 Chiapas (Mexico) earthquake. *Journal of Geophysical Research: Solid Earth*, *124*, 9956–9975. https://doi.org/10.1029/2018JB016765.

Urlaub, M., Petersen, F., Gross, F., Bonforte, A., Puglisi, G., Guglielmino, F., … Kopp, H. (2018). Gravitational collapse of Mount Etna's southeastern flank. *Science Advances*, *4*(10). https://doi.org/10.1126/sciadv.aat9700.

Vaca, S., Vallee, M., Nocquet, J.-M., Battaglia, J., & Regnier, M. (2018). Recurrent slow slip events as a barrier to the northward rupture propagation of the 2016 Pedernales earthquake (Central Ecuador). *Tectonophysics*, *724–725*, 80–92.

Vallee, M., et al. (2013). Intense interface seismicity triggered by a shallow slow slip event in the Central Ecuador subduction zone. *Journal of Geophysical Research - Solid Earth*, *118*, 2965–2981.

Wallace, L. M. (2020). Slow slip events in New Zealand. *Annual Reviews of Earth and Planetary Sciences, 48*, 175–203.

Wallace, L. M., & Beavan, R. J. (2006). A large slow slip event on the Central Hikurangi subduction interface beneath the Manawatu region, North Island, New Zealand. *Geophysical Research Letters, 33*. https://doi.org/10.1029/2006GL026009.

Wallace, L. M., & Beavan, J. (2010). Diverse slow slip behavior at the Hikurangi subduction margin, New Zealand. *The Journal of Geophysical Research, 115*(B12402). https://doi.org/10.1029/2010JB007717.

Wallace, L. M., Beavan, J., Bannister, S., & Williams, C. A. (2012). Simultaneous long- and short-term slow slip events at the Hikurangi subduction margin, New Zealand: Implications for processes that control slow slip event occurrence, duration, and migration. *Journal of Geophysical Research*. https://doi.org/10.1029/2012JB009489.

Wallace, L. M., Hreinsdóttir, S., Ellis, S., Hamling, I., D'Anastasio, E., & Denys, P. (2018). Triggered slow slip and afterslip on the southern Hikurangi subduction zone following the Kaikōura earthquake. *Geophysical Research Letters, 45*. https://doi.org/10.1002/2018GL077385.

Wallace, L. M., Kaneko, Y., Hreinsdottir, S., Hamling, I., Peng, Z., Bartlow, N., ... Fry, B. (2017). Large-scale dynamic triggering of shallow slow slip enhanced by overlying sedimentary wedge. *Nature Geoscience, 10*, 765–770. https://doi.org/10.1038/ngeo3021.

Wallace, L. M., Webb, S. C., Ito, Y., Mochizuki, K., Hino, R., Henrys, S., ... Sheehan, A. (2016). Slow slip near the trench at the Hikurangi subduction zone. *Science, 352*(6286), 701–704. https://doi.org/10.1126/science.aaf2349.

Walter, J. I., Schwartz, S. Y., Protti, M., & Gonzalez, V. (2013). The synchronous occurrence of shallow tremor and verylow frequency earthquakes offshore of the Nicoya peninsula, CostaRica. *Geophysical Research Letters, 40*, 1517–1522. https://doi.org/10.1002/grl.50213.

Wang, K., & Bürgmann, R. (2020). Probing fault frictional properties during afterslip updip and downdip of the 2017 Mw 7.3 Sarpol-e Zahab earthquake with space geodesy. *Journal of Geophysical Research: Solid Earth, 125*(11), e2020JB020319.

Wang, K., & Fialko, Y. (2018). Observations and modeling of coseismic and postseismic deformation due to the 2015 mw 7.8 Gorkha (Nepal) earthquake. *Journal of Geophysical Research: Solid Earth, 123*(1), 761–779.

Wang, K., Hu, Y., & He, J. (2012). Deformation cycles of subduction earthquakes in a viscoelastic Earth. *Nature, 484*(7394), 327–332.

Wang, R., Martin, F. L., & Roth, F. (2003). Computation of deformation induced by earthquakes in a multi-layered elastic crust—FORTRAN programs EDGRN/EDCMP. *Computers & Geosciences, 29*(2), 195–207.

Watanabe, S., Ishikawa, T., Nakamura, Y., et al. (2021). Co- and postseismic slip behaviors extracted from decadal seafloor geodesy after the 2011 Tohoku-oki earthquake. *Earth, Planets and Space, 73*, 162. https://doi.org/10.1186/s40623-021-01487-0.

Watts, A. B., Zhong, S. J., & Hunter, J. (2013). The behavior of the lithosphere on seismic to geologic timescales. *Annual Review of Earth and Planetary Sciences, 41*(1), 443–468.

Wech, A., & Creager, K. (2012). A continuum of stress, strength and slip in the Cascadia subduction zone. *Nature Geoscience*. https://doi.org/10.1038/NGEO1215.

Wei, M., Kaneko, Y., Liu, Y., & McGuire, J. (2013). Episodic fault creep events in California controlled by shallow frictional heterogeneity. *Nature, 6*, 566–570.

Wei, M., Liu, Y., Kaneko, Y., McGuire, J. J., & Bilham, R. (2015). Dynamic triggering of creep events in the Salton trough, Southern California by regional M>5.4 earthquakes constrained by geodetic observations and numerical simulations. *Earth and Planetary Science Letters, 427*. https://doi.org/10.1016/j.epsl.2015.06.044.

Wei, M., McGuire, J. J., & Richardson, E. (2012). A slow slip event in the south Central Alaska Subduction zone and related seismicity anomaly. *Geophysical Research Letters, 39*, L15309. https://doi.org/10.1029/2012GL052351.

Wei, M., Sandwell, D., & Fialko, Y. (2009). A silent M_w 4.7 slip event of October 2006 on the Superstition Hills fault, southern California. *Journal of Geophysical Research, 114*, B07402. https://doi.org/10.1029/2008JB006135.

Weiss, J. R., Walters, R. J., Morishita, Y., Wright, T. J., Lazecky, M., Wang, H., ... Parsons, B. (2020). High-resolution surface velocities and strain for Anatolia from Sentinel-1 InSAR and GNSS data. *Geophysical Research Letters, 47*(17), e2020GL087376.

Wright, T. J., Elliott, J. R., Wang, H., & Ryder, I. (2013). Earthquake cycle deformation and the Moho: Implications for the rheology of continental lithosphere. *Tectonophysics, 609*, 504–523.

Yamashita, Y., H., et al. (2015). Migrating tremor off southern Kyushu as evidence for slow slip of a shallow subduction interface. *Science, 348*(6235), 676–679.

Yang, H., Guo, R., Zhou, J., Yang, H., & Sun, H. (2022). Transient poroelastic response to megathrust earthquakes: A look at the 2015 M_w 8.3 Illapel, Chile, event. *Geophysical Journal International, 230*(2), 908–915.

Yarce, J., Sheehan, A. F., Nakai, J. S., Schwartz, S. Y., Mochizuki, K., Savage, M. K., ... Todd, E. K. (2019). Seismicity at the northern Hikurangi margin, New Zealand, and investigation of the potential spatial and temporal relationships with a shallow slow slip event. *Journal of Geophysical Research, 124*. https://doi.org/10.1029/2018JB017211.

Yokota, Y., & Ishikawa, T. (2020). Shallow slow slip events along the Nankai trough detected by GNSS-A. *Science Advances, 6*(3). https://doi.org/10.1126/sciadv.aay5786.

Zhao, B., Bürgmann, R., Wang, D., Tan, K., Du, R., & Zhang, R. (2017). Dominant controls of downdip afterslip and viscous relaxation on the postseismic displacements following the Mw7.9 Gorkha, Nepal, earthquake. *Journal of Geophysical Research: Solid Earth, 122*(10), 8376–8401.

Zhou, Y., Thomas, M. Y., Parsons, B., & Walker, R. T. (2018). Time-dependent postseismic slip following the 1978 Mw 7.3 Tabas-e-Golshan, Iran earthquake revealed by over 20 years of ESA InSAR observations. *Earth and Planetary Science Letters, 483*, 64–75.

Chapter 6

Earthquake and tsunami early warning with GNSS data

Brendan W. Crowell
Department of Earth and Space Sciences, University of Washington, Seattle, WA, United States

1 Introduction

Early warning, not to be confused with prediction, is the rapid characterization of an event after it has already started but prior to the impacts being felt by the general population. For earthquake early warning (EEW), the origin time, hypocenter, magnitude, and source extent are inferred from a small amount of data after the arrival of P-waves at a few nearby stations. The first EEW system to go live in 1993 was the Mexican Seismic Alert System (SASMEX), which began development following the devastating 1985 Mexico City earthquake (Suárez et al., 2018). SASMEX was originally conceived as a threshold-based system; when a coastal earthquake with a specific body wave magnitude is detected, localities inland of the event are warned. Over the years, SASMEX has evolved to provide local warnings as well. Recently, the SASMEX system gave upwards of 2 min of warning time for the 2017 Mw 8.2 Tehuantepec earthquake. In 2007, the Japan Meteorological Agency (JMA) launched its public earthquake early warning system (Kamigaichi et al., 2009), which has been highly successful in issuing public alerts. Additionally, JMA issues local tsunami warnings based on their rapid earthquake characterization, one of the only local tsunami warning systems in operation globally. Within the United States Geological Survey's (USGS) ShakeAlert system (Given et al., 2018), the EPIC algorithm relies upon the scaling between the peak P-wave displacement (P_d scaling) recorded on strong-motion accelerometers and broadband seismometers, the distance from the source, and the earthquake magnitude (Chung et al., 2019). The FinDer algorithm, also within ShakeAlert, uses interpolated maps of ground motions recorded at accelerometers and performs template matching to determine the source extent and magnitude (Böse et al., 2018). Several next-generation seismic EEW algorithms utilize the ground shaking directly and interpolate forward the predicted ground motions, such as the PLUM algorithm developed for Japan and being applied to ShakeAlert (Cochran et al., 2019; Kodera et al., 2018; Saunders et al., 2022). Machine learning techniques have been proposed to further predict the final ground shaking from small amounts of data, either along individual time series or spatially. Spatially, convolutional neural networks (CNN) that consider the spatial distribution of ground motion and forecast the final observed ground motions have been demonstrated by several groups (Jozinović et al., 2020; Münchmeyer et al., 2021). Recurrent neural networks (RNN) that predict time series properties, such as ground motion, from small chunks of data have been successfully demonstrated with seismic data (i.e., DeepShake; Datta et al., 2022). For a comprehensive overview of existing earthquake warning systems and methodologies, see Allen and Melgar (2019). For tsunami early warning (TEW), the initial magnitude estimates come from M_{wp} (Tsuboi, 2000) or the W-phase (Duputel et al., 2012; Kanamori, 1993), which gives the moment tensor from which to infer fault slip. The information for tsunami early warning comes from a combination of local and global seismic instruments. Furthermore, direct observations of the ocean surface from the National Oceanic and Atmospheric Administration's (NOAA) DART (Deep-ocean Assessment and Reporting of Tsunamis) buoys provide the final validation of tsunami models.

Fig. 6.1 shows a flowchart of all the design elements required for a generalized early warning system. The first two design elements, raw data, and telemetry, generally fall under the authority of seismic and geodetic network operators. For the purposes of early warning, a good telemetry backbone that ensures low latency and high data return is crucial. The monumentation and installation of seismic and geodetic stations should be high-quality, however, low-cost cell phone sensors can also be used to densify networks (e.g., MyShake, Kong et al., 2016). Additionally, the use of distributed acoustic sensing (DAS) fiber optic cables has been explored for EEW, most notably in the offshore region where traditional sensors may be difficult to operate (Farghal et al., 2022). The next three design elements, data analysis, event analysis, and event impact analysis, are the 'algorithmic' elements of an early warning system (i.e., how big is this earthquake, how much

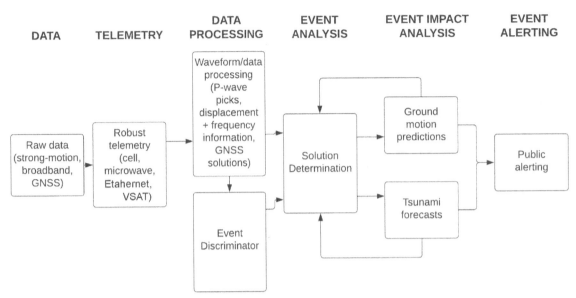

FIG. 6.1 A generalized flowchart for early warning system components. For GNSS, in general displacement estimation occurs at a central processing center. In the future, displacements could be performed at the station, which would subvert the central processing center model.

shaking is expected, should we expect a tsunami). Within the data analysis element, many of the traditional roles of seismic network operations are included such as phase picking and event discrimination, however, with early warning systems, the processes need to be fully automated. The event analysis element includes all the algorithms that take waveform information to say something about the event (magnitude, origin time, slip on the fault, etc.). The event impact design layer uses either the source information or the waveforms to predict the expected ground motions or tsunami wave heights. Some algorithms have been proposed to go between the waveform information and the event impacts without considering the earthquake source. Finally, information about the event is disseminated to the public through a variety of means, either directly to users' phones, email and social media bulletins, radio, TV, or public sirens.

While seismic algorithms do remarkably well at determining origin time, epicenter, and preliminary magnitude, the magnitude estimates tend to saturate for the very largest earthquakes since inertial sensors have difficulty resolving the low-frequency displacements that are diagnostic of total moment release. Sensor rotations and tilts also make it difficult to accurately integrate acceleration into displacement without high-pass filtering, thus reducing the fidelity of the observations (Melgar, Bock, et al., 2013). One of the most often cited magnitude saturation cases is the earthquake, and subsequent, tsunami warnings following the 2011 Mw 9.0 Tohoku-oki earthquake (Hoshiba et al., 2011). While the JMA earthquake early warning system worked well operationally by providing alerts prior to strong shaking in the near field, the magnitude estimates only reached M8.0 by 1 min after origin time where it stayed for the duration of the early warning. Many intermediate and far-field locations did not receive any seismic early warning. The underestimate in magnitude also had a pronounced impact on the ensuing tsunami warning where the maximum wave heights along the Sendai coast were underestimated by many meters, breaching the protective seawalls, and leading to many casualties.

Because of the issue of magnitude saturation, the use of Global Navigation Satellite System (GNSS) data and methods has been proposed. GNSS displacements can track the full spectrum of deformation without going off-scale as occurs with traditional inertial sensors (Grapenthin & Freymueller, 2011; Melgar, Bock, et al., 2013). Within this chapter, I will describe the history of GNSS-based early warning. I will discuss the initial algorithmic developments and the challenges with real-time positioning throughout the years. I will go over best practices for dealing with real-time displacement information and finish with the pathways to operational systems in place today.

2 Real-time GNSS positioning from a historical perspective

There are two primary modes for computing real-time displacements, relative positioning (Dong & Bock, 1989) and precise point positioning (PPP) (Zumberge et al., 1997). Relative positioning relies on performing a double difference between two receivers and all the visible satellites, usually with two frequencies such as L_1 and L_2 for Global Positioning System (GPS) observations. In the double difference, common sources of error are eliminated (receiver and satellite clocks, fractional cycle biases, etc.). If the spacing between the stations is short (<50 km), the tropospheric and ionospheric components are effectively eliminated, and the baseline displacements between the two stations can be easily computed in real-time once the integer values of the phase ambiguities are determined (de Jonge & Tiberius, 1996). Even though this solution works well in real time, it is not computationally scalable to large networks of sites, motions at the reference stations will cause motions in your kinematic stations, and it requires short baselines which might not be feasible in all regions.

Because of these limitations, PPP services have been gaining traction especially as real-time corrections from outside agencies like the International GNSS Service (IGS) are made available. PPP is predicated on obtaining corrections to all the sources of range error between the satellite and receiver and then solving for the displacement at the station. This can be viewed as the classic triangulation problem, like solving for earthquake origin time and hypocenter with P-wave arrivals. Most approaches to PPP typically employ the ionosphere-free combination of pseudorange and carrier-phase observables from multi-frequency receivers to simultaneously estimate position, receiver clock, troposphere delay, and carrier-phase biases for each satellite on an epoch-by-epoch basis. These approaches typically require that initial phase biases be estimated, and subsequent phase cycle slips be flagged, estimated, and fixed as they arise (Gao & Shen, 2002; Kouba & Héroux, 2001; Zumberge et al., 1997). However, most PPP implementations show typical initialization and re-convergence times (after phase breaks) on the order of tens of minutes for 1 s epochs, which significantly degrades the use of PPP for many low-latency applications. To minimize this problem, independent computation of orbits, satellite clock corrections and differential code biases are commonly used to both reduce the number of estimated parameters and increase the precision with which initial phase biases and subsequent cycle slips are fixed. Corrections for a variety of physical processes such as solid earth, ocean, and pole tides, as well as ocean loading effects have also been implemented to further constrain the estimation process and provide greater stability in estimating the correct constant values of biases and phase breaks. Screening the phase and range observables for outliers can further reduce estimation time, but because most PPP algorithms employ combinations of phase and code observables unweighted by formal errors, filter convergence during estimation remains relatively slow, typically taking upwards of tens of minutes and resulting in highly degraded positioning accuracy. An additional PPP implementation, known as PPP-AR (PPP with Ambiguity Resolution), utilizes a regional network to obtain receiver differential code biases and satellite fractional cycle biases, and then applies these corrections in a classical PPP implementation (e.g., Geng et al., 2013). While this improves the positioning quality and can reduce the convergence time to an ambiguity fixed solution, the convergence time is still not instantaneous; reducing the convergence time after the loss of satellite lock for PPP is currently an active area of research.

The first study that demonstrated the possibility of using GNSS time series for tsunami early warning was by Blewitt et al. (2006), using the example of the 2004 Sumatra earthquake. For that study, 30-s sample rate data was processed independently by the GIPSY-OASIS II software (Lichten, 1995) using a full suite of station corrections as well as a 4-day sidereal filter which minimizes issues with multipath error and constellation geometry biases. For the 10 stations closest to the source, the position time series were averaged for the previous 10 min and a 12-min window after shaking would subside (3 min after origin time in their case study). They then used the computed offsets with a fingerprint approach to estimate earthquake magnitude, placing it in a range between M8.7 and M9.3. In contrast, at the same time after origin time, the Pacific Tsunami Warning Center estimated the magnitude to be 8.0; alert issuance in the Indian Ocean was not available at the time of the event, but the underestimate of magnitude would have clearly underestimated the associated impacts of the tsunami (Sardina et al., 2017).

The first study using high-rate, real-time GNSS observations for earthquake early warning was by Crowell et al. (2009). In that study, relative positioning was utilized with the Geodetics, Inc., RTD software (https://geodetics.com/), which provided baseline displacement observations between reference and rover stations and epoch-by-epoch ambiguity resolution was performed. In Crowell et al. (2009), they triangulated the real-time networks using a Delaunay triangulation scheme, and processed each triangle as its own sub-network, parallelizing the problem onto many instances of RTD. The strains within the triangles, computed using the constant strain triangle equations, could be used as a threshold detection criterion, independent of a seismic network. If a detection occurred, the entire network of triangles would be network adjusted to the furthest station away, assuming that station is stationary. At the time of that study, real-time PPP corrections (clocks, orbits, differential code biases) were not good enough for the sub-cm level precision provided through the triangulated sub-network approach. There were several situations that were problematic with the triangulated relative positioning method. First, abnormally small or large triangles, or ones with small or large internal angles, produced anomalously large strains with small baseline displacements. Secondly, motions at the reference station during the network adjustment would get propagated to all stations; this is generally not a problem for earthquake early warning since the characterization of the event is from the initial seconds of the event before surface waves reach the furthest extents of the network. Third, telemetry dropouts would often impact many stations simultaneously, which would break the network into large sub-networks. When attempting to network adjust these solutions, the inter-connectedness was broken, which led to spuriously large displacements that were not true.

Several studies have been performed on the accuracy and noise characteristics of real-time GNSS, showing that even given the limitations above, for the purposes of seismic monitoring over a few minutes time span, the displacement observations can gainfully track M6+ earthquakes in the near field (10s of km). Melgar, Crowell, et al. (2020) performed an analysis of more than 1 year of data across the United States from the Central Washington University Fastlane software. In general, the white noise component on the real-time GNSS displacements over short time windows (<1 min) is roughly 1 cm in each horizontal component and 3–5 cm in the vertical component. These values are reduced when using relative positioning due to the reduction of common error sources, as demonstrated by Genrich and Bock (2006) and Langbein and Bock (2004). The reduction in real-time noise on GNSS is a frontier research area for hazard monitoring and has improved greatly over the past two decades due to improved IGS orbit and clock products, improvements in receiver technology, and the addition of multi-GNSS capabilities.

3 Peak ground displacement

The simplest approach to incorporating GNSS data for early warning is to utilize the peak ground displacements (PGD) recorded on the three-component displacement waveforms. PGD scaling was first proposed by Crowell et al. (2013) using just five earthquakes, the 2011 Mw 9.1 Tohoku-oki, the 2003 Mw 8.3 Tokachi-oki, the 2010 Mw 7.2 El Mayor-Cucapah, and two events during the 2012 Brawley seismic swarm (Mw 5.4 and 5.3). The PGD value is defined as the peak of the Euclidean norm of the three components of motion, North (N), East (E), and Up (U), between the origin time (t_o) and the current time (t_n) such that:

$$PGD(t_n) = max\left(\sqrt{(N(t_0 \to t_n) - N(t_o))^2 + (E(t_0 \to t_n) - E(t_o))^2 + (U(t_0 \to t_n) - U(t_o))^2}\right) \quad (6.1)$$

The principle of PGD scaling is simple: at a given distance, the PGD value will be larger for larger magnitude earthquakes. The scaling law derived by Crowell et al. (2013) is a log–log relationship between the hypocentral distance (R) and PGD, with a general form of:

$$log_{10}(PGD) = A + B * M + C * M * log_{10}(R) \quad (6.2)$$

In Eq. (6.2), the three coefficients, A, B, and C, are determined through least squares and the uncertainty on those coefficients was determined through a bootstrap resampling analysis. The magnitude-dependent attenuation term ($C*M*log_{10}(R)$) is important here and indicates that the attenuation (or slope of the PGD curve) is stronger for larger magnitude earthquakes. If we consider the seismic wave equation, the near, intermediate, and far-field terms have a different attenuation (e.g., Aki & Richards, 2002), -3, -2, and -1 respectively. The near-field term in larger earthquakes will be stronger and may be absent in smaller earthquakes outside of a small distance, so a greater proportion of this higher attenuation term will be present. For example, by Crowell et al. (2013), the coefficient $C = -0.178$, which would make the attenuation -1.6 for a M9 earthquake and -1.2 for a M7 earthquake. PGD scaling, at its core, is a simplified ground motion prediction equation (GMPE), and there is a similarity in the different terms between a traditional GMPE and the PGD scaling law (i.e., magnitude term and magnitude dependent attenuation). In general, GMPEs will relate the magnitude, source terms, site terms, and distance terms to the expected peak ground accelerations, velocities, or spectral accelerations at given periods (PGA, PGV, and SA respectively). GMPEs are often tuned to specific regions to minimize uncertainties in the ground motion predictions. The PGD scaling law is simply an example of a GMPE that is able to compute magnitude through a linear inversion. Table 6.1 shows all the PGD scaling coefficients for different studies.

There have been several analyses of the PGD scaling law with more data, however, the original scaling law still performs well. In Melgar, Crowell, et al. (2015), they analyzed 10 earthquakes between M 6 and 9.1 with 1321 observations in total. They also proposed how to utilize PGD scaling for EEW, with a travel time mask controlling which stations get included in the final magnitude estimate. The travel time mask approach states that only stations that are closer than the travel time mask can be included in the magnitude estimate; this is to ensure that stations have had strong shaking already and thus their PGD values incorporate surface wave observations. There is a trade-off in the choice of travel time mask, too slow and the magnitude takes longer to converge to the correct answer, too fast and stations are included in the magnitude estimate that have not encountered strong shaking. The value that has been chosen for G-FAST is 3 km/s (Crowell et al., 2016). Using a 3 km/s travel time mask, Melgar, Crowell, et al. (2015) showed that the PGD magnitude could reach the true magnitude about halfway to two-thirds of the way through the source-time function of the event, a condition termed as "weak determinism" (Goldberg et al., 2018). In early warning, the concept of determinism refers to an event's magnitude being predetermined and that the characteristics of the nascent stages of rupture are diagnostic

TABLE 6.1 Peak ground displacement scaling coefficients for Eq. (6.1).

Study	Coefficients			Notes
	A	B	C	
Crowell et al. (2013)	−5.013	1.219	−0.178	Seismogeodetic data only
Melgar, Crowell, et al. (2015)	−4.434	1.047	−0.138	A total of 10 earthquakes
Crowell et al. (2016)	−6.687	1.500	−0.214	Exponential distance weighting
Ruhl et al. (2018)	−3.919	1.009	−0.145	A total of 29 earthquakes, used signal-to-noise criterion; original A was −5.919, converting meters to centimeters yields new A value
Goldberg et al. (2021)	−3.841	0.937	−0.127	A total of 33 earthquakes, visually inspected all records

of magnitude. Weak determinism implies that statistically there may be characteristics of the initial rupture that are more likely to lead to a cascading rupture, but longer time windows are required to capture this behavior. GNSS observations capturing the very low-frequency end of ground motions, all the way to the DC offset, are better positioned to observe this stage of the rupture process in the near field.

There are two methods to obtain a single-magnitude estimate for an earthquake: (1) averaging the individual station magnitude estimates at a given time and (2) inverting for a single magnitude by setting up an inverse problem. The approach taken by Crowell et al. (2016) is inverting for a single magnitude for n stations:

$$b = GM \tag{6.3}$$

$$\begin{bmatrix} \log_{10}(PGD_1) - A \\ \log_{10}(PGD_2) - A \\ \vdots \\ \log_{10}(PGD_n) - A \end{bmatrix} = \begin{bmatrix} B + C\log_{10}(R_1) \\ B + C\log_{10}(R_2) \\ \vdots \\ B + C\log_{10}(R_n) \end{bmatrix} M \tag{6.4}$$

Additionally, within the approach taken within G-FAST, exponential distance weighting is employed to give greater value to closer stations. The weighting takes the form of

$$w_i = \exp\left(-\frac{r_{epi,i}^2}{8r_{epi,min}^2}\right) \tag{6.5}$$

where r_{epi} is the epicentral distance and $r_{epi,min}$ is the minimum epicentral distance for all stations in the inversion. The weight matrix in this case is a diagonal matrix of individual station weights and the problem is solved through weighted least squares:

$$WGM = Wb \tag{6.6}$$

Crowell et al. (2016) reregressed the PGD scaling coefficients using the distance weighting in Eq. (6.5). The next large reanalysis of the PGD scaling law was performed by Ruhl et al. (2018), in which they processed 29 earthquakes with 3433 total observations. In that study, the errors of the different PGD scaling laws were shown as a function of distance, which showed dramatically the addition of distance weighting by Crowell et al. (2016) leads to good performance close to the source and poor performance further away. The next reregression and reanalysis of data was performed by Goldberg et al. (2021), who analyzed 2371 records from 33 earthquakes. Differently from Ruhl et al. (2018), every record by Goldberg et al. (2021) was manually inspected to ensure a seismic signal was evident on the records; Ruhl et al. (2018) used a signal-to-noise ratio criterion to determine if records should be included, which explains why more observations are in this dataset. Goldberg et al. (2021) also expanded the PGD relationship to include an anelastic attenuation term and a secondary magnitude term. It should be noted that using higher order terms here makes a simple inversion for magnitude difficult due to nonlinearity, and thus reduces the utility for early warning. Fig. 6.2 shows the natural log residual between the four PGD dataset studies as a function of hypocentral distance. Over time, the distribution of residuals has become tighter, although all four sets of regression coefficients perform similarly, with residuals generally less than 2

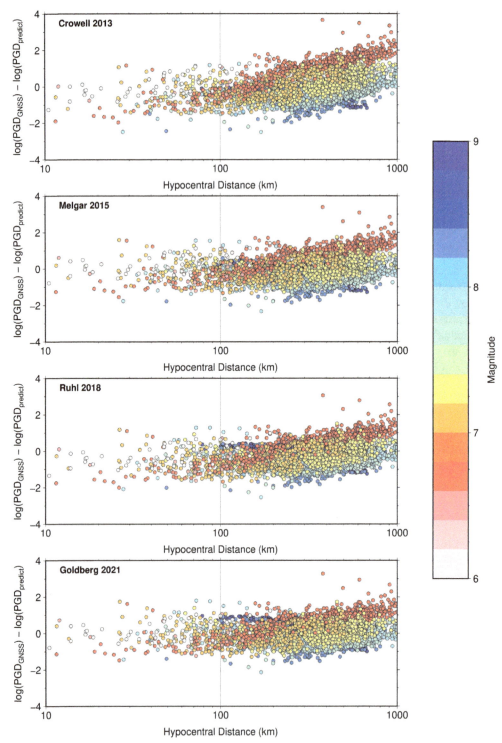

FIG. 6.2 The log residual between the raw PGD values (PGD$_{GNSS}$) from the Goldberg et al. (2021) dataset and the predictions from the different regression studies.

natural log units. At the furthest distances (>500 km), the PGD scaling law tends to underestimate the recorded ground motions; however, it is more likely that the superposition of signal and noise leads to higher PGD values at these distances. The coefficients from Goldberg et al. (2021) improve the bias at far distances by about 0.5 natural log units when compared to Crowell et al. (2013).

Several other studies of PGD scaling have been performed for either a single earthquake or a small subset of events, using the published coefficients from the studies mentioned prior. Crowell, Melgar, and Geng (2018) and Crowell, Schmidt, et al. (2018) performed case studies of three Chilean megathrust events (2010 Mw 8.8 Maule, 2014 Mw 8.2 Iquique, 2015 Mw 8.3 Illapel) and the complex multifault rupture of the 2016 Mw 7.8 Kaikoura earthquake in New Zealand. For the Chilean events (Crowell, Schmidt, et al., 2018), the PGD magnitudes were within 0.2 magnitude units of the true magnitude between 30 and 60 s after the origin time. This is especially impressive since for the Maule earthquake: there were only five stations within 500 km of the epicenter at the time. For the Kaikoura earthquake (Crowell, Melgar, & Geng, 2018), an initial magnitude of 7.3–7.4 is available 18 s after the origin time, which ramps up toward the true magnitude roughly 90 s after the origin time. However, the source time function of the event shows that a significant pulse of moment release is concentrated between 60 and 75 s, so the initial underestimate in PGD scaling reflects the source complexity. Furthermore, including an appropriate grid search for centroid location would yield significantly improved PGD magnitude estimate scatter. This result for the Kaikoura earthquake is further validated by Song et al. (2017), who obtain a magnitude of 7.6 roughly 23 s after the origin time using a smaller subset of stations.

While the data from the 2014 Mw 6.1 Napa, California, earthquake is included in several of the regression analysis studies, at the time, only the regression analysis from Crowell et al. (2013) was available. Melgar, Geng, et al. (2015) retrospectively analyzed the Napa event and found an initial magnitude of 5.6 at 17 s and a final magnitude of 6.05 at 50 s after the origin time. Of note here is that a slower travel-time mask of 1.5 km/s was used rather than the 3.0 km/s used within G-FAST.

Grapenthin et al. (2017) processed stations within 400 km of the 2016 Mw 7.1 Iniskin, Alaska, earthquake using the Track software within the GAMIT/GlobK distribution (Herring et al., 2010). In their analysis, a magnitude of 6.4 is available after 40 s and a magnitude of 7.2 is achieved at 57 s after the origin time. Since the magnitude of the Iniskin earthquake is not very large, the use of baseline observations, given that reference stations are sufficiently far away, does not cause issues with the PGD magnitude estimation. Moreover, the improvement in positioning accuracy from relative positioning allows for small displacements to be resolved, thus reducing the noise level on the derived PGD magnitude.

Hodgkinson et al. (2020) looked at real-time PPP solutions produced by UNAVCO that captured five earthquakes in the Network of the Americas (NOTA) between 2017 and 2019. In their analysis, all their magnitude estimates at 2–3 min after the origin time were within 0.3 magnitude units of the true magnitudes.

Ganas et al. (2020) processed two stations during the 2019 Mw 6.4 Durres, Albania, earthquake and obtained magnitudes of 6.1 and 6.4. In that study, both stations are ~28 km from the source, and the difference in magnitudes can potentially be explained through rupture directivity. No study has explicitly looked at rupture directivity with PGD scaling; however, there are many examples where directivity can be observed easily. Fig. 6.3 shows the map view and log-log PGD-distance plot for the 2016 Mw 7.8 Pedernales, Ecuador, earthquake. In this example, the directivity toward the southwest is evident in the PGD observations, which shows an enhancement of PGD for the seven stations to the southwest of the epicenter and a reduction of PGD for the five stations immediately to the northeast. The other regional stations fall in between these two end members.

Regionally, Ganas et al. (2018) studied a suite of 11 earthquakes around Greece from 1997 to 2017 and computed their own set of regional coefficients and found excellent agreement with all but two events, which had residuals greater than 0.4 magnitude units.

Zang et al. (2022) looked at PGD observations from different multiconstellation PPP solutions for the 2021 Mw 7.4 Maduo, China, earthquake. They found good agreement between all their solutions, and the PGD magnitude reached ~M7.2 40 s after the origin time for most PPP solutions. Furthermore, Fang et al. (2022) looked at the same event but compared the PGD values recorded using the BeiDou B2b corrections against a combination of GPS and BeiDou broadcast solutions, finding good correspondence between the solutions.

On the small end of the scale, Kudlacik et al. (2021) measured PGD at three stations near a mining-related earthquake (M3.7) in the Legnica-Glogow copper mine in Poland, observing displacements between 8.5 and 14.9 mm between 1.3 km and 3.4 km from the hypocenter. When using the Goldberg et al. (2021) regression coefficients, these 3 PGD observations equate to a M4.3 earthquake, larger than the actual magnitude; however, given that none of the PGD studies used any earthquake of this small magnitude to determine the coefficients, it is quite remarkable how close the predicted magnitude is. Moreover, while there are many similarities between anthropogenic and natural earthquakes, there may be differences in the seismic radiation or stress drops between similar moment magnitude events. Also, on the low magnitude end, the original PGD study by Crowell et al. (2013) did include six observations from two events during the 2012 Brawley, California, seismic swarm (e.g., Geng et al., 2013); however, this data was at collocated GNSS and strong-motion stations, and the displacements used were Kalman filtered together (Bock et al., 2011).

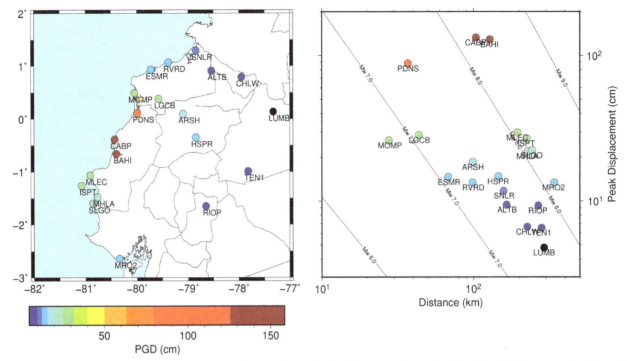

FIG. 6.3 PGD observations for the 2016 Mw 7.8 Pedernales, Ecuador, earthquake. The seven stations to the right of the Mw 8.0 curve on the right are all toward the southwest of the epicenter (see map view), indicating directivity in the rupture. Stations immediately to the northeast of the epicenter *(yellow star)* have a lower predicted magnitude, toward M7. Station PDNS is just south of the epicenter, however, more of the slip front appears to be south of the station; so, PGD is not enhanced.

4 Coseismic-based methods

The static coseismic offsets produced from an earthquake and recorded at high-rate GNSS stations can be used to infer the mechanism of fault rupture and the slip along a fault plane. The offsets are often stable after the period of strong ground shaking, thus complicating the logic required to accurately extract them since the time to stability is based on S and surface wave velocities, the source time function of the earthquake, the local site effects, and the distance to rupture patches. Further complicating the estimation of offsets is the long-period noise due to multipath and constellation geometry that can cause biases in the estimated offsets. Even given the data limitations, using the coseismic offsets from the GNSS position time series remains one of the simplest and accurate ways of obtaining the slip distribution and faulting mechanism of an earthquake in seconds to minutes.

Several methods have been proposed for obtaining the coseismic offsets in real time. The first step is to zero out the time series to either a pre-event average window or to a single point in time prior to the earthquake. For a single point in time, the easiest is to use the origin time of the earthquake that comes from a seismic trigger (as is done with PGD scaling). If data is missing at the origin time, selecting the previously available epoch is often done. For a pre-event window of data, the same seismic origin time can be used, but the user would define a window length to perform an average of the prior epochs of data, which will help to smooth out any potential outliers. However, due to the colored noise structure of GNSS positions, an average of data prior to the event may contain a drift that biases the initialization of the coseismic estimation process. Regardless of the method chosen here, the goal is to have the initial north, east, and up displacements as close to zero as possible so that no erroneous slip or moment is computed.

From here, several different coseismic estimation procedures have been proposed. By Allen and Ziv (2011), a short-term average, long-term average ratio criteria, otherwise known as STA/LTA, was used to determine the onset of shaking and when to allow offsets to be estimated (after making two crossings of a threshold). A similar approach is used by Ohta et al. (2012), which additionally require that the STA minus the LTA is greater than the standard deviation of the LTA by some threshold value of a characteristic function; the maximum of the characteristic function defines when the coseismic offset can be estimated. The method proposed by Ohta et al. (2012) is the same as that is used within the REGARD system under development in Japan (Kawamoto et al., 2017). Crowell et al. (2012) used a simple moving average after the earthquake origin time; they identified a trade-off between the magnitude overrun due to too much shaking and the time to exceed the proper magnitude, which allowed them to settle on a 50-s moving window. The BEFORES algorithm (Minson et al., 2014) uses an exponential moving average instead of a simple moving average; they test an array of decay rates on the exponential moving average and note the lag in the estimated magnitude convergence. In the GlarmS algorithm, the predicted S-wave travel time is used to inform when a window of data can be used (Grapenthin et al., 2014). In G-FAST, a 10-s moving average window is used after a predicted 2-km/s travel time from the earthquake epicenter to the station to remove the effect of shaking in the coseismic estimates (Crowell et al., 2016). As an example of the timing and accuracy of these offset estimation procedures, Fig. 6.4 shows the variance reduction between the estimated offsets and the offset published by the University of Nevada-Reno (UNR) at 20 and 40 s after the origin for the 2019 Mw 7.1 Ridgecrest, California, earthquake for G-FAST, GlarmS, and BEFORES. The UNR offsets represent the best possible near-real-time solution as they are computed using 5-min windows and rapid orbits. Of note, in Fig. 6.4, different stations for which offsets are estimated are presented (stations without ellipses would not have an offset computed at that time for the algorithm). In general, G-FAST is more conservative in when an offset is estimated, but the stations that it does compute offsets for have a higher variance reduction with respect to the UNR offsets. Also of note is that the azimuthal errors in the offset estimation are higher earlier in the earthquake sequence since there will be more shaking at this time.

After coseismic offsets are computed, source modeling can begin. First, a point source representation of the faulting, in the form of a centroid moment tensor (CMT), can be quickly obtained, either using a layered half-space model (Melgar et al., 2012) or a homogeneous half-space (Crowell et al., 2016). For the layered half-space, precomputing Green's functions is necessary since it requires the use of Thomson-Haskell propagator matrices to go from the theoretical deformation at depth to the surface (e.g., Wang et al., 2003). In the homogeneous half-space formulation, Green's functions are computed on the fly since they contain only 18 calculations per iteration. The homogeneous solution is obtained by taking the limit as time goes to infinity in the moment tensor formulation of the seismic wave equation (Hashima et al., 2008). Both the solutions by Melgar et al. (2012) and Crowell et al. (2016) use the deviatoric moment tensor. Fig. 6.5 shows the result of the homogeneous CMT computed with GNSS offsets against the GlobalCMT solution for the 2017 Mw 8.2 Tehuantepec, Mexico, earthquake. The GlobalCMT solution for this has nodal planes of 318/78/-93 and 150/12/-78 for strike/dip/rake and has a source depth of 45 km. The GNSS solution, while slightly off in the nodal planes and a bit deeper at 73 km, is still a fairly accurate representation of the faulting in the earthquake and would be available in about 1–2 min after the event. The deeper depth in the GNSS CMT solution occurs due to the assigning of distributed deformation onto a point source; by pushing the CMT deeper, it is easier to fit the observed offsets. Melgar, Crowell, et al. (2013) proposed a modified version with a line source of CMTs that are regularized through first-order Tikhonov smoothing.

From the derived offsets, it is simple to perform a finite fault slip inversion using static Green's functions (i.e., Okada, 1985). Several different approaches have been taken here, the main difference being in how to choose the appropriate fault orientation/dimensions and the regularization. The slip model will provide the most accurate representation of what occurred during the earthquake, and this information will improve both ground motion and tsunami forecasts. Here, I will outline how three different algorithms determine their slip models: BEFORES (Minson et al., 2014), GlarmS (Grapenthin et al., 2014), and G-FAST (Crowell et al., 2016). For BEFORES, a Bayesian framework is utilized to determine the most likely fault plane, and all potentially orientations are searched over after centering the fault about the hypocenter that is obtained from a seismic EEW algorithm. In BEFORES, regularization is implicit within the Bayesian framework; that is, the choice of priors and the posterior probability of the data guide the inverse problem. The optimal slip model is chosen through maximizing the likelihood. GlarmS utilizes both a fault catalog and the most likely regional fault orientations to determine the slip model. For example, in most of southern California, the orientation of the San Andreas and other parallel faults is roughly northwest (~315–325°); so, GlarmS will generate a fault at this orientation plus use faults that are predefined in the catalog. Regularization in GlarmS is Laplacian and at each time step, the misfit curve (i.e., L-curve) is used to find the optimal smoothness. The optimal slip model in GlarmS is chosen through minimizing the misfit of data residuals. G-FAST generates two fault planes based on the CMT solution generated from the same coseismic offsets, and the one with the lower misfit is chosen. The length and width of the fault planes in both GlarmS and G-FAST are chosen through scaling relationships (e.g., Blaser et al., 2010; Wells & Coppersmith, 1994). For regularization, G-FAST uses an equation that is based on the average values within the Green's function matrix with Laplacian smoothing (Crowell et al., 2012). Fig. 6.6 shows what the slip models for BEFORES, GlarmS, and G-FAST would look like for the 2019 Mw 7.1 Ridgecrest, California, earthquake at 20 and 30 s after the origin time. While there are differences in the discretization and peak slip values for these three algorithms, they all show roughly the same thing, i.e., large slip between 1 and 3 m in the primary epicentral region. Also, note that the offsets used in the slip model are slightly different based on the specific offset estimation scheme within each algorithm (Fig. 6.6).

While not officially demonstrated for early warning, real-time kinematic slip inversions with high-rate GNSS could potentially be done in the future through the precomputing of Green's functions and using 10 s of seconds of data at a time.

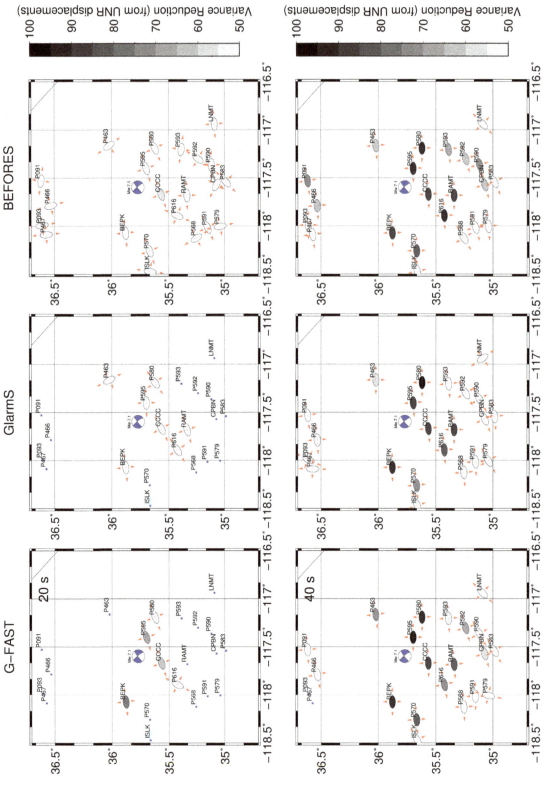

FIG. 6.4 Variance reduction and azimuthal variation between the geodetic algorithm estimated offsets and the University of Nevada, Reno (UNR) offsets at 20 s (*top*) and 40 s (*bottom*) after the origin time. An ellipse that is perfectly horizontal means the displacement azimuth estimated matches the UNR displacement azimuth; any rotation in the ellipse denotes the angle between the two vectors. The *squares* indicate that the station was not yet included in the algorithm at that time.

Earthquake and tsunami early warning with GNSS data **Chapter | 6** 121

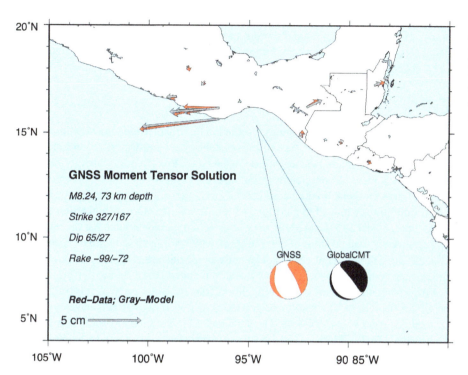

FIG. 6.5 A centroid moment tensor determined from rapid GNSS coseismic offsets for the 2017 Mw 8.2 Tehuantepec, Mexico, earthquake. The *red* vectors show the input data used; the *gray* vectors are the model fits. The GlobalCMT solution, computed through teleseismic records, is shown in the *black* beachball.

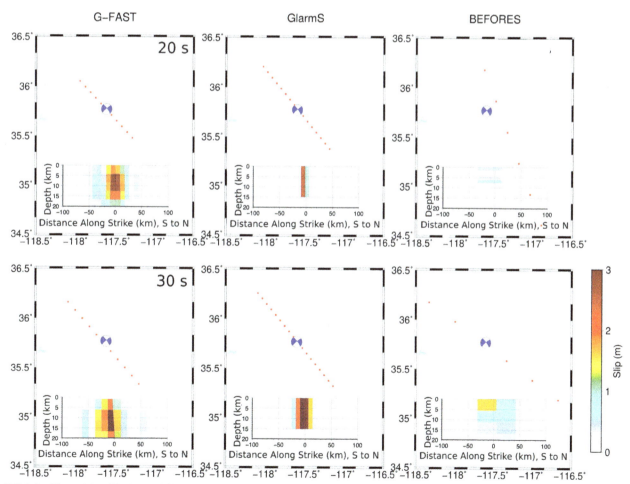

FIG. 6.6 Slip models at 20 s *(top)* and 30 s *(bottom)* after the origin time. The *x*-axis on the slip models is with respect to the epicentral location (indicated by the focal mechanism), with negative distances being south of the epicenter and positive distances north of the epicenter. The *squares* on the maps show the corners of the fault segments at the shallowest depth.

This would only be feasible if the source fault is well known for a specific event, the regularization is predetermined based on prior tests, and multithreading is employed to search for the optimal rupture speed. The timeline of a kinematic inversion is probably too slow to be useful for EEW; however, for TEW, producing a good kinematic slip model in a few minutes after an event would still allow for near-field warning.

5 Algorithm development

Several algorithms have been developed that build on the methodologies presented in this chapter. For an operational algorithm to work successfully, many key elements are required. First, a robust low-latency data pathway is required for real-time GNSS displacements to be properly used in an algorithm. Not only must the data have low latency, which for real-time applications is defined as less than a few seconds, it must be complete with minimal data dropouts. Crowell, Schmidt, et al. (2018) investigated the impact of latency and data dropouts on the potential PGD magnitude and finite fault solutions for several events in Chile and found that these conditions can lead to increased variability in the solutions in the first 10–20 s after the system is triggered. Melgar, Melbourne, et al. (2020) and Melbourne et al. (2020) demonstrated that the telemetry pathways for the Ridgecrest earthquake sequence showed no signs of deterioration, which was an issue for the seismic data telemetry within ShakeAlert (Chung et al., 2020; Stubailo et al., 2021). The reasons for this are complex and beyond the scope of this chapter; however, having multiple redundant data pathways is recommended based on the performance of ShakeAlert during Ridgecrest. Second, a triggering mechanism is required, either from a seismic early warning algorithm or from an algorithm that works on the GNSS data. Several stand-alone geodetic early warning systems have been conceptually demonstrated, although those methods usually use collocated seismic and GNSS stations to address triggering robustness (Goldberg & Bock, 2017; Golriz et al., 2021). Third, algorithms need to have sufficient logic to counteract the high noise level of GNSS observations in real time; over tens of seconds, any GNSS algorithm will report $M > 6$, regardless of earthquake magnitude. Finally, the source information needs to be converted into a useful quantity, such as ground motion or tsunami prediction, and then this information needs to be disseminated to the public while minimizing false positives. In particular, the efficacy of geodetic source information has been shown to be beneficial for ground motion prediction in large earthquakes ($M > 7$) and for events with slip along long faults such as subduction zone events (Ruhl et al., 2017) or large strike-slip system earthquakes (i.e., 2018 Mw 7.8 Kaikoura, Crowell, Melgar, & Geng, 2018). For operational tsunami warnings, dissemination of warnings is generally done on a county/prefecture/state level and provided at certain threshold levels of maximum wave height. Melgar et al. (2016) showed that PGD observations can produce accurate regional tsunami warnings for several large megathrust earthquakes. Williamson et al. (2020) looked at G-FAST solutions for 1300 simulated Cascadia subduction zone earthquakes between M7.5 and 9.5. They found good correspondence between the predicted and true wave heights along the Pacific coast using the existing G-FAST finite faults, and a significant improvement when including a predefined slab model.

Several studies have been performed to show the potential usefulness of geodetic observations to earthquake early warning, and several groups have developed candidate algorithms to be included in the ShakeAlert system (see Murray et al., 2018 for a comprehensive overview). While several other algorithms such as REGARD, BEFORES, and GlarmS are near an operational state, I will focus on the lessons learned from the operationalization process for G-FAST, which provide information for public alerts in ShakeAlert in 2024.

The G-FAST algorithm (Crowell et al., 2016) was officially released in 2016 and contains three core modeling modules: (1) a peak ground displacement determined magnitude, (2) a centroid moment tensor from static offsets, and (3) a finite fault inversion based on the moment tensor results. G-FAST has been running on ShakeAlert development servers since 2018 and is currently undergoing tests for running in the production ShakeAlert system. For data handling, G-FAST continuously reads and ingests 1-Hz displacements over a RabbitMQ broker, which is then converted to Earthworm tracebuf2 streams, which are read whenever there is an event. This data architecture is evolving, and future versions of G-FAST will remove the Earthworm dependency and read directly from a data broker. In order to kick off the modeling modules, an event message must be received, which can come over an ActiveMQ exchange, over a USGS PDL (Product Distribution Layer) feed, or through manual entry of event information. With the trigger information, G-FAST will look for stations that are within the different travel time masks, 3 km/s for PGD and 2 km/s for coseismic offsets. When four stations with data are available from a specific module, messages are broadcast back over an ActiveMQ exchange or stored locally on the machine. Within the PGD output messaging, logic has been added to throttle messages from events where the data is not expected to be above the normal noise levels of real-time displacements. A thorough outline of this logic is provided by Murray et al. (2023). For this, we defined a magnitude sigma that is based on the 99th percentile of time-dependent noise from the real-time solutions, a gross outlier filter, and a requirement that four stations are above certain time-dependent displacement thresholds. For real-time GNSS solutions, the long-period noise leads to an increase in PGD values over time regardless of whether there is an earthquake. The logic we have built into G-FAST attempts to compensate for this issue and to provide a true level of uncertainty that the ShakeAlert system can use. Outside of G-FAST, the ShakeAlert Solution Aggregator working group is modifying the logic that incorporates the magnitude and ground motions from G-FAST, EPIC, FinDer, and PLUM. This is not a trivial issue since seismic algorithms do not know that they are encountering magnitude saturation in real time and the GNSS solutions are considerably noisier than the raw accelerations on strong-motion instruments. The simplest case would be to downweight the seismic and upweight the GNSS solutions for large earthquakes, and vice versa for small earthquakes; but, in a magnitude saturation situation, there is no ground truth to weight magnitude estimates faithfully. The first step that ShakeAlert has taken is to perform a weighted average on the FinDer and G-FAST magnitudes if the FinDer magnitude is greater than the EPIC magnitude. The reported magnitude uncertainty of FinDer is 0.5 magnitude units and G-FAST will be assigned a value from 0.5 to nearly zero if the magnitude estimates at a given time are greater than the 99th percentile value from the noise characteristics (an exponential decay term defines the G-FAST uncertainty). In this scenario, The FinDer magnitude might contribute anywhere from 10% to 50% of the magnitude and G-FAST would provide the rest. Outside of ShakeAlert, G-FAST has been running at NOAA's Center for Tsunami Research since 2020 for the purpose of helping to solve the near-field tsunami warning problem. For G-FAST running at NOAA, data is read from a global RabbitMQ exchange of over 1400 stations processed with Central Washington University's Fastlane software; ShakeAlert is using a subset of this same data. For triggering the G-FAST modeling modules at NOAA, PDL is used, which can transmit many different types of earthquake source information (W-phase, M_{wp}, finite fault, etc.). An additional layer converts the output message of G-FAST to a format that can be ingested by NOAA's SIFT software and run tsunami forecasts rapidly. The messaging from G-FAST was further throttled down to once per minute for ingestion by SIFT since the computational burden of SIFT is large enough to not allow for messages once per second, as is the design within ShakeAlert.

6 Future directions

GNSS-based algorithms provide an independent check on earthquake magnitude, to determine if an earthquake is or is not large. Many scenarios arise where GNSS-based earthquake early warning is useful for the characterization and prediction of large ground motions, most notably subduction zone settings and long earthquake ruptures. Crowell, Melgar, and Geng (2018) noted for the Kaikoura, New Zealand, earthquake that large ground motions recorded toward the northern extent of rupture could be reliably modeled using either a finite fault solution or PGD magnitude from GNSS displacements. PGA at locations toward the end of the rupture during the Kaikoura event exceeded 100% g, but this level of shaking took 40–60 s to manifest, thus presenting an ideal situation for operational early warning.

The different original objectives of GNSS and seismic sensors are also advantageous to early warning systems. Seismic sensors tend to be concentrated in regions with high seismic risks, i.e., near population centers. GNSS stations were installed first to aid in land surveying purposes and, second, to record changes to the dynamic reference frame. The real-time GNSS networks of today grew out of permanent GNSS networks such as SCIGN and PBO that were originally designed to observe regional and larger scale tectonic processes. Because of the original goal of GNSS stations, they tend to be more geographically distributed to capture broad-scale tectonic strains. Since these systems were largely developed independently, the two are sensitive to different types of events and within different tectonic regimes. Grapenthin et al. (2018) noted during the Mw 7.1 Iniskin, Alaska, earthquake in 2016 that GNSS-derived velocities are some of the only constraints on strong ground motions simply due to the different station distributions in the region; Crowell (2021) performed a similar analysis for the four Intermountain western United States earthquakes in 2020. With the ambitious build-out plans for ShakeAlert over the next few years (Given et al., 2018), most GNSS stations along the West Coast will be fully collocated with seismic sensors, at minimum strong-motion accelerometers, and will often have redundant telemetry to ensure system resilience and independence. Many GNSS receivers will also be upgraded to use onboard positioning services that can potentially take in external sources of information, i.e., strong-motion accelerations, and produce a single high-precision broadband displacement waveform (Bock et al., 2011; Geng et al., 2013).

The issues with real-time estimation of displacements have led to several groups proposing using time-differenced, carrier-phase observations of velocity, otherwise known as the variometric approach (Benedetti et al., 2014; Colosimo et al., 2011; Crowell, 2021; Grapenthin et al., 2018; Parameswaran et al., 2023; Shu et al., 2018). With the variometric approach, errors that do not appreciably change over seconds are effectively removed and issues with cycle slips/ambiguity resolution will only cause issues for a single epoch. Crowell et al. (2023) showed that for 5-Hz and greater observations, GNSS velocities match well with seismic ground motion models and can therefore be used to augment seismic warning systems in areas of sparse data coverage. These observations also do not clip for the largest ground motions. Dittmann et al. (2022) demonstrated that a random forest classifier could determine whether GNSS velocities represented a seismic or noise signal, opening the possibility of dynamic blacklisting of GNSS stations used in early warning. Machine learning techniques are also under development to characterize the earthquake source directly from the GNSS time series (i.e., M-LARGE; Lin et al., 2021). This is a promising avenue because it addresses the single, greatest weakness in real-time GNSS displacements, the noise structure.

7 Final thoughts

The field of real-time GNSS for hazard monitoring has only been around since the early 2000s, and is still very much in its infancy; however, it has benefitted greatly from the significant advancements in network seismology over the past half-century. High-rate GNSS time-series analysis is effectively low-frequency seismology, and it uses many of the same techniques that earthquake and tsunami warning systems have employed over the past few decades. In this author's view, the best approach to early warning and rapid response is to holistically use all available datasets and instruments, highlighting their strengths and minimizing their weaknesses. The end goal is to model the full frequency content of the earthquake source as quickly as possible in a fully automated manner; improvements are still necessary for data quality and density, but the future is bright over the coming years.

References

Aki, K., & Richards, P. G. (2002). *Quantitative seismology*.

Allen, R. M., & Melgar, D. (2019). Earthquake early warning: Advances, scientific challenges, and societal needs. *Annual Review of Earth and Planetary Sciences, 47*, 361–388. https://doi.org/10.1146/annurev-earth-053018-060457.

Allen, R. M., & Ziv, A. (2011). Application of real-time GPS to earthquake early warning. *Geophysical Research Letters, 38*, L16310. https://doi.org/10.1029/2011GL047947.

Benedetti, E., Brazanti, M., Biagi, L., Colosimo, G., Mazzoni, A., & Crespi, M. (2014). Global navigation satellite systems seismology for the 2012 Mw 6.1 Emilia earthquake: Exploiting the VADASE algorithm. *Seismological Research Letters, 85*, 649–655. https://doi.org/10.1785/0220130094.

Blaser, L., Kruger, F., Ohrnberger, M., & Scherbaum, F. (2010). Scaling relations of earthquake source parameter estimates with special focus on subduction environment. *Bulletin of the Seismological Society of America, 100*(6), 2914–2926. https://doi.org/10.1785/0120100111.

Blewitt, G., Kreemer, C., Hammond, W. C., Plag, H.-P., Stein, S., & Okal, E. (2006). Rapid determination of earthquake magnitude using GPS for tsunami early warning systems. *Geophysical Research Letters, 33*(11), L11309. https://doi.org/10.1029/2006GL026145.

Bock, Y., Melgar, D., & Crowell, B. W. (2011). Real-time strong-motion broadband displacements from collocated GPS and accelerometers. *Bulletin of the Seismological Society of America, 101*(6), 2904–2925. https://doi.org/10.1785/0120110007.

Böse, M., Smith, D. E., Felizardo, C., Meier, M.-. A., Heaton, T. H., & Clinton, J. F. (2018). FinDer v.2: Improved real-time ground-motion predictions for M 2–M 9 with seismic finite-source characterization. *Geophysical Journal International, 212*, 725–742. https://doi.org/10.1093/gji/ggx430.

Chung, A. I., Henson, I., & Allen, R. M. (2019). Optimizing earthquake early warning performance: ElarmS-3. *Seismological Research Letters, 90A*(2A), 727–743. https://doi.org/10.1785/0220180192.

Chung, A. I., Meier, M.-A., Andrews, J., Böse, M., Crowell, B. W., McGuire, J. J., et al. (2020). ShakeAlert earthquake early warning performance during the 2019 Ridgecrest earthquake sequence. *Bulletin of the Seismological Society of America, 110*(4), 1904–1923. https://doi.org/10.1785/0120200032.

Cochran, E. S., Bunn, J., Minson, S. E., Baltay, A. S., Kilb, D. L., Kodera, Y., et al. (2019). Event detection performance of the PLUM earthquake early warning algorithm in Southern California. *Bulletin of the Seismological Society of America, 109*(4), 1524–1541. https://doi.org/10.1785/0120180326.

Colosimo, G., Crespi, M., & Mazzoni, A. (2011). Real-time GPS seismology with a stand-alone receiver: A preliminary feasibility demonstration. *Journal of Geophysical Research, 116*, B11302. https://doi.org/10.1029/2010JB007941.

Crowell, B. W. (2021). Near-field strong ground motions from GPS-derived velocities for 2020 intermountain Western United States earthquakes. *Seismological Research Letters, 92*(2A), 840–848. https://doi.org/10.1785/0220200325.

Crowell, B. W., Bock, Y., & Melgar, D. (2012). Real-time inversion of GPS data for finite fault modeling and rapid hazard assessment. *Geophysical Research Letters, 39*, L09305. https://doi.org/10.1029/2012GL051318.

Crowell, B. W., Bock, Y., & Squibb, M. B. (2009). Demonstration of earthquake early warning using total displacement waveforms from real-time GPS networks. *Seismological Research Letters, 80*(5), 772–782. https://doi.org/10.1785/gssrl.80.5.772.

Crowell, B., DeGrande, J., Dittmann, T., & Ghent, J. (2023). Validation of peak ground velocities recorded on very-high rate GNSS against NGA-West2 ground motion models. *Seismica, 2*(1). https://doi.org/10.26443/seismica.v2i1.239.

Crowell, B. W., Melgar, D., Bock, Y., Haase, J. S., & Geng, J. (2013). Earthquake magnitude scaling using seismogeodetic data. *Geophysical Research Letters, 40*, 6089–6094. https://doi.org/10.1002/2013GL058391.

Crowell, B. W., Melgar, D., & Geng, J. (2018). Hypothetical real-time GNSS modeling of the 2016 Mw 7.8 Kaikoura earthquake: Perspectives from ground motion and tsunami inundation prediction. *Bulletin of the Seismological Society of America, 108*(3B), 1736–1745. https://doi.org/10.1785/0120170247.

Crowell, B. W., Schmidt, D. A., Bodin, P., Vidale, J. E., Baker, B., Barrientos, S., et al. (2018). G-FAST earthquake early warning potential for great earthquakes in Chile. *Seismological Research Letters, 89*(2A), 542–556. https://doi.org/10.1785/0220170180.

Crowell, B. W., Schmidt, D. A., Bodin, P., Vidale, J. E., Gomberg, J., Hartog, J. R., et al. (2016). Demonstration of the Cascadia G-FAST geodetic earthquake early warning system for the Nisqually, Washington, earthquake. *Seismological Research Letters*, *87*(4), 930–943. https://doi.org/10.1785/0220150255.

Datta, A., Wu, D. J., Zhu, W., Cai, M., & Ellsworth, W. L. (2022). DeepShake: Shaking intensity prediction using deep spatiotemporal RNNs for earthquake early warning. *Seismological Research Letters*, *93*(3), 1636–1649. https://doi.org/10.1785/0220210141.

de Jonge, P., & Tiberius, C. (1996). Integer ambiguity estimation with the lambda method. In *280-284. GPS trends in precise terrestrial, airborne, and spaceborne applications*. https://doi.org/10.1007/978-3-642-80133-4_45.

Dittmann, T., Liu, Y., Morton, Y., & Mencin, D. (2022). Supervised machine learning of high rate GNSS velocities for earthquake strong motion signals. *Journal of Geophysical Research: Solid Earth*, *127*(11), e2022JB024854. https://doi.org/10.1029/2022JB024854.

Dong, D.-. N., & Bock, Y. (1989). Global positioning system network analysis with phase ambiguity resolution applied to crustal deformation studies in California. *Journal of Geophysical Research*, *94*(B4), 3949–3966. https://doi.org/10.1029/JB094iB04p03949.

Duputel, Z., Rivera, L., Kanamori, H., & Hayes, G. (2012). W phase source inversion for moderate to large earthquakes (1990–2010). *Geophysical Journal International*, *189*(2), 1125–1147. https://doi.org/10.1111/j.1365-246X.2012.05419.x.

Fang, R., Lv, H., Hu, Z., Wang, G., Zheng, J., Zhou, R., et al. (2022). GPS/BDS precise point positioning with B2b products for high-rate seismogeodesy: Application to the 2021 Mw 7.4 Maduo earthquake. *Geophysical Journal International*, *231*(3), 2079–2090. https://doi.org/10.1093/gji/ggac311.

Farghal, N. S., Saunders, J. K., & Parker, G. A. (2022). The potential of using fiber optic distributed acoustic sensing (DAS) in earthquake early warning applications. *Bulletin of the Seismological Society of America*, *112*(3), 1416–1435. https://doi.org/10.1785/0120210214.

Ganas, A., Andritsou, N., Kosma, C., Argyrakis, P., Tsironi, V., & Drakatos, G. (2018). A 20-yr database (1997-2017) of co-seismic displacements from GPS recordings in the Aegean area and their scaling with Mw and hypocentral distance. *Bulletin of the Geological Society of Greece*, *52*, 98–130. https://doi.org/10.12681/bgsg.18070.

Ganas, A., Elias, P., Briole, P., Cannavo, F., Valkaniotis, S., Tsironi, V., et al. (2020). Ground deformation and seismic fault model of the M6.4 Durres (Albania) Nov. 26, 2019 earthquakes, based on GNSS/InSAR observations. *Geosciences*, *10*(6), 210. https://doi.org/10.3390/geosciences10060210.

Gao, Y., & Shen, X. (2002). A new method for carrier-phase-based precise point positioning. *Navigation*, *49*, 109–116. https://doi.org/10.1002/j.2161-4296.2002.tb00260.x.

Geng, J., Bock, Y., Melgar, D., Crowell, B. W., & Haase, J. S. (2013). A new seismogeodetic approach applied to GPS and accelerometer observations of the 2012 Brawley seismic swarm: Implications for earthquake early warning. *Geochemistry, Geophysics, Geosystems*, *14*, 2124–2142. https://doi.org/10.1002/ggge.20144.

Genrich, J. F., & Bock, Y. (2006). Instantaneous geodetic positioning with 10–50 Hz GPS measurements: Noise characteristics and implications for monitoring networks. *Journal of Geophysical Research*, *111*, B03403. https://doi.org/10.1029/2005JB003617.

Given, D. D., Allen, R. M., Bodin, P., Cochran, E. S., Creager, K., de Groot, R. M., et al. (2018). *Revised technical implementation plan for the ShakeAlert system – An earthquake early warning system for the west coast of the United States*. U.S. Geol. Surv. Open-File Rept. 2018–1155, 42 pp. https://doi.org/10.3133/ofr20181155.

Goldberg, D. E., & Bock, Y. (2017). Self-contained local broadband seismogeodetic early warning system: Detection and location. *Journal of Geophysical Research - Solid Earth*, *122*, 3197–3220. https://doi.org/10.1002/2016JB013766.

Goldberg, D. E., Melgar, D., Bock, Y., & Allen, R. M. (2018). Geodetic observations of weak determinism in rupture evolution of large earthquakes. *Journal of Geophysical Research - Solid Earth*, *123*(11), 9950–9962. https://doi.org/10.1029/2018JB015962.

Goldberg, D. E., Melgar, D., Hayes, G. P., Crowell, B. W., & Sahakian, V. J. (2021). A ground-motion model for GNSS peak ground displacement. *Bulletin of the Seismological Society of America*, *111*(5), 2393–2407. https://doi.org/10.1785/0120210042.

Golriz, D., Bock, Y., & Xu, X. (2021). Defining the coseismic phase of the crustal deformation cycle with seismogeodesy. *Journal of Geophysical Research: Solid Earth*, *126*, e2021JB022002. https://doi.org/10.1029/2021JB022002.

Grapenthin, R., & Freymueller, J. T. (2011). The dynamics of a seismic wave field: Animation and analysis of kinematic GPS data recorded during the 2011 Tohoku-oki earthquake, Japan. *Geophysical Research Letters*, *38*(18). https://doi.org/10.1029/2011GL048405.

Grapenthin, R., Johanson, I. A., & Allen, R. M. (2014). Operational real-time GPS-enhanced earthquake early warning. *Journal of Geophysical Research - Solid Earth*, *119*, 7944–7965. https://doi.org/10.1002/2014JB011400.

Grapenthin, R., West, M., & Freymueller, J. (2017). The utility of GNSS for earthquake early warning in regions with sparse seismic networks. *Bulletin of the Seismological Society of America*, *107*(4), 1883–1890. https://doi.org/10.1785/0120160317.

Grapenthin, R., West, M., Tape, C., Gardine, M., & Freymueller, J. (2018). Single-frequency instantaneous GNSS velocities resolve dynamic ground motion of the 2016 Mw7.1 Iniskin, Alaska, earthquake. *Seismological Research Letters*, *89*(3), 1040–1048. https://doi.org/10.1785/0220170235.

Hashima, A., Takada, Y., Fukahata, Y., & Matsu'ura, M. (2008). General expressions for internal deformation due to a moment tensor in an elastic/viscoelastic multilayered half-space. *Geophysical Journal International*, *175*, 992–1012. https://doi.org/10.1111/j.1365-246X.2008.03837.x.

Herring, T. A., King, R. W., & McClusky, S. C. (2010). *GAMIT/GLOBK reference manuals, release 10.4*.

Hodgkinson, K. M., Mencin, D. J., Feaux, K., Sievers, C., & Mattioli, G. S. (2020). Evaluation of earthquake magnitude estimation and event detection thresholds for real-time GNSS networks: Examples from recent events captured by the network of the Americas. *Seismological Research Letters*, *91*(3), 1628–1645. https://doi.org/10.1785/0220190269.

Hoshiba, M., et al. (2011). Outline of the 2011 off the Pacific coast of Tohoku earthquake (Mw 9.0) earthquake early warning and observed seismic intensity. *Earth, Planets and Space*, *63*, 547–551. https://doi.org/10.5047/eps.2011.05.031.

Jozinović, D., Lomax, A., Štajduhar, I., & Michelini, A. (2020). Rapid prediction of earthquake ground shaking intensity using raw waveform data and a convolutional neural network. *Geophysical Journal International*, *222*(2), 1379–1389. https://doi.org/10.1093/gji/ggaa233.

Kamigaichi, O., Saito, M., Doi, K., Matsumori, T., Tsukada, S., Takeda, K., et al. (2009). Earthquake early warning in Japan: Warning the general public and future prospects. *Seismological Research Letters, 80*(5), 717–726. https://doi.org/10.1785/gssrl.80.5.717.

Kanamori, H. (1993). W phase. *Geophysical Research Letters, 20*(16), 1691–1694. https://doi.org/10.1029/93GL01883.

Kawamoto, S., Ohta, Y., Hiyama, Y., Todoriki, M., Nishimura, T., Furuya, T., et al. (2017). REGARD: A new GNSS-based real-time finite fault modeling system for GEONET. *Journal of Geophysical Research - Solid Earth, 122*(2), 1324–1349. https://doi.org/10.1002/2016JB013485.

Kodera, Y., Yamada, Y., Hirano, K., Tamaribuchi, K., Adachi, S., Hayashimoto, N., et al. (2018). The propagation of local undamped motion (PLUM) method: A simple and robust seismic wavefield estimation approach for earthquake early warning. *Bulletin of the Seismological Society of America, 108*, 983–1003. https://doi.org/10.1785/0120170085.

Kong, Q., Allen, R. M., Schreier, L., & Kwon, Y.-W. (2016). MyShake: A smartphone seismic network for earthquake early warning and beyond. *Science Advances, 2*(2), 717–718. https://doi.org/10.1126/sciadv.1501055.

Kouba, J., & Héroux, P. (2001). Precise point positioning using IGS orbit and clock products. *GPS Solutions*, 5–12. https://doi.org/10.1007/PL00012883.

Kudlacik, I., Kaplon, J., Lizurek, G., Crespi, M., & Kurpinski, G. (2021). High-rate GPS positioning for tracing anthropogenic seismic activity: The 29 January 2019 mining tremor in Legnica-Glogow Copper District, Poland. *Measurement*, 108396. https://doi.org/10.1016/j.measurement.2020.108396.

Langbein, J., & Bock, Y. (2004). High-rate real-time GPS network at Parkfield: Utility for detecting fault slip and seismic displacements. *Geophysical Research Letters, 31*, L15S20. https://doi.org/10.1029/2003GL019408.

Lichten, S. (1995). *GIPSY-OASIS II: A high precision GPS data processing system and general satellite orbit*.

Lin, J.-T., Melgar, D., Thomas, A. M., & Searcy, J. (2021). Early warning for great earthquakes from characterization of crustal deformation patterns with deep learning. *Journal of Geophysical Research: Solid Earth, 126*, e2021JB022703. https://doi.org/10.1029/2021JB022703.

Melbourne, T. I., Szeliga, W. M., Santillan, V. M., & Scrivner, C. W. (2020). 25-Second determination of 2019 Mw 7.1 Ridgecrest Earthquake Coseismic deformation. *Bulletin of the Seismological Society of America, 110*(4), 1680–1687. https://doi.org/10.1785/0120200084.

Melgar, D., Bock, Y., & Crowell, B. W. (2012). Real-time centroid moment tensor determination for large earthquakes from local and regional displacement records. *Geophysical Journal International, 188*(2), 703–718. https://doi.org/10.1111/j.1365-246X.2011.05297.x.

Melgar, D., Bock, Y., Sanchez, D., & Crowell, B. W. (2013). On robust and reliable automated baseline corrections for strong motion seismology. *Journal of Geophysical Research - Solid Earth, 118*, 1177–1187. https://doi.org/10.1002/jgrb.50135.

Melgar, D., Crowell, B. W., Bock, Y., & Haase, J. S. (2013). Rapid modeling of the 2011 Mw 9.0 Tohoku-oki earthquake with seismogeodesy. *Geophysical Research Letters, 40*, 2963–2968. https://doi.org/10.1002/grl.50590.

Melgar, D., Crowell, B. W., Geng, J., Allen, R. M., Bock, Y., Riquelme, S., et al. (2015). Earthquake magnitude calculation without saturation from the scaling of peak ground displacement. *Geophysical Research Letters, 42*, 5197–5205. https://doi.org/10.1002/2015GL064278.

Melgar, D., Crowell, B. W., Melbourne, T. I., Szeliga, W., Santillan, M., & Scrivner, C. (2020). Noise characteristics of real-time high-rate GNSS positions in a large aperture network. *Journal of Geophysical Research, 125*, e2019JB019197. https://doi.org/10.1029/2019JB019197.

Melgar, D., Geng, J., Crowell, B. W., Haase, J. S., Bock, Y., Hammond, W. C., et al. (2015). Seismogeodesy of the 2014 Mw6.1 Napa earthquake, California: Rapid response and modeling of fast rupture on a dipping strike-slip fault. *Journal of Geophysical Research - Solid Earth, 120*(7), 5013–5033. https://doi.org/10.1002/2015JB011921.

Melgar, D., Melbourne, T. I., Crowell, B. W., Geng, J., Szeliga, W., Scrivner, C., et al. (2020). Real-time high-rate GNSS displacements: Performance demonstration during the 2019 Ridgecrest, CA earthquakes. *Seismological Research Letters, 91*(4), 1943–1951. https://doi.org/10.1785/0220190223.

Melgar, D., et al. (2016). Local tsunami warnings: Perspectives from recent large events. *Geophysical Research Letters, 43*, 1109–1117. https://doi.org/10.1002/2015GL067100.

Minson, S. E., Murray, J. R., Langbein, J. O., & Gomberg, J. S. (2014). Real-time inversions for finite fault slip models and rupture geometry based on high-rate GPS data. *Journal of Geophysical Research - Solid Earth, 119*, 3201–3231. https://doi.org/10.1002/2014JB010622.

Münchmeyer, J., Bindi, D., Leser, U., & Tilmann, F. (2021). The transformer earthquake alerting model: A new versatile approach to earthquake early warning. *Geophysical Journal International, 225*(1), 646–656. https://doi.org/10.1093/gji/ggaa609.

Murray, J. R., Crowell, B. W., Grapenthin, R., Hodgkinson, K., Langbein, J. O., Melbourne, T., et al. (2018). Development of a geodetic component for the U.S. west coast earthquake early warning system. *Seismological Research Letters, 89*(6), 2322–2336. https://doi.org/10.1785/0220180162.

Murray, J. R., Crowell, B. W., Murray, M. H., Ulberg, C. W., McGuire, J. J., Aranha, M. A., et al. (2023). Incorporation of real-time earthquake magnitudes estimated via Peak ground displacement scaling in the ShakeAlert earthquake early warning system. *Bull. Seism. Soc. Am.* https://doi.org/10.1785/0120220181.

Ohta, Y., et al. (2012). Quasi real-time fault model estimation for near-field tsunami forecasting based on RTK-GPS analysis: Application to the 2011 Tohoku-Oki earthquake (M_w 9.0). *Journal of Geophysical Research, 117*, B02311. https://doi.org/10.1029/2011JB008750.

Okada, Y. (1985). Surface deformation due to shear and tensile faults in a half-space. *Bulletin of the Seismological Society of America, 75*(4), 1135–1154.

Parameswaran, R. M., Grapenthin, R., West, M. E., & Fozkos, A. (2023). Interchangeable use of GNSS and seismic data for rapid earthquake characterization: 2021 Chignik, Alaska, Earthquake. *Seismological Research Letters*. https://doi.org/10.1785/0220220357.

Ruhl, C. J., Melgar, D., Geng, J., Goldberg, D. E., Crowell, B. W., Allen, R. M., et al. (2018). A global database of strong-motion displacement GNSS recordings and an example application to PGD scaling. *Seismological Research Letters, 90*(1), 271–279. https://doi.org/10.1785/0220180177.

Ruhl, C. J., Melgar, D., Grapenthin, R., & Allen, R. M. (2017). The value of real-time GNSS to earthquake early warning. *Geophysical Research Letters, 44*(16), 8311–8319. https://doi.org/10.1002/2017GL074502.

Sardina, V., McCreery, C., & Fryer, G. (2017). Compilation and analysis of two decades worth of tsunami bulletins issues by the Pacific tsunami warning center. In *16th World conference on earthquake engineering, paper #193, Santiago de Chile, January 9–13*.

Saunders, J. K., Minson, S. E., Baltay, A. S., Bunn, J. J., Cochran, E. S., Kilb, D. L., et al. (2022). Real-time earthquake detection and alerting behavior of PLUM ground-motion-based early warning in the United States. *Bulletin of the Seismological Society of America*, *112*(5), 2668–2688. https://doi.org/10.1785/0120220022.

Shu, Y., Fang, R., Li, M., Shi, C., Li, M., & Liu, J. (2018). Very high-rate GPS for measuring dynamic seismic displacements without aliasing: Performance evaluation of the variometric approach. *GPS Solutions*, *22*, 121. https://doi.org/10.1007/s10291-018-0785-z.

Song, C., Xu, C.-J., Wen, Y.-M., Yi, L., & Xu, W. (2017). Surface deformation and early warning magnitude of 2016 Kaikoura (New Zealand) earthquake from high-rate GPS observations. *Chinese Journal of Geophysics*, *60*(6), 602–612. https://doi.org/10.1002/cjg2.30071.

Stubailo, I., Alvarez, M., Biasi, G., Bhadha, R., & Hauksson, E. (2021). Latency of waveform data delivery from the Southern California Seismic Network during the 2019 Ridgecrest earthquake sequence and its effect on ShakeAlert. *Seismological Research Letters*, *92*(1), 170–186. https://doi.org/10.1785/0220200211.

Suárez, G., Espinosa-Aranda, J. M., Cuéllar, A., Ibarrola, G., García, A., Zavala, M., et al. (2018). A dedicated seismic early warning network: The Mexican seismic alert system (SASMEX). *Seismological Research Letters*, *89*(2A), 382–391. https://doi.org/10.1785/0220170184.

Tsuboi, S. (2000). Application of Mwp to tsunami earthquake. *Geophysical Research Letters*, *27*(19), 3105–3108. https://doi.org/10.1029/2000GL011735.

Wang, R., Martin, F. L., & Roth, F. (2003). Computation of deformation induced by earthquakes in a multi-layered elastic crust—FORTRAN programs EDGRN/EDCMP. *Computers & Geosciences*, *29*(2), 195–207. https://doi.org/10.1016/S0098-3004(02)00111-5.

Wells, D. L., & Coppersmith, K. J. (1994). New empirical relationships among magnitude, rupture length, rupture width, rupture area, and surface displacement. *Bulletin of the Seismological Society of America*, *84*(4), 974–1002. https://doi.org/10.1785/BSSA0840040974.

Williamson, A. L., Melgar, D., Crowell, B. W., Arcas, D., Melbourne, T. I., Wei, Y., et al. (2020). Toward near-field tsunami forecasting along the Cascadia subduction zone using rapid GNSS source models. *Journal of Geophysical Research - Solid Earth*, *125*, e2020JB019636. https://doi.org/10.1029/2020JB019636.

Zang, J., Wen, Y., Li, Z., Xu, C., He, K., Zhang, P., et al. (2022). Rapid source models of the 2021 Mw 7.4 Maduo, China, earthquake inferred from high-rate BDS3/2, GPS, Galileo and GLONASS observations. *Journal of Geodesy*, *96*, 58. https://doi.org/10.1007/s00190-022-01641-w.

Zumberge, J. F., Heflin, M. B., Jefferson, D. C., Watkins, M. M., & Webb, F. H. (1997). Precise point positioning for the efficient and robust analysis of GPS data from large networks. *Journal of Geophysical Research*, *102*(B3), 5005–5017. https://doi.org/10.1029/96JB03860.

Chapter 7

Measuring volcano deformation with GNSS

Yosuke Aoki

Earthquake Research Institute, The University of Tokyo, Tokyo, Japan

1 Introduction

Volcanic unrest often accompanies pressurization, transport, or both, of magmatic or hydrothermal fluids, resulting in Earth's deformation measured at the surface. While volcano deformation processes have multiple timescales, the shortest of which is less than a second, this chapter only deals with deformation of timescales longer than a few minutes or hours, which can be detected by the Global Navigation Satellite System (GNSS).

Active volcanoes have been recognized to uplift and subside for ~2000 years, and systematic volcano monitoring has been around since the beginning of the 20th century (e.g., Del Gaudio, Aquino, Ricciardi, Ricco, & Scandone, 2010; Di Vito et al., 2016; Dzurisin, 2006). Geodetic monitoring with conventional instruments, such as leveling, triangulation, trilateration, tiltmeters, and strainmeters, have been the main component of volcano monitoring since the dawn of volcano monitoring. However, it gave way to space geodetic techniques, such as GNSS and synthetic aperture radar (SAR), after their emergence in the early 1990s (e.g., Dzurisin, 2006).

Before the emergence of space geodetic techniques, it was technically challenging to monitor active volcanoes in high spatial and temporal resolution. Because leveling, triangulation, and trilateral surveys require substantial human resources, they have issues of temporal resolution. While tiltmeters and strainmeters have high sensitivity, their data often contain drifts due to incomplete coupling between the sensor and host rock (Agnew, 1986).

Because volcanic deformation is often limited to smaller areas than seismic deformation, spatially dense observations are required to capture detailed deformation features. Observations by GNSS and SAR fulfill the requirement. In addition, GNSS can resolve the temporal resolution issues because its continuous measurements can monitor deformation with a temporal resolution of a day or shorter.

Recent volcano deformation studies have been dominated by measurements by GNSS and SAR. GNSS plays a vital role in monitoring volcano deformation because (1) the temporal resolution of GNSS measurements is better than that of SAR, (2) the precision of GNSS measurements is an order of millimeters, while that of SAR measurements is 10 mm or worse, and (3) GNSS can measure three-dimensional (3D) displacements, while SAR usually measures only displacements of the line-of-sight component. Because volcano deformation is often tiny and possesses short timescales, GNSS measurements are indispensable to gain insights into the mechanics of the source of the observed volcano deformation.

Along with monitoring techniques, theoretical developments in deformation modeling have been conducted since the early 20th century. The earliest efforts were to derive analytical expressions of the deformation field, assuming a simple deformation source, such as (de)pressurization of a spherical or spheroidal body embedded in a simple medium, such as a homogeneous, isotropic, and elastic half-space. Recent computational capability development allows us to assume more complex and realistic medium and pressure sources. Because analytical solutions rarely exist in such a situation, the modeling requires extensive numerical techniques, such as the finite element method or boundary element method.

Not only observational but also theoretical studies of fluid transport beneath active volcanoes have advanced volcanology. Because the modeling solely from deformation measurements is purely kinematic without any physical constraints, considering the physics of fluid transport will advance the deformation modeling. However, it is only in the last decade or two that both observational and theoretical insights are combined.

This chapter describes recent progress and perspectives on volcano monitoring with GNSS. Section 2 discusses modeling the observed deformation for (de)pressurization of magmatic or hydrothermal fluids. Section 3 reviews case studies of volcano deformation studies with GNSS. Section 4 introduces unconventional usages of GNSS signals to monitor atmospheric and ionospheric disturbances triggered by volcanic eruptions to discuss insights on the mechanics of volcanic eruptions gained from GNSS measurements. Finally, Section 5 provides a summary of this chapter from the perspectives of volcano monitoring with GNSS.

2 Relating observed deformation to subsurface processes

A goal of volcano deformation monitoring is to understand the location and size of (de)pressurization and any temporal evolution. Therefore, we start by discussing how the surface deformation field, or surface displacements measured by GNSS, is related to the accumulation and migration of magmatic or hydrothermal fluids.

An analytical solution is available if a (de)pressurization point source is embedded in a simple medium, such as homogeneous, isotropic, and elastic half-space. Although simple situations with analytical solutions do not necessarily reconstruct reality in the greatest detail, the availability of analytical solutions helps us efficiently solve an inverse problem to estimate the source properties from the observed deformation field. Adding complexity, such as topography, material heterogeneity, anelasticity of the Earth, or complex geometry of a (de)pressurized body, makes analytical solutions intractable. In that case, we must rely on numerical methods to obtain surface deformation due to the pressure changes. Here we start from a simple problem setting, in which analytical solutions are available, and then move on to more complex and realistic problem settings, in which numerical methods must be invoked to reconstruct the deformation field.

2.1 Governing equations

Although the Earth is spherical, the flat Earth approximation works because the source of volcano deformation is much shallower than Earth's radius. Therefore, here we proceed with the Cartesian coordinate system with the x_1 and x_2 axes in horizontal directions and the x_3 axis in the vertical direction.

The stress field must satisfy the quasistatic equilibrium equation (e.g., Segall, 2010)

$$\nabla \cdot \boldsymbol{\sigma} = 0 \tag{7.1}$$

or

$$\sum_{j=1}^{3} \frac{\partial \sigma_{ij}}{\partial x_j} = 0 \quad (i = 1, 2, 3) \tag{7.2}$$

where $\boldsymbol{\sigma}$ and σ_{ij} denote the stress tensor and its ij component, respectively.

The boundary condition at the surface of the source is (Fig. 7.1)

$$\boldsymbol{n}_s^T \boldsymbol{\sigma} \boldsymbol{n}_s = -\Delta p \tag{7.3}$$

if the overpressure of the source is given, where $\boldsymbol{n}_s$ and Δp represent the normal vector pointing outward of the surface of the source and the overpressure, respectively, and T denotes matrix transpose.

If the boundary condition is given in displacement, rather than the overpressure, quasistatic equilibrium equation (7.1) or (7.2) are given in Navier's equation ignoring the body force as

$$\nabla^2 \boldsymbol{u} + \frac{1}{1-2v} \nabla \nabla \cdot \boldsymbol{u} = \boldsymbol{0} \tag{7.4}$$

where $\boldsymbol{u}$ and v denote the displacement vector and Poisson's ratio, respectively.

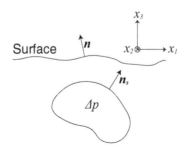

FIG. 7.1 A schematic representation of the boundary condition on a pressure source (Eq. 7.3) and the free surface (Eq. 7.5).

The boundary condition at the free surface is given by (Fig. 7.1)

$$\boldsymbol{\sigma n} = \mathbf{0} \tag{7.5}$$

where $\boldsymbol{n}$ represents the unit vector normal to the surface. The boundary condition on the surface of the half-space can be rewritten as

$$\sigma_{13} = \sigma_{23} = \sigma_{33} = 0. \tag{7.6}$$

Suppose the boundary condition at the source is given in displacements. In that case, the displacements at the surface of an elastic, isotropic, and homogeneous half-space are $4(1 - v)$ times those at the corresponding point within a full-space of the same material property (Davies, 2003). This property makes the derivation of the surface displacement relatively straightforward in some cases.

2.2 Analytical modeling

The simplest analytical model assumes a source embedded in an elastic, isotropic, and homogeneous half-space. Here, we discuss surface displacements by starting from the simplest spherical source and then adding complexity. Here, mathematical details for derivation are omitted; readers should refer to literature cited in this article or Segall (2010, Chapter 7).

2.2.1 Spherical source

One of the simplest model settings is an infinitely small spherical source embedded in an elastic, isotropic, and homogeneous half-space (Fig. 7.2), which is often called the "Mogi model" because Mogi (1958) applied this model to volcano deformation for the first time. In fact, the model itself dates back to Sezawa (1931), Anderson (1937), and Yamakawa (1955). The surface deformation is given by (McTigue, 1987)

$$u_z(r) = \frac{(1-v)a^3 \Delta p}{\mu}\left(1 + \frac{1}{2(-7+5v)}\left(\frac{a}{d}\right)^3\left(1+v + \frac{15d^2(-2+v)}{2(r^2+d^2)}\right) + O\left(\frac{a}{d}\right)^6\right)\frac{d}{(r^2+d^2)^{3/2}} \tag{7.7}$$

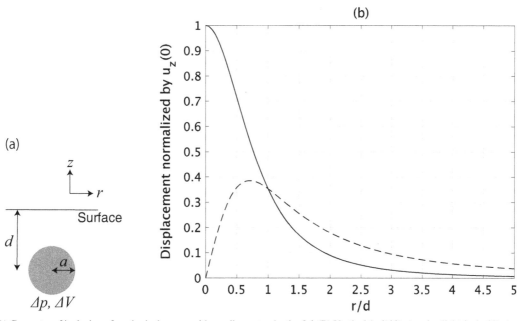

FIG. 7.2 (A) Geometry of inclusion of a spherical source with a radius a at a depth of d. (B) Vertical (*solid line*) and radial (*dashed line*) components of the surface displacements due to the inclusion of a spherical pressure source with an approximation that the radius of the spherical source is much smaller than its depth (Mogi, 1958). The *horizontal axis* denotes the horizontal distance from the source normalized by the depth of the source. The *vertical axis* represents displacements normalized by the vertical displacement right above the source.

$$u_r(r) = \frac{(1-v)a^3 \Delta p}{\mu} \left(1 + \frac{1}{2(-7+5v)}\left(\frac{a}{d}\right)^3\left(1+v + \frac{15d^2(-2+v)}{2(r^2+d^2)}\right) + O\left(\frac{a}{d}\right)^6\right)\frac{r}{(r^2+d^2)^{3/2}}, \qquad (7.8)$$

where $u_z(r)$ and $u_r(r)$ denote displacements in vertical and radial components, respectively, a^3 and d represent the radius and depth of the spherical source, respectively, and μ and v denote rigidity and Poisson's ratio, respectively. $O(a/d)^6$ indicates that the remaining terms are neglected here because they are on the order of $(a/d)^6$ or higher.

If $(a/d)^3 \ll 1$, Eqs. (7.7), (7.8) are rewritten as

$$u_z(r) = \frac{(1-v)a^3 \Delta p}{\mu} \frac{d}{(r^2+d^2)^{3/2}}, \qquad (7.9)$$

$$u_r(r) = \frac{(1-v)a^3 \Delta p}{\mu} \frac{r}{(r^2+d^2)^{3/2}}. \qquad (7.10)$$

Eqs. (7.9), (7.10) can also be written using the volume change $\Delta V = \pi a^3 \Delta p / \mu$ as

$$u_z(r) = \frac{(1-v)\Delta V}{\pi} \frac{d}{(r^2+d^2)^{3/2}}, \qquad (7.11)$$

$$u_r(r) = \frac{(1-v)\Delta V}{\pi} \frac{r}{(r^2+d^2)^{3/2}}. \qquad (7.12)$$

Fig. 7.2 depicts normalized surface displacements in vertical and radial directions as a function of horizontal distance from the source normalized by the source depth. The vertical displacement is maximum right above the source, with decay to half of the maximum at $r = (2^{3/2} - 1)d \sim 0.77d$, while the horizontal displacement is maximum at $r = d/\sqrt{2} \sim 0.71d$. These insights can help roughly estimate the depth of the pressure source from the observed deformation field.

Eqs. (7.9)–(7.12) are under an assumption $(d/a)^3 \ll 1$. Eqs. (7.7), (7.8) show that the contribution of the $(d/a)^3$ is about 10% of the leading term even when $(d/a) = 0.5$, indicating that the approximation is widely applicable (e.g., Taylor, Johnson, & Herd, 2021).

Zhong, Dabrowski, and Jamtveit (2019) investigated the same problem in a different way. They introduced the series expansion of Papkovich-Boussinesq displacement potentials to solve Navier's equation (7.4). While their solution is so tedious to be shown here, it is more general than the solution shown earlier (Eqs. 7.7, 7.8) because it offers not only the surface deformation field but also the deformation field at depth. Also, their solution does not only consider the boundary condition by the volume change or magma overpressure but also incorporates the sliding at the interface between the spherical source and host rock.

2.2.2 Ellipsoidal source

Volcanologists recognized in the 1970s that a pressure source responsible for surface deformation does not always take a spherical shape (e.g., Dieterich & Decker, 1975). Davis (1986) derived an analytical solution of surface displacements due to the pressurization of a spheroidal source with a point-source approximation. Yang, Davis, and Dieterich (1988) relaxed the point-source approximation to derive the surface displacement due to a spheroid with a finite size. Subsequently, Nikkhoo, Walter, Lundgren, and Prats-Iraola (2017) proposed a model to approximate the pressurization of an ellipsoid of arbitrary shape as a combination of three vertically intersected dikes. Amoruso and Crescentini (2013) derived an analytical expression of the surface displacement due to the pressurization of an ellipsoid of arbitrary shape, and Nikkhoo and Rivalta (2023) derived it with more precision.

The pressurization of an elliptical source with a vertical rotational axis gives an axisymmetric deformation field. Fig. 7.3 depicts displacements in this situation. Pressurization of an ellipsoid with a horizontal major axis generates a more localized deformation than pressurization of a spherical body does. Horizontal displacements are relatively smaller than those by pressurization of a sphere with their maximum closer to the source (Fig. 7.3). On the other hand, pressurized ellipsoid with a vertical major axis generates a more widespread deformation with relatively larger horizontal displacements than those by a pressurization of a sphere with their maximum farther from the source (Fig. 7.3).

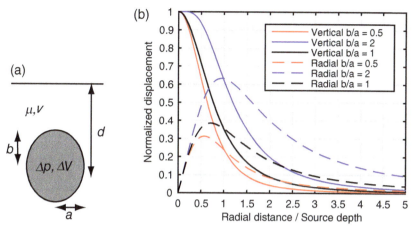

FIG. 7.3 (A) A schematic view of a pressurized ellipsoid. Δp and ΔV denote pressure and volume changes, respectively. d, a, and b represent the depth of the pressurization and major and minor axes of the ellipsoid, respectively. μ and ν are rigidity and Poisson's ratio of the medium, respectively. (B) Vertical (*solid lines*) and radial (*dashed lines*) displacements due to the pressurization of an ellipsoid. *Red and blue lines* depict displacements by pressurizing an ellipsoid with $a = 0.2d$ and $b = 0.1d$, and $a = 0.1d$ and $b = 0.2d$, respectively. For reference, vertical and radial displacements due to the pressurization of a spherical source ($a = b$) are also shown by *black lines*. These displacements are normalized to their maximum vertical displacements. The *horizontal axis* is the radial distance normalized by the depth of the ellipsoidal source.

2.2.3 Dike and sill

Intruded magma favors forming sheets if it is significantly less viscous than the host rock (Rubin, 1993). A vertical or horizontal sheet intrusion is called a dike or sill, respectively. Magma intrusion as a dike or sill is ubiquitous in active volcanoes with low-viscosity magma.

Deriving the precise displacement field due to an intrusion of a dike or a sill is more complicated than it looks, even assuming a homogeneous, elastic, and isotropic half-space. Okada (1985) provided an exact analytical solution of surface displacements with a boundary condition given as the width of the intruded body. While the analytical solution of the surface displacement, given a finite size of the intruded body, is lengthy, that with a point source approximation to a sill is simplified as

$$u_z = \frac{3M_0}{2\pi\mu} \frac{d^3}{(r^2 + d^2)^{5/2}}, \tag{7.13}$$

$$u_r = \frac{3M_0}{2\pi\mu} \frac{rd^2}{(r^2 + d^2)^{5/2}}, \tag{7.14}$$

where M_0 represents the moment of the opening, that is, volume change multiplied by rigidity.

Derivation of surface displacements, given the pressure change of a dike or sill as a boundary condition, is more complex. Fialko, Khazan, and Simons (2001) derived a semianalytical solution of surface displacements due to an inclusion of a penny-shaped crack horizontally embedded in a homogeneous, elastic, and isotropic half-space (Fig. 7.4). Fig. 7.4B shows that a point-source approximation ($d/a = \infty$) gives a more localized displacement field than the inclusion of a penny-shaped crack of finite size. While these two exhibit similar ratios of radial to vertical displacements, the maximum radial displacement with a point-source approximation is closer to the source. In addition, while a sill inclusion exhibits a similar vertical displacement field with an inclusion of a sphere, it exhibits significantly smaller radial displacements, indicating that the observation of both horizontal and vertical displacements is necessary to constrain the shape of the pressure source.

A dike intrusion does not give a point symmetric displacement field (Fig. 7.5); displacements are small along the strike of the dike and large perpendicular to it. In addition, a dike intrusion exhibits subsidence right above the dike and uplifts away from the dike. The near-field tilt is toward the dike, and the far-field tilt is away. When a dike shallows, some points at the surface change the tilt direction. For example, Aoki, Segall, Kato, Cervelli, and Shimada (1999) observed the change of tilt direction of a tiltmeter near the epicenter of a seismic swarm; this observation plays a substantial role in constraining the temporal evolution of the dike propagation during the seismic swarm.

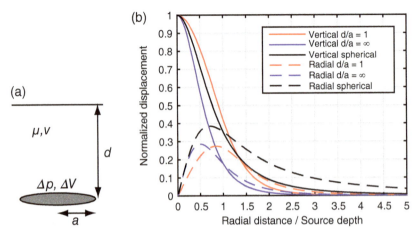

FIG. 7.4 (A) A schematic view of a pressurized sill. Δp and ΔV denote pressure and volume changes, respectively. d and a represent the depth and radius of the sill, respectively. μ and ν are rigidity and Poisson's ratio of the medium, respectively. (B) Vertical (*solid lines*) and radial (*dashed lines*) displacements due to the pressurization of a circular sill. *Red and blue lines* depict displacements by pressurized sills with $d/a = 1$ and $d/a = \infty$, corresponding to a point-source approximation. For reference, vertical and radial displacements due to the pressurization of a sphere are also shown by *black lines*. These displacements are normalized by their maximum vertical displacements. The *horizontal axis* is the radial distance normalized by the depth of the sill.

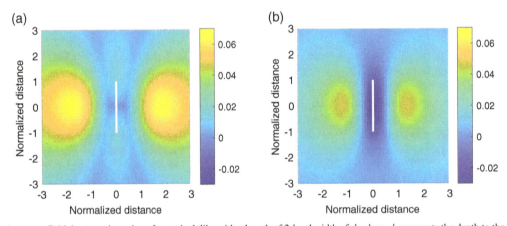

FIG. 7.5 Displacement field due to an intrusion of a vertical dike with a length of $2d$ and width of d, where d represents the depth to the top of the dike. Horizontal and vertical axes are normalized by the depth to the top of the dike. Displacements are normalized by the amount of openings. (A) Horizontal displacement and (B) vertical displacement.

2.2.4 Closed cylinder

A volcanic conduit is often modeled as a cylinder. Some studies (e.g., Bonaccorso & Davis, 1999; Salzer et al., 2014) modeled the observed deformation as the pressurization of a cylinder. Surface displacements due to pressurization of a cylinder of radius a extending from the depth d_1 from d_2 are given by (Segall, 2010)

$$u_z = \frac{a^2 \Delta p}{4\mu} \left(\frac{1 - 2\nu}{(r^2 + c^2)^{1/2}} + \frac{r^2}{(r^2 + c^2)^{3/2}} \right) \Bigg|_{c=d_1}^{c=d_2}, \qquad (7.15)$$

$$u_r = \frac{a^2 r \Delta p}{4\mu} \left(\frac{(3 - 2\nu)c}{r^2(r^2 + c^2)^{1/2}} + \frac{c}{(r^2 + c^2)^{3/2}} \right) \Bigg|_{c=d_1}^{c=d_2}. \qquad (7.16)$$

Fig. 7.6 shows the displacement field associated with the pressurization of a closed cylinder with infinite length ($d_2 = \infty$) embedded in an elastic, homogeneous, and isotropic half-space with $\mu = 20$ GPa and $\nu = 0.25$. Fig. 7.6 shows that pressurization gives uplift and outward displacements everywhere, with its maximum vertical displacement at $r = d_1$ and maximum horizontal displacement at $r \sim 2.02 d_1$.

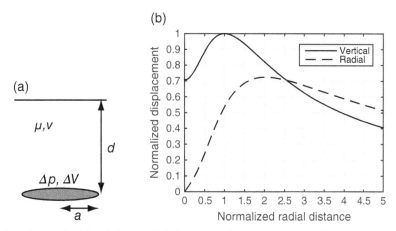

FIG. 7.6 (A) A schematic view of pressurized closed pipe. a and ΔP denote the radius and overpressure of the pipe, respectively. d_1 and d_2 are the depths of the top and bottom of the pipe, respectively. μ and ν represent rigidity and Poisson's ratio of the medium, respectively. (B) Vertical (*solid line*) and radial (*dashed line*) displacements due to pressurization of a closed pipe with an infinite length ($d_2 = \infty$ in Eq. 7.16), the rigidity of 20 GPa, and Poisson's ratio of 0.25. *Horizontal and vertical axes* represent the radial distance normalized by the depth of the top of the pipe and displacements normalized by the maximum vertical displacement, respectively.

2.2.5 Open cylinder

Some volcanoes have an open conduit, where the boundary condition differs from that discussed earlier. The surface deformation due to a pressurized open conduit between depths d_1 and d_2 with a displacement boundary condition (Fig. 7.7A) is given by

$$u_z = a\Delta x \left(\frac{1+2\nu}{2(r^2+c^2)^{1/2}} - \frac{r^2}{2(r^2+c^2)^{3/2}} \right) \bigg|_{c=d_1}^{c=d_2}, \quad (7.17)$$

$$u_r = \frac{ac\Delta x}{r} \left(\frac{1+\nu}{(r^2+c^2)^{1/2}} - \frac{c^2}{2(r^2+c^2)^{3/2}} \right) \bigg|_{c=d_1}^{c=d_2}, \quad (7.18)$$

where Δx denotes the radial displacement of the conduit. Fig. 7.7B shows that the displacement field induced by pressurization of an open conduit is quite different from its closed counterpart (Fig. 7.6B). With $\nu = 0.25$, a pressurization of an

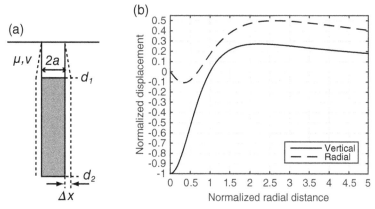

FIG. 7.7 (A) A schematic view of a pressurized open pipe. a and Δx denote the radius of the pipe and the amount of opening, respectively. d_1 and d_2 represent the depths of the top and bottom of the pipe, respectively. μ and ν represent rigidity and Poisson's ratio of the medium, respectively. (B) Vertical (*solid line*) and radial (*dashed line*) displacement due to pressurization of an open pipe with an infinite length ($d_2 = \infty$ in Eqs. 7.17, 7.18) and Poisson's ratio of 0.25. *Horizontal and vertical axes* represent the radial distance normalized by the depth of the top of the pipe and displacements normalized by the subsidence right above the source, respectively.

open conduit subsides right above the conduit. When $d_2 = \infty$, a pressurization induces uplift at radial distance $r > d_1$. The amount of the maximum uplift at $r \sim 2.24d_1$ is $\sim 27\%$ of the maximum subsidence. A pressurization induces inward and outward displacements at $r < 0.69d_1$ and $r > 0.69d_1$, with their maximum at $r \sim 0.30d_1$ and $r \sim 2.64d_1$, respectively.

2.3 Shear on the conduit

Because magma is a viscous fluid, the movement of magma in a volcanic conduit gives shear traction on the conduit wall (Fig. 7.8A). If the distance between the conduit and the observation point is large enough, the shear traction applied to the conduit can be approximated as a single force. This approximation is reasonable because a volcanic conduit is only as wide as a few tens of meters. With this approximation, the surface displacement induced by shear stress τ on the wall of the conduit with radius a at depths between d_1 and d_2 (Fig. 7.8A) is given by (Nishimura, 2009)

$$u_z = \frac{-\tau a}{2\mu}\left(\frac{c}{\sqrt{r^2+c^2}} + (1-2v)\ln\left(\sqrt{r^2+c^2}+c\right)\right)\bigg|_{c=d_1}^{c=d_2}, \tag{7.19}$$

$$u_r = \frac{-\tau ar}{2\mu}\left(\frac{c}{\sqrt{r^2+c^2}} - \frac{1-2v}{\sqrt{r^2+c^2}+c}\right)\bigg|_{c=d_1}^{c=d_2}, \tag{7.20}$$

where μ represents the rigidity of the host rock.

Fig. 7.8B depicts the displacement field with $d_2 = 1.5d_1$ and $d_2 = 4.0d_1$, indicating that the shear traction induces uplift in all areas in both cases. The horizontal displacements are inward near the conduit and the far field while outward at intermediate distances. As in the vertical displacements, the deeper the bottom of the shear traction, the longer the wavelength of the horizontal displacement field.

2.4 Adding material complexities

We have so far discussed surface deformation due to various sources embedded in a homogeneous, elastic, and isotropic half-space. Many studies have employed this assumption because of the simplicity and availability of analytical solutions. This assumption is partially validated because the static ground deformation field is less sensitive to short-wavelength material heterogeneity than the seismic wavefield. Real volcanoes, however, are not homogeneous or flat. Therefore, heterogeneous materials and topography must be considered for more realistic modeling.

The recent development of computational capabilities allows us to compute the deformation field incorporating the full complexity of Earth's elastic and anelastic structure and topography (e.g., Aagaard, Knepley, & Williams, 2013). However, doing so not only takes substantial time to compute the deformation field for some purposes but is also not always effective because the elastic and anelastic parameters beneath the surface are not usually constrained in full resolution. It is, therefore, always important to consider the appropriate complexity to model the deformation field in a given situation.

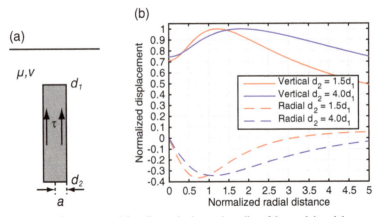

FIG. 7.8 (A) A schematic view of a shear force on a conduit wall. a and τ denote the radius of the conduit and the amount of shear stress applied to the wall, respectively. d_1 and d_2 represent the depths of the top and bottom of the pipe, respectively. μ and v represent rigidity and Poisson's ratio of the medium, respectively. (B) Vertical (*solid lines*) and radial (*dashed lines*) displacements due to the shear force on the conduit wall. *Red and blue lines* depict displacements with the bottom depths of the conduit at $d_2 = 1.5d_1$ and $d_2 = 4.0d_1$. *Horizontal and vertical axes* represent the radial distance normalized by the depth of the top of the conduit and displacements normalized by the maximum vertical displacements, respectively.

2.4.1 Effect of vertical and lateral heterogeneities

The Earth's elastic parameters vary the most with depth because they are most sensitive to pressure. It is thus a good approximation to assume a layered structure (e.g., Dziewonski & Anderson, 1981; Kennett, Engdahl, & Buland, 1995). However, volcanic regions are laterally more heterogeneous than other regions because of various factors, such as magma intrusion and sedimentation of volcanic products.

The deformation field with an assumption of a layered structure is derived by the propagator matrix method (e.g., Aki & Richards, 2002; Gilbert & Backus, 1966 a, 1966 b, pp. 269–282) with a zero frequency limit (Zhu & Rivera, 2002). The propagator matrix connects analytical solutions within each homogeneous layer by satisfying continuity in stress and displacement at material interfaces and by satisfying the boundary condition at the surface.

Analytical solutions are usually only available if the material is laterally heterogeneous. While the surface deformation is derived by numerical methods such as the finite element method, a perturbation solution is available if the perturbation of the elastic constants is small enough (e.g., Cervelli, Kenner, & Segall, 1999; Du, Segall, & Gao, 1994, 1997). As the horizontal variations of elastic constants are usually more minor than the vertical ones, calculating the deformation as a perturbation solution from a layered structure is often valid.

Manconi, Walter, and Amelung (2007) evaluated the effect of material layering on the observed deformation field and concluded that the volume change and the source depth are underestimated when the deformation field is induced by a pressure source embedded in a layered half-space is inverted, assuming a homogeneous half-space. It is because layers of low rigidity or Young's modulus near the surface amplify the deformation. Subsequent studies (e.g., de Zeeuw-van Dalfsen, Pedersen, Hooper, & Sigmundsson, 2012; Langer, Beller, Hirakawa, & Tromp, 2023; Langer, Gharti, & Tromp, 2019; Long & Grosfils, 2009) reached a similar conclusion. Because the accumulated volcanic products make the rigidity of active volcanoes low at shallow depths or make the vertical contrast of rigidity high, it is crucial to consider the vertical contrast of material properties to model volcano deformation. In addition, Masterlark (2007) pointed out that lateral heterogeneity generated by caldera deposits, for example, plays a significant role in modeling volcano deformation.

2.4.2 Effect of viscoelasticity

While the materials at shallow depths can be regarded as purely elastic, materials at depths cannot be purely elastic because of elevated temperature. In tectonic environments, observations of postseismic deformation and postglacial rebound, for example, have identified lower viscosity at the lower crust and upper mantle than the upper crust (e.g., Bürgmann & Dresen, 2008). A viscoelastic layer beneath an elastic half-space also exists in volcanic regions. For example, Sigmundsson et al. (2020), Yamasaki, Sigmundsson, and Iguchi (2020), and Yamasaki, Takahashi, Ohzono, Wright, and Kobayashi (2020) showed that the relaxation of a viscoelastic layer below an elastic half-space of a thickness of 5–10 km can well reconstruct the observed deformation field with a timescale of a few to ~100 years. The estimated elastic thicknesses in those cases are less than those in tectonic environments because of elevated temperature in volcanic regions.

While the earlier discussion is based on a layered structure with no or slight lateral heterogeneity, the presence of a magma reservoir, among other factors, generate lateral heterogeneity in volcanic regions. For example, elevated temperature around a magma reservoir might make materials around the magma reservoir less viscous. Dragoni and Magnanensi (1989) obtained the surface deformation field due to an instantaneous pressurization of a sphere of a radius r_1 embedded in a homogeneous, isotropic, and elastic half-space with a viscoelastic shell of a thickness $r_2 - r_1$ (Fig. 7.9). The surface deformation right after the pressurization is obtained by substituting $a = r_1$ in Eqs. (7.7), (7.8) because a viscoelastic shell behaves as an elastic body. Surface deformation at $t = \infty$ is obtained by substituting $a = r_2$ in Eqs. (7.7), (7.8) because a viscoelastic shell is completely relaxed at $t = \infty$. Because the surface deformation of a pressurized sphere is roughly

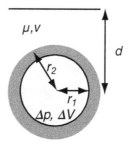

FIG. 7.9 A schematic view of a pressurization of a sphere of radius r_1 with a viscoelastic shell of a thickness of $r_2 - r_1$. Here, the shell is assumed to possess Maxwell rheology.

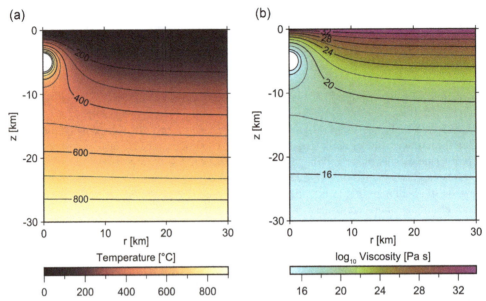

FIG. 7.10 An example distribution of (A) temperature and (B) viscosity, given an intruded magma of 950°C at a depth of 5 km and a geothermal gradient of 30 K/km. The *horizontal and vertical axes* represent the radial distance from the magma reservoir and the depth, respectively. Both the temperature and viscosity field are axisymmetric. *(Modified from Head, M., Hickey, J., Thompson, J., Gottsmann, J., & Fournier, N. (2022, July). Rheological controls on magma reservoir failure in a thermoviscoelastic crust.* Journal of Geophysical Research: Solid Earth, 127(7). https://doi.org/10.1029/2021jb023439.)

proportional to the product of the volume of the sphere and the pressure change (see Eqs. 7.7–7.10) if the sphere is embedded in an elastic, isotropic, and homogeneous half-space, the observed deformation might suggest increased overpressure of a sphere, even if the overpressure does not actually change with time in this case. By the way, the characteristic relaxation time t_R of the sphere in this case is

$$t_R = \frac{3\eta(1-\nu)r_2^3}{\mu(1+\nu)r_1^3}, \qquad (7.21)$$

where η, μ, and ν represent the viscosity of the shell, rigidity, and Poisson's ratio, respectively. Eq. (7.21) indicates that t_R is larger than the relaxation time of the viscoelastic material itself η/μ because $r_1 < r_2$ and $\nu < 0.5$.

The model described earlier is simple, but it can be tuned to be more realistic. First, the background temperature is not homogeneous as Dragoni and Magnanensi (1989), for example, assume. Since the background temperature increases with depth, the real structure beneath a volcano is closer to a viscoelastic half-space overlain by an elastic layer, which is thinner above the magma reservoir by heating (Head, Hickey, Thompson, Gottsmann, & Fournier, 2022; Rucker, Erickson, Karlstrom, Lee, & Gopalakrishnan, 2022, and references therein; Fig. 7.10). As analytical solutions of surface deformation with such complex structures do not exist, numerical methods such as the finite element method are required for the modeling (Head et al., 2022; Rucker et al., 2022, and references therein; Fig. 7.10).

While solving an inverse problem is straightforward if an analytical solution is available, it is not applicable in this case because, as mentioned, no analytical solutions are available and the forward calculation is more time consuming than that with a simpler medium. Nonetheless, some attempts are made to solve an inverse problem in this situation (Currenti et al., 2010).

2.4.3 Effect of topography

Topography is known better than the subsurface structure as it is visible. Now digital elevation maps (DEMs) from the Shuttle Radar Topography Mission and Advanced Spaceborne Thermal Emission and Reflection Radiometer, both with a resolution of 1 arcsecond (~30 m), are globally available (Abrams, Crippen, & Fujisada, 2020; Farr et al., 2007). Some countries have DEMs with finer resolutions; for example, a DEM with resolutions as fine as 10 or 5 m is available in Japan. While it is evident that incorporating detailed topography makes the modeling more precise, it makes the computation more expensive. Therefore, it is essential to understand when topography must be considered and how much detail is required.

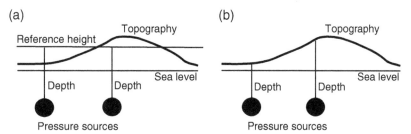

FIG. 7.11 A schematic representation of simple assumptions to consider topography. (A) Reference elevation model in which the depth of the pressure source is defined as the depth with respect to the prescribed reference height. (B) Varying depth model where the depth of the pressure source is defined as the depth from the altitude right above the source. In this case, different depths are defined for two sources with the same distance from sea level.

The simplest way to consider topography is to assume a flat reference surface at a certain altitude (Fig. 7.11A). This method works well only if the topography is relatively gentle. The optimum altitude of the reference flat surface depends on topography and the depth of the pressure source. For example, Cayol and Cornet (1998) favor the reference surface at an altitude of the volcano's summit by simulating surface deformation with an axisymmetric volcano of radius $6d$ and height d and with a spherical source at a depth of $5d$ beneath the summit. However, they pointed out the importance of taking realistic topography into account because this method overestimates the volume change. On the other hand, Williams and Wadge (2000) favor a reference surface between the summit and the average height of the volcano if the pressure source is shallow and a deeper reference surface if the pressure source is deep.

The next simplest way to take topography into account is to translate the depth of the pressure source into the distance from the surface (Fig. 7.11B). Therefore, pressure sources below the summit and flank at the same depth with respect to sea level have different apparent depths in this case. Williams and Wadge (1998) showed that vertical displacements and tilts are consistent with those obtained by the finite element method. However, the volume change is overestimated.

Williams and Wadge (2000) obtained an analytical solution of surface deformation in the presence of topography when the unit vector normal to the surface n can be approximated as $n = (-\partial h/\partial x, -\partial h/\partial y, 1)^T$, where altitude h is a function of location (x, y), and T represents matrix transpose. This study extends previous studies (McTigue & Segall, 1988; McTigue & Stein, 1984) to a 3D problem. Their arguments hold only with the displacement boundary condition of the pressure source and do not hold with the pressure boundary condition.

Then when should we take topography into account? Suppose the depth of the pressure source is greater than the characteristic wavelength of the topography. In that case, the wavelength of stress changes imposed by the pressure source is greater than the characteristic wavelength of the topography. Therefore, the stress change is considered homogeneous at the characteristic length scale of the topography. In this case, the effect of topography on the deformation field is small.

On the other hand, suppose the pressure source is so shallow that the lateral characteristic wavelength of the topography is greater than the source depth. In that case, the surface where stress changes occur can be approximated as locally flat. In this case, too, the effect of topography is small. The earlier argument indicates that the topography affects the surface deformation field when the source depth is similar to the characteristic wavelength of the topography (Segall, 2010, pp. 259–260).

Williams and Wadge (2000) assumes $H/L \ll 1$, where H and L represent characteristics wavelength of the topography, respectively. This approximation holds in the absence of steep topography, and their method is computationally much more inexpensive than numerical methods. However, it does not always give the deformation field that is consistent with numerical methods (e.g., Lungarini, Troise, Meo, & Natale, 2005; Meo, Tammaro, & Capuano, 2008) because it also assumes spatially smooth topography. Ronchin, Geyer, and Martí (2015) demonstrated that the tilt of the topography around the observation point affects the deformation field. This finding indicates that measurements of strain and tilt, spatial differentiation of displacements, are more susceptible to topography. In other words, displacements measured with GNSS observations are less sensitive to the topography than tilt or strain measurements.

2.5 Some caveats

A goal of deformation modeling is to gain insights into the mechanics of the accumulation and migration of magmatic fluids. Therefore, the obtained geometry, the pressure source, and pressure changes should be translated into mechanical insights. This section discusses caveats to discuss volcanological insights through deformation modeling.

2.5.1 Vertical and horizontal displacements

Fig. 7.4B, for example, shows that the vertical displacement field from different source geometries is sometimes similar, indicating that it is impossible to constrain the source geometry solely from vertical displacements and tilt measurements. This problem has been recognized for decades (Dieterich & Decker, 1975).

Fig. 7.4B also shows that even if the vertical displacement field generated by two different pressure sources is similar, the corresponding horizontal displacement field is different, indicating the importance of deformation measurements in three dimensions. GNSS can measure 3D displacements, so it plays a significant role in modeling observed volcano deformation.

2.5.2 Magma compressibility

While many modeling studies assume a spherical magma reservoir for simplicity (e.g., Dragoni & Magnanensi, 1989; Mogi, 1958), we need to note that a spherical magma chamber does not always serve as a simplification of a more complex shape. An obvious reason for this statement is that a sphere is the least compliant geometry in response to pressure changes. The compressibility of a sphere β_s as the least compliant geometry and a penny-shaped crack β_c as the most compliant geometry are

$$\beta_s = \frac{1}{V_s}\frac{dV_s}{d\Delta p} = \frac{3}{4\mu} \tag{7.22}$$

and

$$\beta_c = \frac{1}{V_c}\frac{dV_c}{d\Delta p} = \frac{1}{\Delta p}, \tag{7.23}$$

respectively (McTigue, 1987; Rivalta & Segall, 2008; Tiampo, Rundle, Fernandez, & Langbein, 2000), where V_s, V_c, Δp, and μ denote the volume of the sphere and crack, overpressure, and the rigidity of the host rock, respectively. The compressibility of a spheroid, for example, is between them. Because μ and Δp are on the order of gigapascal and megapascal, respectively, Eq. (7.22) indicates that the compressibility of a magma reservoir can be varied by up to a few orders. In other words, the same volume pressure change results from different overpressures by up to a few orders. Because the observed surface deformation is sensitive to volume changes, this drastic difference in compressibility indicates that it can be difficult to constrain the overpressure, a fundamental parameter to understanding the dynamics of magmatic fluids, solely from deformation measurements, especially from vertical displacements or tilts.

2.5.3 Effect of multiple sources

Many studies employ multiple pressure sources to model the observation. Their deformation cannot be represented as an addition of the contribution from each pressure source *sensu stricto* because the deformation field induced from one source violates the boundary condition at the other source.

Pascal, Neuberg, and Rivalta (2014) demonstrated with the finite element method that a simple addition of the deformation field by a spherical source and a dike or sill is accurate within 5% if the spherical source is located in the strike direction of the dike or sill. They also showed that a simple addition does not work well, or interaction between the pressure sources is not negligible if (1) a spherical source is located in the perpendicular direction to the strike of a sill or dike and (2) the distance between the two sources is less than 4 times the radius of the spherical source. In addition, a simple addition works well if two spherical sources are separated by more than 8 times their radius.

2.5.4 Transcrustal magma reservoir

Many modeling studies assume one or, at most, a few pressure sources to model the observed deformation field. However, a new concept of transcrustal magma reservoir has recently become prevailing (Cashman, Sparks, & Blundy, 2017; Giordano & Caricchi, 2022; Sparks & Cashman, 2017, Fig. 7.12). This concept assumes that the melts are distributed within much of the crust, consisting of mush. This hypothesis is supported by geophysical, geochemical, and petrological evidence (e.g., Jackson, Blundy, & Sparks, 2018; Sparks et al., 2019).

While the transcrustal magma reservoir does not appear to be compatible with a model for which magma reservoirs are embedded in an elastic rock, observations are often wellfit with such a classical model setting. How can we reconcile this inconsistency? Recently, Liao, Soule, and Jones (2018) constructed a mechanical modeling associated with pressurized spherical magma reservoir surrounded by a spherical poroelastic shell embedded in an elastic half-space. They showed

FIG. 7.12 A schematic view of the transcrustal magmatic system. *Color* denotes the distribution of the SiO_2, or proxy to viscosity, where the redder the color is, the higher the viscosity is. The background color is the temperature; the darker the color, the lower the temperature. *(Modified from Gudmundsson, M. T., Jónsdóttir, K., Hooper, A., Holohan, E. P., Halldórsson, S. A., Ófeigsson, B. G., ...Aiuppa, A. (2016, July). Gradual caldera collapse at Bárdarbunga Volcano, Iceland, regulated by lateral magma outflow. Science, 353(6296). https://doi.org/10.1126/science.aaf8988.)*

that deformation continues even after the injection of magma ceased. Mullet and Segall (2022) extended Liao et al. (2018) to the spheroidal magma lens and poroelastic mush (Fig. 7.13) to demonstrate that the surface deformation field fits well with pressurization of the poroelastic mush, rather than the magma lens. They also showed that the model without mush underestimates the magma compressibility by up to 50%.

Liao, Soule, Jones, and Mével (2021) expanded upon the work of Liao et al. (2018) to employ the viscoporoelastic mush. Liao (2022) further extended the previous studies to incorporate thermomechanical coupling between the magma and surrounding rock. They demonstrated that the deformation expected from the model is a combination of short-, mid-, and long-term components from poroelastic diffusion, viscoelastic diffusion, and thermal equilibration, respectively. Incorporating such complexity will allow us to make the modeling more realistic.

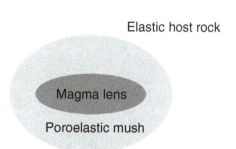

FIG. 7.13 A schematic view of a magma lens surrounded by a poroelastic mush and embedded in the elastic host rock.

3 Observation of volcano deformation

As ground deformation is one of the most fundamental tools for monitoring the activity and understanding the mechanics of the volcano, it is ongoing in many volcanoes. GNSS adds insights into volcanic activity because of its superior temporal resolution and better precision. This section attempts to review ground deformation measurements by GNSS in active volcanoes and introduce insights gained by the GNSS measurements.

Volcanic activity is divided into dormant and active periods; no deformation is observed during the dormant period, and deformation is observed during the active period. The active period is divided into three stages: unrest, erupting, and posteruptive. Unrest often accompanies inflation of the volcano due to magma intrusion into shallow depths. Magma intrusions do not always result in an eruption, but the intruded magma is often arrested without reaching the surface (e.g., Moran, Newhall, & Roman, 2011). Eruptions are often accompanied by deflation of the volcano due to mass loss from the volcano. Eruptions, however, often occur without deflation when the ejected volume is not large enough. Stress changes in the volcano caused by eruptions induce posteruptive deformation. In reality, however, the observed deformation field is diverse because of the diversity of volcanic activity resulting from various factors, including magma composition, the abundance of groundwater, and the explosivity of an eruption. Fig. 7.14 depicts the location of volcanoes discussed in the following.

3.1 Deformation during unrest

Volcanic unrest usually results from the injection of magmatic fluids to shallower depths. The injected magmatic fluids accumulate at a neutral buoyancy depth until the magmatic fluids somehow gain positive buoyancy to intrude toward shallower depths. Magmatic fluids can also migrate horizontally; the cause of the migration is sometimes understood and sometimes not. This section discusses insights gained by measuring ground deformation during volcanic unrest.

3.1.1 Magma accumulation

Most active volcanoes possess magma reservoirs at the crustal level (Fig. 7.12). Especially in volcanoes with intermittent eruptions, ground deformation during much of the unrest period is characterized by volcano inflation due to the injection of magmatic fluids into crustal magma reservoirs.

Because such inflation has a timescale of months to years, monitoring the highest temporal resolution, such as days, is often unnecessary. Therefore, many studies have investigated volcano inflation by SAR (e.g., Chaussard & Amelung, 2014; Chaussard, Amelung, & Aoki, 2013; Pritchard & Simons, 2002). GNSS also plays a significant role in monitoring the ground deformation of active volcanoes because volcano inflation often fluctuates with a timescale shorter than months. Such deformation signals can be recorded only with GNSS.

GNSS has advantages over SAR, at least from two perspectives. First, longer-term records by GNSS than SAR are available in some volcanoes. SAR images from a single constellation are rarely available for over 10 years. For example, Sentinel-1 and ALOS-2 images have been available only since 2014. On the other hand, some volcanoes have GNSS time series for more than two decades (e.g., Bevilacqua et al., 2022; Lisowski, McCaffrey, Wicks, & Dzurisin, 2021; Silverii,

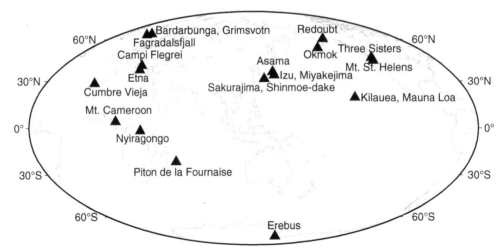

FIG. 7.14 The location of the volcanoes discussed in this chapter.

Montgomery-Brown, Borsa, & Barbour, 2020). Even if a volcano is not erupting, long-term volcano monitoring is vital because the ground deformation associated with its inflation is not constant over time. For example, monitoring with GNSS for two decades revealed an uplift in Campi Flegrei since around 2005 with an acceleration (Bevilacqua et al., 2022) (Fig. 7.15).

The inflation of a volcano sometimes decelerates over time. For example, Lisowski et al. (2021) showed that Three Sisters Volcano has been inflating for at least decades with decaying rates (Fig. 7.16). Also, Silverii et al. (2020) showed that Long Valley Caldera has been inflating since 2011 with decaying rates over time. Interpreting these observed inflations with decreasing rates is not straightforward. The observation can be interpreted as continuing injection of magmatic fluids with decaying rates or more impulsive injections followed by viscoelastic relaxation of the lower crust or an aureole around the magma reservoir.

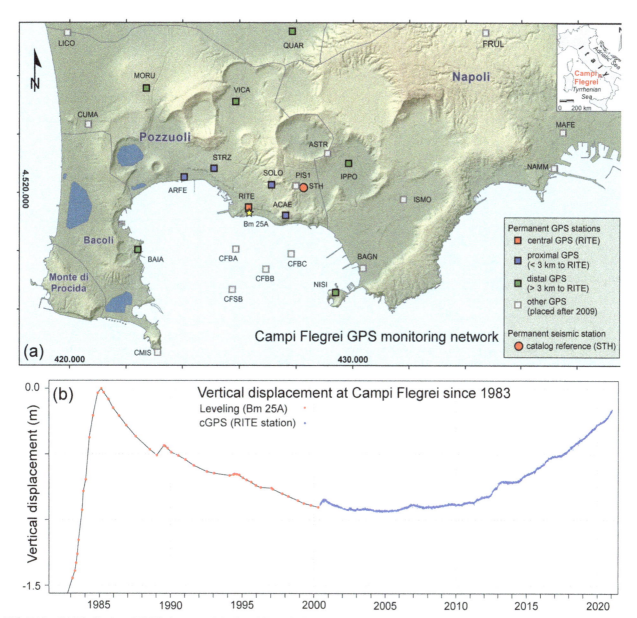

FIG. 7.15 (A) Distribution of GNSS sites around the Campi Flegrei caldera. (B) Time series of the vertical displacement at the RITE GNSS site (*blue*) and the Bm 25A leveling benchmark (*red*) located close to RITE. This time series indicates that the caldera turned from deflation to inflation around 2005, with a recent acceleration of inflation. *(Modified from Bevilacqua, A., Martino, P. D., Giudicepietro, F., Ricciolino, P., Patra, A., Pitman, E. B., ...Neri, A. (2022, November). Data analysis of the unsteadily accelerating GPS and seismic records at Campi Flegrei Caldera from 2000 to 2020. Scientific Reports, 12(1). https://doi.org/10.1038/s41598-022-23628-5.)*

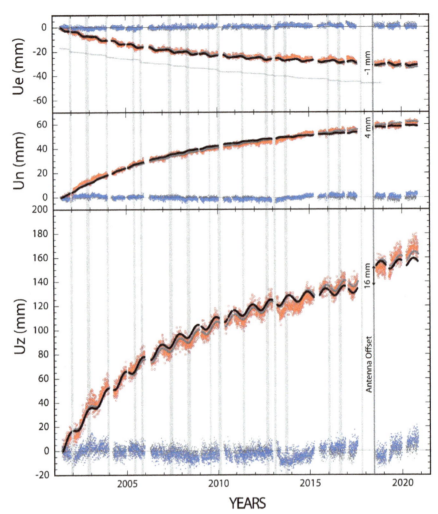

FIG. 7.16 Time series of east, north, and vertical (*top*) daily displacement at the HUSB GNSS site, located about 2.5 km to the north-northwest from the center of inflation of Three Sisters volcano. The displacements (*red*) are corrected for estimated tectonic background movements. The modeled displacement (*black*) includes an exponential fit, annual and semiannual seasonal perturbations, and an antenna offset (the *dark vertical line* in 2018). *Blue lines* denote the difference between the model and the observations. Modeled results and residuals from a log fit to the data are shown in *gray*. *Light gray vertical lines* mark times of slow slip events, and the *sloping light gray line* in the east plot shows estimated offsets from these events superimposed on the volcanic motion. *(Modified from Lisowski, M., McCaffrey, R., Wicks, C. W., & Dzurisin, D. (2021, December). Geodetic constraints on a 25-year magmatic inflation episode near three sisters, Central Oregon. Journal of Geophysical Research: Solid Earth, 126(12). https://doi.org/10.1029/2021jb022360.)*

Second, GNSS can measure displacements where SAR cannot; SAR cannot measure ground deformation in heavily vegetated or snowy regions, but GNSS can. In particular, because many volcanoes get snow around their summit, deformation measurements in such regions by SAR images are available only in summer or unavailable all the time. An extreme example is Erebus, most of which is covered by glaciers all the time except around the vent. Erebus has been monitored by GNSS for over 20 years, and seven continuous GNSS sites presently monitor it. The observed displacement field is well explained by long-term subsidence by viscoelastic relaxation and short-term inflation and deflation cycles by magma intrusions and extrusions or eruptions (Grapenthin et al., 2022).

3.1.2 Vertical magma transport

Magmatic fluids in a magma reservoir move toward the surface if driven by the positive buoyancy force. The intruded magma forms a dike, especially when the fluid's viscosity is low. A low-viscosity fluid can propagate most efficiently as a dike within the host rock with a much higher viscosity than the fluid (Rubin, 1993).

Since vertical dike propagation is ubiquitous in active volcanoes, it has been observed by geodetic instruments in many volcanoes, including Etna (e.g., Bonaccorso & Aloisi, 2021), Piton de la Fournaise (e.g., Peltier et al., 2016), and Asama (e.g., Aoki, Takeo, Ohminato, Nagaoka, & Nishida, 2013) volcanoes. One of the first observations with GPS is associated with the 1989 seismic swarm off the Izu Peninsula, Japan, that resulted in a submarine eruption (Okada & Yamamoto, 1991). Aoki et al. (1999) and Morita, Nakao, and Hayashi (2006) imaged the evolution of the dike propagation by the 1997 and 1998 seismic swarm off the Izu Peninsula, respectively, from GPS data with the aid of other geodetic data, that is, leveling, tiltmeters, and strainmeters, and precise earthquake relocations. Although these seismic swarms did not result in an eruption, imaging dike propagation from geodetic data is not only important for understanding the mechanics of magma transport but also crucial from a practical point of view.

High-viscosity magma cannot propagate as a dike but propagates instead as a thicker geometry. A typical example is Mt. St. Helens. Anderson and Segall (2011) constructed a physical model assuming magma propagation through a pipe from an elliptical magma reservoir. While the propagation mechanics of low-viscosity magma as a dike is extensively studied from observation, experimental, and observational viewpoints, the propagation mechanics of high-viscosity magma appears to need more research.

GNSS measurements play substantial roles in understanding magma propagation because of their superior temporal resolution. An intrusion event sometimes lasts only for a short time. A typical example is the 2015 failed eruption of Sakurajima, in which a rapid magma intrusion of more than 100 m^2/s lasted less than a day (Hotta, Iguchi, & Tameguri, 2016). While SAR images, whose acquisitions are only every few days or more, could not track magma propagation of this event, GNSS can track the evolution of magma propagation even if it lasts less than a day.

3.1.3 Horizontal magma transport

If magma is at neutral buoyancy, it cannot propagate vertically by buoyancy. Instead, it sometimes propagates horizontally out of the magma reservoir. High-viscosity magmas might propagate by expanding their magma reservoir, while low-viscosity magmas propagate as dikes. Here we discuss horizontal dike propagation out of a magma reservoir.

Horizontal dike propagation is often observed in active volcanoes. Observations by GNSS include the 2022 Nyiragongo (Smittarello et al., 2022), 2018 Kī-lauea (e.g., Neal et al., 2019), 2014–15 Bárðarbunga (Sigmundsson et al., 2015), 2007 Piton de la Fournaise (e.g., Chen et al., 2017; Peltier et al., 2015), and 2000 Miyakejima eruptions (Akiyama, Kawabata, & Yoshioka, 2022; Nishimura et al., 2001; Ozawa et al., 2004). They are also observed in a series of eruptions in Kī-lauea (e.g., Cervelli et al., 2002; Neal et al., 2019; Owen, Segall, Lisowski, Miklius, Murray, et al., 2000), Piton de la Fournaise (Peltier, Staudacher, Bachèlery, & Cayol, 2009; Smittarello et al., 2019), and Etna (e.g., Bonaccorso & Aloisi, 2021).

While positive buoyancy as the driving force of vertical dike propagation is conceivable, that of horizontal dike propagation is not straightforward. The horizontal dike propagation is sometimes associated with flank spreading and sometimes not. Relations between the horizontal dike propagation and flank spreading are complex, as will be discussed; stress changes by stationary or transient flank displacements sometimes drive the horizontal dike propagation, and sometimes the dike propagation drives the transient flank displacement. The former and latter are called passive and active dike propagation, respectively.

Kī-lauea, Piton de la Fournaise, and Etna are three typical examples that exhibit horizontal dike propagation and flank spreading (Poland, Peltier, Bonforte, & Puglisi, 2017). These are also among the volcanoes best monitored by geophysical instruments, including GNSS. Also, Mauna Loa exhibits flank spreading by a few tens of millimeters per year (Miklius et al., 1995; Pepe et al., 2018), and Cumbre Vieja and Mt. Cameroon exhibit flank spreading by a few millimeters per year (González, Tiampo, Camacho, & Fernández, 2010; Mathieu, Kervyn, & Ernst, 2011).

GNSS observations demonstrate that the flank spreading of the well-monitored volcanoes is variable. The flank of Kī-lauea, or East Rift Zone, exhibits steady-state seaward displacements of a few tens of millimeters per year (Owen et al., 1995; Owen, Segall, Lisowski, Miklius, Denlinger, & Sako, 2000) with occasional slow-slip transients (e.g., Brooks et al., 2006; Montgomery-Brown et al., 2010; Segall, Desmarais, Shelly, Miklius, & Cervelli, 2006). Geodetic modeling indicates that they are due to slips at the décollement between the base of the volcanic pile and the oceanic crust. While the slow-slip transients and occasional M~5 earthquakes appear to be triggered by dike intrusions, magmatic activity does not affect steady-state flank spreading. This observation suggests that gravity plays a central role in the flank spreading of Kī-lauea. Extensional stress field drives the horizontal dike propagation off the magma reservoir, which triggers slow-slip transients and earthquakes (Brooks et al., 2008; Montgomery-Brown et al., 2010). In other words, the horizontal dike propagation in Kī-lauea is driven passively by flank spreading at least to some degrees. Montgomery-Brown and Miklius (2021) pointed out that the dike intrusion in the East Rift Zone of Kī-lauea volcano is periodic, with a recurrence time of ~8 and ~14 years in the upper and middle East Rift Zone, respectively, in the last four decades, regardless of the magma pressure there. The timing of dike intrusions would not be periodic if magma pressure controls the timing. Therefore, their finding suggests that the flank motion controls the timing of dike intrusions there.

The flank spreading of Piton de la Fournaise has some similarities and differences from that of Kī-lauea. While the flank of both Kī-lauea and Piton de la Fournaise spreads nearly perpendicular to the rift axis, the spreading rate of Piton de la Fournaise is affected by magmatic activity, unlike Kī-lauea. Indeed, GNSS observations show that the flank spreading in Piton de la Fournaise was fast right after the 2007 distal eruption but decayed over time (Chen et al., 2017; Peltier et al., 2015). These observations indicate that the horizontal dike propagation is somewhat active rather than driven by flank spreading. However, details could be better understood because of the sparser deployment of GNSS sites and less intense magmatic activity than Kī-lauea.

Etna has intermediate characteristics between Kī-lauea and Piton de la Fournaise in many ways. The magma supply rate and size of the edifice of Etna is between Kī-lauea and Piton de la Fournaise. Therefore, the style of dike propagation and flank spreading of Etna is expected in between Kī-lauea and Piton de la Fournaise. Geodetic observations, including those with GNSS, exhibit flank spreading in Etna, as in Kī-lauea and Piton de la Fournaise. The style of the flank spreading of Etna has similarities and differences from the other two volcanoes. The most apparent difference is the spatial variability of flank spreading in Etna (e.g., Bonaccorso et al., 2011; Bonforte, Bonaccorso, Guglielmino, Palano, & Puglisi, 2008; Bonforte & Puglisi, 2006). This spatially variable spreading rate reflects extensive faulting in the eastern flank of Etna, unlike the other two volcanoes. The flank spreading rate of Etna is less affected by magmatic activity than that of Piton de la Fournaise (e.g., Houlié, Briole, Bonforte, & Puglisi, 2006; Mattia et al., 2015). Therefore, the horizontal dike propagation in Etna may be controlled by flank spreading as in Kī-lauea. Also, Etna hosts slow slips on the décollement triggered by the dike propagation (Bruno, Mattia, Montgomery-Brown, Rossi, & Scandura, 2017; Mattia et al., 2015) as at Kī-lauea.

Horizontal dike propagation often lacks (apparent) external forces to trigger the propagation. Many horizontal dike propagations associated with, for example, the 2022 Nyiragongo, 2014–15 Bárðarbunga, and 2000 Miyakejima eruptions are without any apparent external stress changes that drive passive magma transport. Therefore, these dike propagations may be caused by active overpressure of the magma reservoir and a lack of vertical buoyancy to drive vertical magma transport.

As horizontal dike propagation results in a loss of magma from the magma reservoir, it can cause a collapse above the magma reservoir. Because a magma reservoir typically sits beneath the summit of a volcano, horizontal dike propagation often results in a summit collapse. The dense deployment of geodetic and seismic instruments allows us to observe the temporal evolution of summit collapses. Recent well-monitored summit collapses include the 2018 Kī-lauea (e.g., Anderson & Johanson, 2022; Anderson et al., 2019), 2014–15 Bárðarbunga (Gudmundsson et al., 2016), 2007 Piton de la Fournaise (e.g., Chen et al., 2017; Peltier et al., 2015), and 2000 Miyakejima eruptions (Munekane, Oikawa, & Kobayashi, 2016).

Geodetic and seismic monitoring has shown that a summit collapse event does not occur at once but takes days to months. Indeed, subsidence associated with the caldera collapse lasted for a few weeks to months associated with the eruptions shown in the previous paragraph. While the summit subsidence often exhibits an exponential decay in a timescale of weeks to months (e.g., Anderson et al., 2019; Gudmundsson et al., 2016), on a shorter timescale, the observed time series is overlain by episodic collapse events. Tiltmeters or strainmeters best monitor each collapse because deformation signals by each collapse are small. However, they are also recorded by kinematic GNSS measurements. As each collapse lasts only on the order of minutes, high-rate GNSS measurements are beneficial to understanding the details of each collapse. Munekane et al. (2016) and Anderson and Johanson (2022) detected each collapse event as an immediate uplift followed by gradual subsidence associated with the 2000 Miyakejima and 2018 Kī-lauea eruptions, respectively (Fig. 7.17). This observation is interpreted as (1) stationary deflation due to the outflow of magma (Fig. 7.18A), (2) the outward displacement due to the faulting on the caldera wall by the subsidence of the roof rock and the pressurization of the magma reservoir by the entrance of the piston column (Fig. 7.18B), and (3) deflation by a gradual depressurization of the magma reservoir (Fig. 7.18C). Fig. 7.18D depicts the time series observed right outside the caldera rim. Munekane et al. (2016) and Segall, Anderson, Pulvirenti, Wang, and Johanson (2020) both favored a nearly vertical caldera wall and a vertically elongated magma reservoir to explain the observation associated with the caldera collapse during the 2000 Miyakejima and 2018 Kī-lauea eruptions, respectively.

3.2 Coeruptive deformation

Since a volcanic eruption results in a mass loss of the magma reservoir, deflation of the volcano is a natural consequence of an eruption. While a volcano does not deflate during an eruption if its size is too small (e.g., Aoki et al., 2013; Mattia et al., 2020), detectable deflation is observed if the size is large enough. The deflation sometimes occurs instantly (e.g., Nakao et al., 2013) and sometimes takes a few months (e.g., Anderson & Johanson, 2022; Bruno et al., 2022; Gudmundsson et al.,

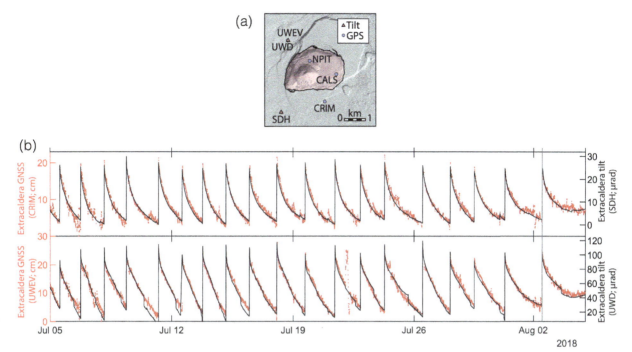

FIG. 7.17 Observation of cyclic uplift and subsidence during the 2018 Kī-lauea eruption. (A) Location of GNSS sites (*blue circles*) and tiltmeters (*red triangles*) around the Kī-lauea caldera. (B) Vertical displacements observed by GNSS (*red dots*) and tiltmeters (*black lines*) at CRIM (*top*) and UWEV (*bottom*). *(Modified from Anderson, K., & Johanson, I. (2022, September). Incremental caldera collapse at Kī-lauea volcano recorded in ground tilt and high-rate GNSS data, with implications for collapse dynamics and the magma system. Bulletin of Volcanology, 84(10). https://doi.org/10.1007/s00445-022-01589-x.)*

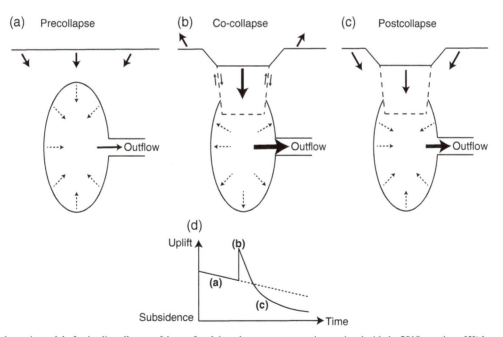

FIG. 7.18 A schematic model of episodic collapses of the roof rock into the magma reservoir associated with the 2018 eruption of Kī-lauea. *Thick arrows* near the surface show ground displacements; *insets* show the corresponding trend of GNSS time series outside the caldera. (A) Precollapse subsidence caused by magma withdrawal. (B) Abrupt subsidence of roof rock along ring faults and associated deformation. (C) Gradual withdrawal of magma between collapse events. (D) GNSS time series outside the caldera rim. Precollapse, cocollapse, and postcollapse stages are denoted by (a), (b), and (c), respectively.

2016). As discussed in Section 3.1.3, a gradual deflation over weeks to months sometimes consists of subevents of a shorter timescale.

It is worth noting that monitoring ground deformation by GNSS, tiltmeters, or strainmeters is useful for eruption warning and assessing volcanic ash hazards. Hreinsdóttir et al. (2014) found that during the 2011 Grívsvötn eruption, the rate of pressure change of the magma reservoir for each eruption derived from GNSS and tilt measurements is proportional to H^{α}, where H denotes the observed height of the eruption column and α represents a constant around 0.25. This relation is consistent with theoretical insights (e.g., Mastin et al., 2009). Their finding indicates that real-time monitoring by GNSS is useful in eruption warning and assessing volcanic ash hazards because the GNSS data can tell us about the flux and height of the eruption column, leading to a more precise assessment of ash hazards.

3.3 Posteruptive deformation

Since mass loss from the magma reservoir due to an eruption perturbs the stress field, surface deformation often continues even after the eruption ends. For example, deflation by the 2011 Shinmoedake eruption in a few days was followed by inflation for a year (Nakao et al., 2013). Li et al. (2021) reported that the summit deflation by the 2014–15 Bárðarbunga eruption that lasted for a few months was followed by inflation for years with deceleration. Reverso et al. (2014) investigated the 1998, 2004, and 2011 eruptions of Grívstötn were followed by inflation with an exponential trend over time. The summit deformation following the 2018 Kī-lauea eruption is more complicated than the Grívstötn case; Wang, Zheng, Pulvirenti, and Segall (2021) observed that the initial subsidence around the summit turned to uplift. While GNSS sites near the summit turned from subsidence to uplift ~100 days after the eruption, those far from the summit subsided for ~300 days after the eruption.

Various models have been proposed to explain the observations. Wang et al. (2021) constructed a physical model to explain the observations associated with the 2018 Kī-lauea eruption by the outflow of the magma from a magma reservoir beneath the summit (Fig. 7.19A). Reverso et al. (2014) proposed a model in which the observed deformation after the 1984, 2004, and 2011 Grívstötn eruptions is driven by the pressure gradient between the deeper and shallower magma reservoir produced by the eruption that reduced the pressure of the shallower magma reservoir (Fig. 7.19B). The pressure gradient drives the magma transport from the deeper to the shallower reservoir. Segall (2016, 2019) added complexity of the model by Reverso et al. (2014) to assume a viscoelastic aureole around the shallower magma reservoir. Got et al. (2017) attempted to explain the same dataset by the damage to the edifice induced by the eruption. These variations of models suggest that uniquely constraining the model requires independent information such as gravity changes, earthquake locations, or gas measurements.

4 Observations of volcanic plumes by GNSS

While GNSS is primarily to measure the position of an antenna, GNSS measurements can extract other information such as the distribution of water vapor in the atmosphere, total electron content (TEC) in the ionosphere, sea level, and snow depth (e.g., Larson, 2019). Among relevant information for volcano monitoring are the ionospheric disturbances manifested by the temporal changes of TEC, atmospheric disturbance, and decay of GNSS signals. TEC changes result from the atmospheric disturbance induced by an explosive eruption that propagates to the ionosphere. Many studies have already reported ionospheric disturbances since a pioneering work by Heki (2006), and such signals may gain more insights into the mechanics of explosive volcanism. Details are left to Chapter 8, which overviews ionospheric disturbance measured by GNSS. This section is, instead, devoted to atmospheric disturbance and decay of GNSS signals associated with volcanic plumes.

4.1 Atmospheric disturbance by volcanic plumes

GNSS signals refract in the presence of water vapor, resulting in fluctuations of the satellite-station range (e.g., Bevis et al., 1992, see also Chapters 9 and 10). The refractivity index N is given as a function of the total atmospheric pressure P_{atm}, partial water vapor pressure P_w, and temperature T as (e.g., Thayer, 1974)

$$N - 1 = \frac{k_1(P_{atm} - P_w)}{T} + \frac{k_2 P_w}{T} + \frac{k_3 P_w}{T^2}, \tag{7.24}$$

where $k_1 = 77.60 \pm 0.009$ K/hPa, $k_2 = 69.4 \pm 2.2$ K/hPa, and $k_3 = (370.1 \pm 1.2) \times 10^3$ K^2/hPa are constants. Volcanic gases and solid particles give negligible perturbations to refractivity (Solheim, Vivekanandan, Ware, & Rocken, 1999).

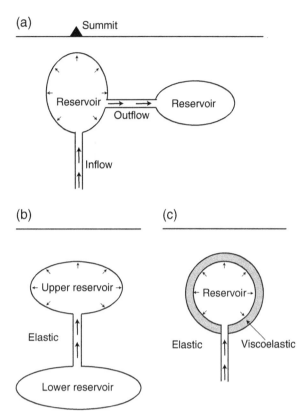

FIG. 7.19 Schematic model driving posteruptive deformation of the (A) 2018 Kīlauea and (B, C) 1998, 2004, and 2011 Grívstötn eruptions, respectively. (A) A schematic representation of magma transport after the 2018 Kīlauea eruption adopted by Wang et al. (2021). The magma reservoir beneath the summit is subject to magma inflow from depth and outflow toward another reservoir at the East Rift Zone. (B) A schematic representation of magma transport after the Grívstötn eruptions proposed by Reverso et al. (2014). A pressure gradient between the upper and lower magma reservoirs generated by mass loss from the upper magma reservoir, or the eruption, drives the magma flow from the lower to the upper reservoir. (C) Same as (B) but proposed by Segall (2016, 2019). The model differs from (B) in that a viscoelastic aureole surrounds the upper reservoir.

Since an ejection of a volcanic plume provides heat and water vapor to the atmosphere, GNSS signals passing through the volcanic plume are strongly refracted, resulting in an apparent extension of the satellite-station range. Therefore, GNSS measurements can add insights into the mechanics of explosive eruptions because they give us a four-dimensional (4D) distribution of water vapor during a volcanic eruption.

Houlié (2005) and Houlié, Briole, Nercessian, and Murakami (2005) were the first studies to detect atmospheric disturbance due to an ejection of the volcanic plume. They detected large misfits between observed and modeled ionosphere-free phase double difference, ϕ_{LC}, given by

$$\phi_{LC} = \phi_{L1} - \frac{f_2}{f_1}\phi_{L2} \tag{7.25}$$

associated with the 2000 Miyakejima and 2004 Mt. St. Helens eruptions. In Eq. (7.25), ϕ_{L1}, ϕ_{L2}, denote L1 and L2 phases, and f_1, and f_2 represent L1 and L2 carrier frequencies (1.57542 and 1.22760 GHz) of the GPS signal, respectively. The observed residuals are up to ~200 mm associated with the 2000 Miyakejima eruption and 140 mm associated with the 2004 Mt. St. Helens eruption, respectively (Houlié, 2005; Houlié et al., 2005, Fig. 7.20A). They successfully mapped the spatial distribution of excess water vapor pressure from the phase anomalies given by Eq. (7.25). The inferred refractivity anomaly of ~200 ppm (Fig. 7.20B) can be explained by a column of water vapor with a radius of ~1 km and an excess temperature of ~40 K, assuming a relative humidity of 100%. This increased temperature is much lower than the temperature of the volcanic jet at the vent. The discrepancy is reconciled either by interpreting that (1) there is a hot part of the jet with a radius of ~100 m or less and (2) the jet is cooled by being entrained by the ambient atmosphere.

The phase delays of GNSS signals can be mapped into the temporal evolution of the shape of the ash plume as a 3D distribution of excess water vapor pressure. For example, from postfit residuals of GNSS signals, Ohta and Iguchi (2015)

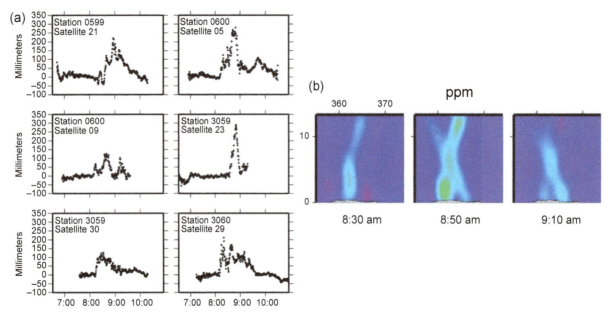

FIG. 7.20 (A) Time series of the ionosphere-free phase double difference (Eq. 7.25) associated with the August 18, 2000 eruption of Miyakejima. (B) An east-west cross-section of the temporal evolution of refractivity anomaly. *(Modified from Houlié, N., Briole, P., Nercessian, A., & Murakami, M. (2005). Volcanic plume above mount St. Helens detected with GPS. Eos, Transactions American Geophysical Union, 86(30), 277. https://doi.org/10.1029/2005eo300001.)*

demonstrated that the height of the plume associated with an eruption of Sakurajima volcano, Japan, in 2012 is about 4000 m, consistent with a visual inspection. They also indicated by comparing the postfit residuals and decays of GNSS signals that while the water-rich component of the plume drifted horizontally at the top of the plume, the ash-rich component stayed about the vent. The following section gives a discussion of the decay of GNSS signals.

An explosive eruption with an ash column reaching the stratosphere often generates volcanic hails. Grapenthin, Hreinsdóttir, and Eaton (2018) argued hail formation at above ~15 km above sea level during the 2011 eruption of Grímsvötn, Iceland, because the partial water pressure within the ash plume there is small, while that below ~15 km above sea level is large. Their argument is consistent with a numerical simulation of ash (e.g., Mastin, 2014). Note that volcanic hails do not generate refraction of GNSS signals.

Postfit residuals of GNSS signals can detect not only large explosive eruptions, as shown above, but also smaller eruptions. For example, Aranzulla, Cannavò, Scollo, Puglisi, and Immè (2013) detected explosive but smaller eruptions of Etna volcano, Italy, on September 4–5, 2007, with a column height of ~2 km, by statistical analysis of postfit residuals of GNSS signals.

As discussed earlier, postfit residuals of GNSS signals allow us to detect explosive eruptions. It must be emphasized that it can be done day and night, regardless of the weather, indicating that systematic monitoring of postfit residuals of the GNSS signals allows us to detect explosive eruptions.

4.2 GNSS signal decay by volcanic plumes

Microwaves radiated from GNSS satellites transmit clouds and raindrops because of their long wavelengths between ~190 and ~250 mm. Nonetheless, their signal strength sometimes decays by passing through volcanic plumes. It was Larson (2013) who found this decay for the first time. Larson (2013) took advantage of the recurrence of GPS satellites every (approximately) one sidereal day to compare the signal-to-noise ratio (SNR) of GNSS signals associated with eruptions of the 2008 Okmok and 2009 Redoubt volcanoes. Because the SNR of GNSS signals is lower when the satellite is at lower elevations, the SNRs from the same elevation angle must be compared. In this regard, GPS is more convenient to assess the signal decay than other satellite constellations because GPS satellites recur every (approximately) one sidereal day, while the recurrence time of GLONASS, for example, is longer. Larson (2013) found that decay of a GPS signal associated with the 2009 eruption of Redoubt Volcano by up to 10 dB-Hz compared with the previous and subsequent 2 days (Fig. 7.21).

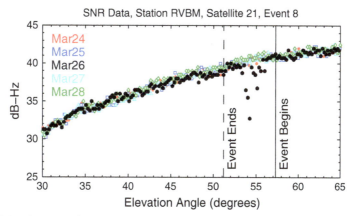

FIG. 7.21 SNR as a function of elevation angle of a satellite between 2 days before and after the event 8 of Mt. Redoubt eruption on March 26, 2009. Because the satellite is downgoing, the elevation of the satellite is higher when the eruption started (*solid line*) than it ended (*dashed line*). A significant decay of SNR during the eruption is visible. *(Modified from Larson, K. M. (2013, June). A new way to detect volcanic plumes.* Geophysical Research Letters, *40(11), 2657–2660. https://doi.org/10.1002/grl.50556.)*

Larson (2013) also found a similar decay associated with the 2008 eruption of Okmok Volcano by up to 15 dB-Hz compared with SNRs 1 day before and 5 days after the eruption (Fig. 7.22). SNRs of the same signals 1 and 2 days after the eruption are substantially lower, leading Larson (2013) to interpret that they are due to the accumulation of volcanic ash above the antenna.

Because SNRs of GNSS signals are recorded in every sample, high-rate GNSS measurements give high-rate SNR changes. Fournier and Jolly (2014) were the first to delineate SNR changes in GPS signals every second. They detected decays of GPS signals associated with two successive eruptions of Tongariro volcano, New Zealand, in 2012. They also found signal decay by a gas-charged debris avalanche and a lateral blast ∼3 km away from the vent.

What causes the decay of GNSS signals? Two mechanisms contribute to the observed signal decay; assuming the signal decay by a particle, absorption dominates when $\chi = \pi D/\lambda$ is less than 1, and scattering dominates otherwise, where D and λ represent the diameter of a particle and the wavelength of the GNSS signal, respectively. Therefore, χ represents the ratio of

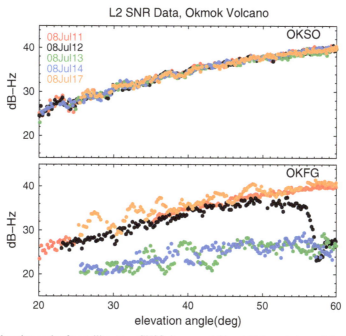

FIG. 7.22 SNR as a function of elevation angle of a satellite at two GNSS sites around Okmok Volcano between 1 day before and 2 days after the eruption on July 12, 2008. While OKSO (*top panel*) does not show any SNR decay, OKFG (*bottom panel*) exhibits SNR decay for 2 days. These SNR decays are probably due to the ash accumulation on the GNSS antenna. *(Modified from Larson, K. M. (2013, June). A new way to detect volcanic plumes.* Geophysical Research Letters, *40(11), 2657–2660. https://doi.org/10.1002/grl.50556.)*

the circumference of the particle to the wavelength of the GNSS signal. As the wavelength of the GNSS signal is between ~190 and ~250 mm, χ equals 1 when D is of between ~60 and ~80 mm. In other words, absorption and scattering are responsible for signal decay by particles of diameter less than and more than 60–80 mm, respectively.

Let us start by considering the contribution of absorption by small particles. The absorption α_a in dB/km is given by (Larson et al., 2017; Solheim et al., 1999)

$$\alpha_a = 10^4 \log e \cdot \frac{6\pi M}{\lambda \rho} \Im\left(\frac{1-\varepsilon}{2+\varepsilon}\right), \tag{7.26}$$

where ρ and M represent the particle density and tephra loading, respectively, in kg/m³. M/ρ is, therefore, the volume fraction of particles. ε denotes the complex dielectric permittivity; lab experiments indicate that the real part of ε is between 5.0 and 6.5, while the imaginary part of ε ranges between −0.05 and −0.30 (e.g., Adams, Perger, Rose, & Kostinski, 1996; Larson et al., 2017; Oguchi, Udagawa, Nanba, Maki, & Ishimine, 2009). $\Im$ takes the imaginary part of the function in parenthesis. These values yield α_a between ~0.007 and 0.05 dB/km with the volume fraction $M/\rho = 4 \times 10^{-3}$. The discussion so far indicates that solely absorption by small particles cannot explain large signal decays of 1–10 dB/km.

While small and electrically neutral particles do not attenuate GNSS signals much, charged particles attenuate GNSS signals. Méndez Harper, Steffes, Dufek, and Akins (2019) showed that charged ash particles with diameters 1–10μm efficiently attenuate GNSS signals by extinction, while particles with diameters of 10 mm or larger decay the strength of GNSS signals by scattering, regardless of the charge on the particles. As large particles do not reach altitudes exceeding a few kilometers from the vent, the observed decay of GNSS signals passing through high altitudes above the vent is solely due to charged small particles. On the other hand, decays of GNSS signals passing through low altitudes are due to a contribution of both large particles and charged small particles.

A combination of the decay of the signal strength and atmospheric disturbance, as discussed earlier, gains more insights into volcanic plumes. Ohta and Iguchi (2015) demonstrated that decays of GNSS signals are observed only above the vent, while the atmospheric delays are also observed away from the vent. This observation indicates that volcanic ashes stay right above the vent without spreading out, while volcanic gases spread out from the vent.

GNSS observations can separate wet and dry (or ice-rich) plumes. Grapenthin et al. (2018) observed decays of GNSS signals passing above the vent at ~25 km above sea level associated with its 2011 eruption of Grímsvötn. As discussed earlier, however, the atmospheric disturbance is limited above the vent at ~15 km above sea level or below. These observations indicate a hail formation at an altitude of ~15 km or above, which is consistent with a numerical simulation of ash dispersal (e.g., Mastin, 2014; Van Eaton et al., 2015).

5 Recommendations

Since the emergence of GPS a few decades ago, GNSS observations have played significant roles in gaining insights into volcanic activity. The main contribution of GNSS is the measurement of surface deformation to gain insights into volcanic processes. In addition, GNSS also significantly contributed to understanding the eruption and plume dynamics by measuring tropospheric and ionospheric disturbances. Here, we discuss future directions to advance volcanology and volcano deformation studies further. The discussion starts from observational aspects and then moves on to modeling aspects.

5.1 Continuing observations

GNSS has monitored some active volcanoes for more than two decades (see Section 3). While it is crucial and well motivated to continue monitoring persistently erupting volcanoes such as Kī-lauea, Piton de la Fournaise, Etna, or Sakurajima, Japan, long-term monitoring of dormant volcanoes is also crucial. While some volcanoes erupt persistently, others erupt only every tens, hundreds, or thousands of years. In such volcanoes, it is crucial to understand the location of magma reservoirs and the accumulation rate of magmatic fluids, which is not always constant. In Campi Flegrei, for example, subsidence between the 1980s and early 2000s was followed by an unsteady acceleration of uplift, continuing until the present (Bevilacqua et al., 2022; Martino, Dolce, Brandi, Scarpato, & Tammaro, 2021). On the other hand, Lisowski et al. (2021) revealed a continuing deceleration of uplift in Three Sisters Volcano, Oregon, The United States.

Fine temporal resolution, which GNSS measurements have, is not required to monitor long-term deformation. Therefore, SAR may be sufficient to monitor the long-term deformation of active volcanoes. Despite the presence of SAR, however, GNSS still plays a significant role in monitoring the long-term deformation of active volcanoes because GNSS measures 3D displacements. In contrast, SAR measures only the line-of-sight distance changes from the satellite to the ground. Also, it is often impossible to measure displacement field from a single SAR constellation because the longevity of an SAR satellite is often less than 10 years.

5.2 Denser observations

The number of GNSS sites has been skyrocketing in the last few decades. Nevada Geodetic Laboratory at University of Nevada Reno obtains more than all available data from more than 17,000 GNSS sites to offer daily coordinates (Blewitt, Hammond, & Kreemer, 2018). Since there are many more GNSS sites from which they do not obtain the data, there would be a total of tens of thousands of GNSS sites on the globe to monitor the Earth's deformation. Since volcano deformation is usually confined within a small area, dense observations are required to understand volcanic activity in detail. Examples seen in Section 3 indicate that the average spacing of GNSS sites needs to be 10 km or less to do so. If a dense deployment of GNSS sites is possible, GNSS measurements can complement measurements by SAR because GNSS measurements have superior temporal resolution than SAR measurements.

A possible way to use more GNSS sites is to use GNSS sites operated by private sectors and those deployed for geodetic purposes. An increasing number of GNSS sites of geodetic quality have been operated by private sectors for various purposes, including automated vehicle operations. In Japan, for example, at least 3300 GNSS sites of sufficient positioning quality have been deployed by private sectors (Ohta & Ohzono, 2022). Combining these GNSS sites with the nationwide GNSS network in Japan, consisting of about 1300 sites, leads to an average spacing of less than 10 km. Because of an expansion of potential applications with GNSS, the number of available GNSS sites will increase in the future, allowing us to monitor surface deformation in high spatial and temporal resolutions.

5.3 Modeling

The recent improvements in computational capabilities allow us to reconstruct the observation with a more complex model, incorporating various factors. These include heterogeneous elastic and rheological structures, realistic topography, and complex geometry of the pressure source. While it is tractable to compute the surface deformation from a given Earth's structure and geometry of the pressure source, the opposite, that is, solving an inverse problem, is not always straightforward. It is particularly the case when the problem is nonlinear, that is, parameters for Earth's structure, the source geometry, or both are unknown. One reason for its difficulty is that solving a forward problem, or computing surface deformation from given parameters, is already computationally expensive. While a few studies already suggested a method to solve an inverse problem in this situation (Huang, Tao, & Shi, 2020; Ronchin, Masterlark, Dawson, Saunders, & Molist, 2017; Trasatti, Giunchi, & Agostinetti, 2008), further improvement is needed to make the computation more tractable.

The last decade or so has witnessed a combination of observational studies of volcano deformation and theoretical studies of the transport of magmatic fluids (e.g., Anderson & Segall, 2011; Wang, Coppess, Segall, Dunham, & Ellsworth, 2022). Because both observational and theoretical studies share the same or at least similar goals from different aspects, it is a natural consequence to attempt to construct a model consistent with observations and theoretical insights. Further methodological developments will lead to more sophisticated modeling and enhanced forecasting capability of volcanic activity.

Acknowledgments

Reviews by Emily Montgomery-Brown and Corné Kreemer improved the manuscript.

References

Aagaard, B. T., Knepley, M. G., & Williams, C. A. (2013). A domain decomposition approach to implementing fault slip in finite-element models of quasi-static and dynamic crustal deformation. *Journal of Geophysical Research: Solid Earth*, *118*(6), 3059–3079. https://doi.org/10.1002/jgrb.50217. June.

Abrams, M., Crippen, R., & Fujisada, H. (2020). ASTER global digital elevation model (GDEM) and ASTER global water body dataset (ASTWBD). *Remote Sensing*, *12*(7), 1156. https://doi.org/10.3390/rs12071156. April.

Adams, R. J., Perger, W. F., Rose, W. I., & Kostinski, A. (1996). Measurements of the complex dielectric constant of volcanic ash from 4 to 19 GHz. *Journal of Geophysical Research: Solid Earth*, *101*(B4), 8175–8185. https://doi.org/10.1029/96jb00193. April.

Agnew, D. C. (1986). Strainmeters and tiltmeters. *Reviews of Geophysics*, *24*(3), 579–624. https://doi.org/10.1029/rg024i003p00579. August.

Aki, K., & Richards, P. G. (2002). *Quantitative seismology*. University Science Books.

Akiyama, T., Kawabata, H., & Yoshioka, S. (2022). Spatiotemporal changes in fault displacements associated with seismovolcanic events in and around Miyakejima and Kozushima in 2000 inferred from GNSS data. *Pure and Applied Geophysics*, *179*(11), 4245–4265. https://doi.org/10.1007/s00024-022-03084-y. July.

Amoruso, A., & Crescentini, L. (2013). Analytical models of volcanic ellipsoidal expansion sources. *Annals of Geophysics*, *56*(4). https://doi.org/10.4401/ag-6441. November.

Anderson, E. M. (1937). The dynamics of the formation of cone-sheets, ring-dykes, and caldron-subsidences. *Proceedings of the Royal Society of Edinburgh*, *56*, 128–157. https://doi.org/10.1017/s0370164600014954.

Anderson, K., & Johanson, I. (2022). Incremental caldera collapse at Kī-lauea volcano recorded in ground tilt and high-rate GNSS data, with implications for collapse dynamics and the magma system. *Bulletin of Volcanology*, *84*(10). https://doi.org/10.1007/s00445-022-01589-x. September.

Anderson, K., & Segall, P. (2011). Physics-based models of ground deformation and extrusion rate at effusively erupting volcanoes. *Journal of Geophysical Research*, *116*(B7). https://doi.org/10.1029/2010jb007939. July.

Anderson, K. R., Johanson, I. A., Patrick, M. R., Gu, M., Segall, P., Poland, M. P., ... Miklius, A. (2019). Magma reservoir failure and the onset of caldera collapse at Kī-lauea volcano in 2018. *Science*, *366*(6470). https://doi.org/10.1126/science.aaz1822. December.

Aoki, Y., Segall, P., Kato, T., Cervelli, P., & Shimada, S. (1999). Imaging magma transport during the 1997 seismic swarm off the Izu Peninsula, Japan. *Science*, *286*(5441), 927–930. https://doi.org/10.1126/science.286.5441.927. October.

Aoki, Y., Takeo, M., Ohminato, T., Nagaoka, Y., & Nishida, K. (2013). Magma pathway and its structural controls of Asama Volcano, Japan. *Geological Society, London, Special Publications*, *380*(1), 67–84. https://doi.org/10.1144/sp380.6. January.

Aranzulla, M., Cannavò, F., Scollo, S., Puglisi, G., & Immè, G. (2013). Volcanic ash detection by GPS signal. *GPS Solutions*, *17*(4), 485–497. https://doi.org/10.1007/s10291-012-0294-4. October.

Bevilacqua, A., Martino, P. D., Giudicepietro, F., Ricciolino, P., Patra, A., Pitman, E. B., ... Neri, A. (2022). Data analysis of the unsteadily accelerating GPS and seismic records at Campi Flegrei Caldera from 2000 to 2020. *Scientific Reports*, *12*(1). https://doi.org/10.1038/s41598-022-23628-5. November.

Bevis, M., Businger, S., Herring, T. A., Rocken, C., Anthes, R. A., & Ware, R. H. (1992). GPS meteorology: Remote sensing of atmospheric water vapor using the global positioning system. *Journal of Geophysical Research*, *97*(D14), 15787. https://doi.org/10.1029/92jd01517.

Blewitt, G., Hammond, W., & Kreemer, C. (2018). Harnessing the GPS data explosion for interdisciplinary science. *Eos*, *99*. https://doi.org/10.1029/2018eo104623. September.

Bonaccorso, A., & Aloisi, M. (2021). Tracking magma storage: New perspectives from 40 years (1980–2020) of ground deformation source modeling on Etna Volcano. *Frontiers in Earth Science*, *9*. https://doi.org/10.3389/feart.2021.638742. March.

Bonaccorso, A., Bonforte, A., Currenti, G., Negro, C. D., Stefano, A. D., & Greco, F. (2011). Magma storage, eruptive activity and flank instability: Inferences from ground deformation and gravity changes during the 1993–2000 recharging of Mt. Etna Volcano. *Journal of Volcanology and Geothermal Research*, *200*(3–4), 245–254. https://doi.org/10.1016/j.jvolgeores.2011.01.001. March.

Bonaccorso, A., & Davis, P. M. (1999). Models of ground deformation from vertical volcanic conduits with application to eruptions of Mount St. Helens and Mount Etna. *Journal of Geophysical Research: Solid Earth*, *104*(B5), 10531–10542. https://doi.org/10.1029/1999jb900054. May.

Bonforte, A., Bonaccorso, A., Guglielmino, F., Palano, M., & Puglisi, G. (2008). Feeding system and magma storage beneath Mt. Etna as revealed by recent inflation/deflation cycles. *Journal of Geophysical Research*, *113*(B5). https://doi.org/10.1029/2007jb005334. May.

Bonforte, A., & Puglisi, G. (2006). Dynamics of the eastern flank of Mt. Etna volcano (Italy) investigated by a dense GPS network. *Journal of Volcanology and Geothermal Research*, *153*(3–4), 357–369. https://doi.org/10.1016/j.jvolgeores.2005.12.005. May.

Brooks, B., Foster, J., Bevis, M., Frazer, L., Wolfe, C., & Behn, M. (2006). Periodic slow earthquakes on the flank of Kī-lauea volcano, Hawai'i. *Earth and Planetary Science Letters*, *246*(3–4), 207–216. https://doi.org/10.1016/j.epsl.2006.03.035. June.

Brooks, B. A., Foster, J., Sandwell, D., Wolfe, C. J., Okubo, P., Poland, M., & Myer, D. (2008). Magmatically triggered slow slip at Kilauea volcano, Hawaii. *Science*, *321*(5893), 1177. https://doi.org/10.1126/science.1159007. August.

Bruno, V., Aloisi, M., Gambino, S., Mattia, M., Ferlito, C., & Rossi, M. (2022). The most intense deflation of the last two decades at Mt. Etna: The 2019–2021 evolution of ground deformation and modeled pressure sources. *Geophysical Research Letters*, *49*(6). https://doi.org/10.1029/2021gl095195. March.

Bruno, V., Mattia, M., Montgomery-Brown, E., Rossi, M., & Scandura, D. (2017). Inflation leading to a slow slip event and volcanic unrest at Mount Etna in 2016: Insights from CGPS data. *Geophysical Research Letters*, *44*(24). https://doi.org/10.1002/2017gl075744. December.

Bürgmann, R., & Dresen, G. (2008). Rheology of the lower crust and upper mantle: Evidence from rock mechanics, geodesy, and field observations. *Annual Review of Earth and Planetary Sciences*, *36*(1), 531–567. https://doi.org/10.1146/annurev.earth.36.031207.124326. May.

Cashman, K. V., Sparks, R. S. J., & Blundy, J. D. (2017). Vertically extensive and unstable magmatic systems: A unified view of igneous processes. *Science*, *355*(6331). https://doi.org/10.1126/science.aag3055. March.

Cayol, V., & Cornet, F. H. (1998). Three-dimensional modeling of the 1983–1984 eruption at Piton de la Fournaise volcano, Réunion Island. *Journal of Geophysical Research: Solid Earth*, *103*(B8), 18025–18037. https://doi.org/10.1029/98jb00201. August.

Cervelli, P., Kenner, S., & Segall, P. (1999). Correction to "Dislocations in inhomogeneous media via a moduli perturbation approach: General formulation and two-dimensional solutions" by Yijun Du, Paul Segall, and Huajian Gao. *Journal of Geophysical Research: Solid Earth*, *104*(B10), 23271–23277. https://doi.org/10.1029/1999jb900229. October.

Cervelli, P., Segall, P., Amelung, F., Garbeil, H., Meertens, C., Owen, S., ... Lisowski, M. (2002). The 12 September 1999 upper east rift zone dike intrusion at Kilauea volcano, Hawaii. *Journal of Geophysical Research: Solid Earth*, *107*(B7). https://doi.org/10.1029/2001jb000602. July, ECV 3-1–ECV 3-13.

Chaussard, E., & Amelung, F. (2014). Regional controls on magma ascent and storage in volcanic arcs. *Geochemistry, Geophysics, Geosystems*, *15*(4), 1407–1418. https://doi.org/10.1002/2013gc005216. April.

Chaussard, E., Amelung, F., & Aoki, Y. (2013). Characterization of open and closed volcanic systems in Indonesia and Mexico using InSAR time series. *Journal of Geophysical Research: Solid Earth*, *118*(8), 3957–3969. https://doi.org/10.1002/jgrb.50288. August.

Chen, Y., Remy, D., Froger, J.-L., Peltier, A., Villeneuve, N., Darrozes, J., ... Bonvalot, S. (2017). Long-term ground displacement observations using InSAR and GNSS at Piton de la Fournaise volcano between 2009 and 2014. *Remote Sensing of Environment*, *194*, 230–247. https://doi.org/10.1016/j.rse.2017.03.038. June.

Currenti, G., Bonaccorso, A., Negro, C. D., Guglielmino, F., Scandura, D., & Boschi, E. (2010). FEM-based inversion for heterogeneous fault mechanisms: Application at Etna volcano by DInSAR data. *Geophysical Journal International, 183*(2), 765–773. https://doi.org/10.1111/j.1365-246x.2010.04769. x. September.

Davies, J. H. (2003). Elastic field in a semi-infinite solid due to thermal expansion or a coherently misfitting inclusion. *Journal of Applied Mechanics, 70*(5), 655–660. https://doi.org/10.1115/1.1602481. September.

Davis, P. M. (1986). Surface deformation due to inflation of an arbitrarily oriented triaxial ellipsoidal cavity in an elastic half-space, with reference to Kilauea volcano, Hawaii. *Journal of Geophysical Research: Solid Earth, 91*(B7), 7429–7438. https://doi.org/10.1029/jb091ib07p07429. June.

Del Gaudio, C., Aquino, I., Ricciardi, G., Ricco, C., & Scandone, R. (2010). Unrest episodes at Campi Flegrei: A reconstruction of vertical ground movements during 1905–2009. *Journal of Volcanology and Geothermal Research, 195*(1), 48–56. https://doi.org/10.1016/j.jvolgeores.2010.05.014. August.

de Zeeuw-van Dalfsen, E., Pedersen, R., Hooper, A., & Sigmundsson, F. (2012). Subsidence of Askja caldera 2000–2009: Modelling of deformation processes at an extensional plate boundary, constrained by time series InSAR analysis. *Journal of Volcanology and Geothermal Research, 213–214*, 72–82. https://doi.org/10.1016/j.jvolgeores.2011.11.004. February.

Di Vito, M. A., Acocella, V., Aiello, G., Barra, D., Battaglia, M., Carandente, A., … Terrasi, F. (2016). Magma transfer at Campi Flegrei Caldera (Italy) before the 1538 AD eruption. *Scientific Reports, 6*(1). https://doi.org/10.1038/srep32245. August.

Dieterich, J. H., & Decker, R. W. (1975). Finite element modeling of surface deformation associated with volcanism. *Journal of Geophysical Research, 80*(29), 4094–4102. https://doi.org/10.1029/jb080i029p04094. October.

Dragoni, M., & Magnanensi, C. (1989). Displacement and stress produced by a pressurized, spherical magma chamber, surrounded by a viscoelastic shell. *Physics of the Earth and Planetary Interiors, 56*(3–4), 316–328. https://doi.org/10.1016/0031-9201(89)90166-0. September.

Du, Y., Segall, P., & Gao, H. (1994). Dislocations in inhomogeneous media via a moduli perturbation approach: General formulation and two-dimensional solutions. *Journal of Geophysical Research: Solid Earth, 99*(B7), 13767–13779. https://doi.org/10.1029/94jb00339. July.

Du, Y., Segall, P., & Gao, H. (1997). Quasi-static dislocations in three dimensional inhomogeneous media. *Geophysical Research Letters, 24*(18), 2347–2350. https://doi.org/10.1029/97gl02341. September.

Dziewonski, A. M., & Anderson, D. L. (1981). Preliminary reference earth model. *Physics of the Earth and Planetary Interiors, 25*(4), 297–356. https://doi.org/10.1016/0031-9201(81)90046-7. June.

Dzurisin, D. (2006). *Volcano deformation: New geodetic monitoring techniques*. Berlin, Heidelberg: Springer. https://doi.org/10.1007/978-3-540-49302-0.

Farr, T. G., Rosen, P. A., Caro, E., Crippen, R., Duren, R., Hensley, S., … Alsdorf, D. (2007). The shuttle radar topography mission. *Reviews of Geophysics, 45*(2). https://doi.org/10.1029/2005rg000183. May.

Fialko, Y., Khazan, Y., & Simons, M. (2001). Deformation due to a pressurized horizontal circular crack in an elastic half-space, with applications to volcano geodesy. *Geophysical Journal International, 146*(1), 181–190. https://doi.org/10.1046/j.1365-246x.2001.00452.x. July.

Fournier, N., & Jolly, A. D. (2014). Detecting complex eruption sequence and directionality from high-rate geodetic observations: The August 6, 2012 Te Maari eruption, Tongariro, New Zealand. *Journal of Volcanology and Geothermal Research, 286*, 387–396. https://doi.org/10.1016/j.jvolgeores.2014.05.021. October.

Gilbert, F., & Backus, G. E. (1966a). Propagator matrices in elastic wave and vibration problems. *Geophysics, 31*(2), 326–332. https://doi.org/10.1190/1.1439771. April.

Gilbert, F., & Backus, G. E. (1966b). To: "Propagator matrices in elastic wave and vibration problems," by Freeman Gilbert and George E. Backus, April, 1966, p. 326–332. *Geophysics, 31*(3), 643. https://doi.org/10.1190/1.1439802. June.

Giordano, G., & Caricchi, L. (2022). Determining the state of activity of transcrustal magmatic systems and their volcanoes. *Annual Review of Earth and Planetary Sciences, 50*(1), 231–259. https://doi.org/10.1146/annurev-earth-032320-084733. May.

González, P. J., Tiampo, K. F., Camacho, A. G., & Fernández, J. (2010). Shallow flank deformation at Cumbre Vieja volcano (Canary Islands): Implications on the stability of steep-sided volcano flanks at oceanic islands. *Earth and Planetary Science Letters, 297*(3–4), 545–557. https://doi.org/10.1016/j.epsl.2010.07.006. September.

Got, J.-L., Carrier, A., Marsan, D., Jouanne, F., Vogfjörd, K., & Villemin, T. (2017). An analysis of the nonlinear magma-edifice coupling at Grimsvötn volcano (Iceland). *Journal of Geophysical Research: Solid Earth, 122*(2), 826–843. https://doi.org/10.1002/2016jb012905. February.

Grapenthin, R., Hreinsdóttir, S., & Eaton, A. R. V. (2018). Volcanic hail detected with GPS: The 2011 eruption of Grímsvötn volcano, Iceland. *Geophysical Research Letters, 45*, 12236–12243. https://doi.org/10.1029/2018GL080317.

Grapenthin, R., Kyle, P., Aster, R. C., Angarita, M., Wilson, T., & Chaput, J. (2022). Deformation at the open-vent Erebus Volcano, Antarctica, from more than 20 years of GNSS observations. *Journal of Volcanology and Geothermal Research, 432*, 107703. https://doi.org/10.1016/j.jvolgeores.2022.107703. December.

Gudmundsson, M. T., Jónsdóttir, K., Hooper, A., Holohan, E. P., Halldórsson, S. A., Ófeigsson, B. G., … Aiuppa, A. (2016). Gradual caldera collapse at Bárdarbunga volcano, Iceland, regulated by lateral magma outflow. *Science, 353*(6296). https://doi.org/10.1126/science.aaf8988. July.

Head, M., Hickey, J., Thompson, J., Gottsmann, J., & Fournier, N. (2022). Rheological controls on magma reservoir failure in a thermo-viscoelastic crust. *Journal of Geophysical Research: Solid Earth, 127*(7). https://doi.org/10.1029/2021jb023439. July.

Heki, K. (2006). Explosion energy of the 2004 eruption of the Asama Volcano, central Japan, inferred from ionospheric disturbances. *Geophysical Research Letters, 33*(14). https://doi.org/10.1029/2006gl026249.

Hotta, K., Iguchi, M., & Tameguri, T. (2016). Rapid dike intrusion into Sakurajima volcano on August 15, 2015, as detected by multi-parameter ground deformation observations. *Earth, Planets and Space, 68*(1). https://doi.org/10.1186/s40623-016-0450-0. April.

Houlié, N. (2005). Sounding the plume of the 18 August 2000 eruption of Miyakejima volcano (Japan) using GPS. *Geophysical Research Letters, 32*(5). https://doi.org/10.1029/2004gl021728.

Houlié, N., Briole, P., Bonforte, A., & Puglisi, G. (2006). Large scale ground deformation of Etna observed by GPS between 1994 and 2001. *Geophysical Research Letters, 33*(2). https://doi.org/10.1029/2005gl024414.

Houlié, N., Briole, P., Nercessian, A., & Murakami, M. (2005). Volcanic plume above Mount St. Helens detected with GPS. *Eos, Transactions American Geophysical Union, 86*(30), 277. https://doi.org/10.1029/2005eo300001.

Hreinsdóttir, S., Sigmundsson, F., Roberts, M. J., Björnsson, H., Grapenthin, R., Arason, P., … Óladóttir, B. A. (2014). Volcanic plume height correlated with magma-pressure change at Grímsvötn volcano, Iceland. *Nature Geoscience, 7*(3), 214–218. https://doi.org/10.1038/ngeo2044. January.

Huang, L., Tao, T., & Shi, Y. (2020). Numerical inversion of magma chamber pressurization in volcanic areas: A case study of Changbaishan volcano. *Journal of Volcanology and Geothermal Research, 395*, 106830. https://doi.org/10.1016/j.jvolgeores.2020.106830. April.

Jackson, M. D., Blundy, J., & Sparks, R. S. J. (2018). Chemical differentiation, cold storage and remobilization of magma in the Earth's crust. *Nature, 564*(7736), 405–409. https://doi.org/10.1038/s41586-018-0746-2. December.

Kennett, B. L. N., Engdahl, E. R., & Buland, R. (1995). Constraints on seismic velocities in the Earth from traveltimes. *Geophysical Journal International, 122*(1), 108–124. https://doi.org/10.1111/j.1365-246x.1995.tb03540.x. July.

Langer, L., Beller, S., Hirakawa, E., & Tromp, J. (2023). Impact of sedimentary basins on Green's functions for static slip inversion. *Geophysical Journal International, 232*(1), 569–580. https://doi.org/10.1093/gji/ggac344. August.

Langer, L., Gharti, H. N., & Tromp, J. (2019). Impact of topography and three-dimensional heterogeneity on coseismic deformation. *Geophysical Journal International, 217*(2), 866–878. https://doi.org/10.1093/gji/ggz060. January.

Larson, K. M. (2013). A new way to detect volcanic plumes. *Geophysical Research Letters, 40*(11), 2657–2660. https://doi.org/10.1002/grl.50556. June.

Larson, K. M. (2019). Unanticipated uses of the global positioning system. *Annual Review of Earth and Planetary Sciences, 47*(1), 19–40. https://doi.org/10.1146/annurev-earth-053018-060203. May.

Larson, K. M., Palo, S., Roesler, C., Mattia, M., Bruno, V., Coltelli, M., & Fee, D. (2017). Detection of plumes at Redoubt and Etna volcanoes using the GPS SNR method. *Journal of Volcanology and Geothermal Research, 344*, 26–39. https://doi.org/10.1016/j.jvolgeores.2017.04.005. September.

Li, S., Sigmundsson, F., Drouin, V., Parks, M. M., Ófeigsson, B. G., Jónsdóttir, K., … Hreinsdóttir, S. (2021). Ground deformation after a Caldera collapse: Contributions of magma inflow and viscoelastic response to the 2015–2018 deformation field around Bárðarbunga, Iceland. *Journal of Geophysical Research: Solid Earth, 126*(3). https://doi.org/10.1029/2020jb020157. March.

Liao, Y. (2022). The roles of heat and gas in a mushy magma chamber. *Journal of Geophysical Research: Solid Earth, 127*(7). https://doi.org/10.1029/2022jb024357. July.

Liao, Y., Soule, S. A., & Jones, M. (2018). On the mechanical effects of poroelastic crystal mush in classical magma chamber models. *Journal of Geophysical Research: Solid Earth, 123*(11), 9376–9406. https://doi.org/10.1029/2018jb015985. November.

Liao, Y., Soule, S. A., Jones, M., & Mével, H. L. (2021). The mechanical response of a magma chamber with poroviscoelastic crystal mush. *Journal of Geophysical Research: Solid Earth, 126*(4). https://doi.org/10.1029/2020jb019395. April.

Lisowski, M., McCaffrey, R., Wicks, C. W., & Dzurisin, D. (2021). Geodetic constraints on a 25-year magmatic inflation episode near three sisters, Central Oregon. *Journal of Geophysical Research: Solid Earth, 126*(12). https://doi.org/10.1029/2021jb022360. December.

Long, S. M., & Grosfils, E. B. (2009). Modeling the effect of layered volcanic material on magma reservoir failure and associated deformation, with application to Long Valley Caldera, California. *Journal of Volcanology and Geothermal Research, 186*(3–4), 349–360. https://doi.org/10.1016/j.jvolgeores.2009.05.021. October.

Lungarini, L., Troise, C., Meo, M., & Natale, G. D. (2005). Finite element modelling of topographic effects on elastic ground deformation at Mt. Etna. *Journal of Volcanology and Geothermal Research, 144*(1–4), 257–271. https://doi.org/10.1016/j.jvolgeores.2004.11.031. June.

Manconi, A., Walter, T. R., & Amelung, F. (2007). Effects of mechanical layering on volcano deformation. *Geophysical Journal International, 170*(2), 952–958. https://doi.org/10.1111/j.1365-246x.2007.03449.x. August.

Martino, P. D., Dolce, M., Brandi, G., Scarpato, G., & Tammaro, U. (2021). The ground deformation history of the Neapolitan Volcanic area (Campi Flegrei Caldera, Somma–Vesuvius Volcano, and Ischia Island) from 20 years of continuous GPS observations (2000–2019). *Remote Sensing, 13*(14), 2725. https://doi.org/10.3390/rs13142725. July.

Masterlark, T. (2007). Magma intrusion and deformation predictions: sensitivities to the Mogi assumptions. *Journal of Geophysical Research, 112*(B6). https://doi.org/10.1029/2006jb004860. June.

Mastin, L., Guffanti, M., Servranckx, R., Webley, P., Barsotti, S., Dean, K., … Waythomas, C. (2009). A multidisciplinary effort to assign realistic source parameters to models of volcanic ash-cloud transport and dispersion during eruptions. *Journal of Volcanology and Geothermal Research, 186*(1–2), 10–21. https://doi.org/10.1016/j.jvolgeores.2009.01.008. September.

Mastin, L. G. (2014). Testing the accuracy of a 1-D volcanic plume model in estimating mass eruption rate. *Journal of Geophysical Research: Atmospheres, 119*(5), 2474–2495. https://doi.org/10.1002/2013jd020604. March.

Mathieu, L., Kervyn, M., & Ernst, G. G. J. (2011). Field evidence for flank instability, basal spreading and volcano-tectonic interactions at Mt Cameroon, West Africa. *Bulletin of Volcanology, 73*(7), 851–867. https://doi.org/10.1007/s00445-011-0458-z. February.

Mattia, M., Bruno, V., Caltabiano, T., Cannata, A., Cannavò, F., D'Alessandro, W., … Salerno, G. (2015). A comprehensive interpretative model of slow slip events on Mt. Etna's eastern flank. *Geochemistry, Geophysics, Geosystems, 16*(3), 635–658. https://doi.org/10.1002/2014gc005585. March.

Mattia, M., Bruno, V., Montgomery-Brown, E., Patanè, D., Barberi, G., & Coltelli, M. (2020). Combined seismic and geodetic analysis before, during, and after the 2018 Mount Etna eruption. *Geochemistry, Geophysics, Geosystems, 21*(9). https://doi.org/10.1029/2020gc009218. September.

McTigue, D. F. (1987). Elastic stress and deformation near a finite spherical magma body: Resolution of the point source paradox. *Journal of Geophysical Research, 92*(B12), 12931–12940. https://doi.org/10.1029/jb092ib12p12931.

McTigue, D. F., & Segall, P. (1988). Displacements and tilts from dip-slip faults and magma chambers beneath irregular surface topography. *Geophysical Research Letters, 15*(6), 601–604. https://doi.org/10.1029/gl015i006p00601. June.

McTigue, D. F., & Stein, R. S. (1984). Topographic amplification of tectonic displacement: Implications for geodetic measurement of strain changes. *Journal of Geophysical Research: Solid Earth, 89*(B2), 1123–1131. https://doi.org/10.1029/jb089ib02p01123. February.

Steffes, P., Dufek, J., & Akins, A. (2019). The effect of electrostatic charge on the propagation of GPS (L-band) signals through volcanic plumes. *Journal of Geophysical Research: Atmospheres, 124*(4), 2260–2275. https://doi.org/10.1029/2018jd029076. February.

Meo, M., Tammaro, U., & Capuano, P. (2008). Influence of topography on ground deformation at Mt. Vesuvius (Italy) by finite element modelling. *International Journal of Non-Linear Mechanics, 43*(3), 178–186. https://doi.org/10.1016/j.ijnonlinmec.2007.12.005. April.

Miklius, A., Lisowski, M., Delaney, P. T., Denlinger, R. P., Dvorak, J. J., Okamura, A. T., & Sakol, M. K. (1995). Recent inflation and flank movement of Mauna Loa Volcano. In *Mauna Loa revealed: Structure, composition, history, and hazards* (pp. 199–205). American Geophysical Union. https://doi.org/10.1029/gm092p0199.

Mogi, K. (1958). Relations between the eruptions of various volcanoes and the deformations of the ground surfaces around them. *Bulletin of the Earthquake Research Institute, University of Tokyo, 36*, 99–134. https://doi.org/10.15083/0000033924.

Montgomery-Brown, E. K., & Miklius, A. (2021). Periodic dike intrusions at Kīlauea volcano, Hawai'i. *Geology, 49*(4), 397–401. https://doi.org/10.1130/g47970.1. December.

Montgomery-Brown, E. K., Sinnett, D. K., Poland, M., Segall, P., Orr, T., Zebker, H., & Miklius, A. (2010). Geodetic evidence for en echelon dike emplacement and concurrent slow slip during the June 2007 intrusion and eruption at Kīlauea volcano, Hawaii. *Journal of Geophysical Research, 115*(B7). https://doi.org/10.1029/2009jb006658. July.

Moran, S. C., Newhall, C., & Roman, D. C. (2011). Failed magmatic eruptions: Late-stage cessation of magma ascent. *Bulletin of Volcanology, 73*(2), 115–122. https://doi.org/10.1007/s00445-010-0444-x. February.

Morita, Y., Nakao, S., & Hayashi, Y. (2006). A quantitative approach to the dike intrusion process inferred from a joint analysis of geodetic and seismological data for the 1998 earthquake swarm off the east coast of Izu Peninsula, Central Japan. *Journal of Geophysical Research: Solid Earth, 111*(B6). https://doi.org/10.1029/2005jb003860. June.

Mullet, B., & Segall, P. (2022). The surface deformation signature of a transcrustal, crystal mush-dominant magma system. *Journal of Geophysical Research: Solid Earth, 127*(5). https://doi.org/10.1029/2022jb024178. May.

Munekane, H., Oikawa, J., & Kobayashi, T. (2016). Mechanisms of step-like tilt changes and very long period seismic signals during the 2000 Miyakejima eruption: Insights from kinematic GPS. *Journal of Geophysical Research: Solid Earth, 121*(4), 2932–2946. https://doi.org/10.1002/2016jb012795. April.

Nakao, S., Morita, Y., Yakiwara, H., Oikawa, J., Ueda, H., Takahashi, H., … Iguchi, M. (2013). Volume change of the magma reservoir relating to the 2011 Kirishima Shinmoe-Dake eruption—Charging, discharging and recharging process inferred from GPS measurements. *Earth, Planets and Space, 65*(6), 505–515. https://doi.org/10.5047/eps.2013.05.017. June.

Neal, C. A., Brantley, S. R., Antolik, L., Babb, J. L., Burgess, M., Calles, K., … Damby, D. (2019). The 2018 rift eruption and summit collapse of Kīlauea volcano. *Science, 363*(6425), 367–374. https://doi.org/10.1126/science.aav7046. January.

Nikkhoo, M., & Rivalta, E. (2023). Surface deformations and gravity changes caused by pressurized finite ellipsoidal cavities. *Geophysical Journal International, 232*(1), 643–655. https://doi.org/10.1093/gji/ggac351. September.

Nikkhoo, M., Walter, T. R., Lundgren, P. R., & Prats-Iraola, P. (2017). Compound dislocation models (CDMs) for volcano deformation analyses. *Geophysical Journal International, 208*(2), 877–894. https://doi.org/10.1093/gji/ggw427. November.

Nishimura, T. (2009). Ground deformation caused by magma ascent in an open conduit. *Journal of Volcanology and Geothermal Research, 187*(3–4), 178–192. https://doi.org/10.1016/j.jvolgeores.2009.09.001. November.

Nishimura, T., Ozawa, S., Murakami, M., Sagiya, T., Tada, T., Kaidzu, M., & Ukawa, M. (2001). Crustal deformation caused by magma migration in the northern Izu Islands, Japan. *Geophysical Research Letters, 28*(19), 3745–3748. https://doi.org/10.1029/2001gl013051. October.

Oguchi, T., Udagawa, M., Nanba, N., Maki, M., & Ishimine, Y. (2009). Measurements of dielectric constant of volcanic ash erupted from five volcanoes in Japan. *IEEE Transactions on Geoscience and Remote Sensing, 47*(4), 1089–1096. https://doi.org/10.1109/tgrs.2008.2008023. April.

Ohta, Y., & Iguchi, M. (2015). Advective diffusion of volcanic plume captured by dense GNSS network around Sakurajima volcano: A case study of the Vulcanian eruption on July 24, 2012. *Earth, Planets and Space, 67*(1). https://doi.org/10.1186/s40623-015-0324-x. September.

Ohta, Y., & Ohzono, M. (2022). Potential for crustal deformation monitoring using a dense cell phone carrier Global Navigation Satellite System Network. *Earth, Planets and Space, 74*(1). https://doi.org/10.1186/s40623-022-01585-7. February.

Okada, Y. (1985). Surface deformation due to shear and tensile faults in a half-space. *Bulletin of the Seismological Society of America, 75*(4), 1135–1154. https://doi.org/10.1785/bssa0750041135. August.

Okada, Y., & Yamamoto, E. (1991). Dyke intrusion model for the 1989 seismovolcanic activity off Ito, Central Japan. *Journal of Geophysical Research, 96*(B6), 10361. https://doi.org/10.1029/91jb00427.

Owen, S., Segall, P., Freymueller, J., Mikijus, A., Denlinger, R., Árnadóttir, T., … Bürgmann, R. (1995). Rapid deformation of the south flank of Kilauea volcano, Hawaii. *Science, 267*(5202), 1328–1332. https://doi.org/10.1126/science.267.5202.1328. March.

Owen, S., Segall, P., Lisowski, M., Miklius, A., Denlinger, R., & Sako, M. (2000). Rapid deformation of Kilauea volcano: Global positioning system measurements between 1990 and 1996. *Journal of Geophysical Research: Solid Earth, 105*(B8), 18983–18998. https://doi.org/10.1029/2000jb900109. August.

Owen, S., Segall, P., Lisowski, M., Miklius, A., Murray, M., Bevis, M., & Foster, J. (2000). January 30, 1997 eruptive event on Kilauea volcano, Hawaii, as monitored by continuous GPS. *Geophysical Research Letters, 27*(17), 2757–2760. https://doi.org/10.1029/1999gl008454. September.

Ozawa, S., Miyazaki, S., Nishimura, T., Murakami, M., Kaidzu, M., Imakiire, T., & Ji, X. (2004). Creep, dike intrusion, and magma chamber deflation model for the 2000 Miyake eruption and the Izu Islands earthquakes. *Journal of Geophysical Research: Solid Earth, 109*(B2). https://doi.org/10.1029/2003jb002601. February.

Pascal, K., Neuberg, J., & Rivalta, E. (2014). On precisely modelling surface deformation due to interacting magma chambers and dykes. *Geophysical Journal International*, *196*(1), 253–278. https://doi.org/10.1093/gji/ggt343. October.

Peltier, A., Beauducel, F., Villeneuve, N., Ferrazzini, V., Muro, A. D., Aiuppa, A., ... Taisne, B. (2016). Deep fluid transfer evidenced by surface deformation during the 2014–2015 unrest at Piton de la Fournaise volcano. *Journal of Volcanology and Geothermal Research*, *321*, 140–148. https://doi.org/10.1016/j.jvolgeores.2016.04.031. July.

Peltier, A., Got, J.-L., Villeneuve, N., Boissier, P., Staudacher, T., Ferrazzini, V., & Walpersdorf, A. (2015). Long-term mass transfer at Piton de la Fournaise volcano evidenced by strain distribution derived from GNSS network. *Journal of Geophysical Research: Solid Earth*, *120*(3), 1874–1889. https://doi.org/10.1002/2014jb011738. March.

Peltier, A., Staudacher, T., Bachèlery, P., & Cayol, V. (2009). Formation of the April 2007 Caldera collapse at Piton de la Fournaise volcano: Insights from GPS data. *Journal of Volcanology and Geothermal Research*, *184*(1–2), 152–163. https://doi.org/10.1016/j.jvolgeores.2008.09.009. July.

Pepe, S., D'Auria, L., Castaldo, R., Casu, F., Luca, C. D., Novellis, V. D., ... Tizzani, P. (2018). The use of massive deformation datasets for the analysis of spatial and temporal evolution of Mauna Loa Volcano (Hawai'i). *Remote Sensing*, *10*(6), 968. https://doi.org/10.3390/rs10060968. June.

Poland, M. P., Peltier, A., Bonforte, A., & Puglisi, G. (2017). The spectrum of persistent volcanic flank instability: A review and proposed framework based on Kīlauea, Piton de la Fournaise, and Etna. *Journal of Volcanology and Geothermal Research*, *339*, 63–80. https://doi.org/10.1016/j.jvolgeores.2017.05.004. June.

Pritchard, M. E., & Simons, M. (2002). A satellite geodetic survey of large-scale deformation of volcanic centres in the central Andes. *Nature*, *418*(6894), 167–171. https://doi.org/10.1038/nature00872. July.

Reverso, T., Vandemeulebrouck, J., Jouanne, F., Pinel, V., Villemin, T., Sturkell, E., & Bascou, P. (2014). A two-magma chamber model as a source of deformation at Grímsvötn volcano, Iceland. *Journal of Geophysical Research: Solid Earth*, *119*(6), 4666–4683. https://doi.org/10.1002/2013jb010569. June.

Rivalta, E., & Segall, P. (2008). Magma compressibility and the missing source for some dike intrusions. *Geophysical Research Letters*, *35*(4). https://doi.org/10.1029/2007gl032521. February.

Ronchin, E., Geyer, A., & Martí, J. (2015). Evaluating topographic effects on ground deformation: Insights from finite element modeling. *Surveys in Geophysics*, *36*(4), 513–548. https://doi.org/10.1007/s10712-015-9325-3. May.

Ronchin, E., Masterlark, T., Dawson, J., Saunders, S., & Molist, J. M. (2017). Imaging the complex geometry of a magma reservoir using FEM-based linear inverse modeling of InSAR data: Application to Rabaul Caldera, Papua New Guinea. *Geophysical Journal International*, *209*(3), 1746–1760. https://doi.org/10.1093/gji/ggx119. March.

Rubin, A. M. (1993). Dikes vs. diapirs in viscoelastic rock. *Earth and Planetary Science Letters*, *119*(4), 641–659. https://doi.org/10.1016/0012-821x(93)90069-l. October.

Rucker, C., Erickson, B. A., Karlstrom, L., Lee, B., & Gopalakrishnan, J. (2022). A computational framework for time-dependent deformation in viscoelastic magmatic systems. *Journal of Geophysical Research: Solid Earth*, *127*(9). https://doi.org/10.1029/2022jb024506. September.

Salzer, J. T., Nikkhoo, M., Walter, T. R., Sudhaus, H., Reyes-Dávila, G., Bretón, M., & Arámbula, R. (2014). Satellite radar data reveal short-term pre-explosive displacements and a complex conduit system at Volcán de Colima, Mexico. *Frontiers in Earth Science*, *2*. https://doi.org/10.3389/feart.2014.00012. June.

Segall, P. (2010). *Earthquake and volcano deformation*. Princeton University Press. https://doi.org/10.1515/9781400833856.

Segall, P. (2016). Repressurization following eruption from a magma chamber with a viscoelastic aureole. *Journal of Geophysical Research: Solid Earth*, *121*(12), 8501–8522. https://doi.org/10.1002/2016jb013597. December.

Segall, P. (2019). Magma chambers: What we can, and cannot, learn from volcano geodesy. *Philosophical Transactions of the Royal Society A: Mathematical, Physical and Engineering Sciences*, *377*(2139), 20180158. https://doi.org/10.1098/rsta.2018.0158. January.

Segall, P., Anderson, K. R., Pulvirenti, F., Wang, T., & Johanson, I. (2020). Caldera collapse geometry revealed by near-field GPS displacements at Kīlauea volcano in 2018. *Geophysical Research Letters*, *47*(15). https://doi.org/10.1029/2020gl088867. August.

Segall, P., Desmarais, E. K., Shelly, D., Miklius, A., & Cervelli, P. (2006). Earthquakes triggered by silent slip events on Kīlauea volcano, Hawaii. *Nature*, *442*(7098), 71–74. https://doi.org/10.1038/nature04938. July.

Sezawa, K. (1931). Relations between the eruptions of various volcanoes and the deformations of the ground surfaces around them. *Bulletin of the Earthquake Research Institute, Tokyo Imperial University*, *9*, 398–406. https://doi.org/10.15083/0000034929.

Sigmundsson, F., Hooper, A., Hreinsdóttir, S., Vogfjörd, K. S., Ófeigsson, B. G., Heimisson, E. R., ... Eibl, E. P. S. (2015). Segmented lateral dyke growth in a rifting event at Bárðarbunga volcanic system, Iceland. *Nature*, *517*(7533), 191–195. https://doi.org/10.1038/nature14111. January.

Sigmundsson, F., Pinel, V., Grapenthin, R., Hooper, A., Halldórsson, S. A., Einarsson, P., ... Yamasaki, T. (2020). Unexpected large eruptions from buoyant magma bodies within viscoelastic crust. *Nature Communications*, *11*(1). https://doi.org/10.1038/s41467-020-16054-6. May.

Silverii, F., Montgomery-Brown, E. K., Borsa, A. A., & Barbour, A. J. (2020). Hydrologically induced deformation in Long Valley Caldera and adjacent Sierra Nevada. *Journal of Geophysical Research: Solid Earth*, *125*(5). https://doi.org/10.1029/2020jb019495. May.

Smittarello, D., Cayol, V., Pinel, V., Peltier, A., Froger, J.-L., & Ferrazzini, V. (2019). Magma propagation at Piton de la Fournaise from joint inversion of InSAR and GNSS. *Journal of Geophysical Research: Solid Earth*, *124*(2), 1361–1387. https://doi.org/10.1029/2018jb016856. February.

Smittarello, D., Smets, B., Barrière, J., Michellier, C., Oth, A., Shreve, T., ... Muhindo, A. S. (2022). Precursor-free eruption triggered by edifice rupture at Nyiragongo Volcano. *Nature*, *609*(7925), 83–88. https://doi.org/10.1038/s41586-022-05047-8. August.

Solheim, F. S., Vivekanandan, J., Ware, R. H., & Rocken, C. (1999). Propagation delays induced in GPS signals by dry air, water vapor, hydrometeors, and other particulates. *Journal of Geophysical Research: Atmospheres*, *104*(D8), 9663–9670. https://doi.org/10.1029/1999jd900095. April.

Sparks, R. S. J., Annen, C., Blundy, J. D., Cashman, K. V., Rust, A. C., & Jackson, M. D. (2019). Formation and dynamics of magma reservoirs. *Philosophical Transactions of the Royal Society A: Mathematical, Physical and Engineering Sciences*, *377*(2139), 20180019. https://doi.org/10.1098/rsta.2018.0019. January.

Sparks, R. S. J., & Cashman, K. V. (2017). Dynamic magma systems: Implications for forecasting volcanic activity. *Elements*, *13*(1), 35–40. https://doi.org/10.2113/gselements.13.1.35. February.

Taylor, N. C., Johnson, J. H., & Herd, R. A. (2021). Making the most of the Mogi model: Size matters. *Journal of Volcanology and Geothermal Research*, *419*, 107380. https://doi.org/10.1016/j.jvolgeores.2021.107380. November.

Thayer, G. D. (1974). An improved equation for the radio refractive index of air. *Radio Science*, *9*(10), 803–807. https://doi.org/10.1029/rs009i010p00803. October.

Tiampo, K., Rundle, J., Fernandez, J., & Langbein, J. (2000). Spherical and ellipsoidal volcanic sources at Long Valley Caldera, California, using a genetic algorithm inversion technique. *Journal of Volcanology and Geothermal Research*, *102*(3–4), 189–206. https://doi.org/10.1016/s0377-0273(00)00185-2. November.

Trasatti, E., Giunchi, C., & Agostinetti, N. P. (2008). Numerical inversion of deformation caused by pressure sources: Application to Mount Etna (Italy). *Geophysical Journal International*, *172*(2), 873–884. https://doi.org/10.1111/j.1365-246x.2007.03677.x. February.

Van Eaton, A. R., Mastin, L. G., Herzog, M., Schwaiger, H. F., Schneider, D. J., Wallace, K. L., & Clarke, A. B. (2015). Hail formation triggers rapid ash aggregation in volcanic plumes. *Nature Communications*, *6*(1). https://doi.org/10.1038/ncomms8860. August.

Wang, T., Zheng, Y., Pulvirenti, F., & Segall, P. (2021). Post-2018 Caldera collapse re-inflation uniquely constrains Kī-lauea's magmatic system. *Journal of Geophysical Research: Solid Earth*, *126*(6). https://doi.org/10.1029/2021jb021803. June.

Wang, T. A., Coppess, K. R., Segall, P., Dunham, E. M., & Ellsworth, W. (2022). Physics-based model reconciles Caldera collapse induced static and dynamic ground motion: Application to Kī-lauea 2018. *Geophysical Research Letters*, *49*(8). https://doi.org/10.1029/2021gl097440. April.

Williams, C. A., & Wadge, G. (1998). The effects of topography on magma chamber deformation models: Application to Mt. Etna and radar interferometry. *Geophysical Research Letters*, *25*(10), 1549–1552. https://doi.org/10.1029/98gl01136. May.

Williams, C. A., & Wadge, G. (2000). An accurate and efficient method for including the effects of topography in three-dimensional elastic models of ground deformation with applications to radar interferometry. *Journal of Geophysical Research: Solid Earth*, *105*(B4), 8103–8120. https://doi.org/10.1029/1999jb900307. April.

Yamakawa, N. (1955). On the strain produced in a semi-infinite elastic solid by an interior source of stress. *Journal of the Seismological Society of Japan*, *8*(2), 84–98. https://doi.org/10.4294/zisin1948.8.2_84.

Yamasaki, T., Sigmundsson, F., & Iguchi, M. (2020). Viscoelastic crustal response to magma supply and discharge in the upper crust: Implications for the uplift of the Aira Caldera before and after the 1914 eruption of the Sakurajima volcano. *Earth and Planetary Science Letters*, *531*, 115981. https://doi.org/10.1016/j.epsl.2019.115981. February.

Yamasaki, T., Takahashi, H., Ohzono, M., Wright, T. J., & Kobayashi, T. (2020). The influence of elastic thickness non-uniformity on viscoelastic crustal response to magma emplacement: Application to the Kutcharo Caldera, Eastern Hokkaido, Japan. *Geophysical Journal International*, *224*(1), 701–718. https://doi.org/10.1093/gji/ggaa440. September.

Yang, X.-M., Davis, P. M., & Dieterich, J. H. (1988). Deformation from inflation of a dipping finite prolate spheroid in an elastic half-space as a model for volcanic stressing. *Journal of Geophysical Research: Solid Earth*, *93*(B5), 4249–4257. https://doi.org/10.1029/jb093ib05p04249. May.

Zhong, X., Dabrowski, M., & Jamtveit, B. (2019). Analytical solution for the stress field in elastic half-space with a spherical pressurized cavity or inclusion containing eigenstrain. *Geophysical Journal International*, *216*(2), 1100–1115. https://doi.org/10.1093/gji/ggy447. October.

Zhu, L., & Rivera, L. A. (2002). A note on the dynamic and static displacements from a point source in multilayered media. *Geophysical Journal International*, *148*(3), 619–627. https://doi.org/10.1046/j.1365-246x.2002.01610.x. March.

Chapter 8

GNSS applications for ionospheric seismology and volcanology

Kosuke Heki

Shanghai Astronomical Observatory, Shanghai, China; Department of Earth and Planetary Sciences, Hokkaido University, Sapporo, Japan

1 Introduction and observation history

The fault dislocation of a large earthquake causes vertical movement of the Earth's surface and excites atmospheric waves. Such waves are also excited by strong volcanic eruptions. They propagate upward and disturb the ionosphere. In typical cases, the disturbance starts shortly after an earthquake/eruption when acoustic and internal gravity waves reach the ionospheric F region. This usually ends as a temporary phenomenon but could be followed by long-lasting disturbances.

Coseismic ionospheric disturbances were first found using Doppler sounding as the vertical oscillation of the ionosphere following the 1968 May Tokachi-oki earthquake (M7.9) (Yuen et al., 1969) and the 1982 March Urakawa-oki (M7.1) earthquake (Tanaka et al., 1984), both in northern Japan. Disturbances related to volcanic eruptions were also found using Doppler sounding shortly after the 1981 eruption of the Mount St Helens volcano, North America (Ogawa et al., 1982). The 1991 eruption of the Pinatubo volcano, the Philippines, was one of the largest eruptions in the 20th century, and Cheng and Huang (1992) and Igarashi et al. (1994) detected traveling ionospheric disturbances in Taiwan and Japan, respectively. It was difficult then to understand these phenomena with such observations of limited temporal and spatial coverage.

Chances of detecting such signals have significantly increased over the last 30 years due to the deployment and densification of continuous receiving stations of global navigation satellite systems (GNSS), such as the Global Positioning System (GPS). Because the primary purpose of GNSS networks is crustal deformation monitoring, they are densely deployed near tectonic plate boundaries. With the phase difference in microwave carriers in different frequencies from GNSS satellites, we can measure changes in ionospheric total electron content (TEC), the number of electrons integrated along the line-of-sight connecting the receiver and the satellite. In this chapter, I review current studies of ionospheric seismology and volcanology based on the "GNSS-TEC" technique.

Coseismic ionospheric disturbance was first detected using GNSS for the 1994 Northridge earthquake ($M_w6.7$), California (Calais & Minster, 1995). A comprehensive study with a dense GNSS array for the 2003 Tokachi-oki earthquake ($M_w8.0$), Japan, revealed various properties of the near-field disturbances, e.g., propagating velocities and directivities (Heki & Ping, 2005). Although ionospheric TEC changes caused by an artificial mine blast were already reported by Calais et al. (1998), those resulting from a real volcanic eruption were first studied using a dense GNSS array by Heki et al. (2006) for the 2004 September eruption of the Asama volcano, central Japan (VEI 2).

Since then, numerous new cases of such disturbances have been reported, as reviewed by Tanimoto et al. (2015), Jin et al. (2018), Astafyeva (2019), and Heki (2021). Here, I present the general methodology of the GNSS-TEC technique in Section 2. Then I discuss the essential contribution of GNSS-TEC observations to seismology (Section 3) and volcanology (Section 4). There I mainly explore how we can infer earthquake magnitudes and volcanic eruption intensities using ionospheric disturbances. Here I do not discuss possible ionospheric changes before earthquakes, which is reviewed by Heki (2021).

2 GNSS-TEC observations

2.1 Phase difference and TEC

Satellites of GPS, the oldest GNSS, transmit microwave signals in the two frequencies 1.57542 GHz (L1) and 1.22760 GHz (L2) from an altitude of ~20,200 km. These signals propagate through the ionosphere, a dispersive media with frequency-dependent delays. The ionosphere ranges in height from ~60 km to beyond 1000 km, and the largest ionization occurs in the F region, ~300 km in altitude. By tracking the differences between the L1 and L2 phases $\Delta(L1-L2)$, we can monitor the temporal changes of TEC along line-of-sight (slant TEC or STEC). The TEC changes are related to the phase difference changes as follows (in TEC unit, i.e., 10^{16} el/m^2).

$$\Delta STEC = (1/40.308) f_1^2 f_2^2 / (f_1^2 - f_2^2) \, \Delta(L1-L2). \tag{8.1}$$

Here, f_1 and f_2 show the frequencies of the two microwave carriers from GNSS, and the phase difference $\Delta(L1-L2)$ should be expressed in meters.

STEC shows U-shaped changes coming from the apparent movement of the satellites in the sky (and consequent changes in the incidence angles of line-of-sights with the ionosphere). Recent GNSS includes geostationary satellites. In that case, STEC does not show such U-shaped changes.

There are an increasing number of ground stations with receivers capable of receiving multi-GNSS. Different systems transmit microwave signals in slightly different frequencies, which should be taken into account when converting the phase differences into TEC. Some GNSS also employ L5 signals (1.17645 GHz in GPS) with a frequency slightly lower than L2. GNSS raw data files are available as RINEX (receiver-independent exchange format) files and can be downloaded from data centers run by, e.g., International GNSS Service (IGS), and University NAVSTAR Consortium (UNAVCO). The typical sampling interval at ground GNSS stations is 30 s. High-rate sampling (e.g., 1 s) is useful for ionospheric seismology/volcanology (e.g., Astafyeva et al., 2013), but is not indispensable. The examples presented in this chapter are all derived from regular 30-s sampling stations.

2.2 From STEC to VTEC

STEC time series of nongeostationary GNSS satellites show U-shaped long-term (hours) changes coming from the satellite motion, in addition to real spatial (e.g., latitudinal difference of ionization) and temporal (e.g., diurnal changes) variations. They also include station- and satellite-dependent biases. Conversion from biased STEC to absolute vertical TEC (VTEC) simplifies TEC behaviors and often makes interpretations easier. The conversion can be done in three steps.

We first remove phase ambiguities by aligning STEC derived from carrier phases with those derived by pseudo-ranges, which have larger noises but no phase ambiguities. Then we correct for satellite and receiver differential code biases (DCBs) available online from the header information of Global Ionospheric Map (GIM) files (Mannucci et al., 1998). If the receiver bias of a station of interest is not available, then we can infer its value by minimizing the scatter of nighttime VTEC at that station, the approach known as "minimum scalloping" (Rideout & Coster, 2006). At last, such bias-free STEC can be converted to absolute VTEC by multiplying it with the cosine of the incidence angle of line-of-sight with ionospheric F region.

The coordinates of the ionospheric piercing points (IPP) of line-of-sight are calculated assuming a thin ionosphere normally at an altitude of 300 km, and the trajectories of their ground projections (subionospheric points, SIP) are plotted on the map to indicate the horizontal position of the observed ionosphere.

2.3 Finding signals related to earthquakes and volcanic eruptions

Ionospheric disturbances by earthquakes and volcanic eruptions are overprinted to other changes, which include regular diurnal changes, spatial changes due to movement of IPP (i.e., crossing of equatorial ionization anomalies), short-term (a few minutes) positive pulses caused by sporadic-E irregularities, sudden drops due to plasma bubbles, large- and medium-scale traveling ionospheric disturbances (LSTID/MSTID), and so on.

Long-term changes can be removed by applying a high-pass filter, easily done by modeling the VTEC (or STEC) changes with polynomials of time (the appropriate degree depends on the time window) and by extracting the departures from these reference curves. Wavelet transformation is also effective in extracting components of certain periods.

Disturbances related to earthquakes and volcanic explosions often appear ~10 min after the events and propagate at several different velocities. The keys to attribute certain TEC changes to earthquakes or eruptions would be (1) if they emerge at the right times and (2) if they propagate outward from the epicenters/volcanoes with prescribed velocities. Then, one would not wrongly interpret irrelevant ionospheric disturbances as those caused by earthquakes and volcanic eruptions. In other words, it is difficult to interpret the observation when we have just one station-satellite pair recording the TEC change.

2.4 Multi-GNSS

Most GNSS-TEC studies have been done using GPS, the first GNSS developed by the United States, and GLONASS, the second GNSS developed by Russia. Now, two other GNSS are operated, namely European Galileo and Chinese BDS (BeiDou Navigation Satellite System). In addition, regional satellite navigation systems are in operation such as QZSS (Quasi-Zenith Satellite System) in Japan and NAVIC (Navigation Indian Constellation) in India. Stations equipped with multi-GNSS receivers are increasing.

BDS, QZSS, and NAVIC partly employ geostationary satellites. By using them, we can keep observing the same satellites continuously and obtain TEC time series without disruptions. This offers a unique opportunity in ionospheric seismology and volcanology. In Section 4, I present an example of ionospheric volcanology taking advantage of the QZSS geostationary satellite.

Another benefit of modern GNSS is the addition of the new carrier frequency L5 to conventional L1 and L2. Its frequency is somewhat lower than L2, and the combination of L1-L5 offers a larger frequency difference than the L1-L2 pair. Now we have three different combinations of carrier phases, L1-L5, L2-L5, and L1-L2. These three have different factors $f_1^2 f_2^2 / (f_1^2 - f_2^2)$ in Eq. (8.1). Heki and Fujimoto (2022) compared TEC time series from these three pairs and separated real TEC changes from noises in GNSS receivers.

3 Ionospheric seismology

3.1 Three different atmospheric waves

An earthquake causes coseismic vertical crustal movement, which excites atmospheric waves (Fig. 8.1). These waves propagate upward and disturb the ionosphere. The disturbance usually starts ~10 min after an earthquake, when the acoustic wave reaches the ionospheric F region. Disturbances caused by direct acoustic waves from epicenters propagate at the sound velocity of the F region height (~0.8 km/s).

Vertical surface motions associated with the passage of the Rayleigh surface waves also excite acoustic waves (Fig. 8.1). They make disturbances propagating much faster (~4 km/s) than direct acoustic waves. They decay less with distance and propagate worldwide. Ionospheric disturbances caused by the 2002 Denali earthquake (M_w7.9), Alaska, were

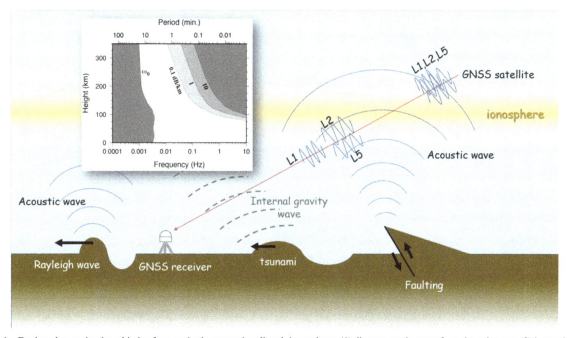

FIG. 8.1 Earthquakes excite three kinds of atmospheric waves that disturb ionosphere: (1) direct acoustic wave from the epicenter, (2) internal gravity wave propagating obliquely upward from the focal area and from propagating tsunami, and (3) acoustic wave excited by propagating Rayleigh surface wave. They are observed with GNSS as TEC changes by receiving microwave signals from satellites in different carrier frequencies. Inset shows the frequency-dependent atmospheric filtering effect of acoustic waves after Blanc (1985).

detected in GNSS stations, California (Ducic et al., 2003), and those caused by the 2004 Sumatra-Andaman earthquake (M_w9.2) have been detected in Japan (Heki, 2021). Disturbances originating from these acoustic waves have typical periods of ~4 min.

Large earthquakes and tsunamis also excite internal gravity waves with longer periods (Fig. 8.1). Clear concentric wave fronts of internal gravity waves were observed after the 2011 Tohoku-oki earthquake (M_w9.0) (Tsugawa et al., 2011). Considering that they emerged above the Japan Sea, the waves would have been excited at the epicenter rather than by propagating tsunamis. Disturbances caused by internal gravity waves typically have periods of ~20 min and propagate with a speed 0.2–0.3 km/s.

3.2 Discriminating the three different waves

Fig. 8.2 shows an example of the coseismic ionospheric disturbance in the 1994 Hokkaido-toho-oki earthquake (M_w8.3) (Astafyeva et al., 2009), where signatures of the three atmospheric waves (Fig. 8.1) are visible. Fig. 8.2A shows the slant TEC time series of GPS satellite 6 (called as G06 here), whose SIP moves northward between stations and the epicenter (Fig. 8.2B). The TEC changes are characterized by the first strong positive anomalies ~10 min after the earthquake, followed by lower amplitude components with slower velocities. We can also recognize faint harmonic oscillation with a period of ~4 min, lasting for a half hour or so within ~500 km from the epicenter.

Here, I perform wavelet transformation using a Mexican hut wavelet to extract certain period components (Heki & Ping, 2005) and show them as a function of time and focal distance in Fig. 8.2C and D. By using a short period wavelet, we can get sharp signatures of Rayleigh waves and harmonic oscillations propagating with a similar speed (Fig. 8.2C). However, direct acoustic wave and internal gravity wave signatures become clearer by using a longer-period wavelet (Fig. 8.2D). There, we can see components propagating at ~4, ~0.8, and ~0.25 km/s, corresponding to three atmospheric waves illustrated in

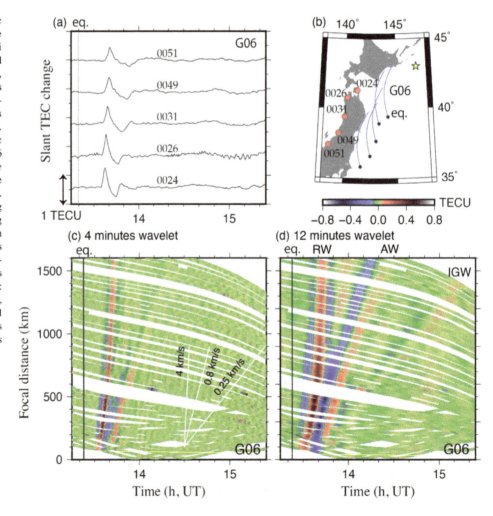

FIG. 8.2 Coseismic ionospheric disturbance associated with the 1994 October Hokkaido-toho-oki earthquake (M_w8.3) that occurred off the east coast of Hokkaido, Japan. (A) Slant TEC time series from five GNSS stations with satellite G06 are drawn as residuals from degree 4 best-fit polynomials. (B) SIP trajectories (ionospheric height assumed as 300 km) of G06 during the period shown in (A). *Blue dots* indicate SIPs at the occurrence time of the earthquake (~13.38 UT). The STEC data were converted using wavelet transformation highlighting components with periods of 4 min (C) and 12 min (D) and drawn as the function of time and focal distance. We could see propagations of three atmospheric waves: Rayleigh surface wave (RW), acoustic wave (AW), and internal gravity wave (IGW) as signatures with different slopes. IGW signals are visible only in (D).

Fig. 8.1. Internal gravity wave signatures are difficult to recognize in the slant TEC time series (Fig. 8.2A) but become clear after wavelet transformation (Fig. 8.2D).

For earthquakes without dense GNSS networks, identification of all the three waves would be difficult. Nevertheless, the first pulses occurring ~10 min after earthquakes are easy to find for most large earthquakes with a few nearby GNSS stations. Although typical periods of disturbances due to acoustic waves are ~4 min, the periods may appear longer for very large earthquakes, e.g., the 2004 Sumatra-Andaman earthquake, because the observed signals are made from acoustic waves excited by multiple segments of the faults with small time lags (Heki et al., 2006). Recently, Bagiya et al. (2023) demonstrated that interferences of such multiple acoustic subpulses may result in azimuthal dependence of waveforms and amplitudes of the disturbance signals.

3.3 Direct acoustic waves from epicenter

Here, I focus on the TEC disturbances generated by direct acoustic waves from the epicenters due to coseismic vertical crustal movements (its signature cannot be separated from Rayleigh wave signatures in near fields, see Fig. 8.2). Indeed, these disturbances have the potential of providing key information on the M_w of earthquakes with their amplitudes relative to background VTEC (Cahyadi & Heki, 2015; Heki, 2021).

In Fig. 8.3, I simulate the propagation of an acoustic wave with a period of 4 min. It propagates upward and is refracted gradually so that it partly propagates horizontally at ionospheric altitudes (ray tracing is illustrated in Fig. 8.3c). Here I assume an N-shaped acoustic pulse using a simple function

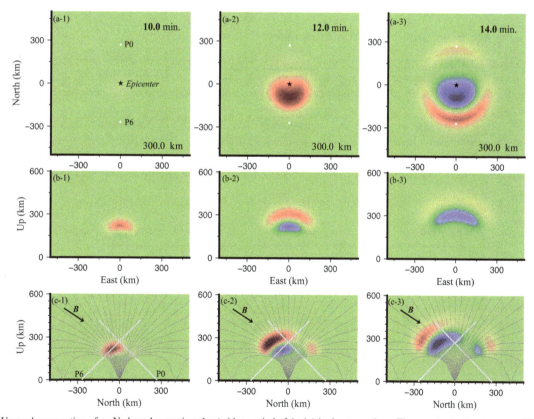

FIG. 8.3 Upward propagation of an N-shaped acoustic pulse (with a period of 4 min) in the atmosphere. The sound velocity structure of US Standard Atmosphere 1976 is assumed and the electron density obeys the Chapman distribution as a function of altitude. Electron density scale is arbitrary (actual intensity depends on M_w and background TEC). *Red* and *blue* indicate positive and negative anomalies, respectively, and green indicates neutral. The geomagnetic field at Fukutoku-Okanoba, Japan, is assumed (declination: −3.0°, inclination: 34.5°). The figure shows three epochs: 10.0 (A1, A2, A3), 12.0 (B1, B2, B3), and 14.0 (C1, C2, C3) min after earthquake. We show horizontal cross sections at altitude 300 km (A1, B1, C1), east-west vertical cross sections (A2, B2, C2), and north-south vertical cross sections (A3, B3, C3). The disturbance front propagates outward with a speed of ~0.8 km/s. The broad southward beam is caused by the interaction of the electron movements with the geomagnetic field. Actual amplitudes of the STEC disturbances strongly depend on the incidence angles of the line-of-sight with the wavefront. TEC signatures at points P0 and P6 *(white circles)* are compared together with those from other points in Fig. 8.4.

$$f(t) = -at \exp\left(\frac{-t^2}{2\sigma^2}\right). \tag{8.2}$$

This function has a maximum and minimum at $t = -\sigma$ and $t = \sigma$, respectively, and I assume 60 s for σ (4 min in period). The parameter a represents the amplitude. This period is close to the acoustic cut-off (Fig. 8.1 inset), and simple ray tracing involves errors due to neglecting gravitational restoring forces. Nevertheless, rough discussions on the line-of-sight geometry dependence of the disturbance amplitudes would be valid.

TEC disturbance amplitudes are sensitive to two factors: geomagnetic fields and line-of-sight geometry (Rolland et al., 2013). Even from stations at the same distance from the epicenter, differences in these factors may result in amplitude contrasts exceeding an order of magnitude. In the F region height, electrons can move only along geomagnetic fields, and this results in the suppression of electron density anomalies in regions where atmospheric particle motion is perpendicular to the field. In the northern hemisphere, this occurs to the north of epicenters (Fig. 8.3C).

The other factor, geometry of line-of-sight and wavefront, matters when we observe electron density anomalies as TEC with line-of-sights penetrating them in various directions. Line-of-sights penetrating only positive parts of the anomalies would show positive TEC anomalies. However, those penetrating both positive and negative parts may show only small changes in TEC (see, e.g., Manta, Occhipinti, Feng, & Hill, 2020). Fig. 8.4 shows synthesized slant TEC time series at seven different stations located along a circle around the epicenter with satellites in northern, eastern, and southern skies. The largest signal occurs when we observe a northern satellite at a southern GNSS station P6 (thick red curve in Fig. 8.5A). If we observe a southern satellite at a northern station P0, the observed signal amplitude is reduced to $\sim$1/4 (thick blue curve in Fig. 8.5C). Arrival times of the TEC anomalies also depend on the geometry.

3.4 Knowing M_w from amplitudes of disturbances

Cahyadi and Heki (2015) and Heki (2021) obtained empirical M_w dependence of the coseismic ionospheric disturbance amplitudes, normalized by background VTEC. Rapid determination of M_w of earthquakes using the coseismic ionospheric disturbance amplitudes, $\sim$10 min after earthquakes, may realize an effective tsunami early warning (e.g., Martire et al., 2023).

In Fig. 8.5, I show normalized amplitudes of 28 earthquakes of M_w from 6.6 to 9.2 compiled by Heki (2021). They show clear coseismic ionospheric disturbances by direct acoustic waves detected by GNSS-TEC. There, the satellite-station pairs showing maximum disturbance amplitudes of individual earthquakes are selected (i.e., pairs with the closest geometry to the bottom curve of Fig. 8.4A). They include two strike-slip earthquakes and two normal fault earthquakes, but all other

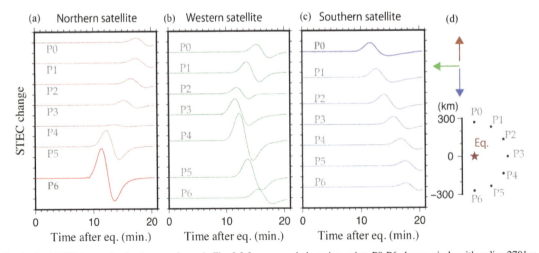

FIG. 8.4 Synthesized TEC time series for the case shown in Fig. 8.3 from ground observing points P0-P6 along a circle with radius 270 km around the epicenter (D). (A), (B), and (C) show the time series when we observe GNSS satellites with elevations and azimuths (45, 0), (45, −90), and (45, 180), respectively. In all the cases, signals get strong when the satellites are in the direction of the epicenter. The *top curve* in (C) and the *bottom curve* in (A), drawn with thick lines, correspond to the line-of-sights shown in Fig. 8.3C.

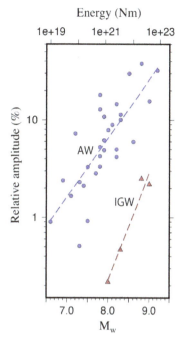

FIG. 8.5 Blue dots compare M_w of the 30 earthquakes given by Heki et al. (2022), with their amplitudes of ionospheric disturbance caused by direct acoustic wave (AW) from epicenters. They are amplitudes of STEC changes and are normalized by background VTEC values. The *dashed line* indicates the best-fit line corresponding to Eq. (8.3). *Red triangles* show amplitudes of internal gravity waves (IGW) by four large earthquakes (1994 Hokkaido-toho-oki, 2003 Tokachi-oki, 2010 Maule, and 2011 Tohoku-oki earthquakes). They obey a different scaling law.

examples are reverse fault earthquakes. Strike-slip earthquakes show somewhat weaker disturbances, but normal fault earthquakes do not show such a tendency (Cahyadi and Heki, 2015).

Heki (2021) considered that these amplitudes (unit: percent) obey a simple law,

$$\log_{10} \text{Amplitude} = a (M_w - 8.0) + b. \tag{8.3}$$

The offset b is the common logarithm of the relative amplitude of a typical $M_w 8$ event. The best-fit line inferred from all the earthquakes (dashed line in Fig. 8.5) has the slope $a \sim 0.602$ and $b \sim 0.804$. The slope is close to 2/3, i.e., the amplitudes increase approximately by two orders of magnitude as M_w increases by three.

Eq. (8.3) enables us to infer M_w by observing STEC oscillation relative to background VTEC, ~ 10 min after earthquakes. The data plotted in Fig. 8.5 still show large scatters around the best-fit line. This is possible because the ideal geometry (like Fig. 8.4A, bottom curve) is not always realized due to insufficient GNSS station coverages. Calibration of the data considering such geometry factors would significantly improve this empirical M_w-amplitude relationship.

3.5 Internal gravity wave signatures

Fig. 8.2 shows the internal gravity wave signature that appeared for the $M_w 8.3$ earthquake, as a component propagating at 0.2–0.3 km/s emerging ~ 40 min after earthquakes. Because we need to confirm its propagating velocity with time-distance plot like Fig. 8.2, we can identify it only when earthquakes are sufficiently large and we have enough number of stations. This was the case for the 2003 Tokachi-oki ($M_w 8.0$), 2010 Maule ($M_w 8.8$), and 2011 Tohoku-oki ($M_w 9.0$) earthquakes, in addition to the 1994 Hokkaido-toho-oki ($M_w 8.3$) shown in Fig. 8.2. I obtained their amplitudes relative to background VTEC. However, I did not correct for the geomagnetic effect considering the angles between the local geomagnetic field and the particle motions.

The M_w dependence of the ionospheric disturbance amplitudes of internal gravity wave is shown in Fig. 8.5. The amplitudes are significantly smaller than those caused by direct acoustic waves. It is interesting that the slope is about twice as steep as acoustic waves. Their amplitudes might be sensitive not only to the area and amount of crustal uplift but also to the duration of faulting (Heki et al., 2022).

4 Ionospheric volcanology

4.1 Two types of ionospheric disturbances by volcanic eruptions

Dautermann, Calais, and Mattioli (2009) and Dautermann, Calais, Lognonné, and Mattioli (2009) reported GNSS-TEC observations of ionospheric disturbance caused by the 2003 eruption of the Soufriére Hills volcano, Montserrat, the Lesser Antilles. TEC changes due to volcanic eruptions were also observed by a dense GNSS network after the 2004 explosion of the Asama volcano, central Japan (Heki, 2006). The TEC signatures of these two eruptions are very different. The former is characterized by ~4 mHz harmonic oscillation of TEC lasting for tens of minutes. The latter, on the other hand, appear as short pulses in TEC ~10 min after the eruption. Since these early reports, more than ten cases of ionospheric disturbances caused by volcanic eruptions have been published. These new cases were found to belong to either of these two types of TEC changes.

The first type is the long-lasting harmonic TEC oscillations ("Type 1" in Fig. 8.6). The interference of upward and downward acoustic waves between the ground surface and the mesopause causes resonant oscillation of the atmosphere in prescribed frequencies (Tahira, 1995). The frequency reflects the vertical atmospheric structure and has various overtones. Such "atmospheric modes" are excited typically by continuous Plinian-type volcanic eruptions.

Such atmospheric modes were first observed with seismometers. Acoustic resonance frequencies of 3.7 and 4.4 mHz were found in background free oscillations of the solid earth (Nishida et al., 2000). These frequency components were found to last >5 h in seismometer records after the 1991 eruption of the Pinatubo volcano, the Philippines (Kanamori & Mori, 1992). Watada and Kanamori (2010) considered that the continuous Plinian eruption of the volcano excited atmospheric resonance, which caused harmonic ground oscillation. Such oscillations were found by GNSS-TEC during the 2003 eruption of the Soufrière Hills (Dautermann, Calais, Lognonné, & Mattioli, 2009; Dautermann, Calais, & Mattioli, 2009), and new cases have been added to the literature since then.

The second type of disturbances ("Type-2" in Fig. 8.6) occur ~10 min after volcanic explosions when acoustic wave pulses reach the ionospheric F region. They have periods of 1–2 min, and their records are characterized by short-term N-shaped impulsive changes as Heki et al. (2006) observed after the 2004 explosion of the Asama volcano.

The volcanic explosivity index (VEI) has been used to measure the intensities of volcanic eruptions (Newhall & Self, 1982). For example, the VEIs of the 2003 Soufrière Hills and the 2004 Asama eruptions are 3 and 2, respectively. In the previous section, I explored the way to use coseismic ionospheric disturbance amplitudes to infer earthquake magnitudes; here, I discuss the feasibility of using ionospheric disturbances as a measure of volcanic eruption intensities. Manta, Occhipinti, Hill, et al. (2020) proposed the Ionospheric Volcanic Power Index (IVPI) and demonstrated its correlation with VEI. However, it is defined using the TEC signal power over a 2-h period and would not be appropriate for short transient type-2 disturbances.

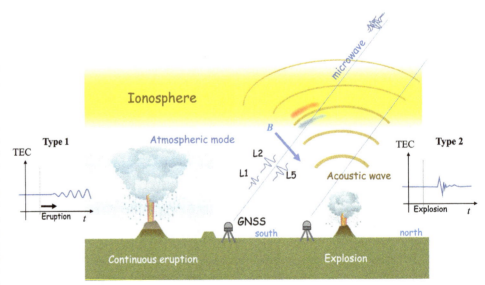

FIG. 8.6 Ionospheric disturbance caused by continuous (Type 1, *left*) and explosive (Type 2, *right*) volcanic eruptions can be detected by differential ionospheric delays of microwave signals of multiple carrier frequencies from GNSS satellites. Strong continuous eruptions sometimes excite atmospheric modes and long-term oscillatory disturbances in ionosphere ("Type 1" disturbance). For explosive eruptions, we often find short-term impulsive disturbances ("Type 2" disturbance) in the ionosphere ~10 min after eruptions. The acoustic wave makes electron density anomalies (pairs of positive and negative anomalies as shown with *red* and *blue* colors in the figure) selectively on the southern side of the volcano (for northern hemisphere cases) due to interaction with geomagnetic fields (*blue arrow*).

On January 15, 2022, the Hunga-Tonga Hunga-Ha'apai submarine volcano, southern Pacific Ocean, erupted. It was the first VEI 6 volcanic eruption after GNSS became our tool, and its ionospheric signatures were beyond a simple categorization into two types. It caused ionospheric disturbances all over the world through the propagation of the Lamb wave that traveled around the world multiple times. The first wave was observed by GNSS receivers worldwide (Themens et al., 2022). In Japan, up to four passages of ionospheric disturbances have been recorded over several days (Heki, 2022). Astafyeva et al. (2022) revealed a complex eruption sequence from near-field TEC observations immediately after the eruption. One of the new remarkable findings is the simultaneous propagation of conjugate anomalies in the northern and southern hemispheres connected by geomagnetic fields (Lin et al., 2022). Another finding is the identification of a unique disturbance that arrived in New Zealand prior to the Lamb wave and propagated ∼0.6 km/s southwestward (Muafiry et al., 2023).

4.2 Type 1 disturbance

In addition to the 2003 eruption of the Soufrière Hills volcano, this type of disturbances has been found in Indonesia, Russia, Chile, and Japan. Harmonic TEC oscillations lasted ∼20 min in the November 5, 2010 eruption of the Merapi volcano (VEI 4), central Java, Indonesia (Cahyadi et al., 2020). Similar oscillations lasted for ∼ 2.5 h during the February 2014 eruption of the Kelud volcano (VEI 4) on eastern Java Island (Nakashima et al., 2016). Both eruptions are Plinian-type continuous eruptions. Shults et al. (2016) found two sequences of similar TEC oscillations in the 2015 April Plinian eruptions of the Calbuco volcano (VEI 4), Chile.

In the northern hemisphere, Shestakov et al. (2021) reported the TEC oscillations lasting for an hour during the 2009 eruption of the Sarychev Peak volcano (VEI 4), Kuril Islands, Russia. The Plinian eruption occurred on August 13, 2021, on a submarine volcano Fukutoku-Okanoba (VEI 4), Japan, and harmonic TEC oscillation was observed for the first time using a geostationary satellite of QZSS (Heki & Fujimoto, 2022). Fig. 8.7 shows four of these past cases, i.e., the 2014 Kelud (Indonesia), 2015 Calbuco (Chile), 2010 Merapi (Indonesia), and 2021 Fukutoku-Okanoba (Japan) eruptions.

Such harmonic TEC oscillations with a period of ∼4 min sometimes occur after large earthquakes, e.g., the 1994 Hokkaido-toho-oki earthquake (Fig. 8.2C). Similar oscillations were also found in the 2004 Sumatra-Andaman earthquake (Choosakul et al., 2009) and the 2011 Tohoku-oki earthquake (Rolland et al., 2011; Saito et al., 2011). However, they propagate outward from the epicenters at a Rayleigh surface wave speed (∼4 km/s) possibly caused by efficient coupling of the Airy phase of the surface wave with the atmosphere (Heki, 2021).

On the other hand, wavefronts of the type-1 disturbances caused by volcanic eruptions propagate at an acoustic wave speed (∼0.8 km/s) (Cahyadi et al., 2020; Heki & Fujimoto, 2022; Nakashima et al., 2016; Shults et al., 2016). Fig. 8.7 presents two curves per eruption, observed by different station-satellite pairs. They slightly lag in time, reflecting the outward propagation of acoustic wave (the SIP of the second pair is somewhat farther from the volcanoes than the first pair).

The time lag between the eruption and the type-1 ionospheric disturbance is not clear because such continuous eruptions do not always have clear onsets (TEC does not show type-2 impulsive changes associated with the onsets, either). In the case of the 2010 Merapi eruption, the TEC oscillations emerged ∼20 min after the "phase-4" Plinian eruption started (Cahyadi et al., 2020). The anomaly also started shortly before 20 min after eruption started (21:04 UT) in the Calbuco case (Shults et al., 2016). Thus, 20 min might be the approximate time for the continuous eruption to excite significant atmospheric modes.

Long continuous TEC records enabled by geostationary satellites are suitable for studying their frequency spectra. Such records are also free from frequency shifts due to the apparent movements of GNSS satellites in the sky. Heki and Fujimoto (2022) were the first to obtain such data for the 2021 Fukutoku-Okanoba eruption. Fig. 8.8C shows that the TEC oscillation was observed in nine satellites of various systems from station 0603 equipped with a multi-GNSS receiver. I selected J07 (geostationary satellite of QZSS), whose slant TEC time series over 5:00–9:30 UT are shown in Fig. 8.8A. We select the 4-h data from 5:20 to 9:20 and estimated their frequency components using the Blackman-Tukey method (Fig. 8.8B).

The four frequency peaks shown in Fig. 8.8B are close to the prescribed frequencies of 3.7, 4.4, 4.8, and 5.4 mHz (periods, 270, 227, 208, and 185 s), with the 4.8 mHz peak somehow weaker than the other three. The two frequencies 3.7 and 4.4 mHz are the atmospheric resonance frequencies detected by seismometers after the 1991 eruption of the Pinatubo volcano (e.g., Kanamori & Mori, 1992). The higher two frequencies are their overtones (Watada & Kanamori, 2010). The power shows a sharp drop for frequencies lower than 3.7 mHz possibly because this frequency is close to the acoustic cut-off of the atmospheric filter (Blanc, 1985).

Fig. 8.8A also suggests the gradual decay of the harmonic oscillation in TEC. The increasing solar zenith angle during this period (local afternoon) also causes natural decay of the signal. However, the observed decay exceeds such natural

FIG. 8.7 Examples of type-1 ionospheric disturbances, harmonic TEC oscillations, caused by the eruptions of the Kelud (Indonesia), Calbuco (Chile), Merapi (Indonesia), and Fukutoku-Okanoba (Japan) volcanoes in 2014, 2015, 2010, and 2021, respectively (Cahyadi et al., 2020; Heki & Fujimoto, 2022; Nakashima et al., 2016; Shults et al., 2016). (A and B) The SIP trajectories for the first three cases and (C) the stationary J07 SIP for the last case. (D) A comparison of the time series of these cases over 1.4h after the onsets of the eruptions. For the 2015 Calbuco and 2010 Merapi cases, the onset times were 21:04 UT and 17:02 UT, respectively. The latter corresponds to the start of "phase 4" of the eruption (Cahyadi et al., 2020). For the 2014 Kelud and 2021 Fukutoku-Okanoba eruptions, exact onset times are unknown and were arbitrarily set as 16:12 UT and 5:03 UT, respectively.

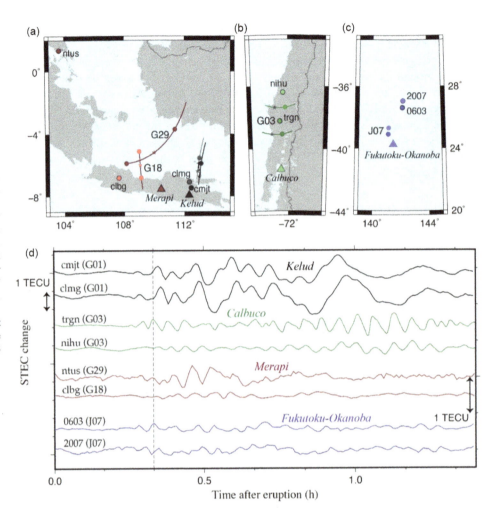

FIG. 8.8 (A) High pass filtered slant TEC time series at the Hahajima (0603) station observed using J07, the QZSS geostationary satellite (Fig. 8.7C). (B) Frequency components of the 4-h data over 5:20–9:20 UT (shown with a *dashed line*) obtained using the Blackman-Tukey method. We see three strong peaks (3.7, 4.4, and 5.4 mHz) and one weaker peak (4.8 mHz) that correspond to known atmospheric modes. (C) A comparison of TEC signals from various satellites observed at station 0603. Alphabets indicate the system (G: GPS, R: GLONASS, E: Galileo, J: QZSS) and the numbers indicate satellite numbers.

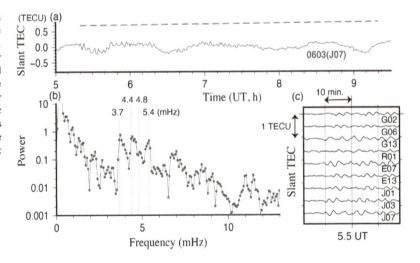

decay and rather reflects the decay of the Plinian eruption itself. Heki and Fujimoto (2022) used a new approach to compare fluctuations in three frequency pairs (L1-L2, L1-L5, L2-L5) and quantified the real decay of ionospheric electron density fluctuations.

Cahyadi et al. (2020) compared the TEC oscillation amplitudes relative to background TEC for three cases: 2010 Merapi, 2014 Kelud, and 2015 Calbuco eruptions. They suggested that such relative TEC oscillation amplitudes are

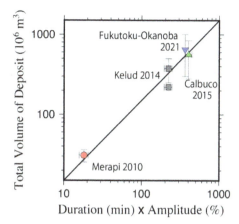

FIG. 8.9 A new value from Fukutoku-Okanoba has been added to Cahyadi et al. (2020). The figure compares the products of the ionospheric disturbance amplitudes and the durations, with the total volume of the deposits inferred from geological approaches. For Fukutoku-Okanoba, I assumed the amplitude 3% of the background VTEC continued for 2 h. Total volume of deposits is inferred assuming the pumice density is the same as water, from the total weight of the deposit of $3–10 \times 10^{11}$ kg (Geological Survey of Japan, 2021).

proportional to the mass eruption rates (MER), and the products of such amplitudes and the duration scale with the total amount of the ejecta. Fig. 8.9 compares the total amount of the ejecta volume from these three eruptions listed by Cahyadi et al. (2020) and a new case from the 2021 Fukutoku-Okanoba eruption (Heki & Fujimoto, 2022).

For the amplitude of TEC oscillations in Fukutoku-Okanoba, the geometry factor has been corrected, i.e., we would have observed ~4 times as strong oscillation if we had a GNSS station to the south of the volcano (Fig. 8.4). After correction, the TEC oscillation amplitude becomes ~3% of the background VTEC, suggesting the MER at the peak time (~5:20 UT) may have reached 5×10^7 kg s^{-1}. If such a strong oscillation continued for 2 h, the product of the amplitude and the duration would become consistent with the total amount of ejecta, inferred as $3–10 \times 10^{11}$ kg by Geological Survey of Japan (2021).

In the future, ionospheric disturbances may provide useful information in near real time. TEC can be monitored from GNSS stations hundreds of kilometers away from the volcano (Fig. 8.7). The amplitudes of the harmonic TEC oscillation could offer a rough estimate of MER of the ongoing eruption within minutes. After the eruptions, we could infer the total volume of deposits from the products of the oscillation amplitudes and the durations. This would be useful where geological approaches are difficult.

4.3 Type 2 disturbance

Type 2 ionospheric disturbances (Fig. 8.6) emerge ~10 min after explosive volcanic eruptions when acoustic wave pulses reach the ionospheric F region. Although they resemble coseismic ionospheric disturbances caused by direct acoustic waves, their periods (1–2 min) are significantly shorter than those related to earthquakes (~4 min). Here, I present five past examples of the type 2 disturbances, all observed in Japan, and discuss their volcanological implications.

Fig. 8.10 illustrates the upward propagation of an N-shaped acoustic pulse. I assumed the function given as Eq. (8.2) (the disturbance period is changed to 80 s, i.e., $\sigma = 20$ s) and repeated the same calculation as in Fig. 8.2. Because the geomagnetic field in the northern hemisphere mid-latitude region was assumed, we see southward directivity. The disturbance front propagates outward with a speed of ~0.8 km/s. Like the earthquake cases, amplitudes of the STEC disturbances depend on the incidence angles of the line-of-sight with the wavefront. A stronger eruption would make a stronger signature in the ionosphere under the same geometry. Hence, GNSS-TEC measurements would help us quantify the explosion intensities.

The intensity of a volcanic explosion has been studied by near-field atmospheric pressure changes associated with the passages of airwaves generated by the eruptions (e.g., Matoza et al., 2019). However, the geometric settings of such sensors and volcanoes vary from volcano to volcano, and the amplitudes of such airwaves cannot be the universal index. Here, I propose using type-2 ionospheric disturbance amplitudes as a new index for explosion intensities. For this purpose, we compare ionospheric TEC responses to five recent explosions from four volcanoes in Japan from 2004 to 2015.

Fig. 8.11 compares the disturbances associated with five eruptions from Cahyadi et al. (2021). The first case is the 2004 September 1 11:02 UT eruption (VEI 2) of the Asama volcano, central Japan, whose ionospheric disturbances were

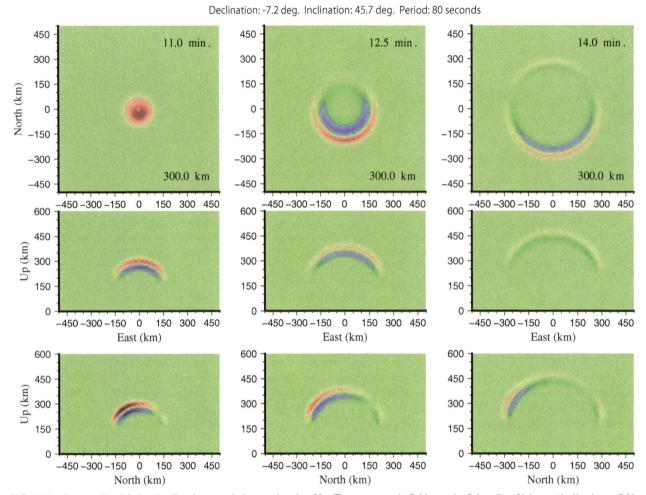

FIG. 8.10 Same as Fig. 8.2, but the disturbance period was reduced to 80 s. The geomagnetic field near the Sakurajima Volcano (declination: −7.2°, inclination: 45.7°) was assumed. The figure shows three epochs: 11.0 *(left)*, 12.5 *(middle)*, and 14.0 *(right)* minutes after the explosion.

reported by Heki et al. (2006). The second case is the October 3, 2009, 07:45 UT eruption (VEI 2) of the Sakurajima volcano, Kyushu. Plume reached a height of ∼3000 m above the caldera rim during this eruption.

The third and the fourth cases are the two VEI 2 explosive eruptions occurred on January 31, 2011, at 22:54 UT, and on February 11, at 02:36 UT, 2011, of the Shin-Moe Volcano, Kyushu. In these eruptions, the plume reached heights of ∼2000 and ∼2500 m above the caldera rim, and the atmospheric pressure changes were 458.5 and 244.3 Pa recorded at a sensor ∼2.6 km southwestward, respectively (Japan Meteorological Agency, 2013).

The last example is the Kuchinoerabu-jima volcano, located ∼100 km to the south of Kyushu. A VEI 3 eruption occurred on May 29, 2015 (00:59 UT). The plume height was ∼9000 m, and pyroclastic flow reached the ocean. This example shows faint harmonic oscillations after the initial N-shape disturbance, suggesting this was a hybrid (of type-1 and -2) eruption.

The raw amplitudes of the five cases are given with red squares in Fig. 8.12. I used background VTEC to normalize such amplitudes following the cases for coseismic ionospheric disturbances (Fig. 8.5). The dark blue squares in Fig. 8.12 express the TEC amplitudes relative to VTEC at the time and location of eruptions from GIM.

Another important factor is the geometry of line-of-sight with the acoustic wavefront (Fig. 8.4). Different line-of-sight geometry with the wavefronts may result in significant differences in disturbance amplitudes. Such geometric factors have been corrected for the five eruption cases by (1) calculating time series with synthesized data using the real geometry of these cases and the "good geometry" (observing a northern satellite from a point 270 km to the south of volcanoes), and (2) converting the observed amplitudes to those under the good geometry. The light blue squares in Fig. 8.12 show such corrected disturbance amplitudes.

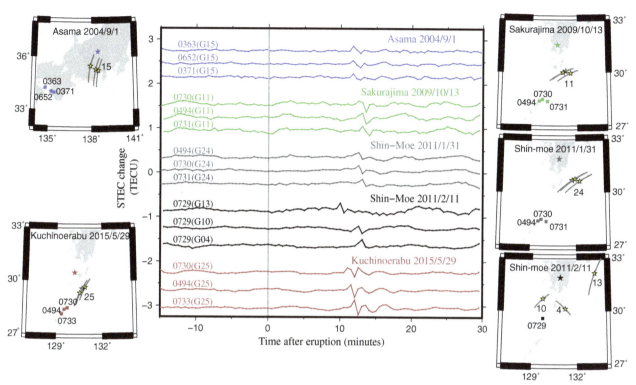

FIG. 8.11 Geometry of volcanoes *(large stars)*, SIP tracks *(gray curves)* and SIP positions at the time of the eruptions *(small yellow stars)*, and GNSS stations *(squares)* for five Vulcanian eruptions in Japan *(left and right)*. The middle panel shows the STEC time series for the three station-satellite pairs for each of the five examples. Long-period changes were removed from STEC by subtracting the best-fit polynomials (degrees 7–9) over 45-min periods. Small disturbances can be seen ~10 min after the eruptions. *(Rewritten after Cahyadi, M. N., Handoko, E. Y., Rahayu, R. W., & Heki, K. (2021), Comparison of volcanic explosions in Japan using impulsive ionospheric disturbances.* Earth, Planets and Space, 73, 228, doi: 10.1186/s40623-021001539-5.*)*

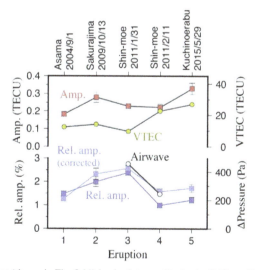

FIG. 8.12 Comparison of the STEC changes (shown in Fig. 8.11) in absolute amplitudes *(red)*. The *yellow circles* show VTEC values at the time and place of the eruptions, calculated using GIM, and *dark blue squares* indicate disturbance amplitudes relative to the background VTEC. These amplitudes were further corrected for geometry factors and are shown in *light blue squares*. For the two eruptions of the Shin-Moe volcano, we compare amplitudes of atmospheric pressure changes detected by the same sensor located ~2.6 km from the volcano.

Explosion intensities can also be studied by measuring amplitudes of airwaves (atmospheric pressure changes). However, different distance of the ground sensors from the volcanoes and different topographic and vegetation conditions make it difficult to compare such intensities for different volcanoes. In contrast, TEC changes focus on upward propagating acoustic waves, and inter-volcano comparison is possible.

Fig. 8.12 includes two cases of atmospheric pressure changes for the January 31 and February 11 explosions of the Shin-Moe volcano, detected using the same sensor (Japan Meteorological Agency, 2013). These two eruptions show similar amplitudes of STEC changes. However, when considering the background VTEC and line-of-sight geometry, the January eruption is about twice as large as the February eruption. This agrees with the ratio of the pressure changes recorded for these two eruptions (458.5 and 244.3 Pa for the January and February eruptions, respectively).

This "within-volcano" comparison provides some support for the validity of using ionospheric disturbances to measure explosion intensities. Deployment of infrasound sensors would remain useful because TEC changes can only be detected for strong volcanic explosions occurring when the number of ionospheric electrons is sufficient (e.g., during daytimes). In fact, there were two explosions of the Shin-Moe volcano (February 1 at 20:25 UT and February 13 at 20:07 UT) with stronger airwaves than the eruption on February 11 at 02:36 UT. However, their ionospheric disturbances were not detected due to the small background VTEC early in the morning (Cahyadi et al., 2021).

Fig. 8.11 suggests that TEC changes caused by the five different volcanic explosions have similar periods ($\sim$80 s), suggesting their origin in the atmospheric properties. Fig. 8.1 inset shows the atmospheric attenuation of acoustic waves with various periods at different altitudes (Blanc, 1985). There, 1.3 min corresponds to the shortest period of the airwaves that reach the ionospheric F region ($\sim$300 km) without large attenuations. Surface infrasound records include stronger powers in periods shorter than 1.3 min for volcanic explosions. However, such components decay before disturbing the ionosphere, resulting in the 1.3 min component outstanding in the TEC records.

Despite the search outside Japan, it has been unsuccessful in adding more clear type-2 cases due to the lack of GNSS stations in appropriate places or the insufficient intensities of the explosions. One exception is the TEC signatures produced by a human-induced explosion in August 2020 in Lebanon (Kundu et al., 2021). I expect that new cases of ionospheric disturbances caused by volcanic explosions will be reported through intensive search efforts in the world.

Acknowledgments

The author acknowledges the support of the Chinese Academy of Sciences, President's International Fellowship Initiative (Grant number 2022VEA0014). He thanks those who produce daily observation data of permanent GNSS networks all over the world. This chapter includes materials from the PhD theses of M.N. Cahyadi and Yuki Nakashima, and the MSc thesis of Ai Matsushita at Hokkaido University.

References

Astafyeva, E. (2019). Ionospheric detection of natural hazards. *Reviews of Geophysics*, *57*, 1265–1288. https://doi.org/10.1029/2019RG000668.

Astafyeva, E., Heki, K., Kiryushkin, V., Afraimovich, E., & Shalimov, S. (2009). Two-mode long-distance propagation of coseismic ionosphere disturbances. *Journal of Geophysical Research*, *114*, A10307. https://doi.org/10.1029/2008JA013853.

Astafyeva, E., Maletckii, B., Mikesell, T. D., Munaibari, E., Ravanelli, M., Coisson, P., et al. (2022). The 15 January 2022 Hunga Tonga eruption history as inferred from ionospheric observations. *Geophysical Research Letters*, *49*, e2022GL098827. https://doi.org/10.1029/2022GL098827.

Astafyeva, E., Rolland, L., Lognonné, P., Khelfi, K., & Yahagi, T. (2013). Parameters of seismic source as deduced from 1 Hz ionospheric GPS data: Case study of the 2011 Tohoku-oki event. *Journal of Geophysical Research*, *118*, 5942–5950. https://doi.org/10.1002/jgra.50556.

Bagiya, M., Heki, K., & Gahalaut, V. K. (2023). Anisotropy of the near-field coseismic ionospheric perturbation amplitudes reflecting the source process: The 2023 February Turkey earthquakes. *Geophysical Research Letters*, *50*, e2023GL103931. https://doi.org/10.1029/2023GL103931.

Blanc, E. (1985). Observations in the upper atmosphere of infrasonic waves from natural or artificial sources – A summary. *Annales Geophysicae*, *3*, 673–687.

Cahyadi, M. N., Handoko, E. Y., Rahayu, R. W., & Heki, K. (2021). Comparison of volcanic explosions in Japan using impulsive ionospheric disturbances. *Earth, Planets and Space*, *73*, 228. https://doi.org/10.1186/s40623-021001539-5.

Cahyadi, M. N., & Heki, K. (2015). Coseismic ionospheric disturbance of the large strike-slip earthquakes in North Sumatra in 2012: M_w dependence of the disturbance amplitudes. *Geophysical Journal International*, *200*, 116–129.

Cahyadi, M. N., Rahayu, R. W., Heki, K., & Nakashima, Y. (2020). Harmonic ionospheric oscillation by the 2010 eruption of the Merapi volcano, Indonesia, and the relevance of its amplitude to the mass eruption rate. *Journal of Volcanology and Geothermal Research*, *405*, 107047. https://doi.org/10.1016/j.jvolgeores.2020.107047.

Calais, E., & Minster, J. B. (1995). GPS detection of ionospheric perturbations following the January 17, 1994, Northridge earthquake. *Geophysical Research Letters*, *22*, 1045–1048. https://doi.org/10.1029/95GL00168.

Calais, E., Minster, J. B., Hofton, M. A., & Hedlin, H. (1998). Ionospheric signature of surface mine blasts from global positioning system measurements. *Geophysical Journal International*, *132*, 191–202.

Cheng, K., & Huang, Y. N. (1992). Ionospheric disturbances observed during the period of Mount Pinatubo eruptions in June 1991. *Journal of Geophysical Research*, *97*(A11), 16,995. https://doi.org/10.1029/92JA01462.

Choosakul, N., Saito, A., Iyemori, T., & Hashizume, M. (2009). Excitation of 4-min periodic ionospheric variations following the great Sumatra-Andaman earthquake in 2004. *Journal of Geophysical Research*, *114*, A10313. https://doi.org/10.1029/2008JA013915.

Dautermann, T., Calais, E., Lognonné, P., & Mattioli, G. S. (2009). Lithosphere-atmosphere-ionosphere coupling after the 2003 explosive eruption of the Soufriere Hills volcano Monserrat. *The Journal of Geophysical Research*, *179*, 1537–1546. https://doi.org/10.1111/j.1365-246X.2009.04390.x.

Dautermann, T., Calais, E., & Mattioli, G. S. (2009). Global positioning system detection and energy estimation of the ionospheric wave caused by the 13 July 2003 explosion of the Soufrière Hills Volcano, Montserrat. *The Journal of Geophysical Research*, *114*, B02202. https://doi.org/10.1029/2008JB005722.

Ducic, V., Artru, J., & Longnonné, P. (2003). Ionospheric remote sensing of the Denali earthquake Rayleigh surface wave. *Geophysical Research Letters*, *30*, 1951. https://doi.org/10.1029/2003GL017812.

Geological Survey of Japan. (2021). *Fukutoku-Oka-no-Ba submarine volcano information*. Geological Survey of Japan. www.gsj.jp/hazards/volcano/fukutokuokanoba/2021/index.html.

Heki, K. (2006). Explosion energy of the 2004 eruption of the Asama volcano, Central Japan, inferred from ionospheric disturbances. *Geophysical Research Letters*, *33*, L14303. https://doi.org/10.1029/2006GL026249.

Heki, K. (2021). Chapter 21: Ionospheric disturbances related to earthquakes. In C. Huang, et al. (Eds.), *Geophysical monograph: vol. 260. Ionospheric dynamics and applications* (pp. 511–526). Wiley/American Geophysical Union. https://doi.org/10.1002/9781119815617.ch21.2021.

Heki, K. (2022). Ionospheric signatures of repeated passages of atmospheric waves by the 2022 Jan. 15 Hunga Tonga-Hunga Ha'apai eruption detected by QZSS-TEC observations in Japan. *Earth, Planets Space*, *74*, 112. https://doi.org/10.1186/s40623-022-01674-7.

Heki, K., Bagiya, M., & Takasaka, Y. (2022). Slow earthquake signatures in coseismic ionospheric disturbances. *Geophysical Research Letters*, *49*, e2022GL101064. https://doi.org/10.1029/2022GL101064.

Heki, K., & Fujimoto, T. (2022). Atmospheric modes excited by the 2021 August eruption of the Fukutoku-Okanoba volcano, Izu-Bonin Arc, observed as harmonic TEC oscillations by QZSS. *Earth, Planets and Space*, *74*, 27. https://doi.org/10.1186/s40623-022-01587-5.

Heki, K., Otsuka, Y., Choosakul, N., Hemmakorn, N., Komolmis, T., & Maruyama, T. (2006). Detection of ruptures of Andaman fault segments in the 2004 great Sumatra earthquake with coseismic ionospheric disturbances. *Journal of Geophysical Research*, *111*, B09313. https://doi.org/10.1029/2005JB004202.

Heki, K., & Ping, J.-S. (2005). Directivity and apparent velocity of the coseismic ionospheric disturbances observed with a dense GPS array. *Earth and Planetary Science Letters*, *236*, 845–855.

Igarashi, K., Kainuma, S., Nishimuta, I., Okamoto, S., Kuroiwa, H., Tanaka, T., et al. (1994). Ionospheric and atmospheric disturbances around Japan caused by the eruption of Mount Pinatubo on 15 June 1991. *Journal of Atmospheric and Terrestrial Physics*, *56*(9), 1227–1234. https://doi.org/10.1016/0021-9169(94)90060-4.

Japan Meteorological Agency (Fukuoka District Meteorological Observatory and Kagoshima Local Meteorological Observatory). (2013). The 2011 eruptive activities of Shinmoedake volcano, Kirishimayama, Kyushu, Japan. *Quarterly Journal of Seismic*, *77*, 65–96 (in Japanese with English abstract).

Jin, S., Jin, R., & Liu, X. (2018). *GNSS atmospheric seismology* (p. 315). Springer. ISBN 978-981-10-3176-2.

Kanamori, H., & Mori, J. (1992). Harmonic excitation of mantle Rayleigh waves by the 1991 eruption of Mount Pinatubo, Philippines. *Geophysical Research Letters*, *19*(7), 721–724. https://doi.org/10.1029/92GL00258/full.

Kundu, B., Senapati, B., Matsushita, A., & Heki, K. (2021). Atmospheric wave energy of the 2020 August 4 explosion in Beirut, Lebanon, from ionospheric disturbances. *Scientific Reports*, *11*, 2793. https://doi.org/10.1038/s41598-021-82355-5.

Lin, J.-T., Rajesh, P. K., Lin, C. C.-H., Chou, M.-Y., Liu, J.-Y., Yue, J., et al. (2022). Rapid conjugate appearance of the giant ionospheric lamb wave signatures in the northern hemisphere after Hunga-Tonga volcano eruptions. *Geophysical Research Letters*, *49*, e2022GL098222. https://doi.org/10.1029/2022GL098222.

Mannucci, A. J., Wilson, B. D., Yuan, D. N., Ho, C. H., Lindqwister, U. J., & Runge, T. F. (1998). A global mapping technique for GPS-derived ionospheric total electron content measurements. *Radio Science*, *33*, 565–582. https://doi.org/10.1029/97RS02707.

Manta, F., Occhipinti, G., Feng, L., & Hill, E. M. (2020). Rapid identification of tsunamigenic earthquakes using GNSS ionospheric sounding. *Scientific Reports*, *10*, 11054. https://doi.org/10.1038/s41598-020-68097-w.

Manta, F., Occhipinti, G., Hill, E. M., Perttu, A., Assinik, J., & Taisne, B. (2020). Correlation between GNSS-TEC and eruption magnitude supports the use of ionospheric sensing to complement volcanic hazard assessment. *Journal of Geophysical Research - Solid Earth*, *126*, e2020JB020726. https://doi.org/10.1029/2020JB020726.

Martire, L. K., Krishnamoorthy, S., Vergados, P., Romans, L., Szilágyi, B., Meng, X., et al. (2023). The GUARDIAN system – A GNSS upper atmospheric real-time disaster information and alert network. *GPS Solutions*, *27*, 32. https://doi.org/10.1007/s10291-022-01365-6.

Matoza, R., Fee, D., Green, D., & Mialle, P. (2019). Volcano infrasound and the international monitoring system. In A. Le Pichon, E. Blanc, & A. Hauchecorne (Eds.), *Infrasound monitoring for atmospheric studies* (pp. 1023–1077). Cham: Springer. ISBN: 978-3-319-75138-2.

Muafiry, I. N., Wijaya, D. D., Meilano, I., & Heki, K. (2023). Diverse ionospheric disturbances by the 2022 Hunga Tonga-Hunga Ha'apai eruption observed by a dense GNSS array in New Zealand. *Journal of Geophysical Research: Space Physics*, *128*, e2023JA031486. https://doi.org/10.1029/2023JA031486.

Nakashima, Y., Heki, K., Takeo, A., Cahyadi, M. N., Aditiya, A., & Yoshizawa, K. (2016). Atmospheric resonant oscillations by the 2014 eruption of the Kelud volcano, Indonesia, observed with the ionospheric total electron contents and seismic signals. *Earth and Planetary Science Letters*, *434*, 112–116. https://doi.org/10.1016/j.epsl.2015.11.029.

Newhall, C. G., & Self, S. (1982). The volcanic explosivity index (VEI) – An estimate of explosive magnitude for historical volcanism. *Journal of Geophysical Research, 87*, 1231–1238.

Nishida, K., Kobayashi, N., & Fukao, Y. (2000). Resonant oscillation between the solid earth and the atmosphere. *Science, 287*, 2244–2246.

Ogawa, T., Kumagai, H., & Sinno, K. (1982). Ionospheric disturbances over Japan due to the 18 May 1980 eruption of Mount St. Helens. *Journal of Atmospheric and Terrestrial Physics, 44*, 863–868.

Rideout, W., & Coster, A. (2006). Automated GPS processing for global total electron content data. *GPS Solutions, 10*, 219–228. https://doi.org/10.1007/s10291-006-0029-5.

Rolland, L. M., Lognonné, P., Astafyeva, E., Kherani, E. A., Kobayashi, N., Mann, M., et al. (2011). The resonant response of the ionosphere imaged after the 2011 off the Pacific coast of Tohoku earthquake. *Earth, Planets and Space, 63*, 853–857.

Rolland, L. M., Vergnolle, M., Nocquet, J.-M., Sladen, A., Dessa, J.-X., Tavakoli, F., et al. (2013). Discriminating the tectonic and non-tectonic contributions in the ionospheric signature of the 2011, Mw7.1, dip–slip Van earthquake, Eastern Turkey. *Geophysical Research Letters, 40*. https://doi.org/10.1002/grl.50544.

Saito, A., Tsugawa, T., Otsuka, Y., Nishioka, M., Iyemori, T., Matsumura, M., et al. (2011). Acoustic resonance and plasma depletion detected by GPS total electron content observation after the 2011 off the Pacific coast of Tohoku earthquake. *Earth, Planets and Space, 63*, 863–867. https://doi.org/10.5047/eps.2011.06.034.

Shestakov, N., Orlyakovskiy, A., Perevalova, N., Titkov, N., Chebrov, D., Ohzono, M., et al. (2021). Investigation of ionospheric response to June 2009 Sarychev Peak Volcano eruption. *Remote Sensing, 13*, 648. https://doi.org/10.3390/rs13040638.

Shults, K., Astafyeva, E., & Adourian, S. (2016). Ionospheric detection and localization of volcano eruptions on the example of the April 2015 Calbuco events. *Journal of Geophysical Research: Space Physics, 121*, 10303–10315. https://doi.org/10.1002/2016JA023382.

Tahira, M. (1995). Acoustic resonance of the atmospheric at 3.7 mHz. *Journal of the Atmospheric Sciences, 52*, 2670–2674.

Tanaka, T., Ichinose, T., Okusawa, T., Shibata, T., Sato, Y., Nagasawa, C., et al. (1984). HF Doppler observations of acoustic wave exhibited by the Urakawa-Oki earthquake on 21 March 1982. *Journal of Atmospheric and Terrestrial Physics, 46*, 233–245.

Tanimoto, T., Heki, K., & Artru-Lambin, J. (2015). Interaction of solid earth, atmosphere, and ionosphere. In G. Schubert (Ed.), *Vol. 4. Treatise on geophysics* (2nd ed., pp. 421–443). Oxford: Elsevier.

Themens, R. T., Watson, C., Žagar, N., Vasylkevych, S., Elvidge, S., McCaffrey, A., et al. (2022). Global propagation of ionospheric disturbances associated with the 2022 Tonga volcanic eruption. *Geophysical Research Letters, 49*, e2022GL098158. https://doi.org/10.1029/2022GL098158.

Tsugawa, T., Saito, A., Otsuka, Y., Nishioka, M., Maruyama, T., Kato, H., et al. (2011). Ionospheric disturbances detected by GPS total electron content observation after the 2011 off the Pacific coast of Tohoku earthquake. *Earth, Planets and Space, 63*, 875–879. https://doi.org/10.5047/eps.2011.06.035.

Watada, S., & Kanamori, H. (2010). Acoustic resonant oscillations between the atmosphere and the solid earth during the 1991 Mt Pinatubo eruption. *The Journal of Geophysical Research, 115*, B12319. https://doi.org/10.1029/2010JB007747.

Yuen, P. C., Weaver, P. F., Suzuki, R. K., & Furumoto, A. S. (1969). Continuous, traveling coupling between seismic waves and the ionosphere evident in May 1968 Japan earthquake data. *Journal of Geophysical Research, 74*, 2256–2264.

Part II

Monitoring climate change with GNSS

Chapter 9

GNSS applications for measuring sea level changes

Rüdiger Haas
Department of Space, Earth and Environment, Chalmers University of Technology, Onsala Space Observatory, Onsala, Sweden

1 Introduction

During the last decades, society has become increasingly aware of climate change and its importance for and effect on sea level changes. In 2020 more than 11% of the world population lived in low-elevation coastal areas, that is, coastal areas below 10 m of elevation, that are hydrologically connected to the sea (IPCC, 2022). It is expected that this population is going to double within the next decades. According to the United Nations, in 2030 about 50% of the world's population will live in coastal areas which are exposed to flooding, storms, and tsunamis (United Nations News, 2021). Accurate and reliable sea level monitoring is thus of high importance for society in an era of ongoing and rapid climate change. For a review on sea level measurements, see, for example, Woodworth (2022).

Traditionally, sea level has been observed with coastal tide gauges. The longest continuous coastal tide gauge records today go back to the 18th century (Ekman, 2003). Coastal tide gauges records are made relative to benchmarks on the land where they are established. Thus, they are potentially affected and disturbed by local deformations, such as tectonic activity, land subsidence, or land uplift due to loading/unloading effects. Repeated geodetic surveys are therefore necessary to monitor the tide gauges with respect to reliable geodetic reference frames that are realized with, for example, GNSS (Schöne, Schön, & Thaller, 2009).

Recently, research has enabled the direct use of GNSS signals to sense the sea surface, see, for example, Martín-Neira (1993). In one or the other way, the GNSS signal reflections are utilized to derive information on the sea level and its changes, and the developed techniques can be applied, for example, using ground-based GNSS equipment at the coast.

In the following chapter we focus on GNSS for coastal sea level measurements, that is, traditional tide gauge measurements that are augmented by GNSS, and sea level observations using reflected GNSS signals.

2 GNSS at traditional tide gauges

Traditional tide gauges measure changes in relative sea level with respect to benchmarks at the coast where the tide gauges are established. The stability of these benchmarks with respect to reliable reference frames has to be monitored repeatedly in order to be able to distinguish between sea level changes and vertical land motion (VLM) (Schöne et al., 2009). Usually, tide gauges are connected to the national height systems of the countries where they are installed, as well as to colocated GNSS stations that are part of the corresponding national GNSS system. These connections are performed repeatedly using geodetic local surveys of the relative position between the reference benchmark referring to the tide gauge to the other corresponding benchmarks. In particular the relative vertical position and its potential variation are of interest, and it is usually determined by precise leveling. The connection to colocated GNSS stations allows for the conversion between the relative sea level observed by the tide gauge and absolute sea level in a global terrestrial reference frame by adding the corresponding VLM derived from the analysis of the GNSS data.

As an example, the left graph in Fig. 9.1 depicts the relative sea level record observed with the traditional tide gauge at the Onsala Space Observatory at the Swedish west coast. This tide gauge started operating in 2015 and is part of the tide gauge network of the Swedish Meteorological and Hydrographical Institute (SHMI) and the data can be retrieved, for example, via SMHI (n.d.) and/or the Permanent Service for Mean Sea Level (PSMSL) (2022). The data set has a temporal resolution of 1 h, and the variability seen is caused by tidal and meteorological effects. Based on the almost 8 years of data recorded so far, an insignificant relative sea level change of 0.1 ± 0.04 mm/year can be observed.

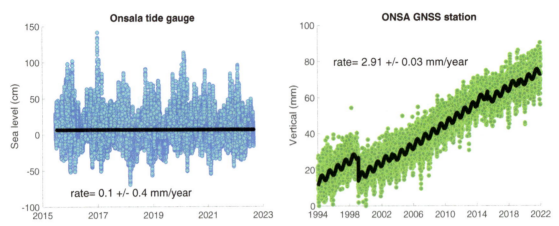

FIG. 9.1 *Left*: Relative sea level as observed with the traditional tide gauge at the Onsala Space Observatory at the Swedish west coast. *Right*: Vertical station component of the GNSS station ONSA at the Onsala Space Observatory. See the text for further explanation and discussion.

The right graph in Fig. 9.1 depicts the VLM of the colocated nearby GNSS station ONSA over the last three decades. This station is part of the International GNSS Service (IGS), and the time series of station positions was produced by the Nevada Geodetic Laboratory (NGL) at the University of Nevada, Reno (Blewitt, Hammond, & Kreemer, 2018) based on GPS data and relates to IGS14, the GPS-only realization of the ITRF2014 (Altamimi, Rebischung, Métivier, & Collilieux, 2016).

From fitting a simple model including offset, drift, a seasonal signal, as well as known discontinuities due to equipment changes, a vertical uplift of 2.91 ± 0.03 mm/year is derived. This VLM is caused by glacial isostatic adjustment (GIA) in the Fennoscandian region, see, for example, Milne et al. (2001) and Lidberg, Johansson, Scherneck, and Milne (2010).

The two sensors, that is, the tide gauge and the GNSS station, are located within less than 550 m distance, and local precise leveling is performed at least once per year to check the stability of the site. So far no relative vertical motion could be detected between the two sensors. From the combination of the vertical uplift of 2.91 ± 0.03 mm/year observed with GNSS and the stable relative sea level (0.1 ± 0.4 mm/year), we thus can conclude that the sea level at the Swedish west coast is rising in an absolute sense by the same amount as the observed land uplift, that is, by 2.91 ± 0.03 mm/year.

Not all tide gauges are as closely colocated to GNSS stations as the one at Onsala. However, there are numerous pairs of tide gauges and GNSS stations worldwide that are still reasonably closely located to each other. As an example, VLM of 546 GPS stations was determined by Gravelle et al. (2023). Furthermore, the technique of GPS imaging was used to derive VLM for more than 2300 tide gauge stations of the PSMSL (Hammond, Blewitt, Kreemer, & Nerem, 2021).

3 Reflected GNSS signals

GNSS signals are transmitted right-hand circular polarized (RHCP) from the satellites and usually RHCP receiving antenna are used for geodetic and geophysical GNSS measurements (Braasch & Scott, 2007). However, when GNSS signals reflect off a surface, the signal polarization is altered, and it is the conductivity and the dielectric properties of the reflecting surface, as well the reflection incident angle, that determine what polarization content the reflected signal will have, see, for example, Katzberg and Garrison (1996) and Zavorotny and Voronovich (2000). In general, in case of small incident angles, the reflected signals are still RHCP dominated, while in the case of large incident angles, the reflected signals are LHCP-dominated. Fig. 9.2 depicts the corresponding reflection coefficients for signals at GNSS L1/E1/B1 frequency (1575.42 MHz) that are reflected off seawater (relative permittivity $\varepsilon_r = 20$, conductivity $\sigma = 4$ S/m).

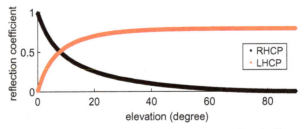

FIG. 9.2 Reflection coefficients for GNSS signals at L1/E1 frequency ($f = 1575.42$ MHz), reflected off seawater ($\varepsilon_r = 20$, $\sigma = 4$).

About three decades ago, the idea to use direct and reflected GNSS signals for mesoscale ocean altimetry was presented by Martín-Neira (1993). The concept was meant to be flown on low earth orbiting satellites where direct incoming RHCP GNSS signals and reflected LHCP signals, after reflection off the ocean surface, should be received. Using this approach and principles of bi-static radar, the sea surface height (SSH) should be determined from the relative delay of the direct and the reflected signals. This concept using two antennas for sea level determination, one upward-looking RHCP and one downward-looking LHCP, is what usually is referred to as GNSS-R. It has also been tested on airborne platforms, for example, Treuhaft, Lowe, Zuffada, and Chao (2001) and Lowe et al. (2002), as well at ground-based coastal installations, for example, Martín-Neira, Caparrini, Font-Rosselló, Lannelongue, and Vallmitjana (2001). Since the LHCP signals are dominant at high elevations (see Fig. 9.2) and received with a dedicated downward-looking antenna, ground-based GNSS-R installations have to be located rather close to the waterline, or even better, directly above the sea surface. Typically, a station with a height difference of 10 m from the mean sea surface and aiming at elevation angles up to 60 degrees, should be located within 5.5 m distance from the waterline.

The effect of signal reflections on standard geodetic GNSS phase measurements with one upward-looking RHCP antenna was described, for example, by Georgiadou and Kleusberg (1988) and Elósegui et al. (1995). The direct incoming and reflected GNSS signals interfere and impact the measurements performed with the GNSS receiver that is connected to the antenna. This is often referred to as multipath and leads to variations in GNSS phase measurements depending on the signal wavelength, satellite elevation, and distance between the GNSS antenna and the reflecting surface. Similarly, also GNSS code and signal-to-noise ratio (SNR) measurements are impacted by these multipath disturbances, see, for example, Bilich, Larson, and Axelrad (2008) and Nievinski and Larson (2014). While often a lot of effort is spent on mitigating multipath effects for more traditional applications of GNSS such as positioning and determination of atmospheric water vapor, see, for example, Elósegui et al. (1995), Jaldehag et al. (1996), and Ning, Elgered, and Johansson (2011), the phenomena can also be used to infer information about the environment surrounding the GNSS station, that is, the reflector and its properties. Thus, coastal GNSS stations with one single upward-looking RHCP antenna that are affected by reflections from the nearby sea surface can be used to monitor sea level, see, for example, Larson, Löfgren, and Haas (2013). Here, the lower elevation range is most interesting, since in this elevation range the reflected signals are still dominantly RHCP (see Fig. 9.2). This approach is often referred to GNSS-IR and such installations can be located further away from the waterline than GNSS-R installations. Typically, a station with a height difference of 10 m from mean sea surface and aiming at elevations angles between 2–10 degrees, should be located within 55 m distance from the waterline.

4 Coastal GNSS-R with two or more antennas

Coastal GNSS-R installations are usually equipped with at least two antennas, one upward-looking RHCP antenna and one downward-looking LHCP antenna. As an example, the left photo in Fig. 9.3 shows the purpose-built GNSS-R installation at the Onsala Space Observatory. The two radome-equipped antennas are mounted on a bar that extends over the sea. Each antenna is connected to one standard off-the-shelf geodetic GNSS receiver with multi-GNSS capability, and analyzed with

FIG. 9.3 GNSS-R installations with two antennas (*left photo*) and three antennas (*right photo*) at the Onsala Space Observatory. See the text for further explanations.

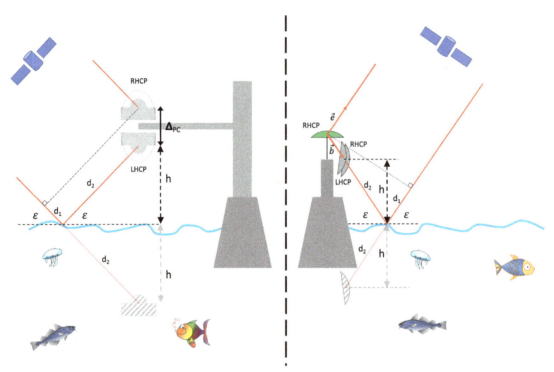

FIG. 9.4 Schematic drawings of the GNSS-R installations with two antennas (*left*) and three antennas (*right*) at the Onsala Space Observatory. See the text for further explanations.

geodetic phase-delay analysis, see, for example, Löfgren, Haas, and Johansson (2011). The data streams from the two antennas can also be used with a receiving and processing system that is built with graphic processing unit-based software radio, see, for example, Hobiger, Haas, and Löfgren (2014).

Other dedicated GNSS-R installations have even more than two antennas. As an example, the right photo in Fig. 9.3 shows the purpose-built GNSS-R installation at the Onsala Space Observatory that is operated in cooperation with GFZ Potsdam. It is equipped with one upward-looking RHCP antenna and two 98 degrees tilted antennas that look at the sea surface, one LHCP and one RHCP. This installation uses a purposely built GNSS receiver with four antenna inputs that allow phase-delay analysis to derive sea level measurements, see, for example, Liu et al. (2017).

The principle of these multiantenna GNSS-R installations is to determine the extra delay that it takes for the GNSS signals that are reflected off the sea surface to reach the downward-looking (or sideward-looking) antenna, compared to the direct signals that reach the upward-looking antenna. This signal delay is related to the height of the upward-looking antenna above the sea surface, see Fig. 9.4 for schematic drawings.

In both cases the downward-looking and sideward-looking antennas experience the same delay w.r.t. to the upward-looking antenna that a mirrored antenna located below the sea surface would experience. This extra delay, expressed in length units, that is, $d = d_1 + d_2$ in Fig. 9.4, is related to the height h of the downward-looking, respectively, sideward-looking antenna, above the sea surface, the satellite elevation ϵ, and the scalar product of the vector $\vec{b}$ between the antenna phase centers and the unit vector $\vec{e}$ toward the satellite:

$$d = d_1 + d_2 = 2\,h\sin\epsilon + \vec{b} \cdot \vec{e} \qquad (9.1)$$

The vector between the antenna phase centers can be determined during the installation from measurements in the local topocentric system between the antenna reference points and taking into account the available information on the antenna phase center with respect to this antenna reference point. While vector $\vec{b}$ is a three-dimensional (3D) topocentric vector for the installation with the sideward-looking antennas, see Fig. 9.3, right, it has just a vertical component Δ_{PC} for the installation with the vertically aligned GNSS antennas with one upward- and one downward-looking antenna, see Fig. 9.3, left. Finally, the height of the downward-looking antenna can be calculated as

$$h = \frac{d}{2\sin\epsilon} - \frac{\Delta_{PC}}{2} \qquad (9.2)$$

For the sideward-looking antenna, the height can be calculated as

$$h = \frac{d - \vec{b} \cdot \vec{e}}{2 \sin \epsilon} \tag{9.3}$$

The extra delay d that is necessary to determine the height above the sea surface can be determined in various ways. For GNSS-R installations with two standard geodetic receivers, one each connected to the RHCP and LHCP antenna, in principle code phase measurements could be used. These are produced individually by the receivers by correlating the received signals with internal noise-free replicas of the GNSS signals. However, since code phase measurements have a precision of several meters, this approach is not sufficient for meaningful coastal sea level observations. An early attempt of using code-phase measurements, though with special receiving equipment and processing, showed a root-mean-square (RMS) error of more than 3 m (Martín-Neira et al., 2001). Instead, carrier-phase measurements need to be used to derive the extra delay that the reflected signals experience. Using two standard geodetic receivers, carrier-phase measurements can be stored in individual RINEX-files and successively postprocessed to derive sea level. It has been shown that both single- and double-difference analyses can be performed to derive sea level (Löfgren, 2014; Löfgren et al., 2011), and an agreement with colocated tide gauge records could be achieved with standard deviations on the level of several centimeters by using single-difference analysis, and subcentimeter by using double-difference analysis (Löfgren, 2014).

Another way of determining the extra delay d is to use interferometric methods with purposely built receivers. One example is the specific GLONASS receiver that was based on graphic processing unit-based software radio, see, for example, Hobiger et al. (2014). With this system, an RMS agreement of better than 2 cm with respect to a colocated traditional tide gauge was achieved for calm weather conditions at the Onsala Space Observatory. Using the previously described three-antenna installation at the Onsala Space Observatory, and utilizing GPS L1 carrier-phase observations, RMS agreement w.r.t. the colocated tide gauge on the order of better than 4.4 cm could be achieved (Liu et al., 2017).

5 Coastal GNSS-IR with single antennas

Coastal GNSS installations that are affected by reflections from nearby sea surface experience interfering multipath disturbances in the recorded code phase, carrier phase, and signal-to-noise ratio (SNR) measurements. Fig. 9.5 depicts examples of such coastal GNSS stations. These stations are standard geodetic GNSS stations and were not installed for

FIG. 9.5 A selection of coastal GNSS installations with some view on the sea surface. The six stations are AC09 (Kayak Island, USA), BRST (Brest, France), GAMB (Rikitea, French Polynesia), HNLC (Honolulu, USA), MARS (Marseille, France), and ROTH (Rothera Station, Antarctica).

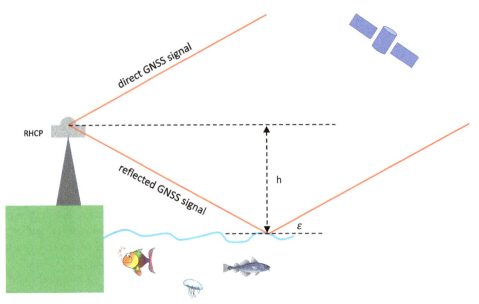

FIG. 9.6 Schematic drawing of a GNSS antenna that is affected by reflections off a nearby sea surface. See the text for further explanation.

the purpose of GNSS-IR. Nevertheless, they offer the potential to additionally provide information on sea level and its variations. Fig. 9.6 provides a schematic drawing of a coastal GNSS station that can be used for GNSS-IR.

It is in particular the SNR measurements that have proven to be useful in determining information about the reflecting surface, see, for example, Anderson (2000) and Larson, Löfgren, and Haas (2013). The recorded SNR data show a variation pattern that is caused by the phase difference between the direct and the reflected GNSS signal. This phase difference changes when the GNSS satellite travels across the local sky and thus causes a periodicity in SNR w.r.t. to elevation angle. This periodicity is superimposed onto the normal SNR trend from the direct GNSS signal due to the rising or setting of the satellite. After detrending, and thus removing the influence of the direct signal, the remaining detrended SNR (dSNR) show a clear oscillation as a function of elevation, reflector distance (or height) h, and signal wavelength λ. It has been shown, see, for example, Larson, Löfgren, and Haas (2013), Löfgren, Haas, and Scherneck (2014), and Strandberg, Hobiger, and Haas (2016), that the dSNR can be described in a simplified form as

$$\mathrm{dSNR} = A \cos\left(\frac{4\pi h}{\lambda} \sin \epsilon + \phi \right) \tag{9.4}$$

Fig. 9.7 depicts examples of dSNR data as a function of $\sin \epsilon$ for a coastal GNSS station. Strong dSNR oscillations below about $\sin \epsilon < 0.3$ are seen, while there is a clear damping envelope visible.

Analyzing dSNR with suitable spectral methods, such as Lomb-Scargle periodogram (LSP), see, for example, Larson, Löfgren, and Haas (2013), allows to derive the dominant multipath frequency, and thus reflector height. Using LSP has the advantage that unevenly sampled data can be analyzed, which is necessary since the dSNR are not evenly sampled with elevation. Usually, suitable azimuth and elevation ranges have to be defined for each station in order to guarantee that the analyzed dSNR variations really are caused by the sea surface, and not by other nearby reflectors. For GNSS-IR, that is, dSNR analysis, low elevation ranges are most suitable, since the reflected signals from low elevations are still predominantly RHCP and thus cause a strong impact on the SNR measurements. For stations with large and fast sea level variations due to, for example, tides, also the temporal height change needs to be considered in the analysis, see, for example, Larson, Löfgren, and Haas (2013) and Löfgren et al. (2014).

Usually, with spectral analysis methods, one reflector height is determined per satellite arc, that is, the temporal resolution of the resulting sea level time series depends on the number of satellite arcs. However, for suitable visibility conditions, that is, large and unobstructed areas of sea surface, long satellite arcs can be split up into smaller subarcs and analyzed separately in order to improve temporal resolution. GPS, GLONASS, Galileo, and Beidou have different orbits, so that the corresponding satellite arcs cover different times, azimuth, and elevation range. Using multi-GNSS data thus improves the temporal resolution of the sea level determination with spectral methods due to dSNR data with different geometries, see, for example, Löfgren and Haas (2014).

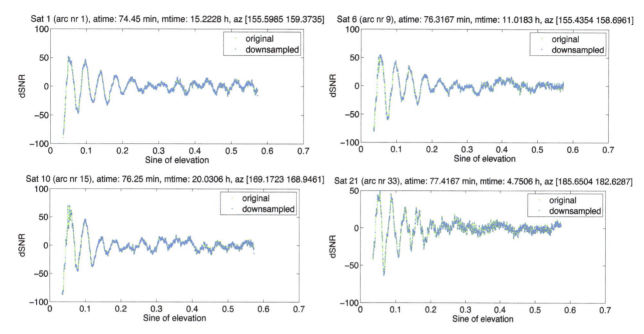

FIG. 9.7 Examples of detrended SNR measurements, that is, dSNR, as a function of $\sin\epsilon$ as observed by a coastal GNSS station. These data sets can be analyzed by, for example, spectral methods, in order to determine the vertical distance to the reflecting surface, that is, the sea level. See the text for further explanation.

A more advance alternative to spectral analysis of dSNR data is to use inverse modeling. With this approach, the dSNR model is extended by including a damping factor related to the variance s^2 of the reflector height to describe surface roughness (Strandberg et al., 2016). The advanced model for dSNR is

$$dSNR = \left(C_1 \sin\left(\frac{4\pi h}{\lambda}\sin\epsilon\right) + C_2 \cos\left(\frac{4\pi h}{\lambda}\sin\epsilon\right)\right) e^{-4k^2 s^2 \sin^2\epsilon} \quad (9.5)$$

Using B-spline functions to model the SSH, and applying nonlinear least squares parameter estimation, the ingredients of the advanced model can be determined. Sea level results with RMS agreement to colocated traditional tide gauge records on the level of less than 1.5 cm could be shown when using multi-GNSS data (Strandberg et al., 2016). The inverse modeling approach is also possible to be used in real-time using a Kalman filter (Strandberg, Hobiger, & Haas, 2019).

Regardless whether the dSNR are analyzed with spectral methods or with the advanced inverse model, the corresponding analyses have to account for differential tropospheric delays (Williams & Nievinski, 2017). Ignoring corresponding corrections, systematic biases can occur, see, for example, Santamaría-Gómez, Watson, Gravelle, King, and Wöppelmann (2015). It was shown by a comparing to colocated tide gauge data that an appropriate handling of the tropospheric corrections for ground-based GNSS-IR is essential in order to avoid seasonal biases (Feng, Haas, & Elgered, 2023).

6 Sensing sea level variability with GNSS

Sea level records derived from GNSS-R or GNSS-IR analyses show, of course, a wide range of variability. Provided long enough records with reasonable temporal resolution, for example, tidal analysis can be performed. This has been shown for various places around the earth, see, for example, Larson, Ray, Nievinski, and Freymueller (2013), Löfgren et al. (2014), Larson, Ray, and Williams (2017), Williams and Nievinski (2017), Tabibi, Geremia-Nievinski, Francis, and van Dam (2020), and Tabibi, Sauveur, Guerrier, Metayer, and Francis (2021). Also nonperiodic phenomena on rather short timescales, such as storms (Peng, Hill, Li, Switzer, & Larson, 2019) and Tsunamis (Larson et al., 2021), have been studied with sea level derived from GNSS-IR.

For climate change studies, long-term absolute sea level records are necessary. However, the sea level records from either GNSS-R or GNSS-IR with spectral analysis or inverse modeling of dSNR data are primarily relative to the GNSS station. Combining these relative sea level results from dSNR with VLM derived from the same GNSS station provides absolute sea level, see, for example, Peng, Feng, Larson, and Hill (2021). These can then compared to, for example, corresponding sea level results from satellite altimetry in coastal areas.

7 Selected highlights

There is a large, and continuously increasing, number of scientific publications that discuss various aspects of GNSS and measurements of sea level changes. The literature that is cited in this chapter is therefore by no means exhaustive. Two interesting examples of work using GNSS in relation to sea level changes are briefly discussed in this section.

Hammond et al. (2021) used more than 13,000 worldwide distributed GPS stations to determine corresponding VLM. The stations had recorded at least 3.5 years of GPS data so that reliable VLM could be determined. Based on geostatistical filtering and interpolation, so-called GPS imaging, the authors then derived VLM for more than 2300 tide gauge stations of the PSMSL. When these resulting VLM results were subtracted from the sea level rates obtained from the tide gauges, the mean rate of sea level rise increased by up to 0.3 mm/year. This rate increase is less than the rate increase due to climate change processes during the last decades.

Peng et al. (2021) used more than 8 years of data of three coastal GNSS stations to derive time series of relative sea level from GNSS-IR analyses of GPS data. The results for relative sea level derived from the GNSS-IR analyses were compared to tide gauge measurements from the nearest tide gauges (between 300 m and 29 km distance). For all three pairs of GNSS-IR and tide gauge sea level time series, linear trends and annual signals were estimated and compared. The results for linear trends agreed within their 1-sigma standard deviations. Also the annual amplitudes and phases agreed reasonably well, though not as well as the linear trends. The remaining residuals after the fit showed correlations coefficients of 0.97 and larger. Furthermore, the sea level results derived from GNSS-IR were compared to SSHs from satellite altimetry. To do so, dynamic atmosphere corrections (DAC) and the VLM of the three GNSS stations were applied to the GNSS-IR results. The DAC products used to correct the GNSS-IR series were identical to the ones used in the satellite altimetry analysis. The SSH results from altimetry are given as gridded results, that is, not colocated with the GNSS-IR stations. The authors compared annual signals (amplitude and phase) and linear trends derived from the GNSS-IR time series (corrected for DAC and VLM), and the gridded SSH from satellite altimetry within 100 km distance from the GNSS stations (for one station within 200 km). The agreement for the pairs of GNSS-IR and satellite altimetry grid point with highest correlation agree within 1-sigma standard deviation for the linear trends. The annual amplitudes and phases show also reasonable agreement, though not as well as the linear trends.

8 Conclusions and outlook

GNSS plays an important role in measurements sea level change. On the one hand, traditional tide gauges are augmented by colocated GNSS stations. This allows, provided that the relative position of the corresponding benchmarks is monitored regularly with high precision and accuracy, to convert the relative sea level records provided by the traditional tide gauges to absolute sea level records. On the other hand, relative and absolute sea levels can be derived directly from coastal GNSS stations. Both dedicated GNSS-R and dedicated or accidental GNSS-IR stations can be used to derive sea level with precision and accuracy on the order of several centimeters to subcentimeters, depending on analysis strategies and local circumstances. Some strategies even allow for real-time analyses. It can thus be envisaged that in the not too far future networks of coastal GNSS stations will be used on a regular basis for sea level monitoring and thus deliver important information for science and society on global change processes and natural hazards.

Acknowledgment

Tonie van Dam and Corné Kreemer provided helpful review comments, which are gratefully acknowledged.

References

Altamimi, Z., Rebischung, P., Métivier, L., & Collilieux, X. (2016). ITRF2014: A new release of the International Terrestrial Reference Frame modeling nonlinear station motions. *Journal of Geophysical Research, 121*, 6109–6131. https://doi.org/10.1002/2016JB013098.

Anderson, K. D. (2000). Determination of water level and tides using interferometric observations of GPS signals. *Journal of Atmospheric and Oceanic Technology, 17*, 1118–1127. https://doi.org/10.1175/1520-0426(2000)017<1118:DOWLAT>2.0.CO;2.

Bilich, A., Larson, K. M., & Axelrad, P. (2008). Modeling GPS phase multipath with SNR: Case study from Salar de Uyuni, Bolivia. *Journal of Geophysical Research*, *113*, B04401. https://doi.org/10.1029/2007JB005194.

Blewitt, G., Hammond, W. C., & Kreemer, C. (2018). Harnessing the GPS data explosion for interdisciplinary science. *Eos*, *99*. https://doi.org/10.1029/2018EO104623.

Braasch, M. S., & Scott, L. (2007). *Signal acquisition and search and antenna polarization*. March/April https://insidegnss-com.exactdn.com/wp-content/uploads/2018/01/MarApr07GNSSsolutionssecure.pdf.

Ekman, M. (2003). *The World's longest sea level series and a winter oscillation index for Northern Europe 1774–2000*. https://www.psmsl.org/products/author_archive/ekman_2003.pdf. Small Publications in Historical Geophysics No. 12.

Elósegui, P., Davis, J. L., Jaldehag, R. T. K., Johansson, J. M., Niell, A. E., & Shapiro, I. I. (1995). Geodesy using the global positioning system: The effects of signal scattering on estimates of site position. *Journal of Geophysical Research*, *100*, 921–993. https://doi.org/10.1029/95JB00868.

Feng, P., Haas, R., & Elgered, G. (2023). A novel tropospheric error formula for ground-based GNSS interferometric reflectometry. *IEEE Transactions on Geoscience and Remote Sensing*. https://doi.org/10.1109/TGRS.2023.3332422.

Georgiadou, P., & Kleusberg, A. (1988). On carrier signal multipath effects in relative GPS positioning. *Manuscripta Geodaetica*, *13*, 172–179.

Gravelle, M., Wöppelmann, G., Gobron, K., Altamimi, Z., Guichard, M., & Herring, T. (2023). The ULR-repro3 GPS data reanalysis and its estimates of vertical land motion at tide gauges for sea level science. *Earth System Science Data*, *15*, 497–509. https://doi.org/10.5194/essd-15-497-2023.

Hammond, W. C., Blewitt, G., Kreemer, C., & Nerem, R. S. (2021). GPS imaging of global vertical land motion for studies of sea level rise. *Journal of Geophysical Research: Solid Earth*, *126*, e2021JB022355. https://doi.org/10.1029/2021JB022355.

Hobiger, T., Haas, R., & Löfgren, J. S. (2014). GLONASS-R: GNSS reflectometry with a frequency division multiple access-based satellite navigation system. *Radio Science*, *49*, 271–282. https://doi.org/10.1002/2013RS005359.

IPCC. (2022). Climate change 2022: Impacts, adaptation, and vulnerability. In H.-O. Pörtner, D. C. Roberts, M. Tignor, E. S. Poloczanska, K. Mintenbeck, & A. Alegría (Eds.), *Contribution of Working Group II to the Sixth Assessment Report of the Intergovernmental Panel on Climate Change*. Cambridge University Press. In Press (Eds.).

Jaldehag, R. T. K., Johansson, J. M., Rönnäng, B. O., Elósegui, P., Davis, J. L., & Shapiro, I. I. (1996). Geodesy using the Swedish permanent GPS network: Effects of signal scattering on estimates of relative site positions. *Journal of Geophysical Research*, *101*, 841–860. https://doi.org/10.1029/96JB01183.

Katzberg, S., & Garrison, J. (1996). Utilizing GPS to determine ionospheric delay over the ocean. *NASA Technical Memorandum*, *4750*, 14 pp.

Larson, K. M., Lay, T., Yamazaki, Y., Cheung, K. F., Ye, L., & Williams, S. D. P. (2021). Dynamic sea level variation from GNSS: 2020 Shumagin earthquake tsunami resonance and Hurricane Laura. *Geophysical Research Letters*, *48*, e2020GL091378. https://doi.org/10.1029/2020GL091378.

Larson, K. M., Löfgren, J., & Haas, R. (2013). Coastal sea level measurements using a single geodetic GPS receiver. *Advances in Space Research*, *51*, 1301–1310. https://doi.org/10.1016/j.asr.2012.04.017.

Larson, K. M., Ray, R. D., Nievinski, F. G., & Freymueller, J. T. (2013). The accidental tide gauge: A GPS reflection case study from Kachemak Bay, Alaska. *IEEE Geoscience and Remote Sensing Letters*, *10*(5), 1200–1204. https://doi.org/10.1109/LGRS.2012.2236075.

Larson, K. M., Ray, R. D., & Williams, S. D. P. (2017). A ten-year comparison of water levels measured with a geodetic GPS receiver versus a conventional tide gauge. *Journal of Atmospheric and Oceanic Technology*. https://doi.org/10.1175/JTECH-D-16-0101.1.

Lidberg, M., Johansson, J. M., Scherneck, H.-G., & Milne, G. A. (2010). Recent results based on continuous GPS observations of the GIA process in Fennoscandia from BIFROST. *Journal of Geodynamics*, *50*, 8–18. https://doi.org/10.1016/j.jog.2009.11.010.

Liu, W., Beckheinrich, J., Semmling, M., Ramatschi, M., Vey, S., & Wickert, J. (2017). Coastal sea-level measurements based on GNSS-R phase altimetry: A case study at the Onsala Space Observatory. *IEEE Transactions on Geoscience and Remote Sensing*, *55*, 5625–5636. https://doi.org/10.1109/TGRS.2017.2711012.

Löfgren, J. S. (2014). Local sea level observations using reflected GNSS signals [Ph.D. dissertation]. Department of Earth and Space Sciences, Chalmers University of Technology, Gothenburg. Retrieved from https://research.chalmers.se/en/publication/191472.

Löfgren, J. S., & Haas, R. (2014). Sea level measurements using multi-frequency GPS and GLONASS observations. *EURASIP Journal on Advances in Signal Processing*, *50*. https://doi.org/10.1186/1687-6180-2014-50.

Löfgren, J. S., Haas, R., & Johansson, J. M. (2011). Monitoring coastal sea level using reflected GNSS signals. *Advances in Space Research*, *47*, 213–220. https://doi.org/10.1016/j.asr.2010.08.015.

Löfgren, J. S., Haas, R., & Scherneck, H.-G. (2014). Sea level time series and ocean tide analysis from multipath signals at five GPS sites in different parts of the world. *Journal of Geodynamics*, *80*, 66–80. https://doi.org/10.1016/j.jog.2014.02.012.

Lowe, S. T., Zuffada, C., Chao, Y., Kroger, P., Young, L. E., & LaBrecque, J. L. (2002). 5-cm-precision aircraft ocean altimetry using GPS reflections. *Geophysical Research Letters*, *29*. https://doi.org/10.1029/2002GL014759.

Martín-Neira, M. (1993). A PAssive Reflectometry and Interferometry System (PARIS): Application to ocean altimetry. *ESA Journal*, *17*, 331–355.

Martín-Neira, M., Caparrini, M., Font-Rosselló, J., Lannelongue, S., & Vallmitjana, C. S. (2001). The PARIS concept: An experimental demonstration of sea surface altimetry using GPS reflected signals. *IEEE Transactions on Geoscience and Remote Sensing*, *39*(1), 142–149.

Milne, G. A., Davis, J. L., Mitrovica, J. X., Scherneck, H.-G., Johansson, J. M., & Vermeer, M. (2001). Space-geodetic constraints on glacial isostatic adjustment in Fennoscandia. *Science*, *291*, 2381–2385. https://doi.org/10.1126/science.1057022.

Nievinski, F. G., & Larson, K. M. (2014). Forward modeling of GPS multipath for near-surface reflectometry and positioning applications. *GPS Solutions*, *18*, 309–322. https://doi.org/10.1007/s10291-013-0331-y.

Ning, T., Elgered, G., & Johansson, J. M. (2011). The impact of microwave absorber and radome geometries on GNSS measurements of station coordinates and atmospheric water vapour. *Advances in Space Research*, *47*, 186–196. https://doi.org/10.1016/j.asr.2010.06.023.

Peng, D., Feng, L., Larson, K. M., & Hill, E. (2021). Measuring coastal absolute sea-level changes using GNSS interferometric reflectometry. *Remote Sensing*, *13*, 4319. https://doi.org/10.3390/rs13214319.

Peng, D., Hill, E. M., Li, L., Switzer, A. D., & Larson, K. M. (2019). Application of GNSS interferometric reflectometry for detecting storm surges. *GPS Solutions, 23*, 47. https://doi.org/10.1007/s10291-019-0838-y.

PSMSL. (2022). *Permanent Service for Mean Sea Level.* https://psmsl.org.

Santamaría-Gómez, A., Watson, C., Gravelle, M., King, M., & Wöppelmann, G. (2015). Levelling co-located GNSS and tide gauge stations using GNSS reflectometry. *Journal of Geodesy, 89*, 241–258. https://doi.org/10.1007/s00190-014-0784-y.

Schöne, T., Schön, N., & Thaller, D. (2009). IGS tide gauge benchmark monitoring pilot project (TIGA), scientific benefits. *Journal of Geodesy, 83*, 3–4. https://doi.org/10.1007/s00190-008-0269-y.

Strandberg, J., Hobiger, T., & Haas, R. (2016). Improving GNSS-R sea level determination through inverse modeling of SNR data. *Radio Science, 51*, 1286–1296. https://doi.org/10.1002/2016RS006057.

Strandberg, J., Hobiger, T., & Haas, R. (2019). Real-time sea-level monitoring using Kalman filtering of GNSS-R data. *GPS Solutions, 23*, 61. https://doi.org/10.1007/s10291-019-0851-1.

Swedish Meteorological and Hydrological Institute (SMHI) (n.d.). Tide gauge data. https://www.smhi.se/data/oceanografi/ladda-ner-oceanografiska-observationer#param=sealevelrh2000,stations=core,stationid=33084.

Tabibi, S., Geremia-Nievinski, F., Francis, O., & van Dam, T. (2020). Tidal analysis of GNSS reflectometry applied for coastal sea level sensing in Antarctica and Greenland. *Remote Sensing of Environment, 248*, 111959. https://doi.org/10.1016/j.rse.2020.111959.

Tabibi, S., Sauveur, R., Guerrier, K., Metayer, G., & Francis, O. (2021). SNR-based GNSS-R for coastal sea-level altimetry. *Geosciences, 11*, 391. https://doi.org/10.3390/geosciences11090391.

Treuhaft, R. N., Lowe, S. T., Zuffada, C., & Chao, Y. (2001). 2-cm GPS altimetry over Crater Lake. *Geophysical Research Letters, 28*, 4343–4346. https://doi.org/10.1029/2001GL013815.

United Nations News. (4 November 2021). https://news.un.org/en/story/2021/11/1104972.

Williams, S. D. P., & Nievinski, F. G. (2017). Tropospheric delays in ground-based GNSS multipath reflectometry—Experimental evidence from coastal sites. *Journal of Geophysical Research Solid Earth, 122*, 2310–2327. https://doi.org/10.1002/2016JB013612.

Woodworth, P. L. (2022). Advances in the observation and understanding of changes in sea level and tides. *Annals of the New York Academy of Sciences, 1516*, 48–75. https://doi.org/10.1111/nyas.14851.

Zavorotny, V. U., & Voronovich, A. G. (2000). Scattering of GPS signals from the ocean with wind remote sensing application. *IEEE Transactions on Geoscience and Remote Sensing, 38*, 951–964. https://doi.org/10.1109/36.841977.

Chapter 10

GNSS application for weather and climate change monitoring

Peng Yuan[a,b], Mingyuan Zhang[c], Weiping Jiang[c], Joseph Awange[d], Michael Mayer[a], Harald Schuh[b,e], and Hansjörg Kutterer[a]

[a]Karlsruhe Institute of Technology, Geodetic Institute, Karlsruhe, Germany, [b]GFZ German Research Centre for Geosciences, Potsdam, Germany, [c]GNSS Research Center, Wuhan University, Wuhan, China, [d]School of Earth and Planetary Sciences, Curtin University, Perth, Australia, [e]Technische Universität Berlin, Institute of Geodesy and Geoinformation Science, Berlin, Germany

1 Introduction

Atmospheric water vapor plays a vital role in the Earth's water and energy cycle. It is a very active atmospheric constituent involved in various weather processes and extreme events, such as heavy rainfall, cyclones, and atmospheric rivers. However, accurate measurement of water vapor remains an issue due to its high spatiotemporal variability.

Water vapor is the dominant greenhouse gas in the Earth's atmosphere. Although water vapor only occupies up to 4% of air, it contributes about 50% of the greenhouse warming effect, larger than the well-known CO_2 and CH_4 (Schmidt et al., 2010). Moreover, there is a positive climate feedback effect between temperature and water vapor. Warmer atmosphere tends to increase water vapor concentrations, and the resulting greenhouse effect will intensify global warming (Held & Soden, 2000). The feedback effect gives rise to challenges, as the warmer and wetter atmosphere is prone to make extreme weather and climate events more frequent and more destructive (Trenberth, 2012). Therefore, evaluating the long-term change of water vapor is critically important in our understanding of climate change.

Tropospheric delay had only been regarded as an error source in GNSS positioning, until Bevis et al. (1992) proposed to employ it for water vapor measurement. This technique, known as GNSS meteorology, has a number of attractive features, such as high temporal resolution (usually from 1 h to 5 min), high precision and accuracy (a few kilogram per square meter or even less), near real-time, all-weather availability, long-term stability, relatively low cost, and easy maintenance. Its horizontal resolution and global coverage have been enhanced with growing number of global (e.g., International GNSS Service, IGS) and regional (e.g., Satellite Positioning Service of the German State Survey, SAPOS) permanent networks established in the past years. The number of tropospheric slant delays at each GNSS station is also increased owing to the multiple GNSS (multi-GNSS) constellations. As a result, the multi-GNSS is capable of enhancing the accuracy and reliability of GNSS tropospheric products compared to a single system (Li et al., 2015). Moreover, it is also capable of reconstructing the four-dimensional (4D and 3D of space and 1D of time) water vapor field using tomography methods (e.g., Flores et al., 2000; Xia et al., 2024). Note that this chapter focuses on the ground-based GNSS water vapor measurement technique, although space-based GNSS radio occultation is also capable of observing water vapor (Wickert et al., 2001).

With three decades of extensive development, ground-based GNSS meteorology has become a well-established technique with diverse applications. For example, GNSS Integrated Water Vapor (IWV) measurements serve as reference data to evaluate the products obtained from other techniques, such as radiosondes (e.g., Wang & Zhang, 2008), microwave radiometers (e.g., Baelen et al., 2005), satellite remote sensing (e.g., Li et al., 2003), and Numerical Weather Models (NWM; e.g., Hagemann et al., 2003). It also contributes to near real-time weather prediction (e.g., Gendt et al., 2004). Today, the earliest installed GNSS stations have accumulated about three decades of observations, which allow for a multiple temporal-scale climatic analysis on their IWV time series, such as annual cycle (e.g., Parracho et al., 2018), interannual variation (e.g., Jiang et al., 2017), and long-term trend (e.g., Yuan, Hunegnaw, et al., 2021). The methodology and advances of GNSS meteorology are comprehensively presented in the books of Boehm and Schuh (2013) and Jones et al. (2020), and its general applications are overviewed by Vaquero-Martínez and Antón (2021). In addition, the role of GNSS meteorology in the monitoring of severe weather and climate events is reviewed by Bonafoni et al. (2019).

2 Data and methods

2.1 Tropospheric delay

The Earth's neutral atmosphere extends from its surface to an altitude of about 50 km. When a GNSS signal passes through the neutral atmosphere, its speed is slower than that in a vacuum. This delay is termed as tropospheric delay in GNSS data processing. The GNSS tropospheric delay cannot be eliminated using multifrequency measurements, as the neutral atmosphere is a nondispersive medium for microwave signals. Therefore, the tropospheric delays are modeled and estimated in GNSS data processing.

The tropospheric delay of a GNSS signal along its propagation path is termed as Slant Total Delay (STD). It can be modeled as follows:

$$\text{STD}(a, e) = mf_H(e) \cdot \text{ZHD} + mf_W(e) \cdot \text{ZWD} + mf_G(e) \cdot (G_N \cdot \cos a + G_E \cdot \sin a) + \varepsilon, \qquad (10.1)$$

where a and e are azimuth and elevation angles, respectively. $mf_H(e)$, $mf_W(e)$, and mf_G are mapping functions for the hydrostatic, wet, and gradient parts, respectively. G_N and G_E are the northern and eastern components of the horizontal gradient vector, respectively. ε is the residual. Tropospheric Zenith Total Delay (ZTD) is divided into two parts: Zenith Hydrostatic Delay (ZHD) and Zenith Wet Delay (ZWD). ZHD is due to various constituents in the atmosphere, whereas ZWD is attributed to water vapor. Despite usually accounting for 90%–100% of ZTD, ZHD can easily be modeled by pressure. However, ZWD is difficult to be modeled due to the strong spatiotemporal variations of water vapor in the atmosphere. Therefore, ZWD and gradient parameters are usually estimated during GNSS data processing, whereas mapping functions and ZHD are determined in advance.

Many mapping functions are developed to map the slant delays into zenith direction, such as Niell Mapping Functions (NMF; Niell, 1996), Isobaric Mapping Functions (IMF; Niell, 2001), Vienna Mapping Function 1 (VMF1; Boehm, Werl, & Schuh, 2006), and Potsdam Mapping Factors (PMF; Zus et al., 2015). VMF1 is one of the most commonly used precise mapping functions developed by the Vienna University of Technology (TUW, https://vmf.geo.tuwien.ac.at/). With elaborations in modeling, VMF1 contributes to enhanced measurement accuracy. The VMF1 is routinely produced by TUW and can be downloaded before each use. For users' convenience, empirical mapping functions have been developed. One of the most famous is Global Mapping Function (GMF; Boehm, Niell, et al., 2006) developed based on VMF1. Now, GMF has been updated and merged into Global Pressure and Temperature 3 (GPT3), and VMF1 has been superseded by VMF3 (Landskron & Böhm, 2018). As an empirical model, GPT3 only includes the average and periodic variations of the mapping functions and other variables. In comparison, VMF3 is operationally updated by using the Numerical Weather Model (NWM), so the interannual and other variations are also modeled. Therefore, VMF3 is superior to GPT3 and is recommended for high-precision GNSS data processing. However, as VMF3 is provided with a time lag of about one day, GPT3 is recommended for real-time applications or in scenarios without Internet availability. In addition to using the mapping functions, the tropospheric delay along the propagation path of GNSS signal can be directly calculated by using ray tracing of NWM, which is more accurate but with higher computational cost (Zhang et al., 2024).

ZHD (in millimeters) can be precisely estimated if the station pressure (P_s in hPa) is available (Saastamoinen, 1972):

$$\text{ZHD} = 2.2768 \frac{P_s}{1 - 0.00266 \cdot \cos(2\varphi) - 0.28 \cdot 10^{-6} H}, \qquad (10.2)$$

where φ and H are the latitude and altitude of the GNSS station, respectively. Note that, H can be geopotential altitude, orthometric height, or ellipsoidal height in different publications, but their difference are negligible for the ZHD estimates in most cases.

GNSS station pressure can be obtained from various sources, such as collocated meteorological sensors, nearby synoptic stations, and NWM. If external data are not available, pressure can be obtained from empirical models such as GPT3 as mentioned above. The GPT3 data are provided in global grids at two optional spatial resolutions of 5° × 5° and 1° × 1° (Landskron & Böhm, 2018). In addition, TUW provides a global ZHD grid product (VZHD) along with VMF1 and VMF3.

Note that the accuracy of a priori ZHD has a significant impact on ZWD and the derived IWV. A positive bias of 1 mm in ZHD can result in a negative bias of 0.12–0.17 kg m^{-2} in the associated IWV. However, on many occasions, a less accurate ZHD estimate has to be employed in GNSS data processing, because its accurate a priori value is unavailable at that time. If so, the ZWD estimate directly obtained from the GNSS data processing is not qualified for accurate water vapor retrieving (Yuan, Blewitt, et al., 2023). Therefore, a usual practice to obtain high-quality ZWD for water vapor measurement is to subtract accurate ZHD from the ZTD estimate. The accurate ZHD estimate can be obtained based on Eq. (10.2) and pressure data from collocated meteorological sensors, nearby synoptic stations, and NWM. However, it is worth mentioning that the inaccurate modeling of ZHD in GNSS data processing can also impact station height estimates (Tregoning & Herring, 2006). Consequently, the accurate a priori ZHD estimates are crucial for high-accuracy GNSS station coordinate solutions.

A number of GNSS tropospheric delay products are publicly available. For example, global GNSS ZTD datasets includes IGS (about 350 stations) https://cddis.nasa.gov/archive/gnss/products/troposphere/zpd/, TIGA (about 500 stations) ftp://ftp.gfz-potsdam.de/GNSS/products/tiga_repro2_tro/, and NGL (more than 22,000 stations) http://geodesy.unr.edu/ (Blewitt et al., 2018). The ZTD products are also available at regional networks, such as EPN https://www.epncb.oma.be/_productsservices/troposphere/ and GURN (Fersch et al., 2022; Yuan & Kutterer, 2021) https://doi.pangaea.de/10.1594/PANGAEA.936134. In addition, GNSS IWV products are available for the NGL and GURN network along with their ZTD products.

2.2 Water vapor retrieval

ZHD (in meters) is calculated as an integration of refractivity (N, unitless) from the altitude of the GNSS station (z_0) to the top of the atmosphere:

$$\text{ZTD} = 10^{-6} \int_{z_0}^{\infty} N dz, \tag{10.3}$$

$$N = k_1 \frac{P_d}{T} Z_d^{-1} + k_2 \frac{P_w}{T} Z_w^{-1} + k_3 \frac{P_w}{T^2} Z_w^{-1}, \tag{10.4}$$

where $k_1 = 77.6890$ K hPa^{-1}, $k_2 = 71.2952$ K hPa^{-1}, and $k_3 = 375{,}463$ K^2 hPa^{-1} are thermodynamic coefficients (Rüeger, 2002). T is temperature in Kelvin. P_d and P_w are the pressure of dry air and water vapor in hectopascals.

Similarly, ZHD and ZWD in meters are expressed as

$$\text{ZHD} = 10^{-6} \int_{z_0}^{\infty} k_1 \frac{R}{M_d} \rho_w dz, \tag{10.5}$$

$$\text{ZWD} = 10^{-6} \int_{z_0}^{\infty} \left(k_2' \frac{P_w}{T} + k_3 \frac{P_w}{T^2} \right) dz, \tag{10.6}$$

where $R = 8.314$ J K^{-1} mol^{-1} is the universal gas constant. ρ_w is the density of water vapor in kilogram per cubic meters. $k_2' = k_2 - k_1 \cdot \frac{M_w}{M_d} = 22.974$ K hPa^{-1}. M_d and M_w are the molar mass of dry air and water vapor with values of 28.964 and 18.015 g mol^{-1} (Iribarne & Godson, 1981), respectively.

Despite an error source in GNSS data processing, GNSS tropospheric delay can be used to retrieve atmospheric water vapor. The absolute amount of atmospheric water vapor can be quantified as IWV, which is defined as the total mass of water vapor in a vertical atmospheric column (in kilogram per square meters):

$$\text{IWV} = \int_{z_0}^{\infty} \rho_w dz. \tag{10.7}$$

In addition to IWV, the term of Precipitable Water Vapor (PWV) is also widely used. PWV is the equivalent liquid water height of atmospheric water vapor. As the density of liquid water is 1000 kg m^{-3}, it is easy to know that the values of IWV in kilogram per square meters and PWV in millimeters are identical. we use the term IWV rather than PWV here, as it is a convention in Europe (Jones et al., 2020).

By applying the ideal gas law and substituting Eq. (10.6) into Eq. (10.7), the IWV can be estimated as follows:

$$\text{IWV} = \frac{10^6 M_w}{(k_2' + k_3/T_m) R} \text{ZWD}, \tag{10.8}$$

where T_m is weighted mean temperature (in Kelvin):

$$T_m = \frac{\int_{z_0}^{\infty} \frac{P_w}{T} dz}{\int_{z_0}^{\infty} \frac{P_w}{T^2} dz} \tag{10.9}$$

Profiles of the meteorological variables can be obtained from radiosonde and NWM. However, only a few GNSS stations are collocated with radiosonde stations. Moreover, despite global coverage, the NWM products usually have time lags from several days to months. Therefore, empirical T_m models have been developed for near real-time IWV retrieving, such as GPT3. The empirical models are based on the analysis of multiyear global accurate T_m estimates obtained from NWM. Alternatively, real-time T_m can be easily calculated according to an empirical relationship with air temperature (T_s, in Kelvin) as proposed by Bevis et al. (1992):

$$T_m = 70.2 + 0.72 \cdot T_s \tag{10.10}$$

The T_m estimates from the empirical model and $T_s - T_m$ relationship are less accurate than those integrated with radiosonde and NWM. From Eq. (10.8) it follows that the relative bias in T_m results in equivalent relative bias in IWV.

Auxiliary meteorological data (e.g., P_s, T_s, and T_m) are crucial for GNSS IWV retrieving. A minority of GNSS stations are equipped with collocated meteorological sensors. The meteorological measurements are provided as RINEX m-file format and archived at IGS data centers: https://igs.org/data-access/. If a collocated meteorological sensor is not available at a GNSS station, the meteorological variables can be estimated based on the nearby synoptic stations. For instance, global synoptic data are provided at Integrated Surface Database (ISD; Smith et al., 2011) at https://www.ncei.noaa.gov/products/land-based-station/integrated-surface-database. Radiosonde data is a good reference for the validation of T_m models and GNSS IWV. Global radiosonde data are accessible at the Integrated Global Radiosonde Archive Version 2 (IGRA2; Durre et al., 2018) at https://www.ncei.noaa.gov/access/metadata/landing-page/bin/iso?id=gov.noaa.ncdc:C00975. Moreover, the Global Climate Observing System (GCOS) Reference Upper-Air Network (GRUAN; https://www.gruan.org/) provides high-quality climatic data of a global network with 30–40 stations. In this network, the GNSS and radiosonde stations are usually collocated so that they are suitable for intercomparison.

Alternatively, the meteorological variables can be obtained from atmospheric reanalysis. The reanalysis provides global or regional gridded meteorological products, and thus it can provide the meteorological estimates anywhere within the coverage area. One of the state-of-the-art global products is the fifth generation European Centre for Medium-Range Weather Forecasts (ECMWF) atmospheric reanalysis (Hersbach et al., 2020) at https://www.ecmwf.int/en/forecasts/datasets/reanalysis-datasets/era5. ERA5 pressure-level product is provided every 1-h at 37 vertical pressure levels with $0.25° \times 0.25°$ horizontal grids. For details on the calculations of the meteorological variables from atmospheric reanalysis, readers are referred to Appendix C by Yuan, Blewitt, et al. (2023). In addition, Global gridded ZHD and T_m products can be obtained from TUW with routine updates. TUW also provides the corresponding empirical models like GPT3 as mentioned above.

In the processing with radiosonde and NWM, readers should bear in mind that their height system is geopotential altitude (H_{gp}, in geopotential meter). The height system is defined based on mean sea level (MSL) with the consideration of gravity variations. The geopotential altitude of a GNSS station with a latitude of φ (in radians) and an orthometric height of H_{or} (in meters) can be estimated as follows (World Meteorological Organization, 2018):

$$H_{gp} = \frac{\gamma_s(\varphi)}{9.80665} \cdot \frac{R(\varphi) \cdot H_{or}}{R(\varphi) + H_{or}}, \tag{10.11}$$

$$\gamma_s(\varphi) = 9.780325 \frac{1 + 1.93185 \times 10^{-3} \cdot \sin^2(\varphi)}{\left(1 - 6.69435 \times 10^{-3} \cdot \sin^2(\varphi)\right)^{0.5}}, \tag{10.12}$$

$$R(\varphi) = \frac{6.378137 \times 10^6}{1.006803 - 6.706 \times 10^{-3} \cdot \sin(\varphi)^2}. \tag{10.13}$$

In this chapter, we used the GNSS ZTD products provided by NGL to present the applications of GNSS meteorology in weather and climate change monitoring. The GNSS data processing strategy is described in http://geodesy.unr.edu/gps/ngl.acn.txt (Accessed 7 November 2022). Moreover, we employed ERA5 pressure level product to obtain P_s and T_m used for the calculation of GNSS IWV.

3 Extreme weather events

Water vapor is of great importance in the prediction, monitoring, and explanation of extreme weather such as heavy rainfall. However, observing the spatiotemporal variations of water vapor in real time is difficult for many traditional techniques, due to their limitations in cost, precision, timeliness, temporal resolution, and availability under severe weather conditions.

Nevertheless, ground-based GNSS provides an effective alternative, and its performance has been extensively improved in the last years. For example, tomographic technique was proposed to reconstruct 4D water vapor field from GNSS STD (e.g., Flores et al., 2000). Jointly using GNSS and InSAR tropospheric observations can improve the spatiotemporal resolution of the water vapor product, as they are complementary in this regard (Alshawaf et al., 2015; Heublein et al., 2019). GNSS meteorology has attracted considerable attention from meteorologists. The GNSS tropospheric products (e.g., ZTD and IWV) have been assimilated into the Weather Research and Forecasting (WRF) model to improve its performance (Wagner et al., 2022). In addition to heavy rainfall, GNSS IWV has also been employed to other extreme weather phenomena such as atmospheric rivers (Wang et al., 2019) and cyclones (Ejigu et al., 2021).

In this section, we present the variations of GNSS IWV during Typhoon Haishen 2020, a Saffir-Simpson Hurricane Scale (SSHS) Category 4 super tropical cyclone. Fig. 10.1 illustrates the daily average GNSS IWV and daily accumulated precipitation maps in the Northwest Pacific during the evolution of the typhoon. Typhoon Haishen 2020 was initiated from a low-pressure area over the southern Philippine Sea and then drifted northward and strengthened to Category 4 on September 3. On September 6, it skimmed over southwestern Japan and made landfall in South Korea with a level of Category 1, triggering heavy rainfall over southwestern Japan and South Korea (Fig. 10.1I). What is striking in the top panel of Fig. 10.1 is that the IWV in southwestern Japan reached the maxima of 60–80 kg m^{-2} with the approach of Haishen on September 6. Then, it rapidly dropped to 30–40 kg m^{-2} with the departure of the typhoon on September 7. By contrast, the IWV in northwestern Japan reached the lowest level of 35–55 kg m^{-2} on September 6, and then increased to 45–55 kg m^{-2} within one day. The results can be explained by horizontal water vapor flux convergence due to the typhoon (Yang et al., 2011).

We selected four GNSS stations along the path of Typhoon Haishen to illustrate the variations of their IWV and precipitation over time (Fig. 10.2). It is observed that all these stations are characterized by several rounds of rainfall due to the movement of spiral rainbands of the typhoon. The IWV peaks at the time of the heaviest rainfall at each station, with a significant reduction after the last rainfall. At station J456 and J462, two peaks of IWV are observed along with the rainfalls in early September 5 and late September 6, respectively. The earlier peak is related to the peripheral circulation of the typhoon before its arrival, and the latter one coincided with the arrival time of the typhoon.

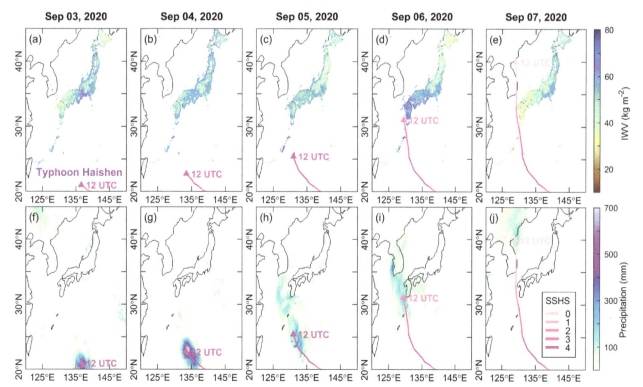

FIG. 10.1 Daily average IWV (A–E) and daily accumulated precipitations (F–J) during Typhoon Haishen 2020. The trail of the typhoon and its Saffir-Simpson Hurricane Scale (SSHS) are provided by The International Best Track Archive for Climate Stewardship (IBTrACS; http://ibtracs.unca.edu/).

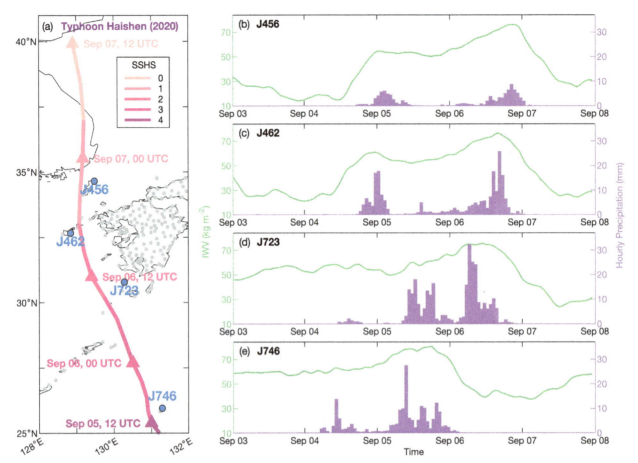

FIG. 10.2 (A) Pathway of Typhoon Haishen 2020. (B–E) GNSS IWV and hourly accumulated precipitations at four stations.

4 Diurnal cycle

The diurnal water vapor cycle is related to many meteo-hydrological processes, such as solar radiation and surface evapotranspiration. Owing to the advantages of high temporal resolution and high accuracy, GNSS meteorology offers a unique tool for the evaluation of diurnal variation of IWV. Dai et al. (2002) analyzed the diurnal IWV signals at 54 GNSS stations in North America and found that the diurnal amplitudes can be as large as $1.8\,\text{kg}\,\text{m}^{-2}$, whereas the amplitudes of semidiurnal cycles are lower than $1\,\text{kg}\,\text{m}^{-2}$. Wang and Zhang (2009) examined the IWV series at hundreds of GNSS stations worldwide and obtained an average peak-to-peak amplitude (PPA) of $0.66\,\text{kg}\,\text{m}^{-2}$ in their diurnal cycles. Diedrich et al. (2016) evaluated the temporal sample errors in the IWV obtained from polar orbiting satellites by taking the diurnal cycle of IWV measured by global GNSS stations into account. Moreover, many regions have also been investigated, such as Japan (Iwasaki & Miki, 2001), Spain (de Galisteo et al., 2011), China (Zhang et al., 2019), Germany (Steinke et al., 2019), Eastern Mediterranean (Ziv et al., 2021), and South Indian Ocean (Lees et al., 2021).

Fig. 10.3 illustrates the month-to-month variations of hourly IWV diurnal cycles at three selected stations, namely INVK (Inuvik, Canada, 68.31°N, 133.53°W, UTC-9), BJFS (Beijing, China, 39.61°N, 115.89°E, UTC+8), and NTUS (Singapore, 1.35°N, 103.68°E, UTC+7), which are located in polar, mid-latitude, and tropical regions, respectively. Their diurnal cycles were calculated based on their IWV in 2001.0–2021.0. The time was adjusted from Coordinated Universal Time (UTC, T_{UTC}) to the respective Local Solar Time (LST, T_{LST}):

$$T_{LST} = T_{UTC} + round\left(\lambda/15°\right) \tag{10.14}$$

where λ (within $-180°$ to $180°$) is the longitude of the station, and *round* is a function to get the nearest integer.

As can be seen from the top panel of Fig. 10.3, the diurnal cycle is quite weak at INVK (Inuvik, Canada, 68.31°N, 133.53°W), which is only observable in summer (June–August) with a PPA of about $0.6\,\text{kg}\,\text{m}^{-2}$ (Fig. 10.3A). The diurnal

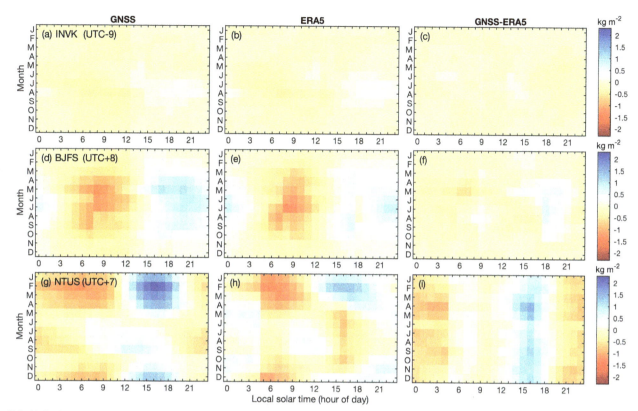

FIG. 10.3 Diurnal cycles of IWV (in kilogram per square meters) from GNSS (A, D, and G), ERA5 (B, E, and H), and their differences (C, F, and I) at three GNSS stations INVK, BJFS, and NTUS, in their respective local time. The diurnal variations were calculated for each month separately based on hourly IWV time series 2001.0–2021.0.

cycle at BJFS (Beijing, China, 39.61°N, 115.89°E) is much more obvious all the time except winter (Fig. 10.3D). The diurnal cycle touches the bottom in the morning and then reaches the peak in the evening, with the largest PPA of 2.3 kg m^{-2} in May. Moreover, the diurnal cycle at NTUS (Singapore, 1.35°N, 103.68°E) is also characterized with strong month-to-month variations (Fig. 10.3G). The cycle is quite strong during the Northeast Monsoon season (December–April). The maximum PPA of 3.6 kg m^{-2} was found in February, which is the dry phase of the Northeast Monsoon season. The IWV values within the diurnal cycle are lower before sunrise and larger in the afternoon, which could be attributed to surface solar heating.

The monthly diurnal cycles modeled by ERA5 are in excellent agreement with those measured by GNSS at INVK, with an average root mean square (RMS) of 0.04 kg m^{-2} for their monthly differences (Fig. 10.3C). However, the monthly PPA estimates at BJFS obtained from ERA5 are 0.5–1.1 kg m^{-2} smaller than those from GNSS in April–October (Fig. 10.3F). What is more striking is their discrepancy at NTUS, where the ERA5-derived PPA values are 1.7 kg m^{-2} smaller than those from GNSS on average (Fig. 10.3I). The results show that the agreement between GNSS and ERA5 in their IWV diurnal cycles can be quite different at specific stations. The discrepancy could be due to representative differences between these two water vapor retrieval techniques or errors in specific techniques.

5 Annual cycle

An annual cycle is a prominent temporal feature in most IWV time series, especially for the regions with active water cycle process such as monsoon. GNSS has been used to evaluate the annual cycle of IWV modeled by atmospheric reanalyzes (Vey et al., 2010; Parracho et al., 2018). Moreover, its potential in the analysis of regional meteo-hydrological characteristics has also been highlighted. One of the first investigations in this regard was conducted by Jiang et al. (2017), who found a 2-month phase lag between the annual signals of GNSS vertical coordinate time series induced by hydrological loading and IWV time series in southwest China and explained it with the water balance equation. Recently, Ferreira et al. (2021) carried out a similar case study in southeast Brazil.

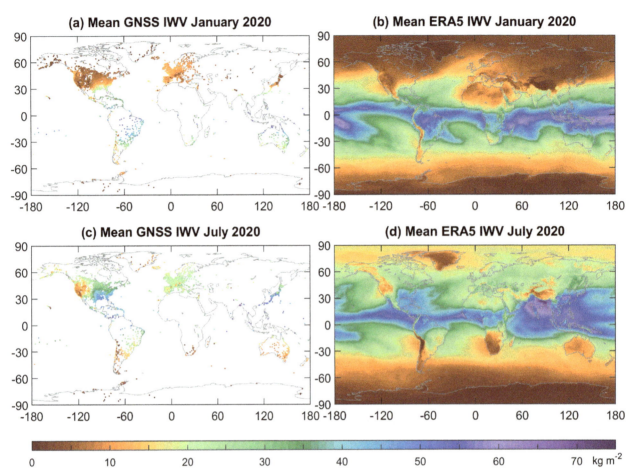

FIG. 10.4 Mean IWV in January (A and B) and July (C and D) 2020 at 5591 stations obtained with GNSS (A and C) and at 0.25° × 0.25° grids on Earth's surface obtained with ERA5 (B and D).

In order to show the seasonal differences in IWV, we compared the monthly mean GNSS IWV in January and July 2020 at 5591 stations worldwide (Fig. 10.4A and C). The seasonal difference is most striking at monsoon climate zones. For example, there are nine GNSS stations located near the coastline of Bangladesh (21.8°N–22.8°N, 89.2°E–89.7°E) characterized by the famous South Asia Monsoon, and their IWV in January and July are 20.9 and 68.7 kg m^{-2} on average, respectively. Moreover, Japan is significantly influenced by East Asia Monsoon, and the IWV at 469 GNSS stations there (21°N–46°N, 122°E–150°E) are 2.5–26.2 (mean value 9.8) and 19.0–56.8 (mean value 44.8) kg m^{-2} in these 2 months, respectively. Obvious seasonal differences are also observed in many other regions, such as Europe, eastern North America, and northern Australia. However, Antarctica only has a low level of IWV in both January (<11 kg m^{-2}) and July (<6 kg m^{-2}), as it is the coldest and driest continent of the Earth. The IWV at inland Antarctica is even lower than that at its coastal regions. This is because the inland Antarctica is characterized by higher altitudes and lower temperatures, and hence the water vapor holding capacity of the air there is weaker. We also show the global IWV maps in these 2 months from ERA5 for comparison (Fig. 10.4B and D). The geographical patterns of the seasonal differences in IWV obtained from ERA5 and GNSS are consistent, indicating that ERA5 has a good performance in modeling the seasonal variations of IWV. ERA5 has a better spatial resolution (0.25° × 0.25°) and global coverage compared to GNSS. Nevertheless, GNSS can serve as a unique tool to validate the IWV product modeled by ERA5.

To illustrate the annual cycle of IWV more explicitly, we selected station TSKB (Japan, 36.11°N, 140.09°E) for which 26 years of data (1995.0–2021.0) are available. The monthly GNSS IWV time series at TSKB is dominated by annual cycle (Fig. 10.5B). We calculated its monthly climatology by averaging the IWV for each month throughout the 26 years (Fig. 10.5A). It can be seen that the climatological averages are larger than 23.7 kg m^{-2} in the monsoon season (May–October) with a maximum of 48.7 in August, whereas they are lower than 16.5 kg m^{-2} in the non-monsoon season (November–April) with a minimum of 6.9 kg m^{-2} in January.

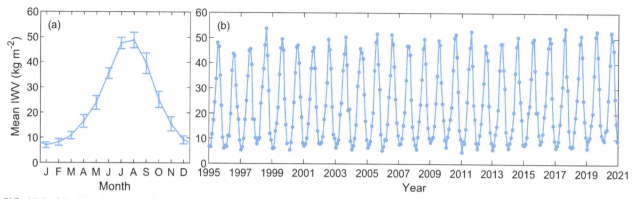

FIG. 10.5 Monthly climatological averages (A) for the GNSS IWV time series in 1995.0–2021.0 (B) at station TSKB (Tsukuba, Japan, 36.11°N, 140.09°E).

6 Interannual variations

Owing to the high-quality long-term GNSS IWV measurements dating back to the 1990s, the interannual variations in water vapor and related climate phenomena have attracted increasing attention. For instance, Jiang et al. (2017) found that the interannual anomalies in both IWV and vertical coordinate time series at the GNSS stations in southwest China are related to droughts.

In addition, it is known that the amount of water vapor at one location can be correlated to the climate phenomenon at a very distant place, even on the other side of the Earth. The correlation pattern is termed as teleconnection, which could be related to large-scale atmospheric circulation anomaly (see Wagner et al., 2021 and references therein). Vey et al. (2009) reported the correlation between the interannual anomaly of GNSS IWV at a coastal station in Australia and El Niño-Southern Oscillation (ENSO). Suparta et al. (2013) analyzed the variations in GNSS IWV over the western Pacific Ocean during the 2011 La Niña event. Wang et al. (2018) carried out a further analysis on the correlation with ENSO by using more coastal stations worldwide with extended time length. In addition, teleconnections of the water vapor over continental inland with more climate indices are also found. Yuan, Van Malderen, et al. (2021) investigated the correlation between the interannual variations of GNSS IWV in Europe and eight climate indices and reported strong correlations with the indices of Arctic Oscillation (AO), North Atlantic Oscillation (NAO), and East Atlantic (EA) pattern as well as a weak correlation with Niño 3.4 ENSO index.

The interannual variations of IWV are usually quantified as monthly climatological anomalies, which are the differences between the IWV time series and monthly climatological averages. Fig. 10.6 shows the monthly GNSS IWV anomalies at station HYDE (India, 17.42°N, 78.55°E). We evaluated the correlations between the IWV anomalies and various climate teleconnection indices, including AO, EA, Northern Oscillation Index (NOI), and Niño 3.4. Although only a weak correlation (0.23) with the EA index is observed for the full GNSS IWV series (Fig. 10.6B), more significant correlations can be found in specific months. For example, the correlation with AO (Fig. 10.6A) is as strong as 0.53 and −0.40 in June–August (JJA) and December–February (DJF), respectively. Similarly, the correlation coefficient with NOI (Fig. 10.6C) is 0.39 and −0.30 in JJA and DJF, respectively. Although station HYDE is located in India, its IWV anomalies in DJF are moderately correlated with Niño 3.4 with a value of 0.30 (Fig. 10.6D). The results indicate that the teleconnections between interannual variations of IWV and climate indices can be planetary-scale and seasonal-dependent.

7 Long-term trends and climate change

Long-term change of IWV is a key indicator of climate change. The atmosphere is able to hold about 7% more water vapor with 1°C rise in temperature, as indicated by the Clausius-Clapeyron Equation (Trenberth et al., 2005). Compared to traditional water vapor measurements from radiosonde, GNSS IWV time series have the advantages of high completeness and good instrumental stability over time, which are crucial for the estimation of their long-term trends. Hence, GNSS has been used to evaluate the IWV trends in a number of global and regional studies. For instance, Nilsson and Elgered (2008) calculated the IWV trends in North Europe with 10 years of GNSS data and obtained estimates from −0.2 to 1.0 kg m^{-2} decade^{-1}. Wang et al. (2016) analyzed global IWV trends obtained from GNSS, radiosonde, and microwave satellite measurements. The authors reported that global IWV were generally increased during 1988–2011, with larger values at nighttime than at daytime. Alshawaf et al. (2017) investigated the climate evolution in Germany by analyzing the trends

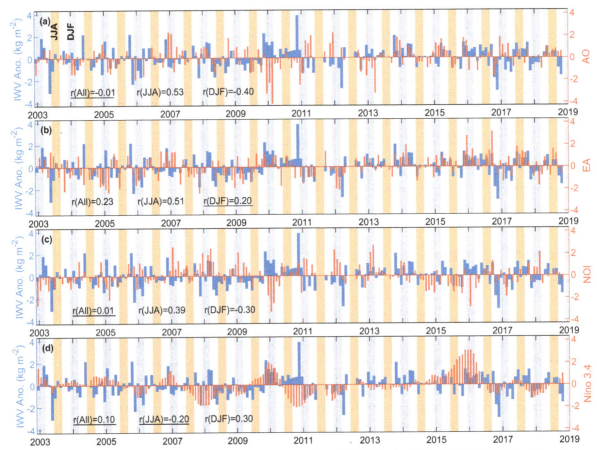

FIG. 10.6 Monthly climatological anomalies of GNSS IWV *(blue bars)* at station HYDE (Hyderabad, India, 17.42°N, 78.55°E) and the teleconnections with respect to climate indices *(red bars)* of Arctic Oscillation (AO, A), East Atlantic pattern (EA, B), Northern Oscillation Index (NOI, C), and Niño 3.4 ENSO index (D). The IWV anomalies and climate indices were standardized with zero mean values and unit standard deviations, respectively. The periods of June–August (JJA) and December–February (DJF) are shown as areas shaded in orange and purple. r(All), r(JJA), and r(DJF) are the Pearson correlation coefficients in the full time series, in JJA, and in DJF, respectively. The correlation coefficients significant at 95% confidence level are underlined.

in temperature and GNSS IWV time series. Parracho et al. (2018) compared global IWV trends from GNSS and two atmospheric reanalyzes (ERA-Interim and MERRA2) and found a fair consistency over the Northern Hemisphere. Yuan, Hunegnaw, et al. (2021) evaluated the feasibility of IWV trends from the state-of-the-art ERA5 reanalysis in Europe with GNSS IWV time series at 109 stations and reported that the ERA5 and GNSS IWV trends are consistent, with magnitudes of 0–1.4 kg m^{-2} decade^{-1}.

Statistical significance of the IWV trend estimate is essential for the interpretation of climate change. Improper noise models (i.e., statistical models) of GNSS IWV time series can under- or overestimate its trend uncertainty. Due to the autocorrelation in many climatic series, the first-order AutoRegressive (AR1) model is believed to be more appropriate than the classical White Noise (WN) model (Tiao et al., 1990). This statement was endorsed by Alshawaf et al. (2018) based on an analysis of the daily GNSS IWV time series in Europe. However, other noise models could be more suitable to describe specific climate conditions and errors related to observation techniques. For example, Yuan, Hunegnaw, et al. (2021) claimed that AutoRegressive Moving Average Model ARMA(1,1) model is superior to AR(1) for the daily IWV time series at stations in Central and South Europe, and the less preferred assumption of AR(1) resulted in overestimations of trend uncertainties by up to 20%. In addition, Power-law (PL) model has been commonly used to describe the statistical properties of many geophysical phenomena (Agnew, 1992). The optimal noise models for a specific IWV time series can be determined by examining its power spectrum and testing with different commonly used models and model combinations, such as WN, AR(1), ARMA(1,1), WN+AR(1), PL, and WN+PL.

Homogenization of climatic series is an important procedure prior to trend analysis. The inhomogeneities in GNSS IWV time series are shown as abrupt shifts at specific epochs, termed as changepoints. The homogenization procedure usually involves identification and estimation of the changepoints as well as evaluation on their statistical significance. The changepoints can be caused by changes in models, strategies, and software packages used for GNSS data processing

(Pacione et al., 2017), changes in local environment, and changes in instrumentation, such as receiver, antenna, and radome of the GNSS station. Detecting the changepoints remains a challenge, as their magnitudes are usually much smaller than the natural variations of IWV. Nevertheless, key information related to the changes in instrumentation is recorded in GNSS station log file. In addition, inspection of associated GNSS coordinate time series is conducive to the detection of undocumented changepoints (Yuan, Hunegnaw, et al., 2021). Reference IWV time series obtained from a nearby GNSS station (Vey et al., 2009) or from other techniques, such as radiosonde, Very Long Baseline Interferometry (VLBI), and Doppler Orbitography and Radiopositioning Integrated by Satellite (DORIS), can also be very helpful. Van Malderen et al. (2020) assessed a number of changepoint detection approaches. The authors emphasized the importance of metadata and the suitability of noise model for the efficiencies of the approaches. Regarding the statistical significances of the changepoints, Ning et al. (2016) employed penalized maximal t-test to account for AR(1) noise. Yuan, Hunegnaw, et al. (2021) reported that the PL-related noise are preferred over the AR(1)-related models when evaluating the changepoints in 76% of the daily (ERA5 minus GNSS) IWV difference time series. Therefore, the significance of the changepoints should be evaluated based on a careful selection of proper noise model.

Fig. 10.7 shows an example of the homogenization approach with the monthly GNSS IWV time series 1999.0–2019.0 at station EUSK (Euskirchen, Germany, 50.67°N, 6.76°E). As can be seen from Fig. 10.7A, the GNSS IWV series is characterized with strong annual and interannual variations, which makes it difficult to detect possible changepoints. Nevertheless, these variations are mostly removed in the IWV difference (GNSS minus ERA5) time series (Fig. 10.7B) where changepoints in the IWV difference series are identified using the RHtestV4 software (Wang, 2008a, 2008b). The software suggests a changepoint in May 2001 with a value of $-0.77\,\mathrm{kg\,m^{-2}}$. As both the GNSS and ERA5 time series may contain changepoints, more information is needed for the determination of its source. For example, the log file of a GNSS station is crucial, as most changepoints in GNSS IWV series are due to device (antenna, radome, and receiver) changes. Vertical coordinate time series are also supportive, as they are sensitive to the device changes (Fig. 10.7C). In particular, it is helpful

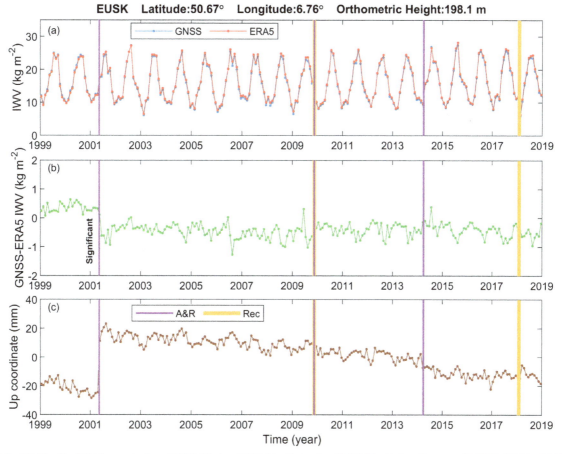

FIG. 10.7 (A) Monthly IWV time series from GNSS *(blue)* and ERA5 *(red)* at station EUSK (Euskirchen, Germany, 50.67°N, 6.76°E). (B) IWV differences (GNSS minus ERA5) time series. (C) Vertical coordinate time series of EUSK. The A&R *(purple)* and Rec *(yellow)* indicate antenna and radome changes and receiver changes, respectively.

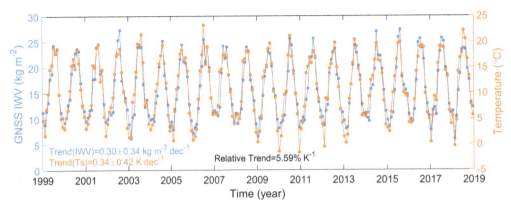

FIG. 10.8 Monthly GNSS IWV *(blue)* and temperature *(orange)* time series at station EUSK.

for the determination of undocumented changepoints. Here, we concluded that the changepoint at EUSK in May 2001 is from GNSS rather than ERA5, as it is supported by the event of antenna and radome change recorded in the GNSS log file as well as an obvious shift in the vertical coordinate time series. We then homogenize the GNSS IWV time series by adding the value of the changepoint to the datapoints before May 2001.

We estimate the linear trend of the homogenized IWV time series and evaluated its statistical significance by using a dedicated time series analysis software Hector (Bos et al., 2013). Annual and semiannual terms are also included in the function model:

$$\text{IWV}(t) = \text{IWV}_R + v \cdot (t - t_R) + \sum_{i=1}^{2} A_i \cdot \sin(\omega_i t + \varphi_i) + \varepsilon, \qquad (10.15)$$

where v is the linear trend of IWV, IWV_R is the IWV at epoch t_R, ε is a vector of residuals, A_i and φ_i ($i = 1, 2$) are the amplitudes and initial phases of annual and semiannual periodic signals, respectively.

In addition, realistic trend uncertainty can only be evaluated with a noise model which can properly represent the statistical properties of the residual IWV time series. We therefore tested the various noise models and their combinations, including WN, Power-Law (PL), WN+PL, AR(1), and ARMA(1,1). For detailed information on the noise models, readers are referred to Yuan, Hunegnaw, et al. (2021 and references therein). The suitability of these alternatives can be evaluated and compared by using Bayesian Information Criterion (BIC, Schwarz, 1978). The model with the smallest BIC value is considered to be optimal. Results show that WN is optimal for the monthly GNSS and ERA5 IWV time series of EUSK, indicating an insignificant autocorrelation in the monthly IWV time series. By contrast, Yuan, Hunegnaw, et al. (2021) concluded that ARMA(1,1) is preferred for the daily IWV time series at EUSK. The difference is most likely to be related with the interval of the time series. Moreover, Fig. 10.7 shows an excellent agreement between the IWV trend estimates from GNSS and ERA5, with values of 0.30 ± 0.34 and $0.32 \pm 0.34 \, \text{kg} \, \text{m}^{-2}$ decade^{-1} at 95% confidence level, respectively. The results are consistent with a recent study by Yuan, Van Malderen, et al. (2023), which shows that ERA5 IWV trends is in better agreement with GPS as compared to many other atmospheric reanalyzes.

Likewise, we selected an optimal noise model for the monthly temperature time series at EUSK and found ARMA(1,1) to be the most suitable. The corresponding trend estimate of temperature is $0.34 \pm 0.42 \, \text{K}$ decade^{-1} at 95% confidence level, which is also insignificant. The results indicate that longer time series are still needed to calculate significant linear trend of IWV and temperature, although 20 years of data have been used in this case study. Nevertheless, the relative trend of GNSS IWV with respect to air temperature is 5.59% K^{-1}, which is close to the theoretical value of 7% K^{-1} (Fig. 10.8).

8 Summary and outlook

This chapter introduced the methodology of ground-based GNSS meteorology and presented several examples on its multiple temporal scale variations and related applications in severe weather event, water cycle, teleconnection, and climate change. The results indicate that GNSS meteorology has a great potential in weather and climate change monitoring.

With the rapid developments of GNSS in multiple GNSS constellations, real-time, high-precision, and low-cost, GNSS meteorology is of great potential for extensive applications. For instance, it has been a unique and effective technique for extreme weather prediction. It is also a valuable data source for climate change studies.

Inter-technique comparisons contribute to a better understanding on the errors in the water vapor estimates from individual techniques. In addition to traditional radiosonde observations, other space geodetic techniques such as VLBI (Teke et al., 2011) and DORIS (Bock et al., 2014) operating in microwave domain can also be used to determine ZWD and the IWV. Their results can be used for mutual validation. In particular, VLBI allows to determine long-term IWV trends due to the long time series available and the high stability of their observing infrastructure. As almost all VLBI stations are equipped with GNSS receivers, the collocation can be used for mutual validation and changepoint detection.

Multi-technique combinations are conducive to more accurate and robust water vapor estimates and related products. As mentioned above, GNSS tropospheric products can enhance the performance of WRF. Moreover, GNSS and Interferometric Synthetic Aperture Radar (InSAR) are complementary in their spatiotemporal resolutions, so that their combinations can enhance the water vapor field.

However, there are still limitations of GNSS meteorology. For example, the spatiotemporal resolutions of the latest mapping functions VMF3 is $1° \times 1°$ and 6-hourly, which is still not sufficient for GNSS to capture local and rapid weather variations, especially during extreme weather events. Further work to enhance the mapping functions mains an imperative.

In addition, the vast amount of GNSS IWV data accumulated over the last three decades brings a lot of opportunities. The big data can be used to develop related machine learning algorithms for weather prediction. With more three decades of data, the concept of GNSS meteorology will be extended to GNSS climatology. However, the big data also bring some challenges. For example, quality control of the GNSS IWV estimates and homogenization of their long-term time series are of great importance.

Acknowledgments

We thank NGL for providing the GPS ZTD products and many institutions for sharing the continuous GPS observations. Sincere thanks are also extended to ECMWF for providing the ERA5 data. This work was funded at KIT with the German Research Foundation (DFG, project number: 321886779), and at GFZ with the German Federal Ministry for the Environment, Nature Conservation, Nuclear Safety and Consumer Protection (BMUV) EKAPEx project (project number: 67KI32002C).

References

Agnew, D. C. (1992). The time-domain behavior of power-law noises. *Geophysical Research Letters, 19*, 333–336. https://doi.org/10.1029/91GL02832.

Alshawaf, F., Balidakis, K., Dick, G., Heise, S., & Wickert, J. (2017). Estimating trends in atmospheric water vapor and temperature time series over Germany. *Atmospheric Measurement Techniques, 10*, 3117–3132. https://doi.org/10.5194/amt-10-3117-2017.

Alshawaf, F., Hinz, S., Mayer, M., & Meyer, F. J. (2015). Constructing accurate maps of atmospheric water vapor by combining interferometric synthetic aperture radar and GNSS observations. *Journal of Geophysical Research: Atmospheres, 120*, 1391–1403. https://doi.org/10.1002/2014JD022419.

Alshawaf, F., Zus, F., Balidakis, K., Deng, Z., Hoseini, M., Dick, G., et al. (2018). On the statistical significance of climatic trends estimated from GPS tropospheric time series. *Journal of Geophysical Research: Atmospheres, 123*, 10967–10990. https://doi.org/10.1029/2018JD028703.

Baelen, J. V., Aubagnac, J.-P., & Dabas, A. (2005). Comparison of near–real time estimates of integrated water vapor derived with GPS, radiosondes, and microwave radiometer. *Journal of Atmospheric and Oceanic Technology, 22*, 201–210. https://doi.org/10.1175/JTECH-1697.1.

Bevis, M., Businger, S., Herring, T. A., Rocken, C., Anthes, R. A., & Ware, R. H. (1992). GPS meteorology: Remote sensing of atmospheric water vapor using the global positioning system. *Journal of Geophysical Research: Atmospheres, 97*, 15787–15801. https://doi.org/10.1029/92JD01517.

Blewitt, G., Hammond, W. C., & Kreemer, C. (2018). Harnessing the GPS data explosion for interdisciplinary science. *EOS, 99*. https://doi.org/10.1029/2018EO104623.

Bock, O., Willis, P., Wang, J., & Mears, C. (2014). A high-quality, homogenized, global, long-term (1993–2008) DORIS precipitable water data set for climate monitoring and model verification. *Journal of Geophysical Research: Atmospheres, 119*, 7209–7230. https://doi.org/10.1002/2013JD021124.

Boehm, J., Niell, A., Tregoning, P., & Schuh, H. (2006). Global mapping function (GMF): A new empirical mapping function based on numerical weather model data. *Geophysical Research Letters, 33*. https://doi.org/10.1029/2005GL025546.

Boehm, J., & Schuh, H. (2013). *Atmospheric effects in space geodesy*. Springer. https://doi.org/10.1007/978-3-642-36932-2.

Boehm, J., Werl, B., & Schuh, H. (2006). Troposphere mapping functions for GPS and very long baseline interferometry from European Centre for Medium-Range Weather Forecasts operational analysis data. *Journal of Geophysical Research: Solid Earth, 111*. https://doi.org/10.1029/2005JB003629.

Bonafoni, S., Biondi, R., Brenot, H., & Anthes, R. (2019). Radio occultation and ground-based GNSS products for observing, understanding and predicting extreme events: A review. *Atmospheric Research, 230*, 104624. https://doi.org/10.1016/j.atmosres.2019.104624.

Bos, M., Fernandes, R., Williams, S., & Bastos, L. (2013). Fast error analysis of continuous GNSS observations with missing data. *Journal of Geodesy, 87*, 351–360.

Dai, A., Wang, J., Ware, R. H., & Hove, T. V. (2002). Diurnal variation in water vapor over North America and its implications for sampling errors in radiosonde humidity. *Journal of Geophysical Research: Atmospheres, 107*, 11–14. https://doi.org/10.1029/2001JD000642.

de Galisteo, J. P. O., Cachorro, V., Toledano, C., Torres, B., Laulainen, N., Bennouna, Y., et al. (2011). Diurnal cycle of precipitable water vapor over Spain. *Quarterly Journal of the Royal Meteorological Society, 137*, 948–958. https://doi.org/10.1002/qj.811.

Diedrich, H., Wittchen, F., Preusker, R., & Fischer, J. (2016). Representativeness of total column water vapour retrievals from instruments on polar orbiting satellites. *Atmospheric Chemistry and Physics, 16*, 8331–8339. https://doi.org/10.5194/acp-16-8331-2016.

Durre, I., Yin, X., Vose, R. S., Applequist, S., & Arnfield, J. (2018). Enhancing the data coverage in the integrated global radiosonde archive. *Journal of Atmospheric and Oceanic Technology, 35*, 1753–1770. https://doi.org/10.1175/JTECH-D-17-0223.1.

Ejigu, Y. G., Teferle, F. N., Klos, A., Bogusz, J., & Hunegnaw, A. (2021). Monitoring and prediction of hurricane tracks using GPS tropospheric products. *GPS Solution, 25*, 76. https://doi.org/10.1007/s10291-021-01104-3.

Ferreira, V., Montecino, H., Ndehedehe, C., Yuan, P., & Xu, T. (2021). The versatility of GNSS observations in hydrological studies. In *GPS and GNSS TECHNOLOGY in Geosciences* (pp. 281–298). Elsevier.

Fersch, B., Wagner, A., Kamm, B., Shehaj, E., Schenk, A., Yuan, P., et al. (2022). Tropospheric water vapor: A comprehensive high-resolution data collection for the transnational Upper Rhine Graben region. *Earth System Science Data, 14*, 5287–5307. https://doi.org/10.5194/essd-14-5287-2022.

Flores, A., Ruffini, G., & Rius, A. (2000). 4D tropospheric tomography using GPS slant wet delays. In *Annales Geophysicae* (pp. 223–234). Springer.

Gendt, G., Dick, G., Reigber, C., Tomassini, M., Liu, Y., & Ramatschi, M. (2004). Near real time GPS water vapor monitoring for numerical weather prediction in Germany. *Journal of the meteorological Society of Japan Ser II, 82*, 361–370.

Hagemann, S., Bengtsson, L., & Gendt, G. (2003). On the determination of atmospheric water vapor from GPS measurements. *Journal of Geophysical Research: Atmospheres, 108*. https://doi.org/10.1029/2002JD003235.

Held, I. M., & Soden, B. J. (2000). Water vapor feedback and global warming. *Annual Review of Energy and the Environment, 25*, 441–475. https://doi.org/10.1146/annurev.energy.25.1.441.

Hersbach, H., Bell, B., Berrisford, P., Hirahara, S., Horányi, A., Muñoz-Sabater, J., et al. (2020). The ERA5 global reanalysis. *Quarterly Journal of the Royal Meteorological Society, 146*, 1999–2049. https://doi.org/10.1002/qj.3803.

Heublein, M., Alshawaf, F., Erdnüß, B., Zhu, X. X., & Hinz, S. (2019). Compressive sensing reconstruction of 3D wet refractivity based on GNSS and InSAR observations. *Journal of Geodesy, 93*, 197–217. https://doi.org/10.1007/s00190-018-1152-0.

Iribarne, J. V., & Godson, W. L. (1981). *Atmospheric thermodynamics*. Springer Science & Business Media.

Iwasaki, H., & Miki, T. (2001). Observational study on the diurnal variation in Precipitable water associated with the thermally induced local circulation over the "Semi-Basin" around Maebashi using GPS data. *Journal of the Meteorological Society of Japan Ser II, 79*, 1077–1091. https://doi.org/10.2151/jmsj.79.1077.

Jiang, W., Yuan, P., Chen, H., Cai, J., Li, Z., Chao, N., et al. (2017). Annual variations of monsoon and drought detected by GPS: A case study in Yunnan, China. *Scientific Reports, 7*, 5874. https://doi.org/10.1038/s41598-017-06095-1.

Jones, J., Guerova, G., Douša, J., Dick, G., de Haan, S., Pottiaux, E., ... van Malderen, R. (Eds.). (2020). *Advanced GNSS tropospheric products for monitoring severe weather events and climate: COST action ES1206 final action dissemination report*. Cham: Springer International Publishing. https://doi.org/10.1007/978-3-030-13901-8.

Landskron, D., & Böhm, J. (2018). VMF3/GPT3: Refined discrete and empirical troposphere mapping functions. *Journal of Geodesy, 92*, 349–360. https://doi.org/10.1007/s00190-017-1066-2.

Lees, E., Bousquet, O., Roy, D., & de Bellevue, J. L. (2021). Analysis of diurnal to seasonal variability of integrated water vapour in the South Indian Ocean basin using ground-based GNSS and fifth-generation ECMWF reanalysis (ERA5) data. *Quarterly Journal of the Royal Meteorological Society, 147*, 229–248. https://doi.org/10.1002/qj.3915.

Li, X., Dick, G., Lu, C., Ge, M., Nilsson, T., Ning, T., et al. (2015). Multi-GNSS meteorology: real-time retrieving of atmospheric water vapor from BeiDou, Galileo, GLONASS, and GPS observations. *IEEE Transactions on Geoscience and Remote Sensing, 53*, 6385–6393. https://doi.org/10.1109/TGRS.2015.2438395.

Li, Z., Muller, J.-P., & Cross, P. (2003). Comparison of precipitable water vapor derived from radiosonde, GPS, and moderate-resolution imaging spectroradiometer measurements. *Journal of Geophysical Research: Atmospheres, 108*. https://doi.org/10.1029/2003JD003372.

Niell, A. E. (1996). Global mapping functions for the atmosphere delay at radio wavelengths. *Journal of Geophysical Research: Solid Earth, 101*, 3227–3246. https://doi.org/10.1029/95JB03048.

Niell, A. E. (2001). Preliminary evaluation of atmospheric mapping functions based on numerical weather models. Physics and Chemistry of the Earth, Part A: Solid Earth and Geodesy. In *26. Proceedings of the first COST action 716 workshop towards operational GPS meteorology and the second network workshop of the international GPS service (IGS)* (pp. 475–480). https://doi.org/10.1016/S1464-1895(01)00087-4.

Nilsson, T., & Elgered, G. (2008). Long-term trends in the atmospheric water vapor content estimated from ground-based GPS data. *Journal of Geophysical Research: Atmospheres, 113*. https://doi.org/10.1029/2008JD010110.

Ning, T., Wickert, J., Deng, Z., Heise, S., Dick, G., Vey, S., et al. (2016). Homogenized time series of the atmospheric water vapor content obtained from the GNSS reprocessed data. *Journal of Climate, 29*, 2443–2456.

Pacione, R., Araszkiewicz, A., Brockmann, E., & Dousa, J. (2017). EPN-Repro2: A reference GNSS tropospheric data set over Europe. *Atmospheric Measurement Techniques, 10*, 1689–1705. https://doi.org/10.5194/amt-10-1689-2017.

Parracho, A. C., Bock, O., & Bastin, S. (2018). Global IWV trends and variability in atmospheric reanalyses and GPS observations. *Atmospheric Chemistry and Physics, 18*, 16213–16237. https://doi.org/10.5194/acp-18-16213-2018.

Rüeger, J. M. (2002). Refractive index formulae for radio waves. In *Proceedings of the FIG XXII international congress, Washington, DC, USA*.

Saastamoinen, J. (1972). Atmospheric correction for the troposphere and stratosphere in radio ranging satellites. *The Use of Artificial Satellites for Geodesy, 15*, 247–251.

Schmidt, G. A., Ruedy, R. A., Miller, R. L., & Lacis, A. A. (2010). Attribution of the present-day total greenhouse effect. *Journal of Geophysical Research: Atmospheres*, *115*. https://doi.org/10.1029/2010JD014287.

Schwarz, G. (1978). Estimating the dimension of a model. *The Annals of Statistics*, *6*(2), 461–464. https://doi.org/10.1214/aos/1176344136.

Smith, A., Lott, N., & Vose, R. (2011). The integrated surface database: Recent developments and partnerships. *Bulletin of the American Meteorological Society*, *92*, 704–708. https://doi.org/10.1175/2011BAMS3015.1.

Steinke, S., Wahl, S., & Crewell, S. (2019). Benefit of high resolution COSMO reanalysis: The diurnal cycle of column-integrated water vapor over Germany. *Meteorologische Zeitschrift*, *165*–177. https://doi.org/10.1127/metz/2019/0936.

Suparta, W., Iskandar, A., Singh, M. S. J., Ali, M. A. M., Yatim, B., & Yatim, A. N. M. (2013). Analysis of GPS water vapor variability during the 2011 La Niña event over the western Pacific Ocean. *Annals of Geophysics*, *56*, R0330. https://doi.org/10.4401/ag-6261.

Teke, K., Böhm, J., Nilsson, T., Schuh, H., Steigenberger, P., Dach, R., et al. (2011). Multi-technique comparison of troposphere zenith delays and gradients during CONT08. *Journal of Geodesy*, *85*, 395–413. https://doi.org/10.1029/JD095iD12p20507.

Tiao, G. C., Reinsel, G. C., Xu, D., Pedrick, J. H., Zhu, X., Miller, A. J., et al. (1990). Effects of autocorrelation and temporal sampling schemes on estimates of trend and spatial correlation. *Journal of Geophysical Research: Atmospheres*, *95*, 20507–20517. https://doi.org/10.1029/JD095iD12p20507.

Tregoning, P., & Herring, T. A. (2006). Impact of a priori zenith hydrostatic delay errors on GPS estimates of station heights and zenith total delays. *Geophysical Research Letters*, *33*. https://doi.org/10.1029/2006GL027706.

Trenberth, K. E. (2012). Framing the way to relate climate extremes to climate change. *Climatic Change*, *115*, 283–290. https://doi.org/10.1007/s10584-012-0441-5.

Trenberth, K. E., Fasullo, J., & Smith, L. (2005). Trends and variability in column-integrated atmospheric water vapor. *Climate Dynamics*, *24*, 741–758. https://doi.org/10.1007/s00382-005-0017-4.

Van Malderen, R., Pottiaux, E., Klos, A., Domonkos, P., Elias, M., Ning, T., et al. (2020). Homogenizing GPS integrated water vapor time series: Benchmarking break detection methods on synthetic data sets. *Earth and Space Science*, *7*, e2020EA001121.

Vaquero-Martínez, J., & Antón, M. (2021). Review on the role of GNSS meteorology in monitoring water vapor for atmospheric physics. *Remote Sensing*, *13*, 2287. https://doi.org/10.3390/rs13122287.

Vey, S., Dietrich, R., Fritsche, M., Rülke, A., Steigenberger, P., & Rothacher, M. (2009). On the homogeneity and interpretation of precipitable water time series derived from global GPS observations. *Journal of Geophysical Research: Atmospheres*, *114*. https://doi.org/10.1029/2008JD010415.

Vey, S., Dietrich, R., Rülke, A., Fritsche, M., Steigenberger, P., & Rothacher, M. (2010). Validation of precipitable water vapor within the NCEP/DOE reanalysis using global GPS observations from one decade. *Journal of Climate*, *23*, 1675–1695. https://doi.org/10.1175/2009JCLI2787.1.

Wagner, T., Beirle, S., Dörner, S., Borger, C., & Van Malderen, R. (2021). Identification of atmospheric and oceanic teleconnection patterns in a 20-year global data set of the atmospheric water vapour column measured from satellites in the visible spectral range. *Atmospheric Chemistry and Physics*, *21*, 5315–5353. https://doi.org/10.5194/acp-21-5315-2021.

Wagner, A., Fersch, B., Yuan, P., Rummler, T., & Kunstmann, H. (2022). Assimilation of GNSS and synoptic data in a convection permitting limited area model: Improvement of simulated tropospheric water vapor content. Frontiers. *Earth Science*, *10*.

Wang, X. L. (2008a). Accounting for autocorrelation in detecting mean shifts in climate data series using the penalized maximal t or F test. *Journal of Applied Meteorology and Climatology*, *47*, 2423–2444. https://doi.org/10.1175/2008JAMC1741.1.

Wang, X. L. (2008b). Penalized maximal F test for detecting undocumented mean shift without trend change. *Journal of Atmospheric and Oceanic Technology*, *25*, 368–384. https://doi.org/10.1175/2007JTECHA982.1.

Wang, J., Dai, A., & Mears, C. (2016). Global water vapor trend from 1988 to 2011 and its diurnal asymmetry based on GPS, radiosonde, and microwave satellite measurements. *Journal of Climate*, *29*, 5205–5222. https://doi.org/10.1175/JCLI-D-15-0485.1.

Wang, M., Wang, J., Bock, Y., Liang, H., Dong, D., & Fang, P. (2019). Dynamic mapping of the movement of landfalling atmospheric rivers over Southern California with GPS data. *Geophysical Research Letters*, *46*, 3551–3559. https://doi.org/10.1029/2018GL081318.

Wang, J., & Zhang, L. (2008). Systematic errors in global radiosonde precipitable water data from comparisons with ground-based GPS measurements. *Journal of Climate*, *21*, 2218–2238. https://doi.org/10.1175/2007JCLI1944.1.

Wang, J., & Zhang, L. (2009). Climate applications of a global, 2-hourly atmospheric precipitable water dataset derived from IGS tropospheric products. *Journal of Geodesy*, *83*, 209–217. https://doi.org/10.1007/s00190-008-0238-5.

Wang, X., Zhang, K., Wu, S., Li, Z., Cheng, Y., Li, L., et al. (2018). The correlation between GNSS-derived precipitable water vapor and sea surface temperature and its responses to El Niño–Southern Oscillation. *Remote Sensing of Environment*, *216*, 1–12. https://doi.org/10.1016/j.rse.2018.06.029.

Wickert, J., Reigber, C., Beyerle, G., König, R., Marquardt, C., Schmidt, T., et al. (2001). Atmosphere sounding by GPS radio occultation: First results from CHAMP. *Geophysical Research Letters*, *28*, 3263–3266. https://doi.org/10.1029/2001GL013117.

World Meteorological Organization. (2018). *Guide to instruments and methods of observation. Measurement of meteorological variables.* (WMO-No. 8).

Xia, P., Peng, W., Yuan, P., & Ye, S. (2024). Monitoring urban heat island intensity based on GNSS tomography technique. *Journal of Geodesy*, *98*, 1–15. https://doi.org/10.1007/s00190-023-01804-3.

Yang, M.-J., Braun, S. A., & Chen, D.-S. (2011). Water budget of typhoon Nari (2001). *Monthly Weather Review*, *139*, 3809–3828. https://doi.org/10.1175/MWR-D-10-05090.1.

Yuan, P., Blewitt, G., Kreemer, C., Hammond, W. C., Argus, D., Yin, X., et al. (2023). An enhanced integrated water vapour dataset from more than 10 000 global ground-based GPS stations in 2020. *Earth System Science Data*, *15*, 723–743. https://doi.org/10.5194/essd-15-723-2023.

Yuan, P., Hunegnaw, A., Alshawaf, F., Awange, J., Klos, A., Teferle, F. N., et al. (2021). Feasibility of ERA5 integrated water vapor trends for climate change analysis in continental Europe: An evaluation with GPS (1994–2019) by considering statistical significance. *Remote Sensing of Environment*, *260*, 112416. https://doi.org/10.1016/j.rse.2021.112416Yuan.

Yuan, P., & Kutterer, H. (2021). Point-scale IWV and zenith total delay (ZTD) derived for 66 stations of the global navigation satellite system (GNSS) Upper Rhine Graben network (GURN). *PANGAEA*. https://doi.org/10.1594/PANGAEA.936134.

Yuan, P., Van Malderen, R., Yin, X., Vogelmann, H., Awange, J., Heck, B., et al. (2021). Characterizations of Europe's integrated water vapor and assessments of atmospheric reanalyses using more than two decades of ground-based GPS. *Atmospheric Chemistry and Physics Discussions*, 1–38. https://doi.org/10.5194/acp-2021-797.

Yuan, P., Van Malderen, R., Yin, X., Vogelmann, H., Jiang, W., Awange, J., et al. (2023). Characterisations of Europe's integrated water vapour and assessments of atmospheric reanalyses using more than 2 decades of ground-based GPS. *Atmospheric Chemistry and Physics*, 23, 3517–3541. https://doi.org/10.5194/acp-23-3517-2023.

Zhang, M., Yuan, P., Jiang, W., Zou, Y., Fan, W., & Wang, J. (2024). A rapid ray tracing method to evaluate the performances of ERA5 and MERRA2 in retrieving global tropospheric delay. *Measurement Science and Technology*, 35. https://doi.org/10.1088/1361-6501/ad1707.

Zhang, W., Zhang, H., Liang, H., Lou, Y., Cai, Y., Cao, Y., et al. (2019). On the suitability of ERA5 in hourly GPS precipitable water vapor retrieval over China. *Journal of Geodesy*. https://doi.org/10.1007/s00190-019-01290-6.

Ziv, Z. S., Yair, Y., Alpert, P., Uzan, L., & Reuveni, Y. (2021). The diurnal variability of precipitable water vapor derived from GPS tropospheric path delays over the Eastern Mediterranean. *Atmospheric Research*, 249, 105307. https://doi.org/10.1016/j.atmosres.2020.105307.

Zus, F., Dick, G., Dousa, J., & Wickert, J. (2015). Systematic errors of mapping functions which are based on the VMF1 concept. *GPS Solution*, 19, 277–286. https://doi.org/10.1007/s10291-014-0386-4.

Chapter 11

Monitoring of extreme weather: GNSS remote sensing of flood inundation and hurricane wind speed

Clara Chew[a] and Chris Ruf[b]

[a]Muon Space, Mountain View, CA, United States, [b]University of Michigan, Ann Arbor, MI, United States

1 GNSS remote sensing for flood inundation mapping

The use of spaceborne GNSS-reflectometry (GNSS-R) for flood inundation mapping is an emerging field that started to gain traction in the scientific community after the launch of cyclone GNSS (CYGNSS) in December 2016. The prime mission of CYGNSS is to retrieve ocean surface wind speed during hurricane intensification events; however, researchers have found that there is a surprising sensitivity of the reflection data to inland surface water. Perhaps the first evidence of this sensitivity came from observations over the Amazon River Basin in South America (similar to those shown in Fig. 11.1).

Small tributaries, some only a few dozen meters wide, appeared to cause an increase in the strength of the surface-reflected signal. In the months that followed, several researchers saw significant changes in the reflected signal before and after notable flooding events. Hurricane Harvey in August 2017 was the first major flooding event to be mapped by CYGNSS (i.e., Chew et al., 2018; Ruf, Chew, et al., 2018), with many others to follow (e.g., Wan et al., 2019; Zhang et al., 2021).

This sensitivity is noteworthy because our current capability to map flood inundation from space is limited by either the temporal repeat period of other satellite systems or by their spectral characteristics. One of the most common ways to map flooding from space is to use optical or multispectral instruments onboard satellites like MODIS or Landsat. These sensors, while providing data at a relatively high spatial resolution (30–500 m) and sometimes high temporal resolution (in the case of MODIS, daily), their data are often obscured by clouds. This introduces a major limitation in using these data during flooding events, particularly in the tropics, where clouds are often ever-present, especially during severe weather events. Vegetation cover, whether it be trees or floating vegetation, also obscures the data, limiting the ability to map flooding beneath vegetation.

Data from microwave sensors, on the other hand, easily penetrate through cloud cover and vegetation, and, in the case of synthetic aperture radar (SAR), can also provide high spatial resolution data on the order of 25 m. However, these missions are typically extremely costly; for example, the joint NASA and ISRO mission NISAR, an L-band SAR, will cost approximately ∼$1.5 billion. Because of the high cost of these missions, rarely is more than a single satellite commissioned, which leads to a poor temporal repeat that is sometimes too long for flood mapping. A single SAR satellite's temporal repeat period is typically between 6 and 12 days, during which time flooding events may have already occurred and receded.

Other microwave sensors (i.e., radiometers) can map inundation faster than SAR instruments, usually on the order of every 2–3 days. However, the spatial footprint of radiometers is quite coarse ($>25 \times 25$ km), often limiting their use for pinpointing regions affected by flooding. There is thus still a need for a remote sensing instrument that can map surface flooding in all-weather conditions, quickly, and at a fine spatial resolution.

Spaceborne GNSS-reflectometry (GNSS-R) instruments could start to fill this need. Because these instruments repurpose L-band signals, they are cheaper to build and launch, which means that constellations of satellites are no longer cost-prohibitive. Constellations of GNSS-R satellites, depending on the number of receivers or number of signals received simultaneously by each receiver, could then provide rapidly updated flood inundation maps at a moderate spatial resolution, no matter the cloud cover extent or time of day.

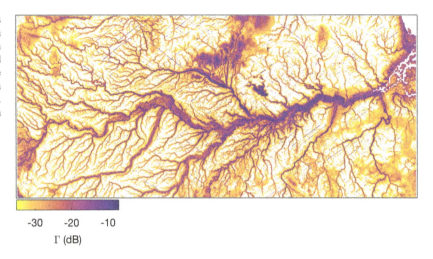

FIG. 11.1 An example of CYGNSS observations of surface reflectivity (Γ) over the Amazon Basin in South America. These observations are a long-term gridded average for the year 2020, in which the grid size is 3×3 km. Areas with no observations are most typically areas with dense vegetation with no or only very little underlying surface water. The highest values of Γ are found in areas with extensive surface water.

There are two main reasons for the sensitivity of surface-reflected GNSS signals to inland water. The first is the high dielectric constant of water at L-band—about 79 at 25°C (Kaatze, 1989). This is much higher than the typical dielectric constant for other natural surfaces like soil—which can vary between about 2–30 depending on its moisture content (Dobson et al., 1985). The dielectric constant of the reflecting surface is one of the main factors determining the amplitude of the reflected signal, with surfaces with higher dielectric constants reflecting the GNSS signal more strongly, and vice versa (Ulaby et al., 1986):

$$R_{lr}(\theta) = \frac{R_{vv}(\theta) - R_{hh}(\theta)}{2} \tag{11.1}$$

where

$$R_{vv}(\theta) = \frac{\varepsilon \cos\theta - \sqrt{\varepsilon - \sin^2\theta}}{\varepsilon \cos\theta + \sqrt{\varepsilon - \sin^2\theta}}, R_{hh}(\theta) = \frac{\cos\theta - \sqrt{\varepsilon - \sin^2\theta}}{\cos\theta + \sqrt{\varepsilon - \sin^2\theta}} \tag{11.2}$$

Here, R is the reflection coefficient for vertical (vv), horizontal (hh), or LHCP (lr) polarizations, θ is the incidence angle, and ε is the complex dielectric constant of the surface.

The second reason is that typically, though certainly not always, inland surface water tends to be less rough than dry land. Together, these two factors are significant because they both determine the surface reflectivity, which is the square of the magnitude of the rough surface reflection coefficient (Bahafza, 2005; De Roo & Ulaby, 1996). While there are several ways to describe the effect of small-scale surface roughness on the reflection coefficient, rougher surfaces decrease the reflection coefficient and vice versa, where roughness refers to the small-scale (cm level) height deviations of the surface. This is different than topographic roughness, which describes larger-scale undulations of the landscape (i.e., hills, valleys, plains, and mountains).

A smooth surface with a high dielectric constant will produce the strongest reflected signal, whereas a rough surface with a low dielectric constant will produce the weakest reflected signal. For CYGNSS, the range of observed reflectivity values exceeds 25 dB.

There is an additional reason why the calm nature of inland surface water is beneficial. The smoother the reflecting surface, the more likely it will be that the reflected signal is coherent. The rougher the reflecting surface, the more likely it will be that the reflected signal is incoherent. Incoherent reflections are diffusely scattered, and the diffuse signal comes from a large sensing footprint. For LEO satellites like CYGNSS, the footprint can be on the order of 25×25 km, which approaches that of a satellite radiometer. Coherent reflections, on the other hand, typically have a spatial footprint on the order of the first Fresnel zone (Katzberg & Garrison, 1996). For LEO GNSS-R satellites, the first Fresnel zone could be as small as 0.3 km^2, assuming an incidence angle of 0°. This, however, assumes no incoherent averaging of the signal; currently, most GNSS-R satellites average the signal over a period of at least 0.5 s. This leads to an along-track smearing of the signal of at least 3.5 km. Thus, although the smallest theoretical footprint of a spaceborne GNSS-R is 0.3 km^2, until the incoherent averaging time period is shortened, this footprint is more realistically $0.3 \times 3.5 \text{ km}^2$, though this assumes a perfectly smooth reflecting surface. The footprint for coherent reflections is thus in between that of SAR or optical measurements (dozens of m to a few hundred m) and that of radiometer observations (a few dozen km). This, combined with a

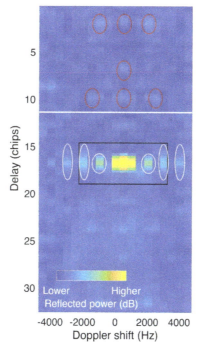

FIG. 11.2 An example of a "full" DDM recorded by CYGNSS, which is an extended version of the "cropped" DDMs that are available for download. The pixels in the cropped DDM are outlined by the *black box*. The *white line* indicates the portion of the DDM that is used for the calculation of the noise—everything above the *white line* is used in the calculation. *Red circles* highlight sidelobes of the autocorrelation function that can appear for coherent reflections. *White circles* indicate "ringing" which is also mostly present in coherent reflections.

potential faster temporal repeat, could make GNSS-R observations complementary to that which already exists for inundation mapping.

The tendency for inland surface water to produce coherent reflections as well as high surface reflectivity has been utilized by researchers for the purpose of retrieval algorithm development for flood and surface water mapping. GNSS-R data are most commonly represented by delay-Doppler maps (DDMs). A DDM is the two-dimensional (2D) cross-correlation between the received (surface reflected) signal and a locally generated replica. The correlation is performed in both delay and Doppler space, and the peak of the cross-correlation occurs at the specular reflection point. Fig. 11.2 shows an example of a DDM recorded by CYGNSS.

From DDMs, several metrics can be calculated. Below is a summary of some of the major pathways for surface water mapping that have been explored, with commentary on each's ability to map flooding specifically. There are two broad categories, with several variations therein:

1.1 Amplitude metrics

"Amplitude metrics" refer to those that quantify the magnitude of the reflected signal in the DDM. The use of these metrics for inundation mapping relies on the assumption that inland surface water reflects a GNSS signal more strongly than un-inundated surfaces.

1. Signal-to-noise ratio (SNR)

One of the first DDM metrics to be explored for the purpose of land surface remote sensing was the signal-to-noise ratio (SNR). SNR is defined as the peak value of the DDM divided by the noise floor, which is usually the mean or median value of the DDM for several rows of delay preceding the reflection. SNR must be corrected for several other factors before it may be used for mapping surface water, such as the GPS transmitted power and transmitted antenna gain, the receiver antenna gain, and the bistatic range. Some studies have also introduced corrections for incidence angle, though this correction has not been standardized across studies.

Several studies noted significant changes in SNR before and after flooding events or in response to seasonal changes in surface water (e.g., Chapman et al., 2022; Chew et al., 2018; Morris et al., 2019). However, since those first studies were published, drawbacks to using SNR for inundation mapping have been noted. One challenge is in the corrections made to SNR. Inaccurate knowledge of the GPS effective isotropic radiated power (GPS EIRP), which is the multiplicative of the transmitted power and transmit antenna gain, can lead to inaccurate interpretation of SNR. After the launch of CYGNSS, researchers discovered that GPS EIRP is not as consistent through time as originally thought (Wang et al., 2019), and short-lived variations in GPS EIRP (called "flex power events") are common, particularly with the later generations of GPS satellites (Steigenberger et al., 2019).

A second challenge of using SNR for inundation mapping is the presence of sidelobes of the autocorrelation function in the noise floor for coherent reflections (Fig. 11.2). As noted above, inland surface water very often produces coherent reflections, and the presence of the sidelobes in the noise floor results in this portion of the DDM no longer being truly noise. As the reflection gets more coherent, the sidelobes grow in tandem with the peak value of the DDM, and the end result of this is a capping of SNR around ~24 dB (Loria et al., 2020). There is thus a ceiling to SNR because of coherency, which could limit its interpretation in some cases.

Finally, the interpretation of SNR is also muddied due to the competing effects of soil moisture and vegetation on the signal. This is described in more detail below.

2. Reflectivity

Surface reflectivity (Γ or sometimes abbreviated as SR) is very similar to SNR, with the exception that it does not consider the noise floor and thus does not suffer from the same ceiling issue. Γ is defined as the peak value of the DDM, with the same corrections as those made in the case of SNR, assuming coherent reflections (De Roo & Ulaby, 1994):

$$\Gamma = \frac{4\pi(R_{ts} + R_{sr})^2}{P^t G^t} \frac{4\pi P^c}{G^r \lambda^2} \tag{11.3}$$

where P^t is the GPS transmitted power, G^t is the transmitted antenna gain, R_{ts} is the distance between the transmitter and the surface, R_{sr} is the distance between the surface and the receiver, P^c is the received power (assumed to be the peak value of the DDM), G^r is the receiver antenna gain, and λ is the GPS L1 signal wavelength.

Obviously, Γ experiences the same problem as SNR in terms of uncertainties associated with imprecise knowledge of GPS EIRP ($P^t G^t$). There is an additional complicating factor that hampers the interpretation of both Γ and SNR, and that is the influence of both soil moisture and vegetation on the reflected signal. Several studies have noted the sensitivity of these two metrics to changes in near-surface soil moisture (e.g., Chew & Small, 2018; Clarizia et al., 2019; Eroglu et al., 2019; Kim & Lakshmi, 2018), and this sensitivity does not disappear just because the metrics are being applied for the purpose of flood mapping. Thus, ignoring the influence of soil moisture on either Γ or SNR could, depending on the area, introduce significant uncertainty into ultimate flooding retrievals.

1.2 Coherency metrics

"Coherency metrics" refers to those that quantify how coherent the reflection is. The use of these metrics for inundation mapping relies on the assumption that inland surface water produces coherent reflections, whereas un-inundated land surfaces do not. Nearly all coherency metrics either exclusively use raw IF data, from which DDMs are ultimately generated, or were trained on raw IF data. Raw IF tracks are not routinely collected by CYGNSS except in special circumstances, and the majority of these collections over land have been made in highly vegetated wetland environments, like the Amazon Basin. Note that the list below is not exhaustive.

1. Entropy

Entropy is an estimate of coherency that can be generated with raw IF data, but not with the standard Level 1 data products produced by CYGNSS, and its use is thus limited to the special raw IF collections. Entropy is calculated by correlating complex waveforms with the PRN code and computing the generalized eigendecomposition of the correlation matrix of N delay waveforms. Von Neumann entropy is applied to the generalized eigenvalues in order to assess the degree of coherence of the scattered signal. Entropy (H) is defined as (Chapman et al., 2022)

$$H = -\sum_{i=1}^{N} p_i \log_N p_i \tag{11.4}$$

where p_i is the normalized eigenvalues.

Entropy is close to zero for strongly coherent reflections and approaches 1 for incoherent reflections. A threshold can then be applied to the entropy values to separate coherent, incoherent, and mixed coherency returns, with coherent returns associated with inland water. In Chapman et al. (2022), a visual examination of two raw IF tracks resulted in threshold values of >0.8 for incoherent reflections, <0.4 for coherent reflections, and 0.4–0.8 for mixed coherency reflections.

2. Power ratio

By Al-Khaldi et al. (2020), raw IF data were used to derive a coherency metric for the standard Level 1 data products from CYGNSS called the power ratio (PR). The power ratio describes the relationship between the average power surrounding the peak power of each DDM relative to the average power in DDM pixels that are not surrounding the peak. PR thus quantifies the amount of spreading in the reflected signal, under the assumption that coherent signals cause less power spread than incoherent ones. Because PR is a ratio of different groups of pixels in the DDM, the standard corrections required by the amplitude metrics are not required. After imposing a threshold of PR ≥ 2 for determining a coherent reflection, Al-Khaldi et al. (2020) concluded that approximately 90% of reflections over the land surface are incoherent, with the vast majority of coherent returns occurring over inland water.

3. Trailing edge slope

The slope of the trailing edge of the reflected signal in the DDM has also been used as an estimate of the signal coherency. By Dong and Jin (2021), an analysis of the distributions of trailing edge slopes for CYGNSS data over land and ocean showed significant differences, with observations over land producing much steeper slopes than over the ocean. The authors attributed these differences to differences in the coherency of the reflections and concluded that 89.6% of CYGNSS observations over the land surface contain a coherent component. Given the disparate conclusions by Al-Khaldi et al. (2020) and Dong and Jin (2021) concerning the occurrence of coherent returns over land, more work is needed in order to come to a community consensus.

After choosing to use either an amplitude or coherency metric, the metric then has to be transformed into some estimate of inundation. Performing this kind of retrieval is of course different than simply noting that there is a sensitivity of CYGNSS data to spatial or temporal differences in inland surface water. Below is a summary of the various approaches that have been tried as well as the advantages and challenges of each one specifically for flood mapping.

1. Thresholding

Imposing a threshold on the data to create a binary map of inundation has been the approach most frequently used to date. This approach involves setting a value such that any data above the value is flagged as water (or a coherent reflection) and any data below the value is flagged as dry land (or an incoherent reflection). Nearly all coherency metrics use a thresholding approach (e.g., Al-Khaldi et al., 2021; Chapman et al., 2022; Dong & Jin, 2021), and several studies using amplitude metrics have also imposed a threshold (e.g., Chew et al., 2018; Morris et al., 2019).

Thresholds are based on observational evidence and are most successful when the data are bimodally distributed with large training datasets. Globally, however, these metrics rarely are bimodally distributed; thresholding algorithms thus work best in regions with highly contrasting landscapes that do produce such a distribution (like the dense forests of the Amazon that are with interspersed surface water). For flood mapping, binary water masks are also most useful when the grid size is fairly small and there is a clear definition of what such a mask binary product represents. In other words, does the binary flag indicate that there is any water in the grid cell, that the grid cell is completely full of water, etc.?

2. Continuous model functions

Using a model function to retrieve fractional inundation extent, as has been employed with radiometer data, has also been used with CYGNSS amplitude metrics (Chew & Small, 2020; Loria et al., 2020). In this approach, observations are transformed into continuous values between 0 and 1, in which 0 (or 0%) indicates the spatial area of interest contains no surface water, and a value of 1 (or 100%) indicates the area of interest is fully inundated. These model functions have the advantage of not requiring empirical observations to determine a threshold. However, as noted by Loria et al. (2020) additional scattering of a coherent signal from outside the first Fresnel zone can introduce fluctuations in amplitude metrics not well captured by model functions. And, as shown by Chew and Small (2020), the sensitivity of amplitude metrics to fractional inundation decreases significantly when more than half of the spatial footprint is flooded, requiring well-calibrated observations, which are difficult given the fluctuations in GPS EIRP described above.

3. Statistical algorithms

Image processing methodologies have also been used on CYGNSS observations to delineate inland water bodies. For example, by Gerlein-Safdi and Ruf (2019), SNR data were transformed into binary surface water maps under the

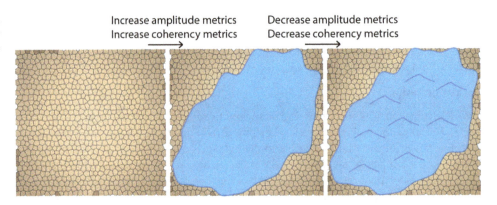

FIG. 11.3 A conceptual example of how amplitude and coherency metrics could evolve over time depending on the water roughness.

assumption that inland surface water produces much higher SNR values than dryland. By calculating the standard deviation of SNR for groups of pixels in time-averaged maps and then using a random walker to segment the image into water/no water maps, Gerlein-Safdi and Ruf (2019) mapped seasonal changes in wetland extent for both Central Africa and the Okavango Delta. Because this methodology requires the aggregation of SNR data over several days or weeks, it is currently unknown if this type of methodology could be also used for flood mapping.

1.3 Current issues

There are several issues that deserve attention concerning challenges in mapping inundation from severe weather events with GNSS-R. For one, understanding how temporally variable water roughness affects CYGNSS inundation retrievals is still an area of active research. Although both amplitude and coherency metrics tend to operate under the assumption that inland surface water is smooth, this is often not the case, particularly when the surface water extent is large. Fig. 11.3 shows a conceptual example of how amplitude and coherency metrics might evolve with time during a flood. Between the first and second panels, the surface floods, increasing both amplitude and coherency metrics. If, however, wind causes surface ripples or even waves, as is depicted in the third panel, then amplitude and coherency metrics will decrease. Depending on the threshold chosen or the model function used, the introduction of surface waves could confuse the retrieval, resulting in either a false-negative or an underestimation of fractional water extent. Although water surface roughness is also a problem for SAR and is by no means unique to GNSS-R, more work is needed to either account for these variations or quantify the uncertainty introduced by them.

The second challenge is the spatial sampling of GNSS-R observations. Unlike traditional remote sensing satellites, which collect data in swaths, GNSS-R data are positioned pseudorandomly over the surface, with tracks of observations eventually achieving full coverage. However, at least with the current constellation of eight CYGNSS receivers, full coverage over the land surface takes several days, if coherent reflections with small spatial footprints are assumed. It is possible that the rise of commercial satellite companies like Spire Global, which aim to launch many CubeSat GNSS-R satellites, could provide full coverage within the typical span of a flooding event. However, this is still a hypothetical proposition.

A third challenge is common to all signals of opportunity missions. As mentioned above, inaccurate knowledge of GPS EIRP remains an issue in the proper calibration of the data. Free signal sources come with the drawback that "we" do not control the source of the signal and are at the whim of those who do. With GNSS-R in particular, the presence of radio frequency interference (RFI) from GPS jammers and spoofers will continue to be a source of noise, and will likely increase in the future. Developing ways to identify and mitigate these issues should be a focus in the GNSS-R community in the coming years.

2 GNSS remote sensing for hurricane wind speed retrieval

GNSS-R observations of the ocean are affected by its surface roughness spectrum, which determines the ocean's bistatic scattering cross section and, hence, the strength of the scattered GNSS signal that is measured (Gleason et al., 2016). The general relationship is given by

$$\sigma_o = \frac{|R|^2}{2\sqrt{\iint_{\kappa<\kappa_*} \kappa_u^2 \Psi(\vec{\kappa}) d^2\kappa \iint_{\kappa<\kappa_*} \kappa_c^2 \Psi(\vec{\kappa}) d^2\kappa}} \qquad (11.5)$$

where σ_o is the scattering cross section, R is the Fresnel reflection coefficient, Ψ is the roughness spectrum of the ocean surface, and κ_u and κ_c are the upwind and crosswind components of the wave number, respectively (Zavorotny & Voronovich, 2000). Integration is carried out with respect to wave number up to a cutoff κ_*. The roughness spectrum below κ_* includes smaller capillary waves which are primarily sensitive to local, instantaneous wind speed, but also includes longer waves which may be affected by other factors. The longer waves are highly correlated with the local winds provided the wind has persisted with the same speed and direction for a sufficiently long time (of order hours). This condition corresponds to a fully developed sea state (Chen-Zhang et al., 2016). In young sea conditions, the longer waves are less well coupled to the local winds but will still affect the scattering cross section measured by a GNSS-R sensor. For this reason, additional information about the sea state is used by GNSS-R retrieval algorithms to accurately determine wind speed in young seas (Pascual et al., 2021).

Fully developed sea conditions are most common in the open ocean away from major storms. One example of an ocean surface wind speed product derived from spaceborne GNSS-R observations is the CYGNSS Science Data Record Fully Developed Seas 10 m-referenced neutral stability-equivalent wind speed (FDS) (Clarizia & Ruf, 2016). Its performance is illustrated in Fig. 11.4, which shows the root-mean-square difference (RMSD) between it and ECMWF reanalysis (ERA5) winds based on a 1-year population of coincident matchups. The RMSD is less than 2 m/s for wind speeds below ~10 m/s and increases at higher winds due to a gradual decrease in the sensitivity of scattering cross section to wind speed (Ruf, Gleason, & McKague, 2018). The figure also shows that the incidence angle of the observation does not have a significant impact on RMSD performance.

The design of GNSS-R sensors benefits from several aspects of the remote sensing technique. Because the transmitted signal is generated by other navigation satellites, the sensor requires much less power than a conventional radar. Because GNSS receivers are widely used for commercial applications, the modified versions used for GNSS-R purposes can be built at a very low cost. And because scattering cross sections in the forward (quasi-specular) direction are typically several orders of magnitude greater than backscatter cross sections, a GNSS-R antenna can be much smaller than a conventional radar antenna. For all of these reasons, it is possible to deploy the sensors on small, low-cost satellites and to deploy a constellation of those satellites for less than the typical cost of a single medium-sized one. An example of this is the CYGNSS constellation of eight GNSS-R satellites (Ruf et al., 2016). Such a large distributed constellation can provide frequent sampling of rapidly changing phenomena such as tropical cyclones. This is illustrated in Fig. 11.5, which shows when and how well CYGNSS sampled every major storm (Category 3 or greater) between 2018 and 2020. In the figure, the red lines are the track of each storm throughout its life cycle, the dots are where there was a CYGNSS overpass, and the

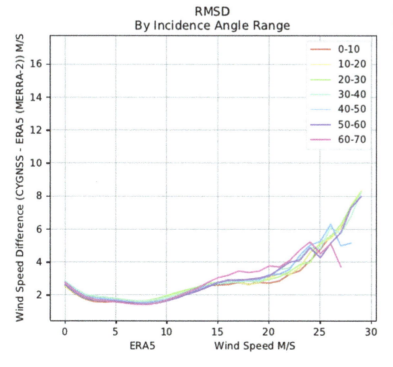

FIG. 11.4 Root-mean-square difference between CYGNSS and coincident ERA5 wind speed for all of the calendar year 2019. Samples are partitioned by the incidence angle of the CYGNSS observation (deg).

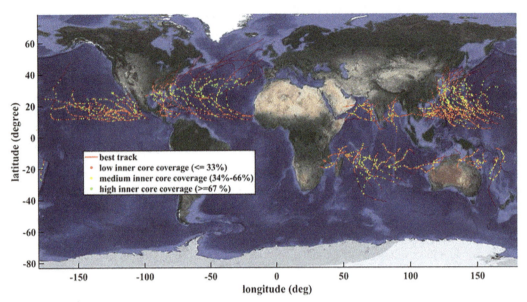

FIG. 11.5 CYGNSS overpasses of all major storms during 2018–2020. *Red lines* are the track of each storm throughout its life cycle, dots are placed where there was a CYGNSS overpass, and the *color* of the dot indicates the density of samples within the inner core of the storm.

color of the dot indicates the density of samples within the inner core of the storm during that overpass. Every storm has been sampled between one and three times a day, which helps monitor and improve the prediction of its development.

CYGNSS wind products have been assimilated into numerical weather prediction models to demonstrate their utility for improving hurricane forecasting. Studies using the NOAA National Centers for Environmental Prediction Hurricane Weather Research and Forecasting (HWRF) regional model with the Gridpoint Statistical Interpolation (GSI) hybrid ensemble three-dimensional variational data assimilation system (3DEnVar) have shown that assimilating CYGNSS data resulted in improved hurricane inner-core structure and surface fluxes. As a result, the track and intensity forecasts for the 2017 Hurricanes Harvey and Irma showed significant improvements (Cui et al., 2019). In a separate independent observing system experiment, CYGNSS was exclusively assimilated into the global model, leading to impacts on HWRF forecasts due to initial conditions (ICs) and lateral boundary conditions (LBCs) for the case of the 2018 Hurricane Michael (Mueller et al., 2021). Fig. 11.6 summarizes the results of this experiment, illustrating the forecast errors for the control case without CYGNSS (HWRF_CTL) and the study case with CYGNSS (HWRF_SPD). While the track forecast remains largely unaffected, a significant reduction in forecast error is observed for both intensity and minimum pressure. Notably, this reduction occurred during the rapid intensification phase of Hurricane Michael's life cycle, suggesting that GNSS-R observations provided crucial information about its initiation.

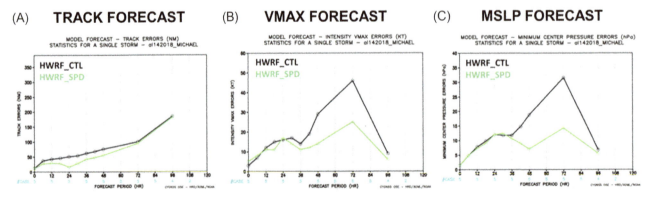

FIG. 11.6 Average HWRF error statistics for (A) track, (B) MSLP, and (C) Vmax for HWRF_CTL *(black)* and HWRF_SPD *(green)*. Sample sizes are indicated in blue as $N-1$, where N represents the number of forecasts.

References

Al-Khaldi, M. M., Johnson, J. T., Gleason, S., Loria, E., O'Brien, A. J., & Yi, Y. (2020). An algorithm for detecting coherence in cyclone global navigation satellite system Mission Level-1 delay-Doppler maps. *IEEE Transactions on Geoscience and Remote Sensing*, 1–10. https://doi.org/10.1109/tgrs.2020.3009784.

Al-Khaldi, M. M., Shah, R., Chew, C., Johnson, J. T., & Gleason, S. (2021). Mapping the dynamics of the south Asian monsoon using CYGNSS's level-1 signal coherency. *IEEE Journal of Selected Topics in Applied Earth Observations and Remote Sensing, 14*, 1111–1119.

Bahafza, B. R. (2005). *Radar systems analysis and design using MATLAB* (2nd ed.). Boca Raton: Chapman and Hall/CRC.

Chapman, B. D., Russo, I. M., Galdi, C., Morris, M., Di Bisceglie, M., Zuffada, C., et al. (2022). Comparison of SAR and CYGNSS surface water extent metrics. *IEEE Journal of Selected Topics in Applied Earth Observations and Remote Sensing, 15*, 3235–3245. https://doi.org/10.1109/JSTARS.2022.3162764.

Chen-Zhang, D. D., Ruf, C. S., Ardhuin, F., & Park, J. (2016). GNSS-R nonlocal sea state dependencies: Model and empirical verification. *Journal of Geophysical Research, Oceans*. https://doi.org/10.1002/2016JC012308.

Chew, C., Reager, J. T., & Small, E. (2018). CYGNSS data map flood inundation during the 2017 Atlantic hurricane season. *Scientific Reports, 8*. https://doi.org/10.1038/s41598-018-27673-x.

Chew, C., & Small, E. (2018). Soil moisture sensing using Spaceborne GNSS reflections: Comparison of CYGNSS reflectivity to SMAP soil moisture. *Geophysical Research Letters, 45*, 4049–4057. https://doi.org/10.1029/2018GL077905.

Chew, C., & Small, E. (2020). Estimating inundation extent using CYGNSS data: A conceptual modeling study. *Remote Sensing of Environment, 246*, 111869.

Clarizia, M. P., Pierdicca, N., Costantini, F., & Floury, N. (2019). Analysis of CYGNSS data for soil moisture retrieval. *IEEE Journal of Selected Topics in Applied Earth Observations and Remote Sensing, 12*, 2227–2235. https://doi.org/10.1109/JSTARS.2019.2895510.

Clarizia, M. P., & Ruf, C. S. (2016). Wind speed retrieval algorithm for the cyclone global navigation satellite system (CYGNSS) mission. *IEEE Transactions on Geoscience and Remote Sensing, 54*, 4419–4432. https://doi.org/10.1109/TGRS.2016.2541343.

Cui, Z., Pu, Z., Tallapragada, V., Atlas, R., & Ruf, C. S. (2019). A preliminary impact study of CYGNSS ocean surface wind speeds on numerical simulations of hurricanes. *Geophysical Research Letters, 46*, 2984–2992. https://doi.org/10.1029/2019GL082236.

De Roo, R. D., & Ulaby, F. T. (1994). Bistatic specular scattering from rough dielectric surfaces. *IEEE Transactions on Antennas and Propagation, 42*, 220–231. https://doi.org/10.1109/8.277216.

De Roo, R. D., & Ulaby, F. T. (1996). A modified physical optics model of the rough surface reflection coefficient. In *IEEE antennas and propagation society international symposium. Baltimore, MD*.

Dobson, M. C., Ulaby, F. T., Hallikainen, M. T., & El-Rayes, M. A. (1985). Microwave dielectric behavior of wet soil-part II: Dielectric mixing models. *IEEE Transactions on Geoscience and Remote Sensing, GE-23*, 35–46. https://doi.org/10.1109/TGRS.1985.289498.

Dong, Z., & Jin, S. (2021). Evaluation of the land GNSS-reflected DDM coherence on soil moisture estimation from CYGNSS data. *Remote Sensing, 13*, 570.

Eroglu, O., Kurum, M., Boyd, D., & Gurbuz, A. C. (2019). High spatio-temporal resolution cygnss soil moisture estimates using artificial neural networks. *Remote Sensing, 11*. https://doi.org/10.3390/rs11192272.

Gerlein-Safdi, C., & Ruf, C. (2019). A CYGNSS-based algorithm for the detection of inland waterbodies. *Geophysical Research Letters, 46*, 12065–12072.

Gleason, S., Ruf, C., Clarizia, M. P., & O'Brien, A. (2016). Calibration and unwrapping of the normalized scattering cross section for the cyclone global navigation satellite system (CYGNSS). *IEEE Transactions on Geoscience and Remote Sensing*. https://doi.org/10.1109/TGRS.2015.2502245.

Kaatze, U. (1989). Complex permittivity of water as a function of frequency and temperature. *Journal of Chemical & Engineering Data, 34*, 371–374.

Katzberg, S. J., & Garrison, J. L. (1996). *Utilizing GPS to determine ionospheric delay over the ocean*. NASA Technical Memorandum TM-4750 (pp. 1–16). 10.1.1.31.3748.

Kim, H., & Lakshmi, V. (2018). Use of cyclone global navigation satellite system (CyGNSS) observations for estimation of soil moisture. *Geophysical Research Letters, 45*, 8272–8282.

Loria, E., O'Brien, A., Zavorotny, V. U., Downs, B., & Zuffada, C. (2020). Analysis of scattering characteristics from inland bodies of water observed by CYGNSS. *Remote Sensing of Environment, 245*, 111825.

Morris, M., Chew, C., Reager, J. T., Shah, R., & Zuffada, C. (2019). A novel approach to monitoring wetland dynamics using CYGNSS: Everglades case study. *Remote Sensing of Environment, 233*, 111417.

Mueller, M., Annane, B., Leidner, M., & Cucurull, L. (2021). Impact of CYGNSS-derived winds on tropical cyclone forecasts in a global and regional model. *Monthly Weather Review*. https://doi.org/10.1175/MWR-D-21-0094.1.

Pascual, D., Clarizia, M. P., & Ruf, C. S. (2021). Improved CYGNSS wind speed retrieval using significant wave height correction. *Remote Sensing*. https://doi.org/10.3390/rs13214313.

Ruf, C. S., Atlas, R., Chang, P. S., Clarizia, M. P., Garrison, J. L., Gleason, S., et al. (2016). New Ocean winds satellite Mission to probe hurricanes and tropical convection. *Bulletin of the American Meteorological Society*. https://doi.org/10.1175/BAMS-D-14-00218.1.

Ruf, C., Chew, C. C., Lang, T., Morris, M., Nave, K., Ridley, A., et al. (2018). A new paradigm in earth environmental monitoring with the CYGNSS small satellite constellation. *Scientific Reports, 8*.

Ruf, C., Gleason, S., & McKague, D. S. (2018). Assessment of CYGNSS wind speed retrieval uncertainty. *IEEE Journal of Selected Topics in Applied Earth Observations and Remote Sensing*. https://doi.org/10.1109/JSTARS.2018.2825948.

Steigenberger, P., Tholert, S., & Montenbruck, O. (2019). Flex power on GPS Block IIR-M and IIF. *GPS Solutions, 23*. https://doi.org/10.1007/s10291-018-0797-8.

Ulaby, F. T., Moore, R., & Fung, A. K. (1986). *Microwave remote sensing, volume 1: Fundamentals and radiometry*. Artech House.

Wan, W., Liu, B., Zeng, Z., Chen, X., Wu, G., Xu, L., et al. (2019). Using CYGNSS data to monitor China's flood inundation during typhoon and extreme precipitation events in 2017. *Remote Sensing, 11*, 854.

Wang, T., Ruf, C., Gleason, S., Block, B., McKague, D., & O'Brien, A. (2019). A real-time EIRP level 1 calibration algorithm for the CYGNSS Mission using the zenith measurements. In *IGARSS 2019–2019 IEEE international geoscience and remote sensing symposium. Yokohama, Japan* (pp. 8725–8728).

Zavorotny, V. U., & Voronovich, A. G. (2000). Scattering of GPS signals from the ocean with wind remote sensing application. *IEEE Transactions on Geoscience and Remote Sensing*. https://doi.org/10.1109/36.841977.

Zhang, S., Ma, Z., Li, Z., Zhang, P., Liu, Q., Nan, Y., et al. (2021). Using CYGNSS data to map flood inundation during the 2021 extreme precipitation in Henan Province, China. *Remote Sensing, 13*, 5181.

Chapter 12

GNSS and the cryosphere

Tonie van Dam[a], Pippa Whitehouse[b], and Lin Liu[c]

[a]Department of Geology and Geophysics, University of Utah, Salt Lake City, UT, United States, [b]Department of Geography, Durham University, Durham, United Kingdom, [c]Earth and Environmental Sciences Programme, Faculty of Science, The Chinese University of Hong Kong, Hong Kong, China

1 Introduction

Sea-level change is driven by a variety of mechanisms operating at different spatial and temporal scales. One of the largest contributions to global mean sea-level rise is the increased mass of water in the oceans due to melting ice from mountain glaciers and the Antarctic and Greenland ice sheets.

Satellite and airborne altimetry and InSAR observations have been invaluable in documenting contemporary changes in the extent and elevation of ice/snow on land (for a recent and thorough review of these observations in the Arctic and Antarctic, see IPCC, 2021). However, these techniques observe volume changes. Converting volume change to a reliable mass change that can be directly linked to sea-level rise, depends on the length of the time series, short-term fluctuations in snow accumulation, compaction rates, and the density of the ice through the ice column (Khan et al., 2016; Morris & Wingham, 2015). Quantifying these parameters over the satellite footprint is often not possible due to the lack of dense in situ observations. Instead, simple physical relationships are assumed, which may introduce errors in the estimates of sea-level change.

In contrast to the volume changes documented by altimetry and InSAR, satellite gravity observations from GRACE and GRACE-FO (G/GFO) provide direct monthly mass change observations over both Greenland and Antarctica, as well as over other significant (>350 km) snow- and ice-covered regions (see Velicogna et al., 2020 and references therein), without the need to apply the intermediate assumptions required by volume-measuring techniques. However, the G/GFO gravity fields represent the total gravity variability from all mass change sources, including ice sheet mass changes and mass variability due to glacial isostatic adjustment (GIA, the viscoelastic response of the solid Earth to glacial unloading over the last several thousand years). This latter contribution to G/GFO mass change is typically modeled and removed from the observations to study present-day mass changes. Estimates of the mass effect of GIA in Greenland using different Earth and ice history models range from 21 to −27 Gton/yr (IMBIE Team, 2020, Extended Data Table 1) compared to surface mass balance values[a] of 125–500 Gton/yr (IMBIE Team, 2020, Extended Data Table 2). The GIA mass effect in Antarctica ranges between 3 and 81 Gton/yr (IMBIE Team, 2018, Extended Data Table 2) compared to surface mass balance values of 1807–2150 Gton/yr (IMBIE Team, 2018, Extended Data Table 1).

Since the mid-1990s, significant progress has been made in the development and use of GNSS to observe, among other things, changes in the cryosphere. In 1995, the first permanent GNSS installations were constructed in the Arctic (Thule Airbase) and in Antarctica (McMurdo Research Station). Since then, GNSS permanent station installations have proliferated in or near many cryospheric regions. Today, POLENET (The Polar Earth Observing Network) maintains over 200 sites, both permanent and temporary, in networks across Greenland (GNET) and Antarctica (ANET). Continental glaciers in Alaska, the Himalayas, New Zealand, Europe, and the Andes are today surrounded by permanent GNSS stations. Advances in geodetic instrumentation, the proliferation of new systems and signals (Glonass, Galileo, and Beidou), and analysis approaches have dramatically improved the quality of the GNSS coordinate time series.

Despite the improvement in the precision of GNSS time series, the recorded surface displacements from sites proximal to the cryosphere are still very difficult to interpret. GNSS observations from the Arctic and Antarctic contain elastic signals due to present-day changes in snow mass and viscoelastic signals due to the past melting of the Pleistocene ice masses, i.e., GIA. Other land-based concentrations of snow and ice are typically found at high elevations, where the elevations

---

a. **Surface mass balance** is the difference between precipitation (rain and snow) that has accumulated on the upper surfaces of glaciers and ice sheets and what has been lost due to melt and eventual runoff and evaporation.

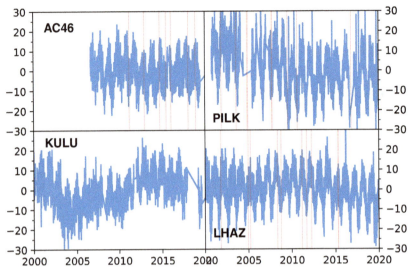

FIG. 12.1 Uplift or vertical coordinate time series from stations in or near cryospheric environments. AC46: Skwentna, Alaska; KULU: Southeast Greenland, near Helheim glacier; PILK: at the foot of the Horace-Walker Glacier in the Sierra Range, West Coast, New Zealand; and LHAZ: Lhasa, Tibet, China. Daily observations are shown in blue. Long-term transients determined using singular spectrum analysis are shown in *red*. The *dotted red vertical lines* indicate the occurrence of nearby earthquakes.

themselves have been created by ongoing tectonic processes such as earthquakes or magmatic processes. These locations are also susceptible to climate-driven displacements at seasonal and transient temporal scales including ice mass change driven by the interaction of glaciers with seasonally varying ocean temperatures.

For example, Fig. 12.1 shows the residual uplift time series from four permanent GNSS stations. The residuals are determined with respect to a linear trend, i.e., the MIDAS-determined velocities in this case (Blewitt et al., 2018). Original data were obtained from the University of Nevada, Reno, Nevada Geodetic Laboratory website. The sites were chosen to illustrate some of the issues associated with time series from cryospheric regions. (Please note that removing the MIDAS trend not only removes the linear tectonic signals but also removes long-term linear uplift/subsidence signals due to surface mass loading and GIA.)

All the time series show a seasonal signal superimposed on nonlinear transient displacements (solid red line). The seasonal signals are likely driven by seasonal variations in cryospheric mass loads but may also include a small contribution from changes in other surface mass loads (atmospheric and nontidal ocean). We also observe data gaps and discontinuities in the time series. Data gaps are common in many GNSS time series from cryospheric regions due to our inability to repair equipment in the offseason. Noting that many ice-covered regions are also found in tectonically active regions (Iceland, Alaska, New Zealand, Himalayas, Cascades, etc.), offsets in the time series due to earthquakes (dotted red lines) or displacements due to magma migration will also be part of the observed signal.

The transient displacements at the stations are shown by the solid red lines in Fig. 12.1. The transients were calculated using singular spectrum analysis (SSA) (Ghil et al., 2002). SSA allows us to decompose the signal into a set of summable components that are grouped together in terms of transient, periodic, and noise signals. The daily GNSS uplift time series were processed using a window width of 700 days. The window of 700 days is chosen to be longer than the number of data points in the annual period under investigation. We summed those components of the SSA decomposition that did not contain any periodic signals (usually the first or the first and second components). At KULU, the transient has been shown to be associated with changes in glacier discharge at the nearby Helheim and Mittegard glaciers (to be discussed in more detail in the section on glacier dynamics). In fact, the transients at all stations are likely related to long-term changes in ocean-atmospheric teleconnections, e.g., the North Atlantic Oscillation (Bevis et al., 2019), that represent long-wavelength mass variability over the planet at long temporal scales (greater than seasonal).

In this chapter, we discuss how GNSS has been used to improve our understanding of cryospheric and geodynamic processes in ice- and/or snow-covered regions. Elastic surface displacements will be discussed in Section 2. The contributions of GNSS to our understanding of Glacial Isostatic Adjustment will be discussed in detail in Section 3. Finally, reflections of the GNSS signal off the snow, ice, or permafrost surface can be used to investigate changes in these quantities. GNSS reflectometry will be discussed in Section 4.

2 Elastic surface displacements

2.1 Theory

The theory to forward model instantaneous surface displacements due to changes in surface mass was presented by Farrell (1972) and Longman (1962, 1963, 1966). In those papers, Green's functions were developed to describe the response of an elastic Earth model to a point load on its surface. Gravitational accelerations, displacements, tilts, and strains can be determined for any surface load by evaluating a convolution sum between Green's functions and a modeled or observed surface mass distribution.

We can determine the uplift at a point due to changes in the spatial redistribution surface mass. If the surface mass is defined as a gridded distribution, over the surface of the Earth, the displacement is calculated as

$$dU = \sum_{i=1}^{n} \sum_{j=1}^{m} dM_{ij} G_{ij}^{u}(\theta) A_{ij} \qquad (12.1)$$

In this equation, the indices i and j represent, respectively, the longitude and latitude of the mass grid cell where n and m represent, respectively, the highest index in the grid; dM_{ij} is the mass at the center of the ith and jth grid cell; $G_{ij}^{u}(\theta)$ is the Green's function that relates the vertical response of the Earth at a point to a load at an angular distance of θ from the load; and A_{ij} is the spherical area of the grid cell. In the case of the cryosphere, $dM_{ij}=0$ for grid cells without ice. Similar equations can be written for the horizontal displacements.

Fig. 12.2 shows the elastic displacement of the Earth's surface caused by removing a uniform disk load of radius 20 km and 1 m equivalent water thickness. Vertical and horizontal displacements are given as a function of the distance to the center of the disk. If you remove a load from the surface of the Earth, the Earth's surface moves upward and away from the center of the load. Uplift is maximum at the center of the disk and decreases rapidly with increasing distance from the center. The horizontal displacements are zero at the center of the load, increase to a maximum at the edge of the load, and then decrease with distance from the edge of the load, however, not as quickly as the vertical displacements.

An example of how Eq. (12.1) can be used to forward model loading in the cryosphere is shown in Fig. 12.3 (Hansen et al., 2021). In the appendix, the reader can find a map of Greenland with the relative locations of these GNSS stations. Fig. 12.3 (black curve) shows the uplift coordinates at three permanent GNSS stations located near the three largest outlet glaciers in Greenland. The uplift in all cases is due to the glacier mass loss to the oceans and the thinning of the ice in the vicinity of the glaciers. In red, is the modeled uplift due to the thinning ice not associated with the glacier dynamics. This effect was modeled using Eq. (12.1), where dM_{ij} the RACMO2.3 surface mass balance model (SMB) (Noël et al., 2018, 2019) convolved with the Green's functions, $G_{ij}^{u}(\theta)$, derived for an elastic Earth model (iasp91 by Wang et al., 2012). The blue lines in the figure show the observed uplift less the effect of regional ice thinning, i.e., corrected for elastic uplift due to SMB. The blue lines represent the uplift at these three GNSS stations due to glacier mass loss on the associated glaciers. The right panels in the figure show the observed elastic uplift due to ice dynamics where a linear + seasonal term has also been removed (Please note that removing a linear trend will also remove the effects of GIA.). The panels on the right-hand side show how the glacier dynamics vary over longer timescales. This example further demonstrates the complexities of interpreting GNSS displacements in terms of ice mass change only.

Koulali et al. (2022) used this approach to assess the elastic loading response to seasonal snowfall across the Southern Antarctic Peninsula. The authors used the load from the RACMO2.3p2 model (van Wessem et al., 2018) as well as surface

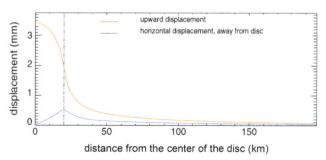

FIG. 12.2 The instantaneous vertical (positive upward) and horizontal (positive away from the disc center) crustal displacements caused by removing a uniform disk load of radius 20 km and 1 m equivalent water thickness. Results are given as a function of the distance to the center of the disc. The vertical dot-dashed line marks the edge of the disk. Results are computed for the Earth model PREM (Dziewonski & Anderson, 1981).

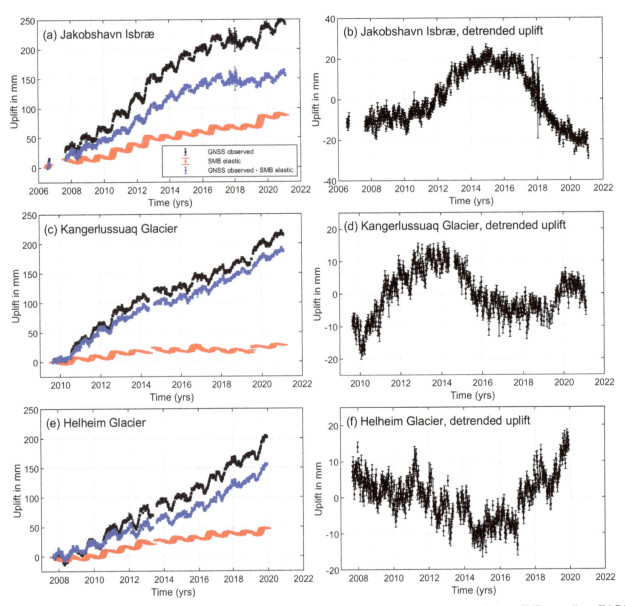

FIG. 12.3 (A) Time series of the weekly uplift solution at KAGA *(black curve)*. Time series of modeled elastic uplift due to SMB anomalies at KAGA (red curve). Observed uplift minus elastic uplift due to SMB *(blue curve)*. (C) Same as (A) but for KUAQ. (E) Same as (A) but for HEL2. (B) Detrended observed uplift minus elastic uplift due to SMB at KAGA. (D) Detrended observed uplift minus elastic uplift due to SMB at KUAQ. (F) Detrended observed uplift minus elastic uplift due to SMB at HEL2. *(Reproduced with permission from Hansen, K., Truffer, M., Aschwanden, A., Mankoff, K., Bevis, M., Humbert, A., et al. (2021). Estimating ice discharge at Greenland's three largest outlet glaciers using local bedrock uplift. Geophysical Research Letters, 48, e2021GL094252, http://doi.org/10.1029/2021GL094252.)*

elevation changes from satellite altimetry assuming an ice density of $850 \, kg/m^3$. Green's functions from the Preliminary Reference Earth Model (PREM; Dziewonski & Anderson, 1981) were used. Koulali et al. (2022) found that the elastic deformation predictions closely reproduced the main features of the vertical displacement variations observed at ten GNSS stations on the Antarctic Peninsula.

In the examples above, it is important to note that the elastic surface displacements that are due to cryospheric mass change are derived from both the surface mass balance of the region as well as dynamic glacier loss. SMB models are available for both Greenland and Antarctica. For a recent comparison of SMB models from Greenland, see Fettweis et al. (2020) and IMBIE Team (2020). For a comparison of models from Antarctica, see Mottram et al. (2021) and IMBIE Team (2018). Glacier volume loss can be obtained from repeat DEMs, satellite altimetry observations, and radar. Before interpreting the GNSS data they need to be corrected for signals that are not related to present-day mass changes. In addition to removing signals associated with surface mass loading effects such as the atmosphere and the oceans, a model for GIA must also be considered.

In comparison to forward modeling the surface mass loads, some authors use the GNSS observed displacements to invert for the surface load, usually at annual periods (Compton et al., 2017; Fu et al., 2015). Taking GNSS 3-D position estimates, the cryospheric load, I, can be determined using

$$I = (G^T W G)^{-1} G^T W d \tag{12.2}$$

where d represents the continuous GNSS position estimates, G are the Green's functions, and W is the GNSS variance matrix. Inversion of surface displacements to determine mass change is most effective when the mass change is localized, and the displacement has an amplitude much larger than the contribution from all other regional loads.

2.2 Half-space loading models

For completeness, we mention the development of elastic half-space models to better understand the effects of short-wavelength cryospheric loads (Pinel et al., 2007). The half-space model has been particularly useful to discriminate between deformation due to ice unloading and magmatic processes at subglacial volcanoes in Iceland. Compton et al. (2017) used the half-space model and GNSS observations to estimate seasonal ice mass variations on Vatnajökull and Mýrdalsjökull/Eyjafjallajökull volcanoes.

2.3 Glacier dynamics

In addition to cryospheric mass loss from surface melting and runoff, dynamically induced ice loss from Greenland's marine-terminating glaciers by calving and submarine melting has accelerated in the last couple of decades. Monitoring the volume changes driven by dynamic glacial discharge requires observations of ice thickness and extent with a high spatial resolution. Converting ice volume change to mass change is not straightforward due to uncertainties in the density of the ice (Kuipers Munneke et al., 2015 discusses the problem of modeling firn compaction and its effects on ice density. Firn is granular snow, on the upper part of a glacier, where it has not yet been compressed into ice).

Khan et al. (2010) used NASA's Airborne Topographic Mapper (ATM) data to model the unloading and rebound caused by the discharge and thinning of Jakobshavn Isbræ, Greenland's largest outlet glacier. This paper is an important contribution to the literature as the authors assess the errors in the GNSS observations due to orbit errors, tropospheric model uncertainties, reference frame uncertainties, and regional mass change effects. They also use different ice histories and viscosity models to quantify the uncertainty of the GIA signal. They determined a common error of 4.2 mm/yr due to uncertainties in the GNSS data processing and GIA. After accounting for these uncertainties in the GNSS data by assessing the variance in the various models and adding this uncertainty to the GNSS uplift observations, the GNSS uplift trend and the predictions from the forward modeling of discharge and thinning were found to agree within the error bars.

Wahr et al. (2013) used the horizontal signals in GNSS displacements to model the mass loss of the Helheim and Midgaard glaciers on the southeast coast of Greenland. Horizontal deformation measurements are 5–10 times more precise than vertical measurements. Surface displacement can be thought of as the sum of the contributions of lots of loads. The vertical displacements from each load add as scalars. The horizontal displacements add as vectors. This allows us to determine the azimuth of the load from the GNSS stations. By comparing the horizontal and vertical displacements recorded at KULU, Wahr et al. (2013) were able to demonstrate that while Helheim was the major contributor to mass loss in the region, there was a significant contribution from Midgaard Glacier located to the northeast of Helheim.

Glacial mass loss is not a steady process. Variations in surface mass balance, ice discharge, and calving front positions all influence glacier mass balance. Zhang et al. (2018) took the work of Khan et al. (2010) further by including horizontal displacements in an analysis of the seasonal and transient components of mass loss on Jakobshavn Isbræ and the surrounding region. To forward model the glacier mass balance the authors used a flux gate approach, that is, they used surface velocity maps to estimate the rate of ice mass passing through a predefined vertical area or gate near the tongue of the glacier. Using multichannel, singular-spectrum analysis, the authors were able to isolate seasonal and transient loading signals in the GNSS up and east components (the north component was not sufficiently robust to separate the transient and seasonal signals). The authors concluded that the transient signals were mainly caused by interannual variations in ice discharge while the seasonal signals were largely influenced by the seasonal retreat and advance of the calving front.

Hansen et al. (2021) used GNSS to investigate discharge at Jakobshavn Isbræ (JI), Kangerlussuaq (KG) Glacier, and Helheim Glacier for the period 2006–2020 (see Fig. 12.3). The authors estimated weekly discharge at the three glaciers using the flux gate approach discussed above. Discharge was then converted to dynamic ice loss by removing a 30-year SMB time series. Glacier thinning was estimated using Airborne Topographic Mapper flights. GNSS uplift time series, corrected for the SMB signal and a long-term velocity (GIA), were then compared to the dynamic ice loss estimates to reveal a 30% increase in annual dynamic ice loss on Helheim Glacier starting in 2017. The authors also found a time lag between the dynamic ice loss and the recorded GNSS uplift. JI and KG have a time lag of 0.87 ± 0.07 years and 0.37 ± 0.17 years, respectively. However, there is no time delay for Helheim Glacier. The authors attributed the lag to the viscous response of the ice to a change in the glacier load. Ice is a viscoelastic material. A change in load induces an elastic response in the ice followed by a viscous response that may take months to years to fully develop depending on the amplitude of the viscosity.

2.4 Geodynamic processes in cryospheric regions

Alaska, Japan, New Zealand, Patagonia, Iceland, British Columbia, and the Antarctic Peninsula are locations where geodynamic and cryospheric processes interact. Separating the displacements in GNSS time series due to changing ice loads from the geodynamic signals or estimating how the stress induced by the cryospheric mass could affect geodynamic processes is necessary.

In Iceland, a decrease in the cryospheric load caused by climate warming would cause the surface to uplift. More importantly, the decreasing surface load would also cause a reduction in pressure at depth, which would increase magma production. Iceland's Vatnajökull ice cap is proximally located to many Icelandic volcanoes. A network of GNSS stations placed around the perimeter of the Vatnajökull ice cap from 1996 to 2004 shows vertical trends of 9–25 mm/yr and horizontal velocities in the range 3–4 mm/yr (Pagli et al., 2007). Viscoelastic modeling of the Earth's response to this unloading (Pagli & Sigmundsson, 2008; Sigmundsson et al., 2010) predicts a minimum rate of magma production of $0.014 \, \text{km}^3/\text{yr}$. Schmidt et al. (2013) extend this analysis to consider the effects of all the shrinking glaciers in Iceland and predicted a rate of magma production of $0.044 \, \text{km}^3/\text{yr}$, three times higher than the rates from Vatnajökull ice cap alone. This additional melt could ultimately extrude at the surface with the equivalent of one Eyjafjallajökull summit eruption about every seventh year.

In Japan, the seasonal components of the GNSS coordinate time series (horizontal and vertical) are coherent in phase. Further, the seasonal E-W amplitudes correlate with the yearly E-W velocities (Murakami & Miyazaki, 2001). The authors demonstrated that the seasonal strain changes affect the seismicity, which also has a seasonal character peaking in March. In a subsequent paper, Heki (2001) estimated the loading due to seasonal snow fall. Heki (2001) tentatively concluded that the snow loads modulate the interseismic stress build up in northeastern Japan.

In the next section, we will discuss recent advances in the use of GNSS data to quantify the ongoing response of the solid Earth to past cryospheric change and hence provide insight into the glacial history and Earth rheology of glaciated and formerly glaciated regions around the world.

3 The viscoelastic response of the Earth to cryospheric change

Glacial Isostatic Adjustment (GIA) is the process by which the solid Earth evolves toward isostatic equilibrium in response to the advance and retreat of ice caps and ice sheets, including those in Greenland and Antarctica, and the former Pleistocene ice sheets of the northern hemisphere. It is an ongoing process, and the largest contemporary deformation signals are found near the centers of former ice sheets, where uplift is surrounded by regions of subsidence (Fig. 12.4). During glacial periods, increased ice loading causes mantle material to be displaced away from glaciated regions to surrounding ice-free regions, creating a forebulge. As the glacial ice melts, and decreases in mass, mantle material beneath the forebulge regions is displaced back toward the former center of ice loading, resulting in adjacent zones of subsidence and uplift. Present-day rates of GIA-related vertical deformation are estimated to approximately range between +20 and −10 mm/yr (Milne & Shennan, 2013), amplitudes well within the measurement precision of GNSS. Measuring this deformation provides insight into past ice sheet change as well as spatial variations in Earth rheology.

GIA modeling is used to predict the evolving solid Earth response to changes in ice and ocean surface loading, while simultaneously solving for the gravitationally consistent redistribution of meltwater across the global ocean (Whitehouse, 2018). Model inputs include an ice-history reconstruction that describes the temporal distribution of ice sheet change over time, often going back to the last glacial maximum (LGM), and an Earth model characterizing the Earth's rheology. These inputs are used to solve the sea-level equation (Farrell & Clark, 1976) and hence predict the long-term response of the solid Earth to cryospheric change. See Whitehouse (2018) for a complete review of GIA modeling.

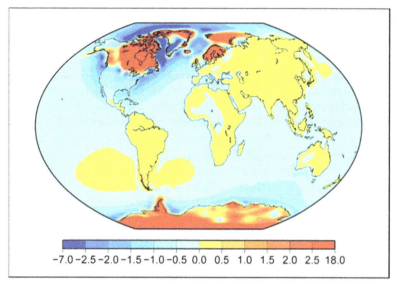

FIG. 12.4 An estimate of the spatial distribution of present-day GIA-related uplift and subsidence predicted using an adaptation of the ICE-5G global deglaciation history (Peltier, 2004) as input to a GIA model (Milne & Shennan, 2013). *(Reproduced with permission from Elsevier License 5327240355569.)*

GIA is a time-decaying process, with deformation rates expected to peak shortly after the disappearance of an ice sheet and decay over multiple millennia (Mitrovica & Peltier, 1993). In locations, such as Canada and Scandinavia, where large-scale ice sheets have long since disappeared, current solid Earth deformation due to GIA will be linear on decadal timescales. This picture becomes more complex in regions that are still ice-covered, and which may have experienced continual ice mass change over many millennia, right up to the present. In this latter scenario, the GIA signal is continually evolving and will reflect the combined response to multiple episodes of ice mass change over a range of spatial and temporal timescales.

In addition to the vertical signal shown in Fig. 12.4, GIA excites a characteristic pattern of horizontal deformation: in the near field, motion is predicted to be directed away from the center of uplift, with magnitudes peaking near the margin of the former ice load (Milne et al., 2001). In the far field, horizontal vectors point toward the former ice sheet (Kreemer et al., 2018). In fact, for a simple radially symmetric viscosity model the horizontal velocity pattern is controlled by viscosities in the lower mantle (Kreemer et al., 2018; Sella et al., 2007). This simple picture breaks down in regions underlain by spatially variable Earth rheology, where the direction and magnitude of deformation will be influenced by regions of high- and low-mantle viscosity, which preferentially resist or accommodate motion, respectively (Hermans et al., 2018; Kaufmann et al., 2005; Kaufmann & Wu, 2002; Whitehouse et al., 2006).

Having described the expected spatial and temporal pattern of GIA-related solid Earth deformation, we now explore what information can be deduced if this pattern can be quantified, for example, via GNSS observations.

3.1 GIA: Three pieces to the puzzle

There are three components to the GIA problem that we wish to understand: the ice load history, the rheology of the solid Earth, and the response of the solid Earth to surface loading. If any two are known, the third can be inferred, with a primary goal being to better understand how ice sheets have evolved during past periods of climate change (Whitehouse, 2018). A secondary goal is to quantify the degree to which Earth deformation may influence ice sheet change (e.g., Gomez et al., 2015; van den Berg et al., 2008). In other words, we seek to answer the question: what is the time-varying response of the solid Earth to a given ice load change, and how might this impact ice sheet dynamics? The glaciological, geological, seismological, and geodetic data used to independently constrain various components of the GIA problem are briefly summarized below.

Ice load history can be pieced together by combining glaciological evidence of past ice extent and thickness (e.g., Batchelor et al., 2019; Bentley et al., 2014; Dalton et al., 2020; Hughes et al., 2016) with ice sheet modeling (e.g., Albrecht et al., 2020; Briggs et al., 2014; Gomez et al., 2013; Lecavalier et al., 2014; Patton et al., 2017; Pittard et al., 2022; Tarasov et al., 2012; Whitehouse, Bentley, & Le Brocq, 2012). In the section "Glacial isostatic adjustment in deglaciated regions," we discuss the degree to which GNSS data may be employed to tune or validate ice sheet reconstructions.

The rheology of the solid Earth can be inferred from late Quaternary records of sea-level change derived from tectonically stable regions (e.g., Lambeck et al., 2014; Peltier et al., 2015). After accounting for processes that alter the height of the sea surface, e.g., perturbations to the geoid or ocean volume change, sea-level records reflect millennial-scale solid Earth deformation in response to regional changes in ice and ocean loading. To accurately constrain Earth rheology using sea-level records, the surface load history must be well known. In regions where the surface load history is poorly known, or long-term records of surface deformation are lacking, independent geophysical evidence, derived from seismic or magnetotelluric data, can be used to constrain the material properties of the solid Earth such as lithosphere thickness or mantle rheology (e.g., ODonnell et al., 2017; Selway et al., 2020).

The final piece of the puzzle—the solid Earth response to surface loading—has already been mentioned when discussing how to infer Earth rheology. Long-term records of sea-level change document the time-evolving nature of the viscoelastic response to cryospheric change, allowing us to place constraints on the relaxation time of the mantle (e.g., McConnell, 1968). However, snapshots of the contemporary rate of Earth deformation derived from GNSS data provide vital, complementary information on ice history and Earth rheology. In particular, GNSS measurements have revolutionized our ability to map the spatial pattern of GIA-related deformation across whole continents (Milne et al., 2004; Sella et al., 2007) at increasingly high levels of precision (Lidberg et al., 2007, 2010), and they have allowed us to document the horizontal response as well as the vertical (Milne et al., 2001; Vardić et al., 2022). Furthermore, combining GNSS data with complementary data sets has enabled us to isolate the elastic and viscous response to past and present cryospheric change in ice-covered areas (e.g., Barletta et al., 2018; Nield et al., 2014; van Dam et al., 2017). This and other applications are discussed in more detail in the section 'Viscous deformation in ice-covered areas.'

3.2 Using GNSS to measure the viscoelastic response to cryospheric change

GNSS observations have been used to quantify surface displacements driven by GIA since the mid-1990s (King et al., 2010) but a number of corrections must be made to isolate the GIA signal. In addition to the elastic signal associated with present-day ice mass change, GNSS time series need to be corrected for signals associated with other types of loading (atmospheric, hydrologic, and nontidal oceanic loading) as well as local signals associated with tectonic motion or fluid input/withdrawal and the subsequent expansion/compaction of underlying sediments. These issues are also discussed in Section 1.

If the processes mentioned above cannot be constrained independently, one option is to draw on our understanding of the characteristic length- and time-scale of each process to interpret the GNSS time series. Deformation associated with GIA is often assumed to be approximately linear over the time span of GNSS measurements, reflecting the fact that the average relaxation time of the upper mantle is a few kiloyears (McConnell, 1968; Mitrovica, 1996). However, other signals that may be essentially linear on decadal timescales include the elastic response to contemporary sea-level rise and signals associated with tectonic processes. The spatial signature of the former is easily identified but separating GIA-related and tectonic signals remains challenging (Drouin et al., 2017; Kreemer et al., 2018; Piña-Valdés et al., 2022; Turner et al., 2020). The assumption of temporal linearity in the GIA signal can also break down, especially in low (mantle)-viscosity, ice-covered regions experiencing ongoing cryospheric change. For example, several recent studies (e.g., Adhikari et al., 2021; Barletta et al., 2018; Khan et al., 2016; Nield et al., 2014) have used GNSS measurements to identify ice-covered regions where mantle relaxation times are on the order of decades. The implication of this is that the GIA signal, i.e., the viscous response to past ice mass change, will be dominated by the response to relatively recent ice mass change and may evolve rapidly over time. This is discussed further in the section "Viscous deformation in ice-covered areas," where we also describe the approach of combining multiple data sets, with different sensitivities, to jointly invert for contemporary ice mass change and the geodetic signature of past ice mass change, building on the approach originally outlined by Wahr et al. (2000).

Even if it is possible to isolate the contemporary GIA signal within GNSS data, there are multiple combinations of ice history and Earth rheology that could result in the observed rate (e.g., Lange et al., 2014). In other words, if we only have one piece of the puzzle, then, there is significant nonuniqueness in solving the GIA problem. This is illustrated in two idealized scenarios, shown in Fig. 12.5 (taken from Whitehouse, 2018). The first (Fig. 12.5A) demonstrates the importance of being able to independently quantify the timing or magnitude of past ice mass change, noting that relatively minor, recent ice loss may yield the same contemporary uplift rate as more significant, much earlier ice loss from a region with the same mantle relaxation time. In the second example (Fig. 12.5B), the two curves represent the situation where the timing of ice loss is known, perhaps constrained by the presence of a dated glacial moraine, but not the magnitude. The same contemporary uplift rate may be explained by relatively minor ice loss from a region where mantle viscosity is high—i.e., the net deformation required to reach isostatic equilibrium is relatively small but the long mantle relaxation time results in

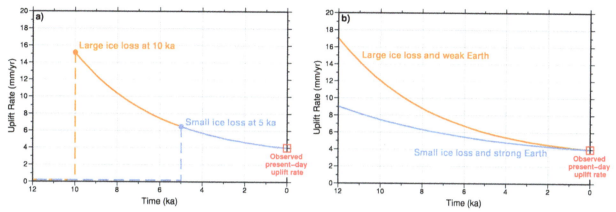

FIG. 12.5 (A) Trade-off between the timing and magnitude of past surface load change for a given Earth rheology: large ice loss at 10 ka can result in the same present-day uplift rate as smaller ice loss at 5 ka. (B) Trade-off between ice load history and Earth rheology: large ice loss combined with a weak rheology can produce the same present-day uplift rate as small ice loss combined with a strong rheology. *(Adapted from Whitehouse, P.L. (2018). Glacial isostatic adjustment modelling: Historical perspectives, recent advances, and future directions. Earth Surface Dynamics, 6(2), pp. 401–429. http://doi.org/10.5194/esurf-6-401-2018.)*

prolonged, subdued rebound rates—or more significant ice loss from a region underlain by low mantle viscosity—i.e., there is greater net rebound to be achieved but fast initial uplift decays rapidly to the contemporary rate. If the ice history and Earth rheology are poorly known, then progress can be made via careful optimization of the distribution of GNSS receivers (Wu et al., 2010), but this is challenging in regions with sparse infrastructure, such as northern Canada, or limited bedrock outcrops, such as Antarctica.

3.3 Horizontal deformation

In the same way that observations of horizontal deformation can improve interpretations of present-day cryospheric mass changes, they are also valuable, but underutilized, in GIA studies seeking to infer past ice history or earth rheology. The small number of studies that adopt this approach is partly due to the need to first remove the signal associated with plate motion, which is poorly defined in plate boundary zones and difficult to constrain in regions experiencing GIA (Ding et al., 2019). The magnitude of the horizontal GIA signal has been shown to be large enough to influence the accuracy of plate motion models determined by GNSS observations (King et al., 2016; Klemann et al., 2008). When constructing global reference frames and plate motion models, GNSS sites are often excluded if they are in a region where the GIA signal is expected to be significant. (Please note that estimating plate rotation while acknowledging that the entire North American plate is affected by GIA was addressed by Kreemer et al. (2018)). However, criteria for exclusion are usually only applied to vertical rates—for example, when constructing ITRF2014, GNSS sites were only excluded if local vertical GIA velocities were predicted to be ≥ 0.75 mm/yr (Altamimi et al., 2017). This approach can lead to bias because horizontal GIA velocities can be significant in areas where vertical rates are small or even negative, and they are expected to persist at magnitudes of several mm/yr in the far-field of the former ice sheets (Klemann et al., 2008).

A more robust approach would involve removing the GIA signal from the GNSS time series prior to reference frame or plate motion estimation. However, the circularity of the problem is clear, and it is exacerbated by significant uncertainty regarding the details of the horizontal GIA velocity field (Vardić et al., 2022). In particular, it has long been known that lateral variations in Earth rheology will significantly influence both the magnitude and the direction of the horizontal GIA signal (Kaufmann et al., 2005), and more recently, it has been shown that GIA-related horizontal velocities can change direction over time, with the timescale dependent on the local mantle viscosity (Hermans et al., 2018). Robustly accounting for these complications remains challenging due to incomplete knowledge of the 3D rheological structure of the Earth (Austermann et al., 2021), and the fact that spatiotemporal records of GIA-related sea-level change or Earth deformation are currently insufficient to perform a formal inversion for Earth structure (Crawford, 2018; Lau et al., 2016, 2018).

For all of the reasons above, horizontal GNSS velocities have largely been neglected in studies of GIA, except in cases where the ice-load history and Earth structure are relatively simple, such as Fennoscandia (Milne et al., 2001). Further efforts should be made to identify the horizontal GIA signal via GNSS measurements because horizontal and vertical deformation fields provide complementary information on past ice sheet change. In addition, and our current inability to robustly quantify intra-plate deformation due to GIA limits our ability to isolate regions of tectonic strain localization, and hence quantify seismic hazard.

3.4 Regions of interest

In the following sections, we review recent applications of GNSS to study cryospheric change in: deglaciated regions, on a global scale, and in still-glaciated regions.

3.4.1 Glacial isostatic adjustment in deglaciated regions

Prior to the establishment of continuous GNSS networks in the 1990s, the spatial pattern of viscoelastic deformation due to past ice loss was reconstructed primarily from records of past sea-level change (Lambeck et al., 1998). By their nature, such records are typically located close to present-day shorelines. The advent of GNSS has enabled us to map the continental-scale GIA signal across Fennoscandia (Johansson et al., 2002; Kierulf et al., 2014, 2021; Lidberg et al., 2010) and North America (Calais et al., 2006; Kreemer et al., 2018; Sella et al., 2007), providing insight into the evolution of former northern hemisphere ice sheets in regions that have limited glaciological constraints on past ice thickness and are far from past or present shorelines.

One of the first applications of GNSS for improving models of GIA was the BIFROST (Baseline Inferences for Fennoscandian Rebound Observations, Sea-level, and Tectonics) project in Sweden and Finland (Johansson et al., 2002). Fennoscandia was covered by some of the greatest ice thicknesses during the Last Glacial Maximum and present-day uplift rates are some of the fastest in the world, peaking at >10 mm/yr (Kierulf et al., 2021). Early comparisons between the observed 3D deformation field and GIA model predictions gave minimum misfits when the ice sheet reconstruction of Lambeck et al. (1998) was combined with an Earth model characterized by a lithospheric thickness of 120 km, an upper mantle viscosity of 8×10^{20} Pa s, and a lower mantle viscosity of 10^{22} Pa s (Milne et al., 2001). This was the first time that model predictions had been compared with both vertical and horizontal GNSS-derived rates of deformation, and it revealed that horizontal and vertical deformation rates have different sensitivities to mantle viscosity, a feature that was exploited in Milne et al. (2004) to refine our understanding of ice history and Earth rheology across Fennoscandia. Although correlations between GNSS-derived observations and GIA model predictions were high in these early studies, systematic differences were evident, partly due to the presence of seasonal and longer-period signals in the GNSS time series (Johansson et al., 2002). Separately, modeling efforts noted that consideration of lateral variations in Earth rheology could perturb predictions of present-day surface deformation by an amount greater than geodetic observational uncertainties (Steffen et al., 2006; Whitehouse et al., 2006). One implication of this is that early inferences of ice history, which were based on the assumption of a radially varying Earth rheology, and which provide a good fit to BIFROST data, when combined with a simplified 1D Earth model (Steffen et al., 2006), are likely to be biased (Whitehouse et al., 2006).

Over the past decade, our ability to constrain the pattern of GIA-related deformation across Fennoscandia has improved with the development of updated and reprocessed 3D GNSS time series (Kierulf et al., 2014, 2021; Lidberg et al., 2010), the inclusion of additional geodetic constraints such as tide gauge data and GRACE data (Hill et al., 2010; Simon et al., 2018), and better estimation of contributions from surface mass loading (Kierulf et al., 2021, 2022; Simon et al., 2018) enabling a more robust assessment of the residual velocity field. This residual deformation is attributed to a combination of GIA model errors and local tectonic processes, confirming and improving on earlier studies that used a smaller geodetic network, and shorter time series, to quantify the GIA signal across northern Europe (Marotta & Sabadini, 2004; Nocquet et al., 2005) and the British Isles (Bradley et al., 2009). The most challenging components of the Fennoscandian geodetic signal to estimate are the short- and long-wavelength signals associated with recent and ongoing ice mass change in Svalbard, the Russian Arctic, and Greenland (Kierulf et al., 2022; Simon et al., 2018). Bedrock GNSS time series from logistically challenging ice-covered regions are often incomplete; they must be corrected for the time-varying elastic signal associated with contemporary ice mass change (Kierulf et al., 2022), and the long-term GIA signal associated with post-LGM ice loss is likely to be overprinted with a poorly constrained signal associated with ice mass change during and since the Little Ice Age. Even outside of currently ice-covered regions, e.g., across the North Sea and British Isles, high-precision geodetic data provide a complex picture that reflects a myriad of contemporary and recent geodynamic processes. If non-GIA signals cannot be accurately constrained or modeled, inferences of past ice sheet change and regional Earth rheology based on the interpretation of geodetic data alone may be biased. One way to resolve this is to incorporate geological-scale, sea-level data, as these provide an independent constraint on the time-decaying Earth's response to past ice loss (Bradley et al., 2011; Simon et al., 2021).

The other main region experiencing ongoing deformation due to the disappearance of a continental-scale ice sheet is North America, the Laurentide Ice Sheet; GNSS coverage of North America is extensive to the south and east of the Great Lakes region, but sites are sparsely distributed across Canada, especially north of ~60°N (Kreemer et al., 2018). In comparison with northern Europe, this sparse geodetic coverage makes it challenging to confirm the location of peak uplift or infer the detailed geometry of past ice loading. Early GNSS studies analyzed ~300 GNSS sites and isolated the position of the hinge line—the transition between uplift in formerly glaciated regions to the north and subsidence in peripheral bulge regions to the south (Calais et al., 2006; Sella et al., 2007)—this is important because it enables us to identify regions where ongoing GIA will contribute to accelerated future sea-level rise on decadal to millennial timescales. An additional motivation for studying intraplate deformation across North America is to quantify the seismic hazard, for example by analyzing residual strain rates after removal of the GIA signal from horizontal GNSS rates. Early studies found a poor fit between GIA model predictions and the long wavelength horizontal velocity field, which was variously attributed to missing complexity in the spatial pattern of ice loading (Calais et al., 2006) or the neglected impact of lateral Earth structure in early GIA models (Sella et al., 2007). This misfit in the horizontal deformation field is confirmed by a more recent study, which draws on observations from >3000 GNSS sites and identifies a poor fit between GIA model predictions and the geodetically constrained spatial pattern of strain rates across North America (Kreemer et al., 2018).

To overcome limitations associated with the spatial distribution of GNSS sites, geodetic observations are usually combined with relative sea-level records when seeking to constrain the glacial history and Earth rheology of North America (Peltier, 2004). Recent improvements in geodetic coverage have permitted data-driven updates to earlier ice sheet reconstructions: Simon et al. (2016) employed a regional tuning approach and found that, compared with the ICE-5G model (Peltier, 2004), a 20%–30% reduction in the thickness of ice in some sectors of the former Laurentide Ice Sheet improved the misfit to vertical GNSS rates by a factor of 9, in close agreement with ice thicknesses in the ICE-6G model (Peltier et al., 2015). However, the details of past ice sheet change across northern Canada are still poorly constrained, leading to significant discrepancies in estimates for the total volume of the Laurentide Ice Sheet during the Last Glacial Maximum (Simms et al., 2019). Recent work has focused on better constraining the spatial details of the GIA signal. For example, Simon et al. (2017) combined GNSS data, GRACE data, and a suite of a priori estimates to determine a data-driven estimate for the present-day GIA signal, which is accompanied by formal uncertainties. Future progress will likely be made by building on this approach of combining data and modeling to yield probabilistic solutions for the past glacial history of North America and other deglaciated regions.

3.4.2 The global-scale response to past cryospheric change

GIA is a global-scale process. Quantifying the GIA signal outside formerly glaciated regions is important for a number of reasons, with the primary motivation being the need to understand how GIA-related vertical deformation will contribute to sea-level change during the coming decades to millennia. An understanding of GIA is also needed when quantifying the global terrestrial reference frame or applying a "GIA correction" to geodetic time series, for example, to isolate the signal associated with contemporary hydrological or cryospheric change.

Due to the viscoelastic nature of the mantle, the isostatic response to past ice sheet change will continue for many centuries, albeit at a decaying rate, influencing future rates of sea-level change and the position of the coastline. In areas formerly covered by ice sheets, such as Canada and Scandinavia, land uplift will impact coastal use due to a shallowing of natural harbors and a reduction in the navigability of some channels. Conversely, in regions adjacent to former ice sheets, such as the US east coast, land subsidence will exacerbate the impacts of future sea-level rise. Even far from the former ice sheets, GIA-related processes such as 'continental levering' (tilting of the lithosphere due to the flooding, and hence loading, of continental shelf regions during postglacial sea-level rise) will influence long-term rates of sea-level change (Walcott, 1972).

In some coastal locations, GNSS measurements quantify the magnitude of the solid Earth contribution to sea-level change (Hammond et al., 2021; Husson et al., 2018; Woppelmann & Marcos, 2016). However, there are many spatial gaps in the coastal GNSS network, and the time series will also record a range of non-GIA signals relating to, e.g., tectonics, hydrology, and sediment compaction (King, Keshin, et al., 2012). These factors will, of course, also contribute to sea-level change, but they will have different temporal and spatial characteristics to the GIA signal. For example, the temporal evolution of non-GIA processes may be nonlinear, with rates varying over months to decades, while the GIA signal will decay smoothly over centuries to millennia, a process that can only robustly be estimated via modeling. In order to quantify the spatial and temporal variability of all the processes contributing to solid Earth deformation, and hence sea-level change, the GIA and non-GIA signals must be separated.

The global GIA field is best estimated by combining data and models. Two main approaches are used: forward modeling and data inversion. The former often makes use of GNSS data to tune model parameters, while the latter draws on a priori information regarding the spatiotemporal characteristics of the GIA signal.

If GNSS data are to be used to test or tune GIA forward models, they must first be corrected for non-GIA effects such as the time-varying elastic signals associated with atmospheric, oceanic, cryospheric, and hydrological surface loading and the accompanying impact of changes to Earth's rotational state. Alternatively, a filter can be applied to identify sites where local effects dominate the time series (Schumacher et al., 2018). Several global-scale GIA models have been tuned to fit GNSS-derived velocity fields (Argus & Peltier, 2010; Peltier et al., 2015). However, when producing a GIA-only velocity field, signals associated with non-GIA effects are often assumed to be linear and, in general, cannot be perfectly removed. Care is, therefore, needed to understand the structure of the uncertainties associated with a GNSS velocity field when using it to tune a GIA forward model.

The statistical properties of a GIA-only velocity field play an important role in studies that use data inversion to quantify the GIA signal. For example, Sha et al. (2019) use a Bayesian hierarchical modeling framework to determine a data-driven estimate for the global GIA signal based on the GIA-only GNSS velocity field of Schumacher et al. (2018) and the output of a single GIA forward model. Caron et al. (2018) extend this approach, drawing on the characteristics of 128,000 GIA forward model runs, and combining these with GNSS velocities and >10,000 estimates of past sea level, in order to quantify the magnitude and uncertainty of the global GIA signal. More commonly, data-driven GIA estimates are derived in combination with efforts to identify the geodetic signature of present-day surface mass change (Jiang et al., 2021; Vishwakarma et al., 2022; Wu et al., 2010). In such 'joint inversion' studies, GNSS observations are not corrected for non-GIA processes; instead, they are combined with additional data sets, such as satellite gravimetry or altimetry, and a priori information on the spatiotemporal signature of all the competing processes is used to separate the signals associated with each of them. Sparse GNSS data coverage limits the capabilities of this approach in some regions (Ziegler et al., 2022) but advantages include the fact that uncertainties on both signals—GIA and present-day surface mass change—are quantified and signals that are currently not typically represented in forward GIA models are revealed. This includes quantification of the relatively short-wavelength GIA signal in regions underlain by anomalously weak Earth rheology and quantification of the GIA signal associated with relatively recent ice mass change, for example during or subsequent to the Little Ice Age (Jiang et al., 2021; Vishwakarma et al., 2022). The long wavelength signature of GIA can also be quantified in global-scale studies, which provides insight into the viscosity of the lower mantle (Jiang et al., 2021).

Not all geodetic studies seek to isolate the GIA signal. Many of them compare observations and predictions with the aim of better understanding the accuracy of GIA models and the nature of the processes contributing to the residual velocity field. For example, Métivier et al. (2020) compare GIA model predictions with vertical velocities provided by the last four International Terrestrial Reference Frame (ITRF) solutions (Altamimi et al., 2002, 2007, 2011, 2016). They demonstrate that the misfit decreases with each ITRF update—reflecting better consideration of GIA effects when creating more recent solutions—but that misfits between the GIA model predictions and the three ITRF2014 velocity fields (velocity fields are produced for 2000, 2005, and 2013) increase over time due to the emergence of a signal associated with recent ice melting, which is not included in the GIA model predictions. Other studies that compare GIA model predictions with global GNSS velocity fields interpret the residuals in terms of surface hydrology (Hammond et al., 2021), plate motion (Vardić et al., 2022), or ocean loading (Frederikse et al., 2019).

Better understanding of the global GIA signal allows us to apply a more accurate 'GIA correction' when seeking to quantify other processes using geodetic data. For example, a globally uniform 'GIA correction' is typically applied to satellite altimetry data when seeking to quantify global mean sea-level change (Cazenave et al., 2018) while a spatially variable 'GIA correction' is applied to GRACE data when seeking to quantify changes in ice mass, ocean mass, or terrestrial hydrology (Cazenave et al., 2018; King, Bingham, et al., 2012; Wang et al., 2022). A "GIA correction" can also be applied to GNSS time series, with the aim being to remove the signal associated with past ice mass change and hence reveal the signal associated with contemporary surface mass change. However, a number of factors must be considered when applying this method in currently ice-covered regions, as discussed in the following section.

3.4.3 Viscoelastic deformation in ice-covered areas

Unlike deglaciated regions, where the ice sheets have long since disappeared and Earth deformation due to GIA is a smoothly decaying signal, the solid Earth response to ice mass change in glaciated regions is a compound signal that reflects contemporary ice mass change, ice mass change that took place thousands of years ago, and everything in between. In this section, we discuss the challenge of untangling this signal and summarize recent advances in our understanding of polar ice sheet change and Earth rheology, as revealed by GNSS.

In glaciated regions, bedrock GNSS observations are used to quantify contemporary ice mass change and/or understand past ice sheet change. Depending on the primary goal, one of three approaches can be used.

1. If the main goal is to quantify contemporary ice mass change, which we assume triggers a time-varying localized elastic response, a "GIA correction" needs to be applied to the GNSS time series (e.g., Khan et al., 2007, 2010; Ludwigsen et al., 2020; Nielsen et al., 2012, 2013; Wake et al., 2016). This correction is a spatially varying, temporally uniform estimate of the present-day rate of uplift/subsidence at each GNSS site due to past ice mass change. Such a correction

can be calculated using a GIA forward model based on two inputs: regional or global ice history constrained by glaciological field constraints and/or ice sheet modeling and an estimate of the local Earth rheology. Due to the filtering effect of the lithosphere and the complex time spectra associated with viscoelastic mantle deformation, it is important to consider the regional ice load history, not just the local ice history, when calculating the GIA correction. The success of this approach depends on our ability to determine the geodetic signal associated with all past ice mass changes.

2. Conversely, if the main goal is to isolate the GIA signal, for example, to distinguish between competing forward GIA models (Fig. 12.6), GNSS time series must be corrected for the elastic signal associated with contemporary ice mass change (e.g., Groh et al., 2012; Hattori et al., 2021; Kappelsberger et al., 2021; Khan et al., 2016, 2008; King et al., 2010; Lange et al., 2014; Martín-Español et al., 2016a, 2016b; Sasgen et al., 2013; Schumacher et al., 2018; Thomas et al., 2011; Wolstencroft et al., 2015). This contemporary signal should be independently constrained, e.g., using satellite altimetry data, and time-dependent (King et al., 2022). As with the above approach, it is assumed that contemporary ice mass change triggers a purely elastic response. Elastic deformation has a much shorter wavelength than viscoelastic deformation, so in this approach it is important to accurately represent the local details of contemporary ice mass change when calculating the correction to be applied to the GNSS data (Spada et al., 2012). It is also important to ensure that the resulting GNSS velocity field is transformed into the same reference frame as the GIA model predictions, typically the center of mass of the solid Earth (CE) (Schumacher et al., 2018). The success of this approach depends on our ability to determine the geodetic signal associated with contemporary ice mass change.

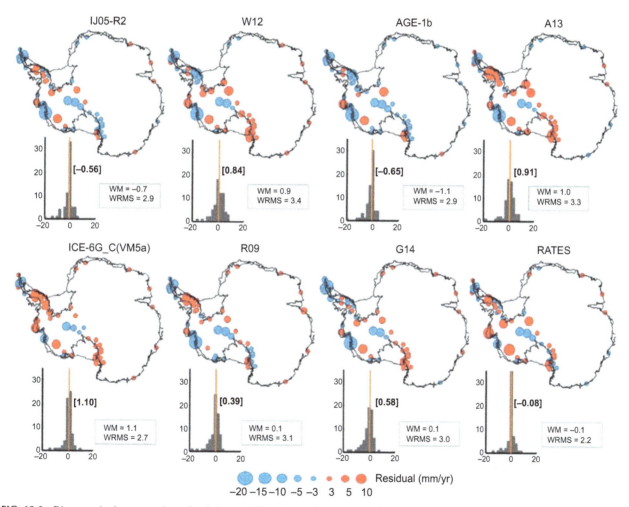

FIG. 12.6 Discrepancies between estimated and observed GIA-related uplift rates at a suite of Antarctic GNSS sites, for a range of GIA models (see Martín-Español et al., 2016b for full details). *Red circles* indicate locations where the estimated GIA rates are greater than the GNSS-observed velocities; *blue circles* indicate the converse. Histograms show the summary statistics; WM = weighted mean, WRMS = weighted root-mean-square, median values are indicated in brackets. *(Reproduced from Martín-Español, A., et al. (2016b). Spatial and temporal Antarctic ice sheet mass trends, glacio-isostatic adjustment, and surface processes from a joint inversion of satellite altimeter, gravity, and GPS data, Journal of Geophysical Research:* Earth Surface, *121(2), pp. 182–200. http://doi.org/10.1002/2015jf003550.)*

3. The third approach involves carrying out a joint inversion for the signals associated with past and present ice mass change (e.g., Engels et al., 2018; Groh et al., 2014; Gunter et al., 2014; Martín-Español et al., 2016a, 2016b; Riva et al., 2009; Sasgen et al., 2017; Schoen et al., 2015). Multiple data sets are required as inputs—e.g., GNSS data, satellite altimetry data, and GRACE data—and the method relies on the fact that different data sets have different sensitivities to the processes being investigated (Velicogna & Wahr, 2002; Wahr et al., 2000). As an example, contemporary ice loss would be recorded as a negative signal in GRACE data, but the solid Earth rebound associated with past ice loss would be recorded as a positive signal in the same data set. This approach relies on the assumption that geodetic signals associated with noncryospheric mass change are negligible or can be removed. An advantage of the method is that signals associated with GIA and contemporary ice mass change are simultaneously quantified. However, errors in the estimation of one signal will lead to bias in the other, and the results do not provide any insight on the cause of the GIA signal, i.e., they do not provide a direct constraint on ice history or Earth rheology.

These three approaches rely on a number of idealized assumptions. In reality, there are several confounding factors, which means that results derived from the implementation of these approaches have many caveats.

In many cases, the "GIA correction" will be poorly known due to uncertainties in ice history, Earth rheology, or both. Attempts to quantify the uncertainty on the GIA signal are often insufficient due to the challenge of rigorously exploring the multidimensional parameter space associated with the model inputs—the recent study of Caron et al. (2018) is an exception to this—and predictions of the GIA signal often disagree significantly (Martín-Español et al., 2016a, 2016b; Whitehouse et al., 2019). A useful review of the reasons for discrepancies in the GIA correction can be found in the supplementary material of Shepherd et al. (2018).

When seeking to quantify the GIA signal in still-glaciated regions, it is important to account for *all* past ice sheet changes, i.e., right up to the present day. However, the centennial-scale history of our remaining ice caps and ice sheets is poorly known prior to the satellite era, largely due to the difficulty of collecting field data that document historical ice-sheet change in logistically challenging locations like Greenland and Antarctica. Continent-scale deglacial reconstructions often assume no ice mass change during the last few millennia (e.g., Whitehouse, Bentley, Milne, et al., 2012) and this is one of the main reasons for misfits between GIA forward model predictions and elastic-corrected GNSS rates in Antarctica (Martín-Español et al., 2016a, 2016b). The inclusion of centennial-scale ice mass change significantly alters the prediction of the GIA signal in Greenland (Adhikari et al., 2021; Simpson et al., 2011; Spada et al., 2012).

Similarly, the details of contemporary ice mass change may be imperfectly known, biasing efforts to isolate the GIA signal (approach 2). Altimetry data provide good temporal coverage of ice elevation change (e.g., Sandberg Sørensen et al., 2018; Schröder et al., 2019) but observations require spatial interpolation and scaling to convert from ice volume change to ice mass change. It is also important to note that GIA-related solid Earth deformation will contaminate altimetry (and gravimetry) observations and an iterative approach may be needed to avoid circularity when seeking to isolate the GIA signal (Kappelsberger et al., 2021).

Data-constrained reconstructions of spatiotemporal ice mass change can be used to determine the time-dependent correction that needs to be applied to GNSS time series to remove the elastic response to contemporary change. Such corrections can be expressed as a time series, but they are often applied assuming a linear (or piecewise linear) rate of change (Martín-Español et al., 2016a, 2016b; Schumacher et al., 2018), which can lead to biased inferences of the GIA signal. Koulali et al. (2022) demonstrate that applying a time-dependent elastic correction to GNSS time series improves the linearity of the residual (GIA) signal (Fig. 12.7).

The most challenging issue to resolve when studying Earth deformation in glaciated regions is the fact that, in regions of weak Earth rheology (e.g., mantle viscosity less than $\sim 10^{19}$ Pa s), ice mass change can trigger a viscous response on the timescale of years to decades (e.g., Nield et al., 2014). This has several implications: (a) ice mass change over recent decades will influence the present-day pattern of deformation, and (b) ice mass change even during the observation period may trigger a viscoelastic response; in other words, we may be dealing with an evolving "GIA signal" rather than a linearly decaying pattern of deformation (Whitehouse, 2018). Due to the difficulty of separating the elastic and viscous signals associated with past and present ice mass change in Alaska, Hu and Freymueller (2019) modeled the evolving response to both as they seek to use GNSS observations to infer regional Earth rheology. In regions where historical ice mass change is poorly known, e.g., Antarctica, simply detrending GNSS time series to remove the "background" GIA signal is problematic because this signal may not be constant in time (e.g., Barletta et al., 2018; Nield et al., 2014; Zhao et al., 2017). It is clear that future geodetic studies will need to account for the possibility of an evolving GIA signal (Chuter et al., 2022; Engels et al., 2018; Samrat et al., 2020).

An additional complication is the fact that, in several areas where GNSS observations have been used to identify regions of low mantle viscosity (e.g., Alaska—Larsen et al., 2005; Iceland—Arnadottir et al., 2009; Patagonia—Dietrich et al., 2010; the northern Antarctic Peninsula—Nield et al., 2014) the very reason for the low viscosity—local tectonics or

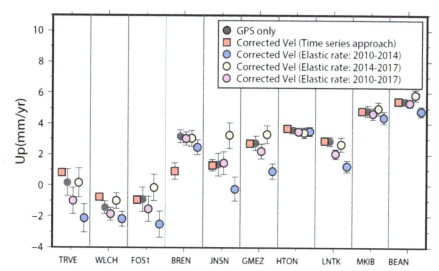

FIG. 12.7 Linear vertical velocities at 10 Antarctic GNSS sites estimated from position time series *(gray)* as well as after correction for present-day elastic deformation due to mass change using a linear *(purple, orange, and blue circles)* or time series *(red squares)* approach. In all cases, the RACMO2.3 (5.5 km resolution) model is used to estimate the elastic signal. Error bars reflect 95% confidence intervals. *(Reproduced from Koulali, A. Whitehouse, P. L., Clarke, P. J., van den Broeke, M. R., Nield, G. A., King, M. A., Bentley, M. J., Wouters, B. & Wilson, T. (2022). GPS-observed elastic deformation due to surface mass balance variability in the Southern Antarctic Peninsula Geophysical Research Letters, 49, http://doi.org/10.1029/2021GL097109.)*

magmatic activity—means that the regional surface velocity field is likely to be contaminated by other processes (e.g., Drouin et al., 2017; Richter et al., 2016).

More generally, consideration of spatial variations in Earth rheology is important when using GNSS records of Earth's deformation to infer Earth's structure. Good spatial distribution of GNSS sites is necessary to be able to determine whether 1D (radially varying) or 3D Earth models provide the optimum approach to modeling GIA (Blank et al., 2021; Marsman et al., 2021). This is problematic in glaciated regions with few bedrock outcrops, but including the analysis of horizontal rates may help overcome some of the challenges associated with inferring complex Earth structure (Powell et al., 2022).

Challenges to overcome in relation to the joint inversion method include irregular or sparse distribution of GNSS sites (Martín-Español et al., 2016a, 2016b), the need to combine data sets with different spatial and temporal coverage or resolution—although, differences in spatial resolution can be an advantage when seeking to separate geophysical signals (Engels et al., 2018)—and the possibility that observations include noncryospheric signals.

3.5 Polar case studies for GIA

In the following section, we summarize progress in understanding cryospheric change in polar regions that has been made possible due to the availability of bedrock GNSS measurements in these regions.

3.5.1 Antarctica

Prior to the establishment of an extensive network of continuous bedrock GNSS sites across Antarctica during the 2007–2008 International Polar Year, it was not possible to validate the GIA models that were being used to correct GRACE data for the effects of past ice mass change. However, the publication of a continent-wide velocity field in 2011 (Thomas et al., 2011) revealed that existing models significantly overestimated the rate of GIA-related deformation at GNSS sites (observed rates were corrected for elastic effects), leading to overestimation of the rate of contemporary ice loss from GRACE. Possible reasons for the misfit included overestimation of the magnitude of post-LGM ice loss, incorrect assumptions about the timing of post-LGM ice loss, or overestimation of upper-mantle viscosity beneath West Antarctica (the location of the greatest misfit).

As GNSS time series have extended, and our ability to quantify contemporary ice mass change has improved, it has become possible to tackle this final issue by using the time-varying solid Earth response recorded by GNSS to infer the third piece of the puzzle, local mantle rheology. For example, Nield et al. (2014) showed that, following the breakup of the Larsen B ice shelf, the timing and magnitude of the increase in uplift recorded at Palmer Station could only be explained in terms of a combined elastic and viscous response, and that upper-mantle viscosity in the region had to be $\sim 10^{18}$ Pa s. The continuous nature of the GNSS time series and the continued evolution of the surface ice load over the past decade have enabled us to identify other regions of West Antarctica that are underlain by weak Earth rheology (Barletta et al., 2018; Samrat et al., 2020, 2021; Zhao et al., 2017). An inability to represent these regions of low mantle viscosity is a significant shortcoming of early continent-wide GIA models and the likely reason for misfits between GIA model predictions and GNSS rates despite improved representation of the ice history (Whitehouse, Bentley, Milne, et al., 2012).

In low-viscosity regions, the solid Earth response to any ice mass change prior to ~2 ka BP will have decayed due to the short relaxation time of the upper mantle, but in locations underlain by higher viscosity, longer term knowledge of the past ice history is needed to accurately model the GIA signal. Misfits between model predictions and GNSS rates in the southern Antarctic Peninsula are likely to be due to incorrect assumptions about the timing or magnitude of Holocene ice mass change (Wolstencroft et al., 2015) while observations of present-day subsidence in the southern Weddell Sea (Bradley et al., 2015) and along a 1000-km stretch of the East Antarctic coast (King et al., 2022) provide strong evidence to support the hypothesis of ice sheet readvance during the Holocene in these regions.

GNSS observations provide powerful constraints on Antarctic ice mass change but caution is needed when using them to directly tune GIA models (e.g., Argus et al., 2014; Ivins et al., 2013) due to the ongoing challenge of rigorously separating the elastic and viscous signals associated with past and present ice mass change.

3.5.2 Greenland

The network of GNSS instruments installed around the coast of Greenland (Fig. 12.8) reveals that observed present-day rates are typically much greater than the rates predicted by GIA models and, hence, the elastic response to contemporary ice mass change is the dominant solid Earth signal (Bevis et al., 2012). Building on this, many studies have used GIA-corrected GNSS rates to document the details of recent ice mass change around Greenland (Khan et al., 2007, 2010; Nielsen et al.,

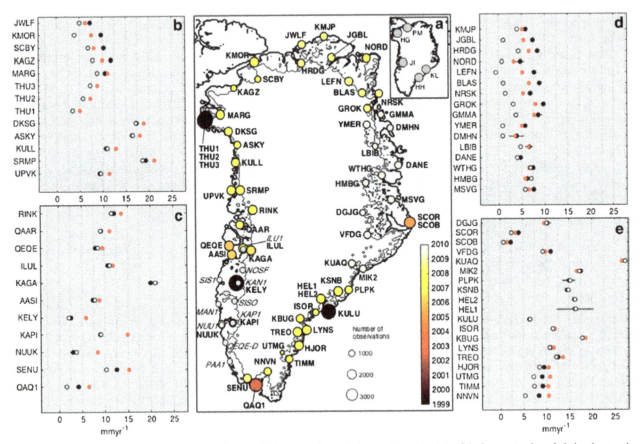

FIG. 12.8 (A) GNSS sites of the Greenland GNET network. Sites are color-coded according to the date of deployment and symbol size denotes the number of daily observations used to calculate the average velocity between deployment and, in most cases, mid-2013. (B)–(E) Mean vertical velocities at 59 GNSS sites in four regions (*black circles* with ±2 standard error). In some cases, the circles obscure the standard error bars. Note that the majority of sites in the network display nonlinear trends in vertical velocity; rates presented in this figure represent the station's average velocity. Also presented in (B)–(E) are the GIA-corrected rates based on the Huy3 *(red)* and ICE-6G_C (VM5a) *(white)* GIA models; see original text for details. *(Reproduced from Wake, L.M., Lecavalier, B.S., & Bevis, M. (2016). Glacial isostatic adjustment (GIA) in Greenland: A review.* Current Climate Change Reports. *http://doi.org/10.1007/s40641-016-0040-z.)*

2012, 2013; Wake et al., 2016), while others have drawn on our improved understanding of contemporary change to apply an elastic correction to GNSS rates, test the accuracy of GIA model predictions, and hence reduce the uncertainty on the GIA correction applied to GRACE (Kappelsberger et al., 2021; Khan et al., 2016, 2008). In locations where both the loading history and the GIA signal are poorly known, GNSS observations can be combined with absolute gravity measurements to jointly constrain the present-day rate of ice mass change and the GIA signal (Wahr et al., 1995). This method has recently been extended by van Dam et al. (2017) and used to infer that the GIA signal at Kulusuk is greater than predicted by current GIA models, likely due to a previous overestimate of mantle viscosity in this region.

In general, determining the Greenland GIA signal via forward modeling is challenging due to the complex ice history of the past few millennia, which has included multiple episodes of retreat and readvance (Lecavalier et al., 2014). Incomplete knowledge of this history has led to large variance in the predicted GIA signal in many regions (Wake et al., 2016). However, the importance of accurately quantifying the GIA signal is likely to decrease in future decades as warming triggers increased ice loss and the elastic signal begins to dominate contemporary deformation to the degree that ice loss estimates are only weakly sensitive to the details of the GIA correction.

It is important to check the robustness of this assumption by investigating whether recent and contemporary ice mass change may trigger a viscous response that would bias the interpretation of GNSS observations, if neglected. By modeling the combined response to recent and contemporary ice mass change and comparing with GNSS rates, Simpson et al. (2011) and Spada et al. (2012) conclude that the contemporary viscous response is likely negligible in most regions. However, both Khan et al. (2016) and Adhikari et al. (2021) use misfits between observed and modeled uplift rates to advance our understanding of Earth rheology across Greenland: the former identifies a region of low mantle viscosity in southeast Greenland, which warrants consideration when interpreting GRACE data, while the latter find evidence for time-dependent viscosity, implying that nonlinear constitutive models of the mantle warrant further investigation (Paxman et al., 2023).

4 GNSS interferometric reflectometry for the cryosphere

4.1 Introduction

Global Navigation Satellite System Reflectometry (GNSS-R) uses GNSS signals reflected from the Earth's surface to estimate ground geophysical properties and their temporal changes. This technique has been tested on terrestrial, air- and space-borne platforms with a wide range of applications in measuring ocean height and surface wind speed, inland water level, near-surface soil moisture, freeze and thaw status, and vegetation water content (e.g., Camps et al., 2016; Chew & Small, 2018; Foti et al., 2015; Li et al., 2018; Ruf et al., 2013).

One technique, GNSS Interferometric Reflectometry (GNSS-IR), is of particular interest to geodesists because it can be readily applied to numerous ground-based, geodetic-level GNSS sites that are operating around the globe. Kristine Larson and collaborators innovated the use of the interference pattern between the direct and reflected GNSS signals to estimate surface soil moisture, snow depth, vegetation growth conditions, and water level changes (Larson et al., 2008, 2009, 2013; Larson & Small, 2014). Comprehensive reviews of GNSS-IR can be found in Larson (2016, 2019). Here, we summarize the applications of GNSS-IR for the cryosphere, with a focus on the mass balance of snow, glaciers, and permafrost, as featured in the growing literature in the last 10 years. A few recent studies have also extended the use of GNSS-IR for estimating sea ice and lake ice thickness (e.g., Ghiasi et al. (2020); Wang et al., 2022; Piña-Valdés et al., 2022), but are not assessed in this article.

4.2 Principles and methodology of GNSS-IR

Global navigation satellites, such as GPS, GLONASS, Galileo, and Beidou, broadcast signals that propagate directly to the ground-based receiver's antenna; whereas some signals reflect off an object, such as ground and snow, before being received by antennas. Fig. 12.9 depicts a typical GNSS-IR geometry of the antenna, signals, and a horizontal and flat snow surface as a reflector. The GNSS monument is anchored into the ground. In conventional positioning-oriented studies, the geocentric vertical and horizontal coordinates accurately estimated by the direct GNSS signals measure how the monument moves with respect to the center of the Earth.

The interference between the direct signal and the reflected signal from a smooth and planar surface, usually a source of noise in positioning applications, becomes a useful observable in GNSS-IR. The interference pattern used by GNSS-IR is imprinted in the signal-to-noise ratio (SNR) data recorded by GNSS receivers. SNR varies as a GNSS satellite rises or sets, and, therefore, it changes as a function of the satellite's elevation angle (e), as expressed below:

$$\text{SNR} = A(e) \sin\left[\frac{4\pi H}{\lambda} \sin e + \phi(e)\right], \tag{12.3}$$

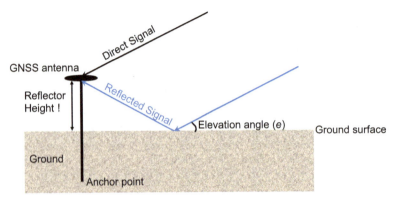

FIG. 12.9 Scheme diagram of a typical GNSS-IR geometry. Here, a flat horizontal surface reflects GNSS signals that interfere with the direct signals.

where A and ϕ are the amplitude and phase of the SNR oscillation, respectively, mainly depending on surface reflectivity, surface roughness, and antenna gain pattern; H is the height of the GNSS receiver antenna's phase center above the reflecting surface, commonly referred to as the reflector height; λ is the wavelength of the GNSS signal (e.g., GPS L1 signal $\lambda = 0.19029$ m, GPS L2 signal $\lambda = 0.24438$ m).

Noting that Eq. (12.3) describes how the SNR oscillates with the sine of the elevation angle (e), the reflector height is related to the oscillation frequency via the relationship

$$H = \frac{f\lambda}{2} \tag{12.4}$$

which is, in turn, used to estimate many cryospheric variables such as snow depth, snow compaction, ice mass balance, and freeze/thaw movements, as detailed in the next few sections. The phase is mainly determined by the soil moisture in the layer of 0–5 cm depth (Larson et al., 2010; Chew et al., 2014), while the amplitude is sensitive to the moisture content in the soil and vegetation (Larson et al., 2008; Small et al., 2010).

The computer program of Roesler and Larson (2018) and the latest updated python package gnssrefl (https://github.com/kristinemlarson/gnssrefl) are recommended for GNSS-IR data processing. Here, we only briefly summarize the key steps. For any given SNR series corresponding to a rising/setting satellite track at low elevation angles (e.g., 5–30 degrees), one needs to first fit and remove a low-order polynomial to obtain the residuals. Then, the Lomb-Scargle Periodogram analysis is applied to the residuals to derive its frequency spectrum. One can then identify the dominant frequency of SNR oscillations from the peak power and subsequently estimate the reflector height by Eq. (12.4).

The footprint of GNSS-IR, determined by the size of the first Fresnel reflection zone, is of the order of 1000 m^2 and increases with the reflector height. In other words, cryospheric variables retrieved by GNSS-IR are averaged over a local region that lies between in situ point measurements and remote sensing observations.

4.3 Snow depth

Estimating snow depth is arguably the most mature cryosphere application of GNSS-IR. The idea is straightforward and illustrated in Fig. 12.10A: using the reflector height above a snow-free bare ground surface (usually an averaged value for summer and early fall) as the initial reference, the difference between the reflector height above the snow surface and the snow-free reference distance gives the snow depth on a given day.

One of the pioneering GNSS-IR snow-depth studies was conducted by Larson et al. (2009) at a GPS site near Boulder, Colorado. By comparing the snow depths estimated by GNSS-IR and measured by ultrasonic snow-depth sensors and by hand during two spring snowstorms, they proved the concept and feasibility of this new method to turn existing GNSS sites into snow sensors.

A few follow-up studies further improved and extended GNSS-IR-based snow studies. For instance, McCreight and Small (2014) estimated daily snow depth, density, and snow water equivalent (SWE) at 155 GPS sites in the western US and validated their products against in situ measurements at 18 sites. The snow bulk density was estimated using the regression model of McCreight and Small (2014) and the product of snow depth and bulk density gives SWE. The validation results of McCreight and Small (2014) showed that near real-time GNSS-based snow depth and SWE estimates gave root mean square errors on the order of 10 and 4 cm, respectively, accurate enough for many applications such as water-resource management and regional climate modeling. Recent technical efforts to combine observables from multiple

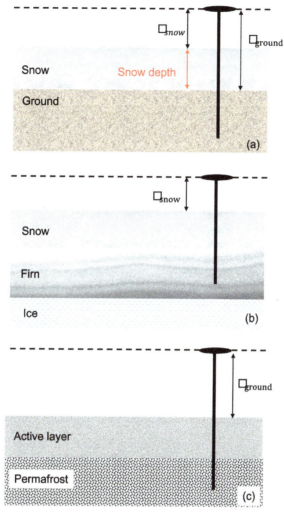

FIG. 12.10 A ground-based GNSS receiver estimates reflector height on a snow surface (A), over a glacier/ice sheet (B), and in a permafrost area (C). The diagrams are not to scale.

GNSS satellites has moved the field toward improved precision and better azimuth coverage (e.g., Tabibi et al., 2017; Zhang et al., 2020).

4.4 Ice mass balance

GNSS sites installed on glaciers and ice sheets, originally used to measure ice flow, can also use the GNSS-IR-retrieved snow surface elevations to place constraints on ice mass balance estimates. Fig. 12.10B shows the geometry of a GNSS monument anchored within the firn layer on a flat ice surface. Using the methods described above, GNSS-IR retrieves the surface elevation changes that are attributed to snow accumulation (resulting in increasing elevation), firn compaction, and surface melting (decreasing elevation). Larson et al. (2015) were the first to propose a method that combines the change in the vertical position of the GNSS antenna with the GNSS-IR-derived snow surface elevations. They tested their method at three sites located in interior Greenland where the observations were used together to validate firn densification models and surface mass balance models.

Shean et al. (2017) and Siegfried et al. (2017) extended a similar idea to GNSS networks in West Antarctica, the former on the floating ice shelf of Pine Island Glacier and the latter on the Whillans and Mercer ice streams. In addition to estimating surface mass balance from the surface elevation changes derived from GNSS-IR, Shean et al. (2017) also combined the GNSS-IR surface elevation with GNSS horizontal velocities, surface mass balance model output, and a dynamic firn model to infer basal melting rates beneath ~1 km of ice. Their estimated basal melting rates ranged from 2 to 40 m/yr, showing good agreement with in situ measurements and revealing higher rates in regions undergoing strong longitudinal extension. Siegfried et al. (2017) validated the GNSS-IR-retrieved reflector height using in situ field measurements and reported an accuracy of 0.02 m, thus providing an effective estimate of surface mass balance.

4.5 Freeze and thaw movements in permafrost areas

Permafrost is defined as ground that remains at or below 0°C for at least two consecutive years. On top of the permafrost is the active layer that undergoes seasonal freeze and thaw. The ground surface in permafrost areas subsides and uplifts annually due to water-ice phase changes. Superimposed on the seasonal cycle, ground surface elevation also changes interannually and secularly (decades or longer) because of permafrost degradation/aggradation and the associated changes in subsurface thermal, hydrological, and mechanical properties.

Liu and Larson (2018) first demonstrated the use of reflected signals received by a GNSS site in northern Utqiagvik (Alaska) to retrieve daily changes in ground surface elevation over 12 thaw seasons from 2004 to 2015. The geometry and the underlying frozen ground layers are illustrated in Fig. 12.10C. Because the anchor position is frozen into the permafrost and the monument is rigid, the temporal changes in reflector height in snow-free conditions are exactly opposite to the elevation changes due to permafrost and active layer dynamics. Liu and Larson (2018) quantified the uncertainties of the retrieved elevation changes as a few centimeters. The magnitude of this uncertainty level is greater than the expected magnitude of day-to-day variations in surface elevation. Therefore, this type of measurement is unreliable for revealing the changes at daily intervals. Still, the study obtained a 12-year time series of daily changes that can resolve temporal variations at seasonal, interannual, and secular timescales. The results revealed a regular thaw-season subsidence whose magnitude varied between 1.1 and 7.5 cm and a small yet statistically significant subsidence trend of 0.26 ± 0.02 cm/yr, which can be used to estimate the loss of ground ice due to permafrost degradation. In 2016, the first snow-in event at the site occurred in November, much later than the other years, thus allowing Hu et al. (2018) to retrieve daily elevation changes spanning the thaw to freezing season. Fig. 12.11A and B presents the estimated snow depth and ground elevation changes on snow-free days, respectively. Once the ground is covered by snow, GNSS-IR gives snow-depth variations that are at least one order of magnitude greater than the frozen ground elevation changes. Fig. 12.11B shows the cyclic elevation changes characterized by thaw subsidence and frost heave.

In a detailed study focusing on the Canadian Arctic, Zhang et al. (2020) used GNSS-IR to retrieve vertical surface movement at five sites (Alert, Resolute Bay, Repulse Bay, Baker Lake, and Iqaluit) to investigate multiyear linear trends. The results revealed that the ground surface underwent subsidence at all sites except Repulse Bay; the subsidence trends were attributed to near-surface permafrost degradation induced by warming air temperatures.

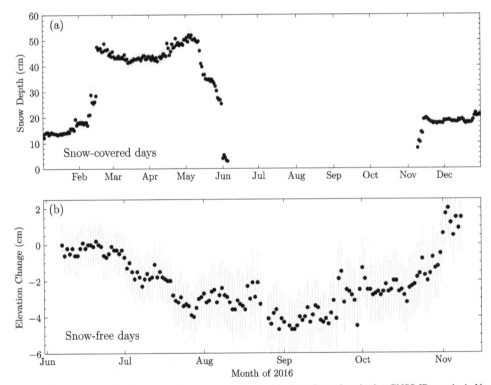

FIG. 12.11 Daily time series of snow depth (A) and elevation changes on snow-free days (B) retrieved using GNSS-IR at a site in Northern Alaska (ID: SG27, coordinates: 71°19′22″ N, 156°36′37″ W). The ground surface underwent thaw subsidence from June to early September, followed by frost heave. *(Data source: Hu, Y., Liu, L., Larson, K.M., Schaefer, K.M., Zhang, J., & Yao, Y. (2018). GPS interferometric reflectometry reveals cyclic elevation changes in thaw and freezing seasons in a permafrost area (Barrow, Alaska). Geophysical Research Letters, 45(11), 5581–5589.)*

4.6 Summary

In the last decade, we have witnessed the innovative use of reflected GNSS signals at existing geodetic-level sites to document and understand various cryospheric properties and processes such as snow depth, snow-water equivalence, snow densification, firn compaction, surface mass balance, basal mass balance, frost heave, and thaw settlement. Despite the fact that most GNSS sites in cold regions were not designed for estimating the variables summarized in this section, GNSS-IR turns noises into signals that provide continuous, long-term, and, in many cases, unprecedented records. At each site, GNSS-IR also offers useful spatial coverage with a footprint of about 1000 m^2, complementary to point and remote-sensing measurements. As both the geodesy and cryosphere communities have realized the potential of GNSS reflectometry, more discoveries are waiting to be made, which will improve our quantitative understanding of the cryosphere in the ever-warming climate.

Appendix

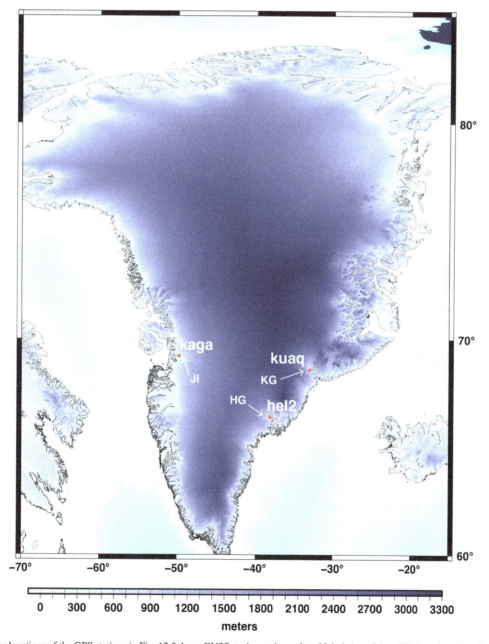

Figure showing the locations of the GPS stations in Fig. 12.3. kaga GNSS station at the outlet of Jakobshavn Isbræ (JI), hel2 GNSS station at the outlet of Helheim Glacier (HG), and kuaq GNSS station at the outlet of Kangerlussuaq Glacier (KG).

References

Adhikari, S., et al. (2021). Decadal to centennial timescale mantle viscosity inferred from modern crustal uplift rates in Greenland. *Geophysical Research Letters*, 48, e2021GL094040. https://doi.org/10.1029/2021GL094040.

Albrecht, T., Winkelmann, R., & Levermann, A. (2020). Glacial-cycle simulations of the Antarctic ice sheet with the parallel ice sheet model (PISM)-part 2: Parameter ensemble analysis. *The Cryosphere*, 14(2). https://doi.org/10.5194/tc-14-633-2020.

Altamimi, Z., Collilieux, X., & Métivier, L. (2011). ITRF2008: An improved solution of the international terrestrial reference frame. *Journal of Geodesy*, 85(8), 457–473. https://doi.org/10.1007/s00190-011-0444-4.

Altamimi, Z., Sillard, P., & Boucher, C. (2002). ITRF2000: A new release of the international terrestrial reference frame for earth science applications. *Journal of Geophysical Research: Solid Earth*, 107(B10). https://doi.org/10.1029/2001JB000561. ETG 2-1-ETG 2-19.

Altamimi, Z., et al. (2007). ITRF2005: A new release of the international terrestrial reference frame based on time series of station positions and earth orientation parameters. *Journal of Geophysical Research*, 112(B9), B09401. https://doi.org/10.1029/2007JB004949.

Altamimi, Z., et al. (2016). ITRF2014: A new release of the international terrestrial reference frame modeling nonlinear station motions. *Journal of Geophysical Research: Solid Earth*, 121(8), 6109–6131. https://doi.org/10.1002/2016JB013098.

Altamimi, Z., et al. (2017). ITRF2014 plate motion model. *Geophysical Journal International*, 209(3), 1906–1912. https://doi.org/10.1093/gji/ggx136.

Argus, D. F., & Peltier, W. R. (2010). Constraining models of postglacial rebound using space geodesy: A detailed assessment of model ICE-5G (VM2) and its relatives. *Geophysical Journal International*, 181(2). https://doi.org/10.1111/j.1365-246X.2010.04562.x.

Argus, D. F., et al. (2014). The Antarctica component of postglacial rebound model ICE-6G_C (VM5a) based on GPS positioning, exposure age dating of ICE thicknesses, and relative sea level histories. *Geophysical Journal International*, 198(1), 537–563. https://doi.org/10.1093/Gji/Ggu140.

Arnadottir, T., et al. (2009). Glacial rebound and plate spreading: Results from the first countrywide GPS observations in Iceland. *Geophysical Journal International*, 177(2), 691–716. https://doi.org/10.1111/j.1365-246X.2008.04059.x.

Austermann, J., et al. (2021). The effect of lateral variations in earth structure on last Interglacial Sea level. *Geophysical Journal International*, 227(3), 1938–1960. https://doi.org/10.1093/gji/ggab289.

Barletta, V. R., et al. (2018). Observed rapid bedrock uplift in Amundsen Sea embayment promotes ice-sheet stability. *Science*, 360(6395), 1335. https://doi.org/10.1126/science.aao1447.

Batchelor, C. L., et al. (2019). The configuration of northern hemisphere ice sheets through the quaternary. *Nature Communications*, 10(3713). https://doi.org/10.1038/s41467-019-11601-2.

Bentley, M. J., et al. (2014). A community-based geological reconstruction of Antarctic ice sheet deglaciation since the last glacial maximum. *Quaternary Science Reviews*, 100, 1–9. https://doi.org/10.1016/j.quascirev.2014.06.025.

Bevis, M., Harig, C., Khan, S. A., Brown, A., Simons, F. J., Willis, M., et al. (2019). Accelerating changes in ice mass within Greenland, and the ice sheet's sensitivity to atmospheric forcing. *PNAS*, 116. https://doi.org/10.1073/pnas.1806562116.

Bevis, M., et al. (2012). Bedrock displacements in Greenland manifest ice mass variations, climate cycles and climate change. *Proceedings of the National Academy of Sciences of the United States of America*, 109(30). https://doi.org/10.1073/pnas.1204664109.

Blank, B., et al. (2021). Effect of lateral and stress-dependent viscosity variations on GIA induced uplift rates in the Amundsen Sea embayment. *Geochemistry, Geophysics, Geosystems*, 22(9). https://doi.org/10.1029/2021GC009807.

Blewitt, G., Hammond, W. C., & Kreemer, C. (2018). Harnessing the GPS data explosion for interdisciplinary science. *Eos*. https://doi.org/10.1029/2018EO104623.

Bradley, S. L., et al. (2009). Glacial isostatic adjustment of the British Isles: New constraints from GPS measurements of crustal motion. *Geophysical Journal International*, 178(1), 14–22. https://doi.org/10.1111/j.1365-246X.2008.04033.x.

Bradley, S. L., et al. (2011). An improved glacial isostatic adjustment model for the British Isles. *Journal of Quaternary Science*, 26(5), 541–552. https://doi.org/10.1002/Jqs.1481.

Bradley, S. L., et al. (2015). Low post-glacial rebound rates in the Weddell Sea due to late Holocene ice-sheet readvance. *Earth and Planetary Science Letters*, 413, 79–89. https://doi.org/10.1016/j.epsl.2014.12.039.

Briggs, R. D., Pollard, D., & Tarasov, L. (2014). A data-constrained large ensemble analysis of Antarctic evolution since the Eemian. *Quaternary Science Reviews*, 103, 91–115. https://doi.org/10.1016/j.quascirev.2014.09.003.

Calais, E., et al. (2006). Deformation of the north American plate interior from a decade of continuous GPS measurements. *Journal of Geophysical Research: Solid Earth*, 111(6). https://doi.org/10.1029/2005JB004253.

Camps, A., Park, H., Pablos, M., Foti, G., Gommenginger, C. P., Liu, P.-W., & Judge, J. (2016). Sensitivity of GNSS-R spaceborne observations to soil moisture and vegetation. *IEEE Journal of Selected Topics in Applied Earth Observations and Remote Sensing*, 9(10), 4730–4742.

Caron, L., et al. (2018). GIA model statistics for GRACE hydrology, cryosphere, and ocean science. *Geophysical Research Letters*, 45(5), 2203–2212. https://doi.org/10.1002/2017gl076644.

Cazenave, A., et al. (2018). Global sea-level budget 1993-present. *Earth System Science Data*, 10(3), 1551–1590. https://doi.org/10.5194/essd-10-1551-2018.

Chew, C. C., & Small, E. E. (2018). Soil moisture sensing using spaceborne GNSS reflections: Comparison of CYGNSS reflectivity to SMAP soil moisture. *Geophysical Research Letters*, 45(9), 4049–4057.

Chew, C. C., Small, E. E., Larson, K. M., & Zavorotny, V. U. (2014). Effects of near-surface soil moisture on GPS SNR data: Development of a retrieval algorithm for soil moisture. *IEEE Transactions on Geoscience and Remote Sensing*, 52(1), 537–543.

Chuter, S. J., et al. (2022). Mass evolution of the Antarctic peninsula over the last 2 decades from a joint Bayesian inversion. *The Cryosphere*, 16(4). https://doi.org/10.5194/tc-16-1349-2022.

Compton, K., Bennett, R. A., Hreinsdóttir, S., van Dam, T., Bordoni, A., Barletta, V., et al. (2017). Short-term variations of Icelandic ice cap mass inferred from cGPS coordinate time series. *Geochemistry, Geophysics, Geosystems, 18*. https://doi.org/10.1002/2017GC006831.

Crawford, O. (2018). Quantifying the sensitivity of post-glacial sea level change to laterally varying viscosity. *Geophysical Journal International, 214*(2), 1324–1363. https://doi.org/10.1093/gji/ggy184.

Dalton, A. S., et al. (2020). An updated radiocarbon-based ice margin chronology for the last deglaciation of the north American ice sheet complex. *Quaternary Science Reviews, 234*.

Dietrich, R., et al. (2010). Rapid crustal uplift in Patagonia due to enhanced ice loss. *Earth and Planetary Science Letters, 289*(1–2), 22–29. https://doi.org/10.1016/J.EPSL.2009.10.021.

Ding, K., et al. (2019). Glacial isostatic adjustment, intraplate strain, and Relative Sea level changes in the eastern United States. *Journal of Geophysical Research: Solid Earth, 124*(6). https://doi.org/10.1029/2018JB017060.

Drouin, V., et al. (2017). Deformation in the northern volcanic zone of Iceland 2008–2014: An interplay of tectonic, magmatic, and glacial isostatic deformation. *Journal of geophysical research: Solid Earth, 122*(4). https://doi.org/10.1002/2016JB013206.

Dziewonski, A. M., & Anderson, D. L. (1981). Preliminary reference earth model. *Physics of the Earth and Planetary Interiors, 25*. https://doi.org/10.1016/0031-9201(81)90046-7.

Engels, O., et al. (2018). Separating geophysical signals using GRACE and High-resolution data: A case study in Antarctica. *Geophysical Research Letters, 45*(22). https://doi.org/10.1029/2018GL079670.

Farrell, W. (1972). Deformation of the earth by surface loads. *Reviews of Geophysics, 10*, 761–797.

Farrell, W. E., & Clark, J. A. (1976). On postglacial sea level. *Geophysical Journal of the Royal Astronomical Society, 46*(3), 647–667.

Fettweis, X., Hofer, S., Krebs-Kanzow, U., Amory, C., Aoki, T., Berends, C. J., ... Zolles, T. (2020). GrSMBMIP: Intercomparison of the modelled 1980–2012 surface mass balance over the Greenland Ice Sheet. *The Cryosphere, 14*, 3935–3958. https://doi.org/10.5194/tc-14-3935-2020.

Foti, G., Gommenginger, C., Jales, P., Unwin, M., Shaw, A., Robertson, C., & Roselló, J. (2015). Spaceborne GNSS reflectometry for ocean winds: First results from the UK TechDemoSat-1 mission. *Geophysical Research Letters, 42*(13), 5435–5441.

Frederikse, T., Landerer, F. W., & Caron, L. (2019). The imprints of contemporary mass redistribution on local sea level and vertical land motion observations. *Solid Earth, 10*(6), 1971–1987. https://doi.org/10.5194/se-10-1971-2019.

Fu, Y., Argus, D. F., & Landerer, F. W. (2015). GPS as an independent measurement to estimate terrestrial water storage variations in Washington and Oregon. *Journal of Geophysical Research - Solid Earth, 120*. https://doi.org/10.1002/2014JB011415.

Ghiasi, Y., Duguay, C. R., Murfitt, J., van der Sanden, J. J., Thompson, A., Drouin, H., & Prévost, C. (2020). Application of GNSS interferometric reflectometry for the estimation of lake ice thickness. *Remote Sensing, 12*(17), 2721.

Ghil, M., Allen, M. R., Dettinger, M. D., Ide, K., Kondrashov, D., Mann, M. E., et al. (2002). Advanced spectral methods for climatic time series. *Reviews of Geophysics, 40*. https://doi.org/10.1029/2000RG000092.

Gomez, N., Pollard, D., & Holland, D. (2015). Sea-level feedback lowers projections of future Antarctic ice-sheet mass loss. *Nature Communications, 6*, 8798. https://doi.org/10.1038/ncomms9798.

Gomez, N., Pollard, D., & Mitrovica, J. X. (2013). A 3-D coupled ice sheet-sea level model applied to Antarctica through the last 40 ky. *Earth and Planetary Science Letters, 384*, 88–99.

Groh, A., et al. (2012). An investigation of Glacial Isostatic adjustment over the Amundsen Sea sector, West Antarctica. *Global and Planetary Change, 98–99*(December 2012), 45–53. https://doi.org/10.1016/j.gloplacha.2012.08.001.

Groh, A., et al. (2014). Assessing the current evolution of the Greenland ice sheet by means of satellite and ground-based observations. *Surveys in Geophysics, 35*(6). https://doi.org/10.1007/s10712-014-9287-x.

Gunter, B. C., et al. (2014). Empirical estimation of present-day Antarctic glacial isostatic adjustment and ice mass change. *The Cryosphere, 8*(2), 743–760. https://doi.org/10.5194/tc-8-743-2014.

Hammond, W. C., et al. (2021). GPS imaging of global vertical land motion for studies of sea level rise. *Journal of Geophysical Research: Solid Earth, 126*(7). https://doi.org/10.1029/2021JB022355.

Hansen, K., Truffer, M., Aschwanden, A., Mankoff, K., Bevis, M., Humbert, A., et al. (2021). Estimating ice discharge at Greenland's three largest outlet glaciers using local bedrock uplift. *Geophysical Research Letters, 48*, e2021GL094252. https://doi.org/10.1029/2021GL094252.

Hattori, A., et al. (2021). GNSS observations of GIA-induced crustal deformation in Lützow-Holm Bay, East Antarctica. *Geophysical Research Letters, 48*(13). https://doi.org/10.1029/2021GL093479.

Heki, K. (2001). Seasonal modulation of interseismic strain buildup in northeastern Japan driven by snow loads. *Science, 293*.

Hermans, T. H. J., van der Wal, W., & Broerse, T. (2018). Reversal of the direction of horizontal velocities induced by GIA as a function of mantle viscosity. *Geophysical Research Letters, 45*(18). https://doi.org/10.1029/2018GL078533.

Hill, E. M., et al. (2010). Combination of geodetic observations and models for glacial isostatic adjustment fields in Fennoscandia. *Journal of Geophysical Research: Solid Earth, 115*, B07403. https://doi.org/10.1029/2009jb006967.

Hu, Y., & Freymueller, J. T. (2019). Geodetic observations of time-variable glacial isostatic adjustment in Southeast Alaska and its implications for earth rheology. *Journal of Geophysical Research: Solid Earth, 124*(9). https://doi.org/10.1029/2018JB017028.

Hu, Y., Liu, L., Larson, K. M., Schaefer, K. M., Zhang, J., & Yao, Y. (2018). GPS interferometric reflectometry reveals cyclic elevation changes in thaw and freezing seasons in a permafrost area (Barrow, Alaska). *Geophysical Research Letters, 45*(11), 5581–5589.

Hughes, A. L. C., et al. (2016). The last Eurasian ice sheets – a chronological database and time-slice reconstruction, DATED-1. *Boreas, 45*(1), 1–45. https://doi.org/10.1111/bor.12142.

Husson, L., Bodin, T., Spada, G., Choblet, G., & Kreemer, C. (2018). Bayesian surface reconstruction of geodetic uplift rates: Mapping the global fingerprint of glacial isostatic adjustment. *Journal of Geodynamics, 122*, 25–40. https://doi.org/10.1016/j.jog.2018.10.002.

IMBIE Team. (2018). Mass balance of the Antarctic ice sheet from 1992 to 2017. *Nature, 558*. https://doi.org/10.1038/s41586-018-0179-y.

IMBIE Team. (2020). Mass balance of the Greenland ice sheet from 1992 to 2018. *Nature, 579*. https://doi.org/10.1038/s41586-019-1855-2.

IPCC. (2021). In V. Masson-Delmotte, P. Zhai, A. Pirani, S. L. Connors, C. Péan, S. Berger, ... B. Zhou (Eds.), *Climate change 2021: The physical science basis. Contribution of working group I to the sixth assessment report of the intergovernmental panel on climate change.* Cambridge, United Kingdom and New York, NY, USA: Cambridge University Press. https://doi.org/10.1017/9781009157896.

Ivins, E. R., et al. (2013). Antarctic contribution to sea-level rise observed by GRACE with improved GIA correction. *Journal of Geophysical Research: Solid Earth, 118*(6), 3126–3141. https://doi.org/10.1002/jgrb.50208.

Jiang, Y., et al. (2021). Assessing global present-day surface mass transport and glacial isostatic adjustment from inversion of geodetic observations. *Journal of Geophysical Research: Solid Earth, 126*(5). https://doi.org/10.1029/2020JB020713.

Johansson, J. M., et al. (2002). Continuous GPS measurements of postglacial adjustment in Fennoscandia – 1. Geodetic results. *Journal of Geophysical Research-Solid Earth, 107*(B8).

Kappelsberger, M. T., et al. (2021). Modeled and observed bedrock displacements in north-East Greenland using refined estimates of present-day ice-mass changes and densified GNSS measurements. *Journal of Geophysical Research: Earth Surface, 126*(4). https://doi.org/10.1029/2020JF005860.

Kaufmann, G., & Wu, P. (2002). Glacial isostatic adjustment in Fennoscandia with a three-dimensional viscosity structure as an inverse problem. *Earth and Planetary Science Letters, 197*(1–2), 1–10. https://doi.org/10.1016/S0012-821x(02)00477-6.

Kaufmann, G., Wu, P., & Ivins, E. R. (2005). Lateral viscosity variations beneath Antarctica and their implications on regional rebound motions and seismotectonics. *Journal of Geodynamics, 39*(2), 165–181.

Khan, S. A., Sasgen, I., Bevis, M., van Dam, T., Bamber, J. L., Wahr, J., et al. (2016). Geodetic measurements reveal similarities between post-last glacial maximum and present-day mass loss from the Greenland ice sheet. *Science Advances, 2*. https://doi.org/10.1126/sciadv.1600931.

Khan, S. A., Wahr, J., Stearns, L. A., Hamilton, G. S., van Dam, T., Larson, K. M., et al. (2007). Elastic uplift in Southeast Greenland due to rapid ice mass loss. *Geophysical Research Letters, 34*, L21701. https://doi.org/10.1029/2007GL031468.

Khan, S. A., et al. (2008). Geodetic measurements of postglacial adjustments in Greenland. *Journal of Geophysical Research: Solid Earth, 113*(2). https://doi.org/10.1029/2007JB004956.

Khan, S. A., et al. (2010). Spread of ice mass loss into Northwest Greenland observed by GRACE and GPS. *Geophysical Research Letters, 37*(6). https://doi.org/10.1029/2010GL042460.

Kierulf, H. P., et al. (2014). A GPS velocity field for Fennoscandia and a consistent comparison to glacial isostatic adjustment models. *Journal of Geophysical Research: Solid Earth, 119*(8), 6613–6629. https://doi.org/10.1002/2013jb010889.

Kierulf, H. P., et al. (2021). A GNSS velocity field for geophysical applications in Fennoscandia. *Journal of Geodynamics, 146*. https://doi.org/10.1016/j.jog.2021.101845.

Kierulf, H. P., et al. (2022). Time-varying uplift in Svalbard—An effect of glacial changes. *Geophysical Journal International, 231*(3), 1518–1534. https://doi.org/10.1093/gji/ggac264.

King, M. A., Bingham, R. J., et al. (2012). Lower satellite-gravimetry estimates of Antarctic Sea-level contribution. *Nature, 491*(7425), 586–589. https://doi.org/10.1038/nature11621.

King, M. A., Keshin, M., et al. (2012). Regional biases in absolute sea-level estimates from tide gauge data due to residual unmodeled vertical land movement. *Geophysical Research Letters, 39*, L14604. https://doi.org/10.1029/2012gl052348.

King, M. A., Watson, C. S., & White, D. (2022). GPS rates of vertical bedrock motion suggest late Holocene ice-sheet readvance in a critical sector of East Antarctica. *Geophysical Research Letters, 49*(4). https://doi.org/10.1029/2021GL097232.

King, M. A., Whitehouse, P. L., & van der Wal, W. (2016). Incomplete separability of Antarctic plate rotation from glacial isostatic adjustment deformation within geodetic observations. *Geophysical Journal International, 204*, 324–330.

King, M. A., et al. (2010). Improved constraints on models of glacial isostatic adjustment: A review of the contribution of ground-based geodetic observations. *Surveys in Geophysics, 31*(5), 465–507. https://doi.org/10.1007/s10712-010-9100-4.

Klemann, V., Martinec, Z., & Ivins, E. R. (2008). Glacial isostasy and plate motion. *Journal of Geodynamics, 46*(3–5), 95–103. https://doi.org/10.1016/j.jog.2008.04.005.

Koulali, A., Whitehouse, P. L., Clarke, P. J., van den Broeke, M. R., Nield, G. A., King, M. A., et al. (2022). GPS-observed elastic deformation due to surface mass balance variability in the Southern Antarctic Peninsula. *Geophysical Research Letters, 49*. https://doi.org/10.1029/2021GL097109.

Kreemer, C., Hammond, W. C., & Blewitt, G. (2018). a robust estimation of the 3-D intraplate deformation of the North American Plate from GPS. *Journal of Geophysical Research: Solid Earth, 123*(5), 4388–4412. https://doi.org/10.1029/2017JB015257.

Kuipers Munneke, P., Ligtenberg, S. R. M., Noël, B. P. Y., Howat, I. M., Box, J. E., Mosley-Thompson, E., ... van den Broeke, M. R. (2015). Elevation change of the Greenland Ice Sheet due to surface mass balance and firn processes, 1960–2014. *The Cryosphere, 9*, 2009–2025. https://doi.org/10.5194/tc-9-2009-2015.

Lambeck, K., Rouby, H., Purcell, A., & Sambridge, M. (2014). Sea level and global ice volumes from the Last Glacial Maximum to the Holocene. *PNAS, 111*(43), 15296–15303. https://doi.org/10.1073/pnas.1411762111.

Lambeck, K., Smither, C., & Johnston, P. (1998). Sea-level change, glacial rebound and mantle viscosity for northern Europe. *Geophysical Journal International, 134*(1), 102–144.

Lange, H., et al. (2014). Observed crustal uplift near the southern Patagonian icefield constrains improved viscoelastic earth models. *Geophysical Research Letters, 41*(3), 805–812. https://doi.org/10.1002/2013gl058419.

Larsen, C. F., et al. (2005). Rapid viscoelastic uplift in Southeast Alaska caused by post-little ice age glacial retreat. *Earth and Planetary Science Letters, 237*(3–4), 548–560. https://doi.org/10.1016/j.epsl.2005.06.032.

Larson, K. M. (2016). GPS interferometric reflectometry: applications to surface soil moisture, snow depth, and vegetation water content in the western United States. *Wiley Interdisciplinary Reviews: Water, 3*(6), 775–787.

Larson, K. M. (2019). Unanticipated uses of the global positioning system. *Annual Review of Earth and Planetary Sciences, 47*(1), 19–40.

Larson, K. M., Braun, J. J., Small, E. E., Zavorotny, V. U., Gutmann, E. D., & Bilich, A. L. (2010). GPS multipath and its relation to near-surface soil moisture content. *IEEE Journal of Selected Topics in Applied Earth Observations and Remote Sensing, 3*(1), 91–99.

Larson, K. M., Gutmann, E. D., Zavorotny, V. U., Braun, J. J., Williams, M. W., & Nievinski, F. G. (2009). Can we measure snow depth with GPS receivers? *Geophysical Research Letters, 36*(17), L17502.

Larson, K. M., Ray, R. D., Nievinski, F. G., & Freymueller, J. T. (2013). The accidental tide gauge: A GPS reflection case study from Kachemak Bay, Alaska. *IEEE Geoscience and Remote Sensing Letters, 10*(5), 1200–1204.

Larson, K. M., & Small, E. E. (2014). Normalized microwave reflection index: A vegetation measurement derived from GPS networks. *IEEE Journal of Selected Topics in Applied Earth Observations and Remote Sensing, 7*(5), 1501–1511.

Larson, K. M., Small, E. E., Gutmann, E. D., Bilich, A. L., Braun, J. J., & Zavorotny, V. U. (2008). Use of GPS receivers as a soil moisture network for water cycle studies. *Geophysical Research Letters, 35*(24), L24405.

Larson, K. M., Wahr, J., & Munneke, P. K. (2015). Constraints on snow accumulation and firn density in Greenland using GPS receivers. *Journal of Glaciology, 61*(225), 101–114.

Lau, H. C. P., et al. (2016). Inferences of mantle viscosity based on ice age data sets: Radial structure. *Journal of Geophysical Research: Solid Earth, 121*(10), 6991–7012. https://doi.org/10.1002/2016jb013043.

Lau, H. C. P., et al. (2018). Inferences of mantle viscosity based on ice age data sets: The bias in radial viscosity profiles due to the neglect of laterally heterogeneous viscosity structure. *Journal of Geophysical Research: Solid Earth, 123*(9), 7237–7252. https://doi.org/10.1029/2018JB015740.

Lecavalier, B. S., et al. (2014). A model of Greenland ice sheet deglaciation constrained by observations of relative sea level and ice extent. *Quaternary Science Reviews, 102*, 54–84. https://doi.org/10.1016/j.quascirev.2014.07.018.

Li, W., Cardellach, E., Fabra, F., Ribó, S., & Rius, A. (2018). Lake level and surface topography measured with spaceborne GNSS-reflectometry from CYGNSS mission: Example for the Lake Qinghai. *Geophysical Research Letters, 45*(24).

Lidberg, M., et al. (2007). An improved and extended GPS-derived 3D velocity field of the glacial isostatic adjustment (GIA) in Fennoscandia. *Journal of Geodesy, 81*(3), 213–230. https://doi.org/10.1007/s00190-006-0102-4.

Lidberg, M., et al. (2010). Recent results based on continuous GPS observations of the GIA process in Fennoscandia from BIFROST. *Journal of Geodynamics, 50*(1), 8–18. https://doi.org/10.1016/j.jog.2009.11.010.

Longman, I. M. (1962). A Green's function for determining the deformation of the earth under surface mass loads, 1. Theory. *Journal of Geophysical Research, 67*, 845.

Longman, I. M. (1963). A Green's function for determining the deformation of the earth under surface mass loads, 2. Computations and numerical results. *Journal of Geophysical Research, 68*, 485.

Longman, I. M. (1966). Computation of love numbers and load deformation coefficients for a model earth. *Geophysical Journal International, 11*, 133–137.

Ludwigsen, C. A., et al. (2020). Vertical land motion from present-day deglaciation in the wider Arctic. *Geophysical Research Letters, 47*(19). https://doi.org/10.1029/2020GL088144.

Marotta, A. M., & Sabadini, R. (2004). The signatures of tectonics and glacial isostatic adjustment revealed by the strain rate in Europe. *Geophysical Journal International, 157*(2), 865–870. https://doi.org/10.1111/j.1365-246X.2004.02275.x.

Marsman, C. P., et al. (2021). The impact of a 3-D earth structure on glacial isostatic adjustment in Southeast Alaska following the little ice age. *Journal of Geophysical Research: Solid Earth, 126*(12). https://doi.org/10.1029/2021JB022312.

Martín-Español, A., et al. (2016a). An assessment of forward and inverse GIA solutions for Antarctica. *Journal of Geophysical Research: Solid Earth, 121*(9), 6947–6965. https://doi.org/10.1002/2016jb013154.

Martín-Español, A., et al. (2016b). Spatial and temporal Antarctic ice sheet mass trends, glacio-isostatic adjustment, and surface processes from a joint inversion of satellite altimeter, gravity, and GPS data. *Journal of Geophysical Research: Earth Surface, 121*(2), 182–200. https://doi.org/10.1002/2015jf003550.

McConnell, R. K. (1968). Viscosity of mantle from relaxation time spectra of isostatic adjustment. *Journal of Geophysical Research, 73*(22), 7089–7105. https://doi.org/10.1029/JB073i022p07089.

McCreight, J. L., & Small, E. E. (2014). Modeling bulk density and snow water equivalent using daily snow depth observations. *The Cryosphere, 8*, 521–536.

Métivier, L., Altamimi, A., & Rouby, H. (2020). Past and present ITRF solutions from geophysical perspectives. *Advances in Space Research, 65*(12). https://doi.org/10.1016/j.asr.2020.03.031.

Milne, G. A., & Shennan, I. (2013). Isostasy: Glaciation-induced sea-level change. In S. A. Elias, & C. J. Mock (Eds.), *Encyclopedia of quaternary science* (pp. 452–459). Oxford: Elsevier.

Milne, G. A., et al. (2001). Space-geodetic constraints on glacial isostatic adjustment in Fennoscandia. *Science, 291*(5512), 2381–2385.

Milne, G. A., et al. (2004). Continuous GPS measurements of postglacial adjustment in Fennoscandia: 2. Modeling results. *Journal of Geophysical Research: Solid Earth, 109*(B2), B02412. https://doi.org/10.1029/2003jb002619.

Mitrovica, J. X. (1996). Haskell [1935] revisited. *Journal of Geophysical Research: Solid Earth, 101*(B1), 555–569. https://doi.org/10.1029/95jb03208.

Mitrovica, J. X., & Peltier, W. R. (1993). A new formalism for inferring mantle viscosity based on estimates of post glacial decay times: Application to RSL variations in N.E. Hudson Bay. *Geophysical Research Letters, 20*(20). https://doi.org/10.1029/93GL02136.

Morris, E. M., & Wingham, D. J. (2015). Uncertainty in mass-balance trends derived from altimetry: A case study along the EGIG line, Central Greenland. *Journal of Glaciology, 61*. https://doi.org/10.3189/2015JoG14J123.

Mottram, R., Hansen, N., Kittel, C., van Wessem, J. M., Agosta, C., Amory, C., et al. (2021). What is the surface mass balance of Antarctica? An intercomparison of regional climate model estimates. *The Cryosphere, 15*(8). https://doi.org/10.5194/tc-15-3751-2021.

Murakami, M., & Miyazaki, S. (2001). Periodicity of strain accumulation detected by permanent GPS array: Possible relationship to seasonality of major earthquakes' occurrence. *Geophysical Research Letters, 28,* 2983–2986. https://doi.org/10.1029/2001GL013015.

Nield, G. A., et al. (2014). Rapid bedrock uplift in the Antarctic Peninsula explained by viscoelastic response to recent ice unloading. *Earth and Planetary Science Letters, 397,* 32–41. https://doi.org/10.1016/j.epsl.2014.04.019.

Nielsen, K., et al. (2012). Crustal uplift due to ice mass variability on Upernavik Isstrøm, West Greenland. *Earth and Planetary Science Letters, 353–354.* https://doi.org/10.1016/j.epsl.2012.08.024.

Nielsen, K., et al. (2013). Vertical and horizontal surface displacements near Jakobshavn Isbræ driven by melt-induced and dynamic ice loss. *Journal of Geophysical Research: Solid Earth, 118*(4). https://doi.org/10.1002/jgrb.50145.

Nocquet, J. M., Calais, E., & Parsons, B. (2005). Geodetic constraints on glacial isostatic adjustment in Europe. *Geophysical Research Letters, 32*(6). https://doi.org/10.1029/2004GL022174.

Noël, B., van de Berg, W. J., van Wessem, J. M., van Meijgaard, E., van As, D., Lenaerts, J. T. M., et al. (2018). Modelling the climate and surface mass balance of polar ice sheets using RACMO2 – Part 1: Greenland (1958–2016). *The Cryosphere, 12,* 2018. https://doi.org/10.5194/tc-12-811-2018.

Noël, B., van Kampenhout, L., van de Berg, W. J., Lenaerts, J. T. M., Wouters, B., van den Broeke, M. R., et al. (2019). Brief communication: CESM2 climate forcing (1950–2014) yields realistic Greenland ice sheet surface mass balance. *The Cryosphere Discussions,* 1–17. https://doi.org/10.5194/tc-2019-209.

ODonnell, J. P., et al. (2017). The uppermost mantle seismic velocity and viscosity structure of central West Antarctica. *Earth and Planetary Science Letters, 472,* 38–49.

Pagli, C., & Sigmundsson, F. (2008). Will present day glacier retreat increase volcanic activity? Stress induced by recent glacier retreat and its effect on magmatism at the Vatnajökull ice cap, Iceland. *Geophysical Research Letters, 35.* https://doi.org/10.1029/2008GL033510.

Pagli, C., Sigmundsson, F., Pedersen, R., Einarsson, P., Arnadottir, T., & Feigl, K. L. (2007). Crustal deformation associated with the 1996Gj.lp subglacial eruption, Iceland: InSAR studies in affected areas adjacent to the Vatnajökull ice cap. *Earth and Planetary Science Letters, 259,* 24–33. https://doi.org/10.1016/j.epsl.2007.04.019.

Patton, H., et al. (2017). Deglaciation of the Eurasian ice sheet complex. *Quaternary Science Reviews, 169,* 148–172.

Paxman, G. J. G., et al. (2023). Inference of the timescale-dependent apparent viscosity structure in the upper mantle beneath Greenland. *AGU Advances, 4*(2). https://doi.org/10.1029/2022av000751.

Peltier, W. R. (2004). Global glacial isostasy and the surface of the ICE-age earth: The ICE-5G (VM2) model and GRACE. *Annual Review of Earth and Planetary Sciences, 32,* 111–149.

Peltier, W. R., Argus, D. F., & Drummond, R. (2015). Space geodesy constrains ICE age terminal deglaciation: The global ICE-6G_C (VM5a) model. *Journal of Geophysical Research: Solid Earth, 120*(1), 450–487. https://doi.org/10.1002/2014jb011176.

Piña-Valdés, J., et al. (2022). 3D GNSS velocity field sheds light on the deformation mechanisms in Europe: Effects of the vertical crustal motion on the distribution of seismicity. *Journal of Geophysical Research: Solid Earth, 127*(6). https://doi.org/10.1029/2021JB023451.

Pinel, V., Sigmundsson, F., Sturkell, E., Geirsson, H., Einarsson, P., Gudmundsson, M. T., & Högnadóttir, T. (2007). Discriminating volcano deformation due to magma movements and variable surface loads: Application to Katla subglacial volcano, Iceland. *Geophysical Journal International, 169*(1), 325–338. https://doi.org/10.1111/j.1365-246X.2006.03267.x.

Pittard, M. L., et al. (2022). An ensemble of Antarctic deglacial simulations constrained by geological observations. *Quaternary Science Reviews, 298,* 107800. https://doi.org/10.1016/j.quascirev.2022.107800.

Powell, E., et al. (2022). The robustness of geodetically derived 1-D Antarctic viscosity models in the presence of complex 3-D viscoelastic earth structure. *Geophysical Journal International, 231*(1). https://doi.org/10.1093/gji/ggac129.

Richter, A., et al. (2016). Crustal deformation across the southern Patagonian icefield observed by GNSS. *Earth and Planetary Science Letters, 452,* 206–215. https://doi.org/10.1016/j.epsl.2016.07.042.

Riva, R. E. M., et al. (2009). Glacial isostatic adjustment over Antarctica from combined ICESat and GRACE satellite data. *Earth and Planetary Science Letters, 288*(3–4), 516–523. https://doi.org/10.1016/j.epsl.2009.10.013.

Ruf, C., Unwin, M., Dickinson, J., Rose, R., Rose, D., Vincent, M., & Lyons, A. (2013). CYGNSS: Enabling the future of hurricane prediction. *IEEE Geoscience and Remote Sensing Magazine, 1,* 52–67.

Samrat, N. H., et al. (2020). Reduced ice mass loss and three-dimensional viscoelastic deformation in northern Antarctic peninsula inferred from GPS. *Geophysical Journal International, 222*(2). https://doi.org/10.1093/gji/ggaa229.

Samrat, N. H., et al. (2021). Upper mantle viscosity underneath northern Marguerite Bay, Antarctic peninsula constrained by bedrock uplift and ice mass variability. *Geophysical Research Letters.* https://doi.org/10.1029/2021GL097065.

Sandberg Sørensen, L., et al. (2018). 25 years of elevation changes of the Greenland ice Sheet from ERS, Envisat, and CryoSat-2 radar altimetry. *Earth and Planetary Science Letters, 495.* https://doi.org/10.1016/j.epsl.2018.05.015.

Sasgen, I., et al. (2013). Antarctic ice-mass balance 2003 to 2012: Regional reanalysis of GRACE satellite gravimetry measurements with improved estimate of glacial-isostatic adjustment based on GPS uplift rates. *The Cryosphere, 7*(5), 1499–1512. https://doi.org/10.5194/tc-7-1499-2013.

Sasgen, I., et al. (2017). Joint inversion estimate of regional glacial isostatic adjustment in Antarctica considering a lateral varying earth structure (ESA STSE project REGINA). *Geophysical Journal International, 211*(3), 1534–1553. https://doi.org/10.1093/gji/ggx368.

Schmidt, P., Lund, B., Hieronymus, C., Maclennan, J., Àrnadòttir, T., & Pagli, C. (2013). Effects of present-day deglaciation in Iceland on mantle melt production rates. *Journal of Geophysical Research - Solid Earth, 118.* https://doi.org/10.1002/jgrb.50273.

Schoen, N., et al. (2015). Simultaneous solution for mass trends on the West Antarctic ice sheet. *The Cryosphere, 9,* 805–819.

Schröder, L., et al. (2019). Four decades of Antarctic surface elevation changes from multi-mission satellite altimetry. *The Cryosphere, 13*(2). https://doi.org/10.5194/tc-13-427-2019.

Schumacher, M., et al. (2018). A new global GPS data set for testing and improving modelled GIA uplift rates. *Geophysical Journal International, 214*(3). https://doi.org/10.1093/gji/ggy235.

Sella, G. F., et al. (2007). Observation of glacial isostatic adjustment in "stable" North America with GPS. *Geophysical Research Letters, 34*(2), L02306.

Selway, K., et al. (2020). Magnetotelluric constraints on the temperature, composition, partial melt content, and viscosity of the upper mantle beneath Svalbard. *Geochemistry, Geophysics, Geosystems, 21*(5). https://doi.org/10.1029/2020GC008985.

Sha, Z., et al. (2019). Bayesian model–data synthesis with an application to global glacio-isostatic adjustment. *Environmetrics, 30*(1). https://doi.org/10.1002/env.2530.

Shean, D. E., Christianson, K., Larson, K. M., Ligtenberg, S. R. M., Joughin, I. R., Smith, B. E., ... Holland, D. M. (2017). GPS-derived estimates of surface mass balance and ocean-induced basal melt for Pine Island Glacier ice shelf, Antarctica. *The Cryosphere, 11*, 2655–2674.

Siegfried, M. R., Medley, B., Larson, K. M., Fricker, H. A., & Tulaczyk, S. (2017). Snow accumulation variability on a West Antarctic ice stream observed with GPS reflectometry, 2007–2017. *Geophysical Research Letters, 44*.

Sigmundsson, F., Pinel, V., Lund, B., Fabien, A., Pagli, C., Geirsson, H., et al. (2010). Climate effects on volcanism: Influence on magmatic systems of loading and unloading from ice mass variations, with examples from Iceland. *Philosophical Transactions of the Royal Society A, 368*. https://doi.org/10.1098/rsta2010.0042.

Simms, A. R., et al. (2019). Balancing the last glacial maximum (LGM) sea-level budget. *Quaternary Science Reviews, 205*, 143–153. https://doi.org/10.1016/j.quascirev.2018.12.018.

Simon, K. M., Riva, R. E. M., & Vermeersen, L. L. A. (2021). Constraint of glacial isostatic adjustment in the North Sea with geological relative sea level and GNSS vertical land motion data. *Geophysical Journal International, 227*(2), 1168–1180. https://doi.org/10.1093/gji/ggab261.

Simon, K. M., et al. (2016). A glacial isostatic adjustment model for the central and northern Laurentide ice sheet based on relative sea level and GPS measurements. *Geophysical Journal International, 205*(3), 1618–1636. https://doi.org/10.1093/gji/ggw103.

Simon, K. M., et al. (2017). A data-driven model for constraint of present-day glacial isostatic adjustment in North America. *Earth and Planetary Science Letters, 474*, 322–333. https://doi.org/10.1016/j.epsl.2017.06.046.

Simon, K. M., et al. (2018). The glacial isostatic adjustment signal at present day in northern Europe and the British Isles estimated from geodetic observations and geophysical models. *Solid Earth, 9*(3). https://doi.org/10.5194/se-9-777-2018.

Simpson, M. J. R., et al. (2011). The influence of decadal- to millennial-scale ice mass changes on present-day vertical land motion in Greenland: Implications for the interpretation of GPS observations. *Journal of Geophysical Research, 116*, B02406. https://doi.org/10.1029/2010jb007776.

Small, E. E., Larson, K. M., & Braun, J. J. (2010). Sensing vegetation growth with GPS reflections. *Geophysical Research Letters, 37*, L12401. https://doi.org/10.1029/2010GL042951.

Spada, G., et al. (2012). Greenland uplift and regional sea level changes from ICESat observations and GIA modelling. *Geophysical Journal International, 189*(3), 1457–1474. https://doi.org/10.1111/j.1365-246X.2012.05443.x.

Steffen, H., Kaufmann, G., & Wu, P. (2006). Three-dimensional finite-element modeling of the glacial isostatic adjustment in Fennoscandia. *Earth and Planetary Science Letters, 250*(1–2), 358–375. https://doi.org/10.1016/j.epsl.2006.08.003.

Tabibi, S., Geremia-Nievinski, F., & van Dam, T. (2017). Statistical comparison and combination of GPS, GLONASS, and multi-GNSS multipath reflectometry applied to snow depth retrieval. *IEEE Transactions on Geoscience and Remote Sensing, 55*(7), 3773–3785.

Tarasov, L., et al. (2012). A data-calibrated distribution of deglacial chronologies for the north American ice complex from glaciological modeling. *Earth and Planetary Science Letters, 315*, 30–40. https://doi.org/10.1016/j.epsl.2011.09.010.

Thomas, I. D., et al. (2011). Widespread low rates of Antarctic glacial isostatic adjustment revealed by GPS observations. *Geophysical Research Letters, 38*, L22302. https://doi.org/10.1029/2011GL049277.

Turner, R. J., Reading, A. M., & King, M. A. (2020). Separation of tectonic and local components of horizontal GPS station velocities: A case study for glacial isostatic adjustment in East Antarctica. *Geophysical Journal International, 222*(3). https://doi.org/10.1093/gji/ggaa265.

van Dam, T., et al. (2017). Using GPS and absolute gravity observations to separate the effects of present-day and Pleistocene ice-mass changes in south East Greenland. *Earth and Planetary Science Letters, 459*, 127–135.

van den Berg, J., et al. (2008). Effect of isostasy on dynamical ice sheet modeling: A case study for Eurasia. *Journal of Geophysical Research, 113*(B5), B05412. https://doi.org/10.1029/2007jb004994.

van Wessem, J. M., van de Berg, W. J., Noel, B. P. Y., van Meijgaard, E., Amory, C., Birnbaum, G., et al. (2018). Modelling the climate and surface mass balance of polar ice sheets using RACMO2-part 2: Antarctica (1979–2016). *The Cryosphere, 12*(4). https://doi.org/10.5194/tc-12-1479-2018.

Vardić, K., Clarke, P. J., & Whitehouse, P. L. (2022). A GNSS velocity field for crustal deformation studies: The influence of glacial isostatic adjustment on plate motion models. *Geophysical Journal International, 231*(1), 426–458. https://doi.org/10.1093/gji/ggac047.

Velicogna, I., Mohajerani, Y., Geruo, A., Landerer, F., Mouginot, J., Noel, B., et al. (2020). Continuity of ice sheet mass loss in Greenland and Antarctica from the GRACE and GRACE follow-on missions. *Geophysical Research Letters, 47*, e2020GL087291. https://doi.org/10.1029/2020GL087291.

Velicogna, I., & Wahr, J. (2002). A method for separating Antarctic postglacial rebound and ice mass balance using future ICESat geoscience laser altimeter system, gravity recovery and climate experiment, and GPS satellite data. *Journal of Geophysical Research, B10*. https://doi.org/10.1029/2001JB000708.

Vishwakarma, B. D., et al. (2022). Separating GIA signal from surface mass change using GPS and GRACE data. *Geophysical Journal International, 232*(1), 537–547. https://doi.org/10.1093/gji/ggac336.

Wahr, J., Dazhong, H., & Trupin, A. (1995). Predictions of vertical uplift caused by changing polar ice volumes on a viscoelastic earth. *Geophysical Research Letters, 22*(8), 977–980. https://doi.org/10.1029/94gl02840.

Wahr, J., Khan, S. A., van Dam, T., Liu, L., Angelen, J. H., van den Broeke, M. R., et al. (2013). The use of GPS horizontals for loading studies, with applications to northern California and Southeast Greenland. *Journal of Geophysical Research: Solid Earth, 118*(4). https://doi.org/10.1002/jgrb.50104.

Wahr, J., Wingham, D., & Bentley, C. (2000). A method of combining ICESat and GRACE satellite data to constrain Antarctic mass balance. *Journal of Geophysical Research, 105*(B7), 16279–16294.

Wake, L. M., Lecavalier, B. S., & Bevis, M. (2016). Glacial isostatic adjustment (GIA) in Greenland: A review. *Current Climate Change Reports*. https://doi.org/10.1007/s40641-016-0040-z.

Walcott, R. I. (1972). Past sea levels, eustasy and deformation of the earth. *Quaternary Research, 2*(1), 1–14.

Wang, H., Xiang, L., Jia, L., Jiang, L., Wang, Z., Hu, B., et al. (2012). Load love numbers and Green's functions for elastic earth models PREM, iasp91, ak135, and modified models with refined crustal structure rom crust 2.0. *Computers and Geosciences, 49*. https://doi.org/10.1016/j.cageo.2012.06.022.

Wang, H., et al. (2022). GRACE-based estimates of groundwater variations over North America from 2002 to 2017. *Geodesy and Geodynamics, 13*(1), 11–23. https://doi.org/10.1016/j.geog.2021.10.003.

Whitehouse, P. L. (2018). Glacial isostatic adjustment modelling: Historical perspectives, recent advances, and future directions. *Earth Surface Dynamics, 6*(2), 401–429. https://doi.org/10.5194/esurf-6-401-2018.

Whitehouse, P. L., Bentley, M. J., & Le Brocq, A. M. (2012). A deglacial model for Antarctica: Geological constraints and glaciological modelling as a basis for a new model of Antarctic glacial isostatic adjustment. *Quaternary Science Reviews, 32*, 1–24.

Whitehouse, P. L., Bentley, M. J., Milne, G. A., et al. (2012). A new glacial isostatic adjustment model for Antarctica: Calibrated and tested using observations of relative sea-level change and present-day uplift rates. *Geophysical Journal International, 190*(3), 1464–1482. https://doi.org/10.1111/j.1365-246X.2012.05557.x.

Whitehouse, P., et al. (2006). Impact of 3-D earth structure on Fennoscandian glacial isostatic adjustment: Implications for space-geodetic estimates of present-day crustal deformations. *Geophysical Research Letters, 33*(13), L13502.

Whitehouse, P. L., et al. (2019). Solid earth change and the evolution of the Antarctic ice sheet. *Nature Communications, 10*. https://doi.org/10.1038/s41467-018-08068-y.

Wolstencroft, M., et al. (2015). Uplift rates from a new high-density GPS network in Palmer Land indicate significant late Holocene ice loss in the southwestern Weddell Sea. *Geophysical Journal International, 203*, 737–754. https://doi.org/10.1093/gji/ggv327.

Woppelmann, G., & Marcos, M. (2016). Vertical land motion as a key to understanding sea level change and variability. *Reviews of Geophysics, 54*(1), 64–92. https://doi.org/10.1002/2015rg000502.

Wu, X. P., et al. (2010). Simultaneous estimation of global present-day water transport and glacial isostatic adjustment. *Nature Geoscience, 3*(9), 642–646. https://doi.org/10.1038/ngeo938.

Zhang, B., Zhanga, E., Liu, L., Khan, S. A., vanDam, T., Yaob, Y., et al. (2018). Geodetic measurements reveal short-term changes of glacial mass near Jakobshavn Isbræ (Greenland) from 2007 to 2017. *EPSL, 503*. https://doi.org/10.1016/j.epsl.2018.09.029.

Zhang, J., Liu, L., & Hu, Y. (2020). Global Positioning System interferometric reflectometry (GPS-IR) measurements of ground surface elevation changes in permafrost areas in northern Canada. *The Cryosphere, 14*(6), 1875–1888.

Zhao, C., et al. (2017). Rapid ice unloading in the Fleming glacier region, southern Antarctic peninsula, and its effect on bedrock uplift rates. *Earth and Planetary Science Letters, 473*, 164–176.

Ziegler, Y., et al. (2022). Can GPS and GRACE data be used to separate past and present-day surface loading in a data-driven approach? *Geophysical Journal International, 232*(2), 884–901. https://doi.org/10.1093/gji/ggac365.

Chapter 13

The role of GNSS monitoring in landslide research

Halldór Geirsson and Þorsteinn Sæmundsson
Institute of Earth Sciences, Science Institute, University of Iceland, Reykjavík, Iceland

1 Introduction

Landslides and their secondary effects are one of the important natural hazards that humans need to deal with. During 2004–2016, more than 55,000 people lost their lives due to landslides, and monetary losses are estimated to be over 20 billion USD annually (Sim et al., 2022). With increasing urbanization, building land may grow to cover known active landslide areas in some places, e.g., in the Funu landslide in the Democratic Republic of the Congo (Fig. 13.1), posing a risk to the inhabitants (Dille et al., 2022). With changing climate, several factors that affect landslide triggering and motion are of importance. The first factor is increased extreme rainfall events, a well-known trigger for landslide motion. The IPCC report states that the frequency and intensity of heavy precipitation events have likely already increased on a global scale over the majority of land masses (IPCC, 2023). It is further predicted that heavy precipitation will generally become more intense and more frequent, rendering overall increasing landslide hazards. Glacial and permafrost thawing also has an effect on landslide hazards. The high-latitude and high-elevation thawing of permafrost due to warming climate acts to destabilize slopes (e.g., Morino et al., 2019, 2021; Patton et al., 2019; Sæmundsson et al., 2018). Furthermore, the retreat of glaciers leaves steeply carved ("U-shaped") valleys that are prone to failure (Ben-Yehoshua et al., 2022; Kos et al., 2016; Lacroix et al., 2022; Sæmundsson et al., 2022). It is important to note that these triggering and destabilization mechanisms may not only affect shallow landslides and soil slumps but also large-scale deep-seated landslide bodies.

Landslide hazards often fall within the realm of cascading hazards. The abovementioned example of heavy rainfall-triggered landslides is one example, however, landslides can also fall into or occur at the bottom of water bodies, causing devastating tsunamis. Examples of this include a landslide into Lake Askja, Iceland, in 2014, causing a tsunami with run-up reaching approximately 80m (Gylfadóttir et al., 2017). The largest known tsunami run-up in recent decades exceeds 500m, in Lituya Bay in southeast Alaska (Fritz et al., 2009). In this event, a magnitude 8.3 earthquake in 1958 triggered a ~30 Mm^3 landslide into a narrow fjord. Several volcanic islands around the globe are susceptible to large-scale slope failure and thus have the potential to create tsunamis large enough to cause damage at across the ocean basins; for example, at the Canary Islands (Masson et al., 2002). Fully submarine landslides originating from continental shelves can also bring large tsunamis, for example, the Storegga landslides off the coast of Norway 8100 years ago (Bugge et al., 1988). Another well-known example of cascading hazards involving landslides and density currents in general comes from Mount St. Helens, where the cryptodome growth triggered a landslide, which released pressure on the magmatic system, causing an explosive eruption and a tsunami in Spirit Lake (Harris, 1988). Indeed, rockfall and collapse, i.e., landslides, from active domes is a well-known triggering effect for explosive eruptions (e.g., Voight & Elsworth, 2000).

The deformation rate of landslides is highly variable: from several millimeters per year for slow-creeping landslides, up to ~100 km/h for collapsing landslides and mudflows. The spatial scales also span orders of magnitude: from individual boulders to whole mountainsides. In general, landslide motion is quite variable in space and time and thus challenging to monitor. Some landslides have been observed to accelerate motion before catastrophic failure (e.g., Scoppettuolo et al., 2020), while many landslides show periods of acceleration and deceleration in relation to precipitation (see examples below). The complexities of landslide motion highlight the importance of continuous deformation monitoring to give warnings for large-scale motion and to understand the nature of these complex deformations.

There are several techniques in addition to GNSS that can be used to measure landslide motion. These include Interferometric Synthetic Aperture Radar (InSAR); ground-based SAR; wire extensometers; leveling equipment; triangulation; various forms of electronic distance measurements [EDM; total stations; terrestrial laser scanners (TLS)]; fiber optics;

FIG. 13.1 Houses crowding the Funu landslide in Bukavu, Democratic Republic of the Congo. *(Figure courtesy of Antoine Dille.)*

optical imaging; and inclinometers (tiltmeters). All these techniques have their advantages and disadvantages. Below, we review the basics of landslide terminology, reflect on the technical practicalities of GNSS measurements in landslides, introduce several specific case studies, and finally give a perspective on the usefulness of GNSS for landslide research in relation to other available deformation techniques.

2 Landslide motion and landslide types

Landslide is a commonly used term for discrete slope movements of rock, debris, or soil (Crozier, 1989), but can also be referred to as mass wasting, mass movements, or slope failures (e.g., Sæmundsson et al., 2022). Landslides refer in general to the process of gravitational downslope movement of rock, debris, or soil (e.g., Crozier, 2002; Hermanns, 2018; Hutchinson, 1968) and can take place in different environmental settings. The topographic terrain that best facilitates landslides is steep, however, observations show that landslides can also occur in areas of very gentle slope (<2°), in which cases a specific weak layer is needed to facilitate the movement (e.g., Gatter et al., 2021). Landslides and their secondary hazards, such as glacial lake outburst floods, tsunami waves, and river damming by large rock slope failures, can threaten human life and damage infrastructure (Crozier & Glade, 2005; Geertsema et al., 2009; Haque et al., 2016; Kjekstad & Highland, 2009; Oppikofer et al., 2020; Petley, 2012; Strom & Korup, 2006).

Mass movements at landslides are conventionally divided into slow or rapid movements, which can either develop over years, decades, centuries, or even longer, or occur near-instantaneously. Slow moving landslides, also termed creeping landslides, with appreciable thickness, are termed deep-seated gravitational slope deformation (DSGSD; e.g., Dramis & Sorriso-Valvo, 1995; Lacroix, Handwerger, & Bièvre, 2020). The instantaneously failing landslides are often referred to as high-magnitude, low-frequency processes (Hungr, 2007) and may possibly have high impact. DSGSD is usually not considered to be a direct hazard due to their slow rate of movement; however, parts of the moving mass can accelerate and cause catastrophic failures such as rock avalanches or rockslides (Agliardi et al., 2012).

Triggering factors for landslide motion can be of various origin. Additional input of water such as intense or prolonged precipitation or rapid snowmelt is a common triggering factor (e.g., Hungr et al., 2014; Oliveira et al., 2022; Sæmundsson et al., 2003). Hydrologic triggering affects landslides in two ways; by increasing the net mass of the landslide, and by increasing the pore pressure, which in turn lowers normal stress on weak sliding planes and affects friction coefficients (e.g., Agliardi et al., 2020; Bogaard & Greco, 2016). Erosion, such as undercutting of slopes by coastal waves, river erosion, or glacial erosion, gradually causes oversteepening and promotes landslide failure (e.g., Ballantyne, 2002; Ben-Yehoshua et al., 2022; McColl & Davies, 2013). Hence large floods or large waves can be very effective landslide triggers, often accompanied by heavy precipitation. Large- and intermediate-magnitude earthquakes routinely trigger landslides and rockfall (e.g., Meunier et al., 2007; Shreve, 1966), and landslides are common in volcanic areas because of steep

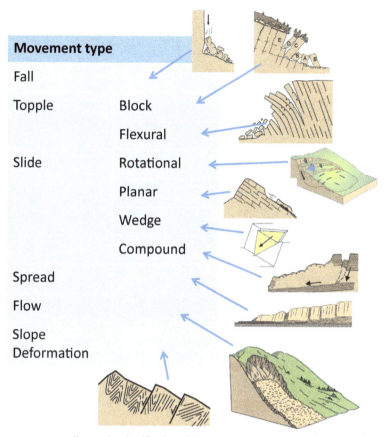

FIG. 13.2 Landslide movement types according to the classification of Hungr et al. (2014). *(From Hermanns, R.L. (2018). Landslides. In: P. T. Bobrowsky & B. Marker (Eds.),* Encyclopedia of engineering geology, *183-1.)*

topography and weak materials, such as unconsolidated ash layers (e.g., Rault et al., 2022). Humans can also directly trigger landslides through, e.g., construction, mining, or deforestation (Glade, 2003). Perhaps the largest human contribution to landslide triggering is through climate change (Haque et al., 2019; IPCC, 2023), which affects rainfall patterns, glacial retreat, and thawing of permafrost (Patton et al., 2019; Czekirda et al., 2019; Sæmundsson et al., 2018; Morino et al., 2019, 2021; Svennevig et al., 2022, 2023).

Landslides have been classified in numerous ways, but here the widely used classification of Varnes (1978), with later updates of Hungr et al. (2014), will be introduced. The Varnes (1978) classification is based on the dominating type of movement: fall, topple, slide, spread, flow, and complex motion (Fig. 13.2), and the type of material involved in the movement: i.e., bedrock, debris, or soils. In the updated classification (Hungr et al., 2014), several changes were made such as using the engineering properties of materials (rock, clay, mud, silt, sand, gravel, boulders, debris, peat, and ice) rather than "debris" and "soils" in the original classification, and the movement type "slope deformation" was introduced (Fig. 13.2). Several handbooks have been published regarding classification of landslides based on the Varnes classification and their associated hazards (e.g., Highland & Bobrowsky, 2008). A detailed description of movement types and materials involved can be seen in Hungr et al. (2014). It is interesting to note that although landslide motion is generally downward and outward, there can be various partitioning and spatiotemporal scales of rotation versus translation. Furthermore, many of the examples introduced in Fig. 13.2 are two-dimensional (2D) representations, while the structure and mode of movement can vary laterally across landslides. Thus, placing GNSS instruments in landslides to measure their motion usually requires careful planning.

3 GNSS landslide equipment and data processing

Active landslides tend to be broken up or show initial signs of movement such as fractures in rocks or soil (Fig. 13.2). Thus, installing GNSS equipment in active landslides can be challenging. The access to sites of most interest can be dangerous because of steep slopes and overhanging landslide scarps with loose material. In many cases, the safest access is via helicopters. However, in some cases, the access can be fairly easy, for example, if the purpose is to monitor road movements that cross active landslides, or if building infrastructure already exists in the active landslide area.

FIG. 13.3 An example of setup of a permanent GNSS station in a slow-creeping landslide in the Almenningar region, north Iceland. The mast on the right, driven approximately 2 m into the ground, holds the GNSS antenna. Guy wires are used to reduce vibration, which eventually might loosen the pole. The mast is deliberately high to avoid winter burial in snow. The mast on the left holds a solar panel, battery, GNSS receiver, and communication equipment. *(Photo: Þorsteinn Sæmundsson.)*

Due to the broken-up nature of many landslides, it can be tricky to install quality monuments for the GNSS antenna: by definition landslides are unstable! In some cases, individual boulders can be used for monuments, drilling, and cementing in short pins with threads to hold the GNSS antenna. Sometimes it may be feasible to build concrete pillars or anchored rods/pipes drilled (e.g., PBO-style monuments) or driven in by hammering (Fig. 13.3). In some cases, it may be sufficient to lay the monument on the ground and cover foundations by piles of rock. For any such monuments set in loose gravels, it is good practice to measure routinely the tilt angle and tilt direction of the monument, for example, using a carpenter's bubble level and a compass, to be able to separate local rotational movement of the monument from the movement of the landslide.

Since landslide motion spans orders of magnitude, there are a range of GNSS equipment that can be used. For the slowest-creeping landslides, one will want to use instruments and monuments comparable to standard equipment at Continuously Operating Reference Stations (CORS). For faster-moving landslides, it may be sufficient to run cheaper instruments and lower-grade antennas. Measurements can be campaign-style (e.g., Brückl et al., 2006; Dille et al., 2022) or continuous. Some of the site locations may be such that there is a significant probability of the loss of the instrument, which favors the installation of more expandable products. Furthermore, the spatial variability of landslide motion may favor a higher number of lower-quality stations compared to a few high-quality stations for the same cost (Šegina et al., 2020). Some researchers have used single-frequency instruments (e.g., Squarzoni et al., 2005; Zuliani et al., 2022) or even code-only receivers, which compute and send coarse position results automatically as opposed to streaming raw satellite observations for real-time or postprocessing. It is, however, the experience of the authors that the initial purchase price of the instruments is only a part of the total budget when operating permanent stations in harsh environments for many years.

All GNSS instruments need some electrical power to operate. Typical CORS-like GNSS instruments use approximately 2–5 W of power (Berglund, 2016; see also https://kb.unavco.org/category/development-and-testing/power-test-reports/90/). Power-saving options for the GNSS receivers include: tracking fewer GNSS systems; disabling Wi-Fi/Bluetooth options where available; tracking L1 only (Zuliani et al., 2022; L1 has a higher signal/noise ratio than L2 and thus requires less amplification to track); use periodic power cycling, i.e., only turning the receiver on for a few minutes or hours each day or other periodicities. However, for effective real-time monitoring of deformation, it is essential to stream data continuously. For short-term installations (days), it is sufficient to run the instruments off car batteries, however, for permanent installations solar panels and/or wind generators need often to be used since few landslides have mains-powered infrastructure. The conditions for operating such solar and wind power vary from place to place. In low- to mid-latitudes, solar energy is often sufficient year-round, while for high-latitude installations large battery packs and solar panels are needed to survive through the dark winters.

The GNSS data need to be transmitted. The most rudimentary method is manual transfer of data [manual file transfer protocol (mftp)]. Most permanent stations, especially those used for monitoring purposes, have some form of communication equipment, for example, cellular modems, radio links, WiFi network extenders, satellite modems, or LoRa network equipment. Each of these equipment in turn costs subscription plans and require power. The power consumption varies but typically is in the range 1–5 W. It is important to design landslide equipment to function with as little intervention as possible. Nothing is more daunting than traveling for hours or days only to do a simple instrument reboot. However, care must be taken with backup systems and fail-safe engineering, as these can also malfunction and cause problems.

The data processing for landslide GNSS data follows conventional pathways. For postprocessing of dual-frequency GNSS data, one can use commercial or academic software such as GAMIT/GLOBK, GIPSY-X, BERNESE, PRIDE PPP-AR, or vendor-specific software. Typically, 24-h solutions are produced, however, the potential high rate of deformation at landslides also calls for high-resolution epoch-wise solutions and high receiver sampling rates to study or detect sudden changes in movements. Long-term postprocessed GNSS time series are useful to understand the dynamics of landslide motion—see examples in Section 4.

For many landslide applications, real-time processing of data is very important. The tricky part of GNSS real-time processing is that one wants to allow the position solutions to capture large movements immediately, but yet give stable results such that few or no outliers occur in the solution. This may sound simple, but is rather challenging to acquire in practice and spurious outliers often occur (Fig. 13.4). Common GNSS processing algorithms apply filters, which suppress to some degree the allowed deviation of the latest data point from previous values. There are several strategies for real-time analysis of GNSS data. The classic strategy is to have a local base station(s) and perform double-difference processing of data, using software such as GAMIT/TRACK, RTKLIB, or vendor-specific software. Having a nearby base station (less than a few km) can help considerably with the quality of the results (e.g., Wang, 2011). Running in parallel a single station precise-point-positioning (PPP) algorithm, or combined PPP-RTK algorithms generally improves the robustness and quality of the position monitoring (Huang et al., 2023). For the monitoring of sudden motion it can be useful to integrate GNSS receivers with accelerometers (e.g., Grapenthin et al., 2014; Notti et al., 2020). This approach has dual benefits: the accelerometers provide an individual measure of large motion; and they also help to eliminate positioning outliers, which would require corresponding accelerations if real (Jing et al., 2022).

Occasionally, landslides are located in regions where other deformation processes can also cause deformation. Consider for example unstable slopes of inflating or deflating volcanoes or slopes in plate boundary and earthquake regions. In these cases, it can become challenging to disentangle the signals, and eventually the deformation processes may be interlinked, such as if earthquakes are triggering landslide motion. Fortunately, landslides have usually a much smaller footprint than, say, earthquake deformation. The deformation from isolated events can thus often be estimated by considering a spatially larger set of deformation data that can be modeled or interpolated and subtracted from the overall signal to highlight only the landslide deformation.

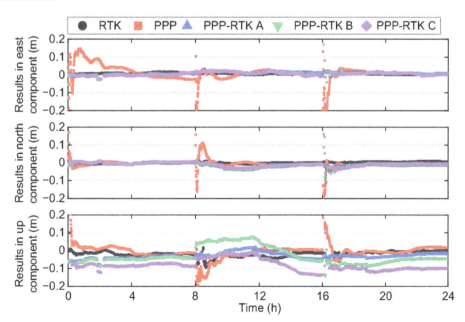

FIG. 13.4 GNSS time series from five different processing approaches. The parameter filter is reset every 8 h to reflect the monitoring performance of each approach. *(Huang, G., Du, S., & Wang, D. (2023). GNSS techniques for real-time monitoring of landslides: A review. Satellite Navigation, 4(1), 5.)*

4 Case studies

There are numerous landslides around the world which have continuous or episodic GNSS instruments for monitoring and research of deformation. Below, we explore lessons from four GNSS-instrumented landslides.

4.1 Åknes, Norway

The Åknes landslide is a slow-moving landslide in gneiss bedrock in western Norway, with the potential of generating a tsunami if it fails catastrophically into the fjord below. The landslide is steep (30°–35°) and has an area of ~0.7 km². Parts of the landslide have been observed to move at up to 8 cm/yr and the landslide mass is at least 100 m thick (Blikra et al., 2012). The Åknes landslide has a multiparametric monitoring network installed, including 10 GNSS stations, two lasers with reflectors, 30 prisms surveyed automatically by a total station, several extensometers and crackmeters, and seven boreholes with biaxial inclinometers and piezometers. Ground-based InSAR has been applied intermittently since 2005. Additionally, a network of six seismometers and a borehole-mounted string of eight seismometers are installed, as well as a weather station (Pless et al., 2021).

The 10 GNSS stations are equipped with Trimble R9 receivers and Trimble Zephyr 2 or Trimble Zephyr 3 antennas. Data are processed at 15 min, 4 h, and 12 h intervals, for increasing accuracy. The GNSS time series show interannual variability (Fig. 13.5) that correlates generally between stations such that higher rates are observed at most stations in the same year. The episodic faster creep episodes are driven by intense precipitation episodes and snowmelt (Aspaas et al., 2022).

4.2 Almenningar, Iceland

The Almenningar landslide region in northern Iceland is composed of several deep-seated gravitational slow-moving rockslides that are undercut by the ocean (Sæmundsson et al., 2007). The region spans approximately 6 km across the landslides, including three regions that display most movement. A road was established over the landslide area in 1968 to connect

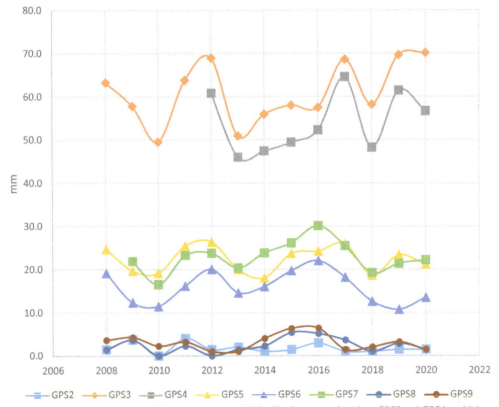

FIG. 13.5 Annual total displacement of GNSS stations in the Aknes landslide. The fastest moving sites, GPS3 and GPS4 are highest on the landslide. *(From Pless, G., Blikra, L. H., & Kristensen, L. (2021). Possibility for using drainage as mitigation to increase the stability of the Åknes rock-slope instability, Stranda in western Norway. Norges vassdrags- og energidirektorat (NVG), report no. 22/2021, 74 pp.)*

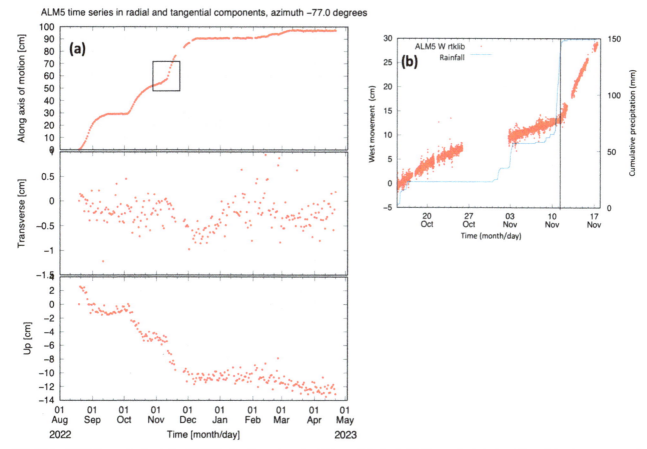

FIG. 13.6 GNSS time series for station ALM5 in the Almenningar landslide, north Iceland. (A) Time series in horizontally radial and transverse, and vertical components. Each dot is from GIPSY/OASIS II PPP processing of 24-h data. (B) Zoom-in on boxed area in (A), showing 5-min solutions from real-time processing using RTKLIB and a local (<2 km) reference station. Gaps in data are due to problems with the RTKLIB processing. Blue curve shows cumulative precipitation from a nearby weather station. Vertical black line highlights abrupt velocity change.

better one of the remote towns of Iceland, Siglufjörður. Shortly after the installation of the road, creep was noted along certain road segments such that periodic road maintenance is needed. Developing fractures have formed on both sides of the road, at approximately 60 m above sea level, raising concerns of safety. The deformation speed of the three main landslide masses is quite variable, as observed with both long-term campaign GPS and optical image pixel tracking (Ninuson, 2023; Sæmundsson et al., 2007). The maximum annual speed is approximately 1 m/yr, as observed by campaign GNSS measurements (Geirsson et al., 2023).

In 2022, a network of nine GNSS stations was installed in the landslide area for monitoring and research purposes. One of the stations is deliberately installed in a stable region to act as a local reference station. Each station is equipped with a Septentrio Mosaic-X5 receiver, an Ardusimple antenna, a cellular modem, 100 W solar panel, and a 100 Ah battery (Fig. 13.3; Geirsson et al., 2023). The data are streamed over a cellular network into RTKLIB for real-time processing at 5-min intervals (Fig. 13.6B). The data are also postprocessed in 24-h bins using GIPSY/OASIS II (Fig. 13.6A).

The GNSS time series at Almenningar show distinct deformation episodes, lasting days to weeks (Fig. 13.6A). The dominating motion at most stations is horizontal, with a stable azimuth over time at most stations. The deformation episodes vary in their temporal behavior, and show overall a strong correlation with cumulative rainfall over several days to weeks (Fig. 13.6B; Geirsson et al., 2023). Previous monitoring of the landslide was done using annual GPS campaigns, hence the improvement in temporal resolution is eye-opening for the dynamics of the landslide area.

4.3 El Yunque, Puerto Rico

In the El Yunque forest, a small (roughly 60 m × 60 m) landslide offered the opportunity of combining continuous GPS measurements and terrestrial laser scanner (TLS) observations of landslide deformation. Two rover stations (Topcon

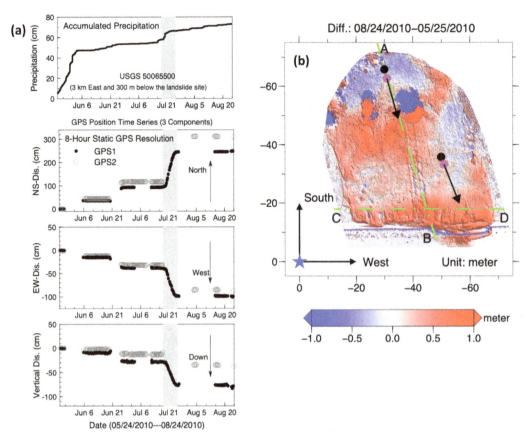

FIG. 13.7 (A) Time series of precipitation and Landslide motion in the El Yunque forest, Puerto Rico. (B) Map view of vertical displacements from a TLS survey, showing large- and small-scale slumps, as well as the location of the GPS stations. *(From Wang, G., Philips, D., Joyce, J., & Rivera, F. (2011). The integration of TLS and continuous GPS to study landslide deformation: A case study in Puerto Rico.* Journal of Geodetic Science, *1(1), 25–34.)*

GB1000 receivers and PG-A1 antennas) were processed using a base station located at 2 km distance and running a Trimble NetR8 GPS receiver with a Zephyr Geodetic 1 antenna. Processing was done using the Topcon Tools software package (Wang et al., 2011). The GPS stations record displacements of about a meter during 3 months in 2010 (Fig. 13.7A). The displacements are clearly triggered by rainfall, however, it is interesting to note that the amplitude of motion does not correspond directly with the amount of rainfall; small rainfall can also trigger movement of the landslide. The application of the TLS shows in detail the spatial distribution of motion (Fig. 13.7B), highlighting the strength of combining different deformation techniques.

4.4 Cà Lita, Italy

The Cà Lita landslide is located in the northern Apennines in Italy. The landslide is approximately 3 km long, 1.5 km wide, and has an elevation drop of approximately 400 m. The landslide is classified as a complex landslide, and has shown periodic dekameter-scale reactivation. In 2002–2004, the landslide resumed activity (Cervi et al., 2012). In 2016, another reactivation period occurred, in which continuous GNSS measurements were started, recording over 50 m of displacement although the initial phase of the sliding was not caught (Fig. 13.8; Mulas et al., 2020a, 2020b). Two different types of GNSS instrumentation have been tested here: high-end Leica receivers with a base-station and low-cost Emlid receivers (Mulas et al., 2020b). The Leica set used a dual frequency base and two L1-rovers. The Emlid receivers (type Reach RS; L1 receivers) have onboard computers running specific modules of the RTKLIB software. The Emlid master station broadcasts corrections over long-range radio (LoRa), which the rovers use to compute positions at 5 Hz sampling rates, transferred in 1-h-long files and median values used for a daily coordinate solution (Fig. 13.8; Mulas et al., 2020b). Data from the Leica receivers are processed using the Leica Spider software (Mulas et al., 2020a). The landslide was further reactivated in late 2017–2018 and in 2019. During 2019, the largest displacements so far occurred, reaching up to 70 m

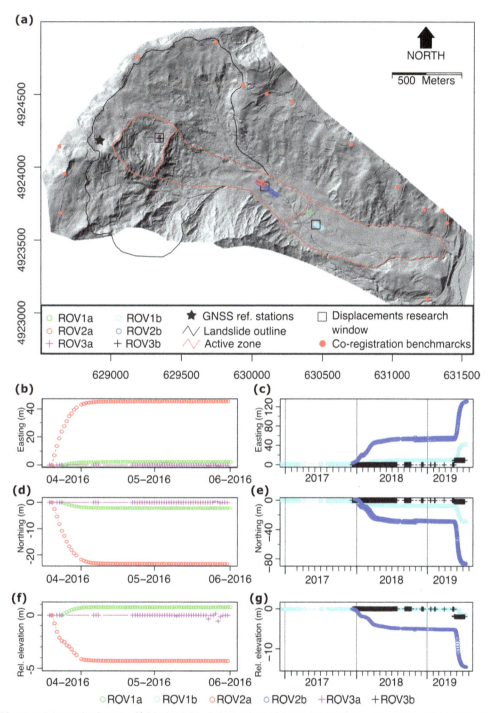

FIG. 13.8 GNSS map and time series from the Cà Lita landslide, Italy. (A) Map of the station locations and landslide outline. (B, D, F) show time series from March 24 to July 15, 2016. (C, E, G) show time series from November 2016 to July 2019. Station ROV3b uses an Emlid receiver, the rest use Leica receivers. *(From Mulas, M., Ciccarese, G., Truffelli, G., & Corsini, A. (2020a). Integration of digital image correlation of Sentinel-2 data and continuous GNSS for long-term slope movements monitoring in moderately rapid landslides. Remote Sensing, 12(16), 2605.)*

in the center part of the landslide over approximately 1-month long period. The highest speeds reached 9.6 m in 12 h, and a downslope propagation of a movement wave could be observed in the data. The motion is attributed to weak flysch rock masses (submarine turbditite sequences) and hydrologic triggering (Cervi et al., 2012). It is interesting to see for this case study that for such large-scale motion and short baselines, the monitoring and high-rate data acquisition can be made in a cost-effective way.

5 Perspective of GNSS and other landslide deformation methods

As there exist several techniques to measure landslide deformation, it is interesting to compare and contrast the advantages and disadvantages of each technique.

GNSS instruments can be relatively inexpensive and allow continuous (up to 50 Hz) observations of 3D displacements. Long time series can be valuable for studying landslide deformation. However, GNSS instruments only observe the deformation at individual points.

InSAR provides excellent spatial coverage, especially the higher-resolution satellites such as TERRASAR-X. The measurement is one-dimensional (1D) in the line of sight of the satellite, however, multiple look angles can be applied to separate horizontal and vertical deformation. Due to polar orbits, there is little sensitivity to north-south-oriented motion. Steep topography commonly found at landslides can prove troublesome in data processing. Furthermore, too large motion between acquisitions causes decorrelation and loss of information. The temporal sampling interval is usually on the order of days to weeks, but microsatellites such as ICEYE may allow higher temporal resolution (Ignatenko et al., 2020). Ground-based InSAR, with high sampling rates, are useful if a good lookout of the landslide can be found, albeit expensive to operate.

Extensometers are simple devices that measure tension over at most a few meters, typically installed over faults or cracks that need to be monitored especially. The measurement is 1D length change. Results can be streamed in real time. Frost and snow cover can affect the measurements.

Using a total station, where mirrors installed in the landslide are surveyed automatically at repeated time intervals, provides an efficient way to monitor landslide movement. This technique bypasses the need for powered infrastructure in the landslide. However, this approach requires a clear line of sight. Thus, total stations have limitations in cloudy conditions and rainy periods, when the probability for landslide triggering is generally higher. Terrestrial laser scanning (TLS) and other optical imaging monitoring techniques are also limited by cloud cover.

Inclinometers or tiltmeters are useful to detect sudden or large motion. They can be surface-mounted, or in shallow boreholes, with the borehole installation being less sensitive to temperature-related errors. These instruments offer high sampling rate and real-time data streaming.

Strain gauges, in the sense of angular strain, are sometimes installed as vertical strings and can be useful to detect the motion across bedded interfaces (e.g., Petrisor et al., 2013). However, too much displacement can cut the cable apart, and if the string does not reach the bedding interface then no deformation is recorded.

From the list above, one can gauge that GNSS instrumentation is most practical for measuring motion of slow-to-fast creeping landslides, or bodies of landslides that may change motion quickly. GNSS is not practical for mudflows or fast-moving density currents. In practice, it is usually best to combine several techniques; i.e., use GNSS for temporal resolution and long-term deformation; use tension gauges for specific cracks or faults that may be of interest or require specific monitoring; and apply InSAR, TLS, or optical image differences to acquire spatial density of measurements. Because of the spatially and temporally complex nature of landslide movement (Fig. 13.2), it is often a good idea to install several GNSS stations in each landslide.

Several landslide bodies have multiparametric geophysical instrumentation, and in addition to deformation measurements may include hydrometers, seismometers, accelerometers, geo-electrical measurements, soil moisture, and meteorological measurements. These state-of-the-art networks, where GNSS measurements are an important cornerstone, allow monitoring and research of the landslides from multiple viewpoints, overall improving understanding and hazard monitoring of landslides.

Acknowledgments

We thank the Icelandic Road and Coastal Administration for funding landslide research and monitoring in the Almenningar region. RTKLIB processing and data streaming was developed by Jóhanna Malen Skúladóttir, Institute of Earth Sciences, University of Iceland. We thank Antoine Dille for supplying Fig. 13.1 and valuable comments on the manuscript.

References

Agliardi, F., Crosta, G. B., & Frattini, P. (2012). Slow rock slope deformations. In J. J. Clague, & D. Stead (Eds.), *Landslides types, mechanisms and modeling* (pp. 207–221). Cambridge University Press.

Agliardi, F., Scuderi, M. M., Fusi, N., & Collettini, C. (2020). Slow-to-fast transition of giant creeping rockslides modulated by undrained loading in basal shear zones. *Nature Communications*, *11*(1), 1352.

Aspaas, A., Lacroix, P., Kristensen, L., Etzelmüller, B., & Renard, F. (2022). What causes transient deformations in the Åknes landslide, Norway? In *EGU general assembly conference abstracts* (pp. EGU22-1718).

Ballantyne, C. K. (2002). Paraglacial geomorphology. *Quaternary Science Reviews, 21*, 1935–2017.

Ben-Yehoshua, D., Sæmundsson, Þ., Helgason, J. K., Belart, J. M. C., Sigurðsson, J. V., & Erlingsson, S. (2022). Paraglacial exposure and collapse of glacial sediment: The 2013 landslide onto Svínafells-jökull, Southeast Iceland. *Earth Surface Processes and Landforms, 47*, 2612–2627.

Berglund, H. (2016). *Septentrio PolaRx5 power consumption. UNAVCO power test report* (p. 396). Retrieved from https://kb.unavco.org/kb/pdf-841.html.

Blikra, L. H., Fasani, G. B., Esposito, C., Lenti, L., Martino, S., Pecci, M., ... Diederichs, M. (2012). The Aknes rockslide, Norway. In J. J. Clague, & D. Stead (Eds.), *Landslides: types, mechanisms and modeling* (pp. 323–335).

Bogaard, T. A., & Greco, R. (2016). Landslide hydrology: From hydrology to pore pressure. *Wiley Interdisciplinary Reviews: Water, 3*(3), 439–459.

Brückl, E., Brunner, F. K., & Kraus, K. (2006). Kinematics of a deep-seated landslide derived from photogrammetric, GPS and geophysical data. *Engineering Geology, 88*(3–4), 149–159.

Bugge, T., Belderson, R. H., & Kenyon, N. H. (1988). The Storegga slide. *Philosophical Transactions of the Royal Society of London. Series A, Mathematical and Physical Sciences, 325*(1586), 357–388.

Cervi, F., Ronchetti, F., Martinelli, G., Bogaard, T. A., & Corsini, A. (2012). Origin and assessment of deep groundwater inflow in the Ca' Lita landslide using hydrochemistry and in situ monitoring. *Hydrology and Earth System Sciences, 16*, 4205–4221.

Crozier, M. (1989). *Landslides: Causes, Consequences and Environment*. London: Helm252. https://doi.org/10.1111/j.1745-7939.1989.tb01143.x.

Crozier, M. (2002). Landslides. In *Applied geography* (pp. 111–122). Routledge.

Crozier, M. J., & Glade, T. (2005). Landslide hazard and risk: Issues, concepts and approach. In *Landslide hazard risk* (pp. 1–40).

Czekirda, J., Westermann, S., Etzelmüller, B., & Johannesson, T. (2019). Transient modelling of permafrost distribution in Iceland. *Frontiers in Earth Science, 7*, 130. https://doi.org/10.3389/feart.2019.00130.

Dille, A., Dewitte, O., Handwerger, A. L., d'Oreye, N., Derauw, D., Ganza Bamulezi, G., ... Kervyn, F. (2022). Acceleration of a large deep-seated tropical landslide due to urbanization feedbacks. *Nature Geoscience, 15*(12), 1048–1055.

Dramis, F., & Sorriso-Valvo, M. (1995). Deep-seated gravitational slope deformations, related landslides and tectonics. *International Journal of Rock Mechanics and Mining Sciences & Geomechanics, 5*, 203A.

Fritz, H. M., Mohammed, F., & Yoo, J. (2009). Lituya Bay landslide impact generated mega-tsunami 50th anniversary. In *Tsunami science four years after the 2004 indian ocean tsunami: part ii: observation and data analysis* (pp. 153–175).

Gatter, R., Clare, M. A., Kuhlmann, J., & Huhn, K. (2021). Characterisation of weak layers, physical controls on their global distribution and their role in submarine landslide formation. *Earth-Science Reviews, 223*, 103845.

Geertsema, M., Highland, L., & Vaugeouis, L. (2009). Environmental impact of landslides. In *Landslides–disaster risk reduction* (pp. 589–607). Springer.

Geirsson, H., Sæmundsson, Þ., Malen Skúladóttir, J., & Jónasson, N. (2023). Seasonal and precipitation-triggered movements of the Almenningar and Tungnakvíslarjökull landslides, Iceland, monitored by low-cost GNSS observations. In *EGU general assembly conference abstracts* (pp. EGU-13439).

Glade, T. (2003). Landslide occurrence as a response to land use change: A review of evidence from New Zealand. *Catena, 51*, 297–314.

Grapenthin, R., Johanson, I. A., & Allen, R. M. (2014). Operational real-time GPS-enhanced earthquake early warning. *Journal of Geophysical Research: Solid Earth, 119*(10), 7944–7965.

Gylfadóttir, S. S., Kim, J., Helgason, J. K., Brynjólfsson, S., Höskuldsson, Á., Jóhannesson, T., ... Løvholt, F. (2017). The 2014 Lake Askja rockslide-induced tsunami: Optimization of numerical tsunami model using observed data. *Journal of Geophysical Research: Oceans, 122*(5), 4110–4122.

Haque, U., Blum, P., Da Silva, P. F., Andersen, P., Pilz, J., Chalov, S. R., ... Poyiadji, E. (2016). Fatal landslides in Europe. *Landslides, 13*, 1545–1554.

Haque, U., Da Silva, P. F., Devoli, G., Pilz, J., Zhao, B., Khaloua, A., & Glass, G. E. (2019). The human cost of global warming: Deadly landslides and their triggers (1995–2014). *Science of the Total Environment, 682*, 673–684.

Harris, S. L. (1988). *Fire mountains of the west: The Cascade and mono Lake volcanoes* (p. 454). Missoula, MT: Mountain Press Pub. Co.

Hermanns, R. L. (2018). Landslides. In P. T. Bobrowsky, & B. Marker (Eds.), *Encyclopedia of engineering geology*. 183-1.

Highland, L. M., & Bobrowsky, P. (2008). The landslide handbook—A guide to understanding landslides: Reston, Virginia, U.S. *Geological Survey Circular, 1325*. 129 pp.

Huang, G., Du, S., & Wang, D. (2023). GNSS techniques for real-time monitoring of landslides: A review. *Satellite Navigation, 4*(1), 5.

Hungr, O. (2007). Dynamics of rapid landslides. In *Progress in landslide science* (pp. 47–57). Springer.

Hungr, O., Leroueil, S., & Picarelli, L. (2014). The Varnes classification of landslide types, an update. *Landslides, 11*, 167–194.

Hutchinson, J. N. (1968). Field meeting on the coastal landslides of Kent: 1–3 July 1966. *Proceedings of the Geologists Association, 79*, 227–237.

Ignatenko, V., Laurila, P., Radius, A., Lamentowski, L., Antropov, O., & Muff, D. (2020). ICEYE microsatellite SAR constellation status update: Evaluation of first commercial imaging modes. In *IGARSS 2020; IEEE international geoscience and remote sensing symposium* (pp. 3581–3584). IEEE.

IPCC. (2023). Climate change 2023: Synthesis report. Contribution of working groups I, II and III to the sixth assessment report of the intergovernmental panel on climate change. In H. Lee, & J. Romero (Eds.), *Core writing team* (pp. 35–115). Geneva, Switzerland: IPCC. https://doi.org/10.59327/IPCC/AR6-9789291691647.

Jing, C., Huang, G., Zhang, Q., Li, X., Bai, Z., & Du, Y. (2022). GNSS/accelerometer adaptive coupled landslide deformation monitoring technology. *Remote Sensing, 14*(15), 3537.

Kjekstad, O., & Highland, L. (2009). Economic and social impacts of landslides. *Landslides – Disaster Risk Reduction*, 573–587.

Kos, A., Amann, F., Strozzi, T., Delaloye, R., von Ruette, J., & Springman, S. (2016). Contemporary glacier retreat triggers a rapid landslide response, great Aletsch glacier, Switzerland. *Geophysical Research Letters, 43*(24), 12–466.

Lacroix, P., Belart, J. M., Berthier, E., Sæmundsson, Þ., & Jónsdóttir, K. (2022). Mechanisms of landslide destabilization induced by glacier-retreat on Tungnakvíslarjökull area, Iceland. *Geophysical Research Letters, 49*(14), e2022GL098302.

Lacroix, P., Handwerger, A. L., & Bièvre, G. (2020). Life and death of slow-moving landslides. *Nature Reviews Earth & Environment, 1*(8), 404–419.

Masson, D. G., Watts, A. B., Gee, M. J. R., Urgeles, R., Mitchell, N. C., Le Bas, T. P., & Canals, M. (2002). Slope failures on the flanks of the western Canary Islands. *Earth-Science Reviews, 57*(1–2), 1–35.

McColl, S. T., & Davies, T. R. H. (2013). Large ice-contact slope movements: Glacial buttressing, deformation and erosion. *Earth Surface Processes and Landforms, 38*, 1102–1115.

Meunier, P., Hovius, N., & Haines, A. J. (2007). Regional patterns of earthquake-triggered landslides and their relation to ground motion. *Geophysical Research Letters, 34*, L20408.

Morino, C., Conway, S. J., Balme, M. R., Helgason, J. K., Sæmundsson, T., Jordan, C., … Argles, T. (2021). The impact of ground-ice thaw on landslide geomorphology and dynamics: Two case studies in northern Iceland. *Landslides, 18*, 2785–2812.

Morino, C., Conway, S. J., Sæmundsson, T., Helgason, J. K., Hillier, J., Butcher, F. E. G., … Argles, T. (2019). Molards as an indicator of permafrost degradation and landslide processes. *Earth and Planetary Science Letters, 516*, 136–147.

Mulas, M., Ciccarese, G., Truffelli, G., & Corsini, A. (2020a). Integration of digital image correlation of Sentinel-2 data and continuous GNSS for long-term slope movements monitoring in moderately rapid landslides. *Remote Sensing, 12*(16), 2605.

Mulas, M., Ciccarese, G., Truffelli, G., & Corsini, A. (2020b). Displacements of an active moderately rapid landslide—A dataset retrieved by continuous GNSS arrays. *Data, 5*(3), 71.

Ninuson, E. A. (2023). *Displacement measurements of three slow moving landslides at Almenningar, North Iceland. A feature tracking application*. MS thesis University of Iceland. 106 pp.

Notti, D., Cina, A., Manzino, A., Colombo, A., Bendea, I. H., Mollo, P., & Giordan, D. (2020). Low-cost GNSS solution for continuous monitoring of slope instabilities applied to Madonna Del Sasso sanctuary (NW Italy). *Sensors, 20*(1), 289.

Oliveira, E. D. P., Acevedo, A. M. G., Moreira, V. S., Faro, V. P., & Kormann, A. C. M. (2022). The key parameters involved in a rainfall-triggered landslide. *Water, 14*, 3561.

Oppikofer, T., Hermanns, R. L., Jakobsen, V. U., Böhme, M., Nicolet, P., & Penna, I. (2020). Forecasting dam height and stability of dams formed by rock slope failures in Norway. In *Natural hazards and Earth system sciences discussion* (pp. 1–24).

Patton, A. I., Rathburn, S. L., & Capps, D. M. (2019). Landslide response to climate change in permafrost regions. *Geomorphology, 340*, 116–128.

Petley, D. (2012). Global patterns of loss of life from landslides. *Geology, 40*, 927–930.

Petrisor, D., Fosalau, C., Zet, C., & Damian, C. (2013). Measurement of landslide displacement and orientation using strain gauges based on amorphous magnetic microwires. In *Proceedings of the 19th IMEKO TC4 symposium advances in instrumentation and sensors interoperability, July, Barcelona, Spain* (pp. 655–660).

Pless, G., Blikra, L. H., & Kristensen, L. (2021). *Possibility for using drainage as mitigation to increase the stability of the Åknes rock-slope instability, Stranda in western Norway*. Norges vassdrags- og energidirektorat (NVG), report no. 22/2021, 74 pp.

Rault, C., Thiery, Y., Chaput, M., Reninger, P. A., Dewez, T. J. B., Michon, L., et al. (2022). Landslide processes involved in volcano dismantling from past to present: The remarkable open-air laboratory of the cirque de Salazie (Reunion Island). *Journal of Geophysical Research: Earth Surface, 127*, e2021JF006257.

Sæmundsson, Þ., Morino, C., & Conway, S. J. (2022). Mass-movements in cold and polar climates. In J. F. Shroder (Ed.), *Treatise on geomorphology* (pp. 350–370). Elsevier.

Sæmundsson, Þ., Morino, C., Helgason, J. K., Conway, S. J., & Pétursson, H. G. (2018). The triggering factors of the Móafellshyrna debris slide in northern Iceland: Intense precipitation, earthquake activity and thawing of mountain permafrost. *Science of the Total Environment, 621*, 1163–1175.

Sæmundsson, Þ., Petursson, H. G., & Decaulne, A. (2003). Triggering factors for rapid mass movements in Iceland. In *1. International conference on debris-flow hazards mitigation: mechanics, prediction, and assessment, proceedings* (pp. 167–178).

Sæmundsson, T., Petursson, H. G., Kneisel, C., & Beylich, A. (2007). Monitoring of the Tjarnardalir landslide, in central North Iceland. In *Vol. 23. First north American landslide conference, Vail Colorado* (pp. 1029–1040). AEG Publication.

Scoppettuolo, M. R., Cascini, L., & Babilio, E. (2020). Typical displacement behaviours of slope movements. *Landslides, 17*(5), 1105–1116.

Šegina, E., Peternel, T., Urbančič, T., Realini, E., Zupan, M., Jež, J., & Auflič, M. J. (2020). Monitoring surface displacement of a deep-seated landslide by a low-cost and near real-time GNSS system. *Remote Sensing, 12*(20), 3375.

Shreve, R. L. (1966). Sherman landslide, Alaska. *Science, 154*, 1639–1643.

Sim, K. B., Lee, M. L., & Wong, S. Y. (2022). A review of landslide acceptable risk and tolerable risk. *Geoenvironmental Disasters, 9*(1), 3.

Squarzoni, C., Delacourt, C., & Allemand, P. (2005). Differential single-frequency GPS monitoring of the La Valette landslide (French Alps). *Engineering Geology, 79*(3–4), 215–229.

Strom, A., & Korup, O. (2006). Extremely large rockslides and rock avalanches in the Tien Shan Mountains, Kyrgyzstan. *Landslides, 3*, 125–136.

Svennevig, K., Hermann, R., Keiding, M., Binder, D., Citterlo, M., Dahl-Jensen, T., … Voss, P. (2022). A large frozen debris avalanche entraining warming permafrost ground—The June 2021 Assapaat landslide, West Greenland. *Landslides, 19*, 2549–2567.

Svennevig, K., Koch, J., Keiding, M., & Leutzenburg, G. (2023). Assessing the impact of climate change to landslides using public data, a case study from Vejle, Denmark. *Natural Hazards and Earth System Sciences Discussions*, 1–26. https://doi.org/10.5194/nhess-2023-68.

Varnes, D. J. (1978). Slope movement types and processes. *Transportation Research Board, Special Report, 176*, 11–33.

Voight, B., & Elsworth, D. (2000). Instability and collapse of hazardous gas-pressurized lava domes. *Geophysical Research Letters*, *27*(1), 1–4.

Wang, G. (2011). GPS landslide monitoring: Single base vs. network solutions—A case study based on the Puerto Rico and Virgin Islands permanent GPS network. *Journal of Geodetic Science*, *1*(3), 191–203.

Wang, G., Philips, D., Joyce, J., & Rivera, F. (2011). The integration of TLS and continuous GPS to study landslide deformation: A case study in Puerto Rico. *Journal of Geodetic Science*, *1*(1), 25–34.

Zuliani, D., Tunini, L., Di Traglia, F., Chersich, M., & Curone, D. (2022). Cost-effective, single-frequency GPS network as a tool for landslide monitoring. *Sensors*, *22*(9), 3526.

Chapter 14

Climate- and weather-driven solid Earth deformation and seismicity

Roland Bürgmann[a,b], Kristel Chanard[c], and Yuning Fu[d]

[a]Berkeley Seismological Laboratory, University of California Berkeley, Berkeley, CA, United States, [b]Department of Earth and Planetary Science, University of California Berkeley, Berkeley, CA, United States, [c]Institut de Physique du Globe de Paris, Université Paris Cité, CNRS, IGN, Paris, France, [d]School of Earth, Environment and Society, Bowling Green State University, Bowling Green, OH, United States

1 Introduction

The Earth's surface represents the interface of the lithosphere, the outer layer of the solid Earth, with the cryosphere, hydrosphere, and atmosphere. Surface processes, including the hydrological cycle and the erosion, transport, and deposition of sediments, reflect the effect of climate and weather on the solid Earth. Tectonics and crustal deformation, in turn, can affect climate and weather by reshaping Earth's topography. Thus, the lithosphere and the atmosphere may interact over a range of timescales, from billions of years (associated with the compositional evolution of the atmosphere, the formation and growth of continents, and plate tectonics) to seconds (associated with storm events, earthquakes, and volcanic eruptions). Mankind has long been fascinated by the possibility of atmospheric and hydrological processes (i.e., climate, weather, and the hydrological cycle) affecting solid Earth deformation processes in the brittle lithosphere, including earthquakes that are generally understood to mostly be a result of plate tectonics. For good reasons, the idea of "earthquake weather," that is, a particular weather pattern or season in which earthquakes preferably occur, has generally been considered an urban myth (e.g., Fallou, Marti, Dallo, & Corradini, 2022).

Recent advances in geophysical observations, including the Global Navigation Satellite System (GNSS) have revealed significant deformation associated with climate and weather (e.g., White, Gardner, Borsa, Argus, & Martens, 2022), which is superimposed on plate tectonic and earthquake cycle deformation. Processes involved in this atmospheric forcing of solid Earth deformation include the redistribution of surface loads via sediments, water and ice, near-surface pressure and temperature changes in the atmosphere, and varying fluid pressures in the shallow subsurface (Fig. 14.1). Deformation associated with changes in surface loads and poroelastic and thermoelastic eigenstrains associated with changes in subsurface fluid pressure and temperature will also perturb the state of stress on faults at depth. Depending on how close such faults are to their critical failure stress and to what degree the climate-driven stress changes encourage or discourage failure, we may thus expect some effect on the distribution of earthquakes in space and time.

Tectonic stressing rates in plate boundary zones are on the order of 1–100 kPa/yr, and substantially less in intracontinental regions. Earthquakes typically cause a drop in stress of about 1–100 MPa, but the absolute ambient stress is rather poorly known, with an upper bound provided by the frictional strength of faults (Byerlee, 1978). Atmospheric processes drive hydrological and sedimentary mass transport and produce changes in temperature and fluid pressure in the shallow Earth, which modify subsurface stress over a wide range of spatial and temporal scales (Fig. 14.2). The addition or removal of a meter-thick layer of water or ice on the Earth's surface adds or subtracts a vertical load of about 10 kPa and the most extreme atmospheric pressure drops associated with tropical cyclones can also approach 10 kPa. Thus, while stresses caused by atmospheric and hydrological processes are generally small compared to those from the unrelenting forces associated with plate tectonics, they can cause substantial variations in stressing rates and may thus change the distribution of seismicity in space and time. Of course, lake- or sea-level changes of many tens of meters or the growth and retreat of glaciers and ice sheets many hundreds of meters thick will have an even greater effect on solid Earth deformation and stress. Here, we review recent advances in our understanding of climate- and weather-driven solid Earth deformation, stress, and seismicity. We consider timescales ranging from a hundred thousand years (associated with ice age cycles) to days (associated with daily weather events). We find that while most earthquakes are the result of tectonic processes, the occurrence of a small but significant number of earthquakes appears to have been accelerated, delayed, or modulated by atmospheric and hydrological processes.

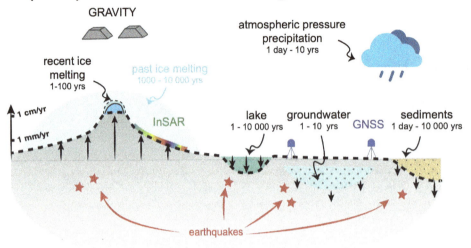

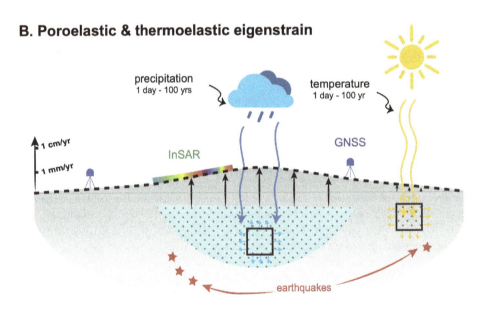

FIG. 14.1 Deformation and stresses due to climate-driven surface loading and poro- and thermoelastic eigenstrain. (A) Schematic representation of the effect of various processes involving past and present surface and subsurface loading, including glacial isostatic adjustment (GIA) in response to unloading of the last glacial maximum ice cap, recent ice melting, lake loading, groundwater, precipitation, atmospheric pressure, and sediments. (B) Schematic representation of the effect of poroelastic and thermoelastic eigenstrain induced by fluctuations in groundwater level and surface temperature. Geodetic measurements by GNSS, InSAR, and space-based gravity illuminate these processes.

2 Observing and modeling climate-driven deformation, stress, and seismicity

2.1 Measuring climate-driven deformation

The solid Earth deforms in response to climate-driven spatial and temporal changes in surface mass load, including terrestrial water storage, ice, snow, and sediment transport through landsliding and erosion (Fig. 14.1A). Additionally, temperature variations near the Earth's surface and pressure fluctuations in groundwater are associated with thermoelastic and poroelastic eigenstrain in the shallow Earth, which also causes the solid Earth to deform (Fig. 14.1B). Using space geodetic observations, it becomes possible to comprehensively capture and constrain models of these mass load changes and deformation processes (Fig. 14.1).

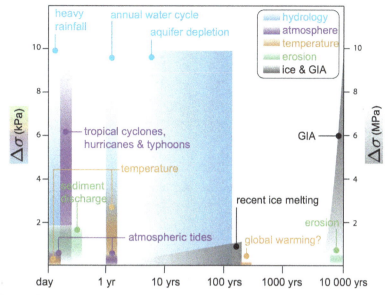

FIG. 14.2 Timescales of climate-driven surface processes and associated stress variations. Note the substantially higher peak stress changes of several MPa associated with ice sheet loading and glacial isostatic adjustment (GIA, right y-axis), compared to atmospheric, hydrological, erosional, and thermally induced stress changes of a few kPa that occur on shorter timescales (left y-axis).

Measuring climate-driven mass load changes has been enabled by the launch of the Gravity Recovery and Climate Experiment (GRACE; 2002–17) and GRACE Follow-On (GRACE-FO, 2018-in orbit) satellite missions which introduced a unique observable: monthly global measurements of Earth's space and time-varying gravity field at a spatial resolution of 300–400 km (Landerer et al., 2020; Tapley, Bettadpur, Ries, Thompson, & Watkins, 2004). By combining GRACE-derived mass loads with a (visco-) elastic Earth model through the load Love numbers theory, it is possible to compute the loading-induced deformation of the Earth (Cathles, 1975; Farrell, 1972), which can be compared to independent GNSS measurements.

Climate-induced deformation of the Earth's surface can be directly measured using geodetic observations (Fig. 14.3). GNSS stations measure positions and changes in position at millimeter-level accuracy. Time series of GNSS stations have been used to investigate the, primarily elastic, solid Earth's response to global (e.g., Blewitt, Lavallée, Clarke, & Nurudinov, 2001; Dong, Fang, Bock, Cheng, & Miyazaki, 2002; van Dam et al., 2001) and regional seasonal and long-term variations in continental water storage, and within the atmosphere and ocean. Regional examples include seasonal hydrological loading deformation in the Amazon River basin (e.g., Bevis et al., 2005; Davis, Elósegui, Mitrovica, & Tamisiea, 2004; Knowles, Bennett, & Harig, 2020), the Nepal Himalaya (e.g., Chanard, Avouac, Ramillien, & Genrich, 2014; Fu & Freymueller, 2012), snow cover in Japan (Heki, 2004), and Iceland (Compton, Bennett, & Hreinsdóttir, 2015; Grapenthin, Sigmundsson, Geirsson, Árnadóttir, & Pinel, 2006). In the western United States, snow and rainfall cause strong seasonal deformation, reaching up to a few millimeters and a few centimeters in horizontal and vertical annual amplitude, and drought periods induce multiyear signals of similar amplitudes (Fig. 14.3; e.g., Amos et al., 2014; Argus, Fu, & Landerer, 2014; Borsa, Agnew, & Cayan, 2014; Fu, Argus, & Landerer, 2015; Ouellette, de Linage, & Famiglietti, 2013). In addition to hydrological loads, the change in atmospheric pressure loading causes deformation of the Earth (Darwin, 1882; van Dam & Wahr, 1987). van Dam, Blewitt, and Heflin (1994) analyzed daily GNSS time series and concluded that atmospheric loading displacements account for up to 24% of the total variance in GNSS height series.

At the global scale (Chanard, Fleitout, Calais, Rebischung, & Avouac, 2018) and in regions of large-scale hydrological, atmospheric, and nontidal oceanic mass variations (Davis, Elósegui, Mitrovica, & Tamisiea, 2004; van Dam & Wahr, 2007) captured by GRACE/-FO, predicted surface displacements using a purely elastic spherical and layered Earth models (Dziewonski & Anderson, 1981) are in good agreement with GNSS observations (Fig. 14.3D). However, discrepancies at the regional scale exist and are unlikely to be related to the choice of Earth's mechanical properties (Chanard, Fleitout, Calais, Barbot, & Avouac, 2018). These discrepancies may be due to the high noise level of GRACE/-FO products (e.g., Chen et al., 2022) or its coarse spatial resolution (~350 km) compared to GNSS observations. Moreover, as GRACE measures integrated mass variations, distinguishing between sources of mass change at the Earth's surface and within its

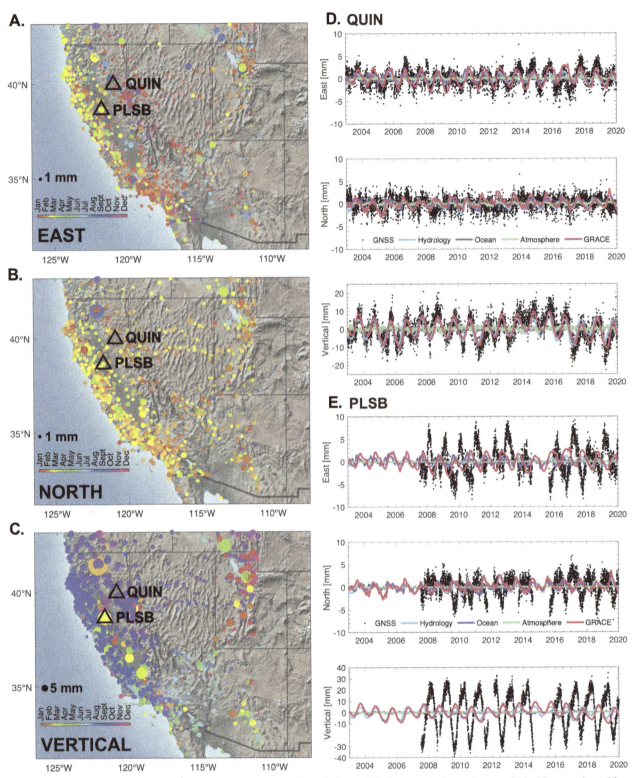

FIG. 14.3 Amplitude and phase of maximum (A) east, (B) north, and (C) vertical annual displacements in the southwestern United States, estimated from GNSS time series processed by the Nevada Geodetic Laboratory (Blewitt, Hammond, & Kreemer, 2018) aligned to the IGS14 reference frame (Altamimi, Rebischung, Métivier, & Collilieux, 2016). Black triangles highlight GNSS stations QUIN and PLSB, located in the Sierra Nevada mountain range and above the Sacramento aquifer system, respectively. Detrended daily east, north, and vertical position time series at these two GNSS stations are shown in (D) and (E) with deformation models derived from, in blue, hydrological (GLDAS; Rodell et al., 2004), in green, atmospheric (ERA5; Hersbach et al., 2020), in purple, nontidal oceanic (ECCO; Wunsch et al., 2009), and, in red, GRACE/-FO load (Gauer, Chanard, & Fleitout, 2023).

interior is critical to correctly infer climate-driven deformation from gravity measurements (e.g., Chen et al., 2022). Environmental models, including hydrological, atmospheric, and nontidal oceanic loads, provide sources of deformation independently and present higher spatiotemporal resolution but, similar to GRACE-derived models, their combination can explain only up to 50% of the observed seasonal vertical displacements and up to 20% of the horizontal ones (Fig. 14.3D; e.g., Li, van Dam, Li, & Shen, 2016).

Greater details of surface loading can be mapped by inverting ground deformation recorded by dense GNSS networks. For example, using the EarthScope plate boundary observatory (PBO) GNSS data, climate-related terrestrial water storage variations have been quantified at high spatial resolution (tens of kilometers) in California (Amos et al., 2014; Argus et al., 2017; Argus, Fu, & Landerer, 2014; Johnson, Fu, & Bürgmann, 2017a), the Pacific Northwest (Fu, Argus, & Landerer, 2015), and across the western United States (Borsa, Agnew, & Cayan, 2014; Enzminger, Small, & Borsa, 2018). These studies captured both annual cycles of water load and multiyear variations associated with regional drought conditions. Milliner et al. (2018) extended this technique to load variations over the course of a few days and estimated the total water mass released by Hurricane Harvey at the Texas Gulf Coast. Yet, the method is limited to well-instrumented regions, with a resolution that depends on the (often uneven) interstation distances and may not capture the full load complexity (Wahr et al., 2013). Moreover, results may be biased by other geophysical signals or systematic errors polluting GNSS observations (Chanard, Métois, Rebischung, & Avouac, 2020; Larochelle et al., 2022).

Other geophysical signals recorded by GNSS stations include poroelastic deformation driven by fluctuations in groundwater level and thermoelastic strain due to spatiotemporal variations in near-surface temperature (Fig. 14.4). When surface hydrological loading increases, it can also lead to downward fluid infiltration and diffusion, causing a rise in pore pressure, and Coulomb stress. It can also enhance hydrofracturing in favorably oriented cracks. While surface hydrological loading has been established as the primary source of GNSS seasonal observations, the contribution of poroelastic processes to surface deformation has received renewed attention in recent years. The decametric resolution and millimeters/year accuracy of Interferometric Synthetic Aperture Radar (InSAR) make it a powerful tool to monitor localized surface subsidence in response to groundwater depletion (e.g., Amelung, Galloway, Bell, Zebker, & Laczniak, 1999; Chaussard, Wdowinski, Cabral-Cano, & Amelung, 2014; Ojha, Shirzaei, Werth, Argus, & Farr, 2018). GNSS provides valuable complementary observations to monitor poroelastic deformation, offering a better temporal resolution, a more accurate separation of horizontal and vertical displacements, and the potential to identify errors specific to each geodetic technique (Larochelle et al., 2022). However, while surface loading models generally agree with observations of nonlinear motion at GNSS sites anchored to bedrock (Fig. 14.3D), they fail at predicting the amplitude and phase of site motion when located on or around groundwater basins (Fig. 14.3E). Additionally, fluctuations in surface temperature that can span tens of degrees annually cause the shallow Earth to deform thermoelastically, with differential motions of up to several millimeters (Ben-Zion & Allam, 2013; Prawirodirdjo, Ben-Zion, & Bock, 2006; Xu et al., 2017). Poroelastic and thermoelastic processes are now well recognized, but their modeling remains challenging and further research is needed to fully comprehend and model the link between changes in groundwater, temperature, and surface deformation.

2.2 Modeling climate-driven deformation and stress

Mechanical models are needed to quantitatively relate surface loads to Earth deformation and stress (Fig. 14.4). In the 3D, semi-infinite, homogeneous elastic solid approximation, analytical solutions exist for strain and stresses caused by a surface point load (Boussinesq, 1885). These solutions have been used to model stresses on faults induced by seasonal hydrology (Bettinelli et al., 2008) or erosion (Steer, Simoes, Cattin, & Shyu, 2014). The solution of deformation by surface loads on a more realistic spherical elastic Earth model was provided by Farrell (1972) following the method of Longman (1962, 1963). Load Love numbers and the resulting Green's functions can be computed based on a spherical, radially stratified Earth model (e.g., Wang et al., 2012), such as the PREM (Preliminary Reference Earth model, Dziewonski & Anderson, 1981). The loading-induced stress changes in the Earth can then be calculated combining strain and the elastic moduli of the Earth. Many programs have been developed to compute elastic strain induced by surface loading in the spherical Earth approximation. For example, SPOTL is designed to calculate ocean tidal loading (Agnew, 1997) but can be utilized to calculate deformation induced by surface loads (e.g., Borsa, Agnew, & Cayan, 2014). STATIC1D (Pollitz, 1996) can be used to model strain and stress from a surface load force (e.g., Johnson, Fu, & Bürgmann, 2017a, 2017b). LoadDef models elastic deformation by surface mass loading for any planetary body (Martens, Rivera, & Simons, 2019). ISSM-SESAW is especially designed for cryospheric loading problems of ice and glaciers (Adhikari, Ivins, & Larour, 2016) and REAR is another simulation tool to model loading deformation by cryospheric loads (Melini, Gegout, Spada, & King, 2014). These spherical models can be extended to the viscoelastic domain for Newtonian rheologies, including Maxwell and Burgers

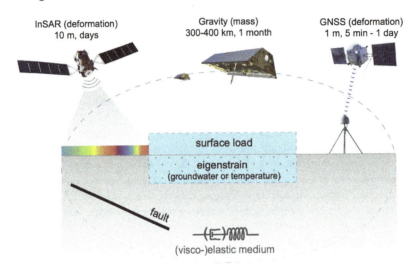

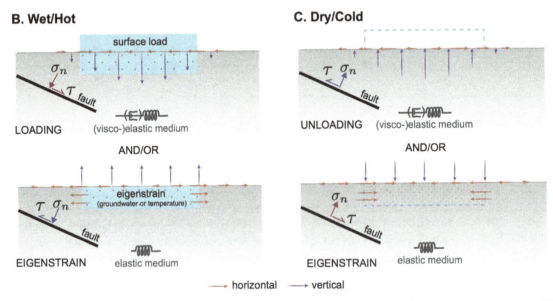

FIG. 14.4 (A) Schematic illustration of space-geodetic observations and models of hydrologically induced deformation and stress changes. (B) During a wet season, an increase of near-surface water load (surface and groundwater, ice and snow) causes 3D loading displacements and also applies normal (σ_n) and shear (τ) stress changes on a nearby fault, thus inducing Coulomb stress changes (CS $= \tau + \mu\,\sigma_n$, where μ is the fault friction coefficient; Stein, 1999). An increase of subsurface groundwater also results in an expansion and elevated pore fluid pressure, which may also lead to Coulomb stress change on an adjacent fault. Similarly, zones of high temperature produce thermoelastic eigenstrains and stresses. (C) The opposite scenario occurs during a dry/cold period.

bodies, which are necessary to investigate deglaciation-induced strain and stress fields through time (Craig, Calais, Fleitout, Bollinger, & Scotti, 2016; Lambeck & Purcell, 2003; Michel & Boy, 2022; Pollitz, Wech, Kao, & Bürgmann, 2013; Wu, 1999; Wu & Johnston, 2000). In addition to analytical solutions, programs relying on the finite element method, such as Pylith (Aagaard, Knepley, & Williams, 2017), have been developed to compute more detailed loading-induced strain and stress of the Earth by surface forces, but are often limited to a smaller region (Brandes, Steffen, Steffen, & Wu, 2015; Hampel, Hetzel, Maniatis, & Karow, 2009; Lund, Schimdt, & Hieronymus, 2009).

Models also exist that allow for computing deformation and stress from subsurface fluid pressure changes caused by the infiltration of surface water. These include simple point pressure sources (Mogi, 1958), solutions of finite strain cuboid sources in a homogeneous half-space (Barbot, Moore, & Lambert, 2017), analytical solutions for an unconfined aquifer with heterogeneous elastic properties (Larochelle et al., 2022), and numerical solutions of the poroelastic problem

(Wang & Kümpel, 2003). In the upper few kilometers of the crust, strain and stress changes from the poroelastic response to groundwater level changes greatly exceed those from hydrological surface loading, but rapidly decay with increasing depth (e.g., Hu & Bürgmann, 2020; Miller, 2008).

Poroelasticity and thermoelasticity are based on the same mathematical framework (Biot, 1956; Rice & Cleary, 1976; Tsai, 2011). The problem of thermoelastic eigenstrain and stresses has been relatively well studied, particularly due to its importance in material sciences. Berger (1975) provided a solution to describe thermoelastic strain in a 2-D homogeneous elastic half-space, induced by spatially and temporally periodic surface temperature variations. Ben-Zion and Leary (1986) extended this solution to the case of an elastic half-space covered by a thin layer of unconsolidated incompetent material, and the model was further developed and applied by Tsai (2011). The thermoelastic problem has only recently been tackled for geophysical applications in a spherical frame, which is more suited for large-wavelength or global studies (Fang, Dong, & Hager, 2014).

The Coulomb failure criterion has been widely applied to quantify the stress condition required for a fault to rupture. Coulomb stress change can be simplified as $\Delta CS = \Delta\tau + \mu\Delta\sigma_e$, where $\Delta\tau$ is the shear stress change, $\Delta\sigma_e = \Delta\sigma_n - \Delta p$ is the effective normal-stress change (i.e., the difference between the changes in normal stress σ_n and pore pressure Δp), and μ is the fault friction coefficient (Stein, 1999). Positive Coulomb stress change is considered to promote fault failure and negative Coulomb stress change suppresses failure. Under undrained conditions, the change in pore pressure due to a change in normal stress is given by $\Delta p = B\Delta\sigma_n$, where B is the Skempton coefficient (and typically ranges between 0.5 and 0.9). Thus, $\Delta CS = \Delta\tau + \mu\Delta\sigma_n(1-B)$. Because of fluid flow, the induced pore pressure changes are time-dependent and it becomes important to consider the timescale for which a poroelastic stress change model is applied (e.g., Cocco & Rice, 2002).

2.3 Documenting earthquake triggering and modulation

The idea that climate might influence seismicity dates back to at least the fourth century B.C., when Aristotle proposed that winds trapped in subterranean caves generated earthquakes (Missiakoulis, 2008). Since the early 20th century, studies have suggested that climate-driven transient and oscillatory stresses may be sufficient to trigger or modulate seismicity given the slow buildup of stress along active faults (Fig. 14.5A; Drake, 1912; Sayles, 1913; Taber, 1914). However, establishing correlations between climate forcing and seismicity is challenging because it requires quantitative or statistical methods, high-quality data sets, and extended observation periods. A number of studies have investigated temporal changes in the statistics of recorded seismicity, such as earthquake frequency, seismic moment rate, or the b-value of the Gutenberg-Richter earthquake frequency-magnitude distribution, in relation to climatic events or periodic forcing (e.g., Zhai et al., 2021).

Several statistical approaches, continuous or discrete in time, have been shown to be effective in detecting periodicities in seismic catalogs. The Schuster test (Schuster, 1897) and its recent extension the Schuster spectrum (Fig. 14.5B; Ader & Avouac, 2013) use the relative discrete timing of seismic events for statistical analysis. The probability that the temporal distribution of events is periodic at a given frequency is compared to the null hypothesis, where the event distribution arises from a uniform seismicity rate. Other approaches exist, based on the use of time series analysis and periodogram statistics, such as the unifrequential Schuster periodogram (Schuster, 1898) or its multifrequential extension and generalization, the multifrequential periodogram (Dutilleul, Johnson, Bürgmann, Wan, & Shen, 2015), which can detect overlapping signals with fractional frequencies. However, periodograms require a trigonometric model hypothesis. Hsu et al. (2021) use the empirical mode decomposition to separate seismicity rate time series into signals with different timescales (intrinsic mode functions), focusing on annual and interannual variations. If the period of the external forcing is known (e.g., annual climate cycles), simple stacks or histograms of event occurrences are also commonly used (Fig. 14.5A; e.g., Bollinger et al., 2007; Craig, Chanard, & Calais, 2017; Taber, 1914).

All these methods are subject to the implicit mathematical assumption of independent earthquake occurrences, and therefore can be biased by the occurrence of clustered aftershocks. As a result, statistical analysis of periodicities generally requires first declustering the seismic catalog through procedures relying on judgment-based parameter choices (e.g., Reasenberg, 1985; Zaliapin & Ben-Zion, 2020; Zhuang, Ogata, & Vere-Jones, 2002), which may impact the inference of periodicities (Dutilleul, Johnson, Bürgmann, Wan, & Shen, 2015; Ueda & Kato, 2019).

In addition to applying statistical techniques to the temporal distribution of earthquakes, Johnson et al. (2017a, 2020) quantified the percentage of earthquakes that exceed the background seismicity rate with regard to climate-driven stress perturbations that encourage slip on known faults or enhance the background stress field to assess seismic modulation. Similarly, Craig, Chanard, and Calais (2017) investigated correlations between seismicity rates and hydrology-driven stress variations on faults computed from satellite gravity measurements. These methods take faulting style and orientation into consideration when determining whether or not climatic forcing increases or decreases stress and seismic activity on faults.

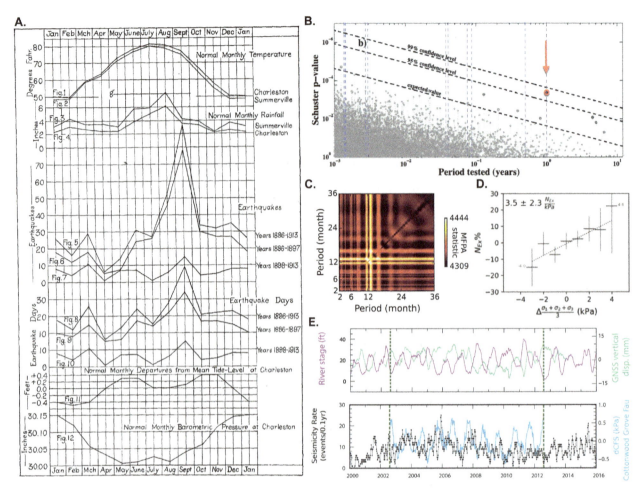

FIG. 14.5 Examples of methods to detect seismic modulation by climate forcing. (A) Seasonal variations in temperature, precipitation, number of earthquakes and days with earthquakes (to reduce aftershock bias), oceanic and atmospheric pressure in South Carolina over the 1886–1913 period. (B) Seasonal periodicity of the $M_b \geq 4$ mid-crustal Nepal seismic events from a Schuster spectrum of the 1965–2008 ISC seismic catalog. (C) Seasonal periodicity of seismicity from the 1988–2002 Central San Andreas Fault near Parkfield seismic catalog inferred from contour maps of the multifrequential periodogram analysis in the bifrequential case. Here, the lighter the color, the higher the statistical significance of the detected periodicity (Dutilleul, Johnson, Bürgmann, Wan, & Shen, 2015). (D) Percent excess seismicity of the 2000–2016 southern Alaska seismic catalog assuming a 3-month lagged response to hydrologically driven differential stress variations, suggesting a delayed modulation mechanism (Johnson, Fu, & Bürgmann, 2020). (E) Correlation between the New Madrid region seismicity rate (black), variations in the Mississippi river stage, local GNSS vertical displacement, and calculated Coulomb failure stress variations from GRACE loading variations on the Cottonwood grove fault (Craig, Chanard, & Calais, 2017). ((A) Figure from Taber, S. (1914). Seismic activity in the Atlantic coastal plain near Charleston, South Carolina. Bulletin of the Seismological Society of America, 4 (3), 108–160. (B) Figure from Ader, T. J., & Avouac, J. P. (2013). Detecting periodicities and declustering in earthquake catalogs using the Schuster spectrum, application to Himalayan seismicity. Earth and Planetary Science Letters, 377, 97–105.)

3 Deformation and seismicity from changing climate and weather

3.1 Ice age climate cycles

Starting about 2.6 million years ago, Earth's climate has been dominated by ice age cycles accompanied by the emplacement and retreat of vast ice sheets and associated global sea-level changes of over 100 m. In addition to ice sheets reaching far across high-latitude and elevated continental regions, large lakes at lower latitudes have also undergone correlated filling cycles. This cyclic global water redistribution, as well as associated erosion, transport, and deposition of sediments, led to substantial changes in loading at the Earth's surface and stress under and surrounding continental ice sheets and lakes (e.g., Gilbert, 1890; Johnston, 1987), as well as in near-coastal regions (e.g., Luttrell & Sandwell, 2010). Since the end of the last ice age ~12,000 years ago, the Earth has been adjusting to the associated removal of vast ice sheets and sea-level rise through glacial isostatic adjustment (GIA), which, due to the time-dependent viscoelastic relaxation of the Earth's mantle, is an ongoing process evident in GNSS-measured deformation, GRACE measured gravity changes, and coastal sea-level variations captured by raised shorelines and tide gauges (e.g., Caron et al., 2018; Kierulf et al., 2021; Milne et al., 2001; Sella et al., 2007; Tamisiea, Mitrovica, & Davis, 2007; van Dam, Whitehouse, & Liu, 2024). Depending on the tectonic setting and associated lateral variation of the lithospheric thickness and effective mantle viscosity, the amplitude, duration, rate, and sign of GIA-induced deformation and associated stress changes are highly variable (e.g., Steffen, Wu, Steffen, & Eaton, 2014; Vachon et al., 2022; Wu, Steffen, Steffen, & Lund, 2021, and references therein).

Stress changes induced by GIA can reach more than 10 MPa near, and in the forebulge of, the continental ice-sheet load zones (Fig. 14.6), reaching deep into the lithosphere (e.g., Vachon et al., 2022; Wu & Johnston, 2000; Wu, Steffen, Steffen, & Lund, 2021). As these stress changes can reach the range of typical earthquake stress drops, they are capable of triggering earthquakes on faults suitably oriented with respect to the induced stress field or arrest slip on unfavorably oriented faults (e.g., Steffen & Steffen, 2021). Large prehistoric earthquake ruptures and current seismicity in Scandinavia, Greenland,

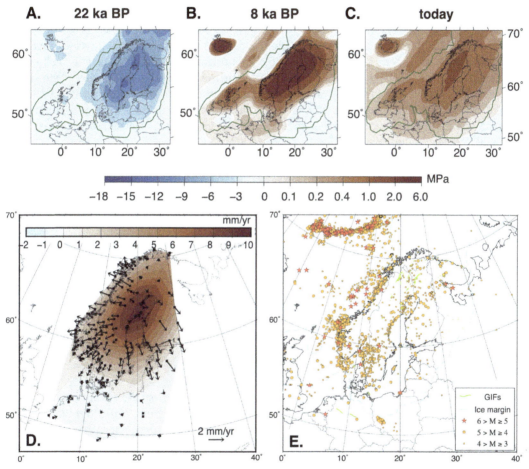

FIG. 14.6 Stress, surface deformation, and earthquakes associated with glacial isostatic adjustment in Fennoscandia. Change in Coulomb stress at 2.5 km depth from viscoelastic model calculation at (A) Last Glacial Maximum (22 ka). (B) Soon after the ice sheet has completely melted (8 ka). (C) Present day. The green solid line shows the maximum ice sheet extent at Last Glacial Maximum. The numerical model assumes a background thrust-faulting stress regime with a N150°E-oriented maximum horizontal stress (see Wu, Steffen, Steffen, & Lund, 2021 for details). Assuming zero Coulomb stress before glaciation, thrust faults were stabilized during maximum ice sheet loading, strongly encouraged to slip immediately following unloading, and continue to be loaded due to ongoing viscoelastic rebound. (D) Current vertical and horizontal surface velocities measured with GNSS. (E) Glacially induced faults (GIFs) (green lines, from Munier et al., 2020) and recent (M > 3) seismicity. The dashed blue lines show the ice margin at Last Glacial Maximum. ((C) Modified from Wu, P., Steffen, R., Steffen, H., & Lund, B. (2021). Glacial isostatic adjustment models for earthquake triggering. In H. Steffen, O. Olesen, & R. Sutinen (Eds.), Glacially-triggered faulting (pp. 383–401). Cambridge: Cambridge University Press. https://doi.org/10.1017/9781108779906.029 and Patrick Wu (pers. comm., 2022). (D) Modified from Kierulf, H .P., Steffen, H., Barletta, V. R., Lidberg, M., Johansson, J., Kristiansen, O., & Tarasov, L. (2021). A GNSS velocity field for geophysical applications in Fennoscandia. Journal of Geodynamics, 146, 101845 and Holger Steffen (pers. comm.). (E) Modified from Kierulf, H. P., Steffen, H., Barletta, V. R., Lidberg, M., Johansson, J., Kristiansen, O., & Tarasov, L. (2021). A GNSS velocity field for geophysical applications in Fennoscandia. Journal of Geodynamics, 146, 101845 and Holger Steffen (pers. comm.).)

North America, the European Alps, and around Antarctica have been linked to these large and enduring climate-driven stress changes (e.g., Ivins, James, & Klemann, 2003; Steffen, Olesen, & Sutinen, 2021; Steffen et al., 2020; Vachon et al., 2022; Wu & Johnston, 2000; Wu, Steffen, Steffen, & Lund, 2021, and references therein). In the near field, triggered earthquake activity likely was most intense during and shortly after the latest pulse of deglaciation (e.g., Fig. 14.6B), but enhanced seismicity in the previously glaciated areas appears to continue into the present, both in North America and in Fennoscandia (Fig. 14.6E). Glacial-load stresses may also suppress earthquake activity if they counteract the applied tectonic stress field, moving faults further away from critical failure stress conditions (e.g., Johnston, 1987). Activity in the forebulge during loading and ice-sheet advance is also observed (e.g., Pisarska-Jamroży et al., 2018; Steffen & Steffen, 2021; Štěpančíková et al., 2022). Štěpančíková et al. (2022) documented a case of fault-slip acceleration during the peak glaciation phase in the Late Pleistocene on a fault in central Europe ~150 km from the ice sheet margin. This active period was followed by a cessation of activity throughout the Holocene, consistent with stress-change conditions in the forebulge, favoring slip during loading, rather than unloading.

Glacially triggered earthquakes generally reflect the combined effect of plate tectonic stresses and glacially induced isostatic stresses. It is often difficult to ascertain a direct relationship between GIA and earthquake occurrence, and thus some caution is warranted in assessing proposed triggering relationships, especially for events at large distances from the former ice sheet margin (e.g., Steffen, Olesen, & Sutinen, 2021; Wu & Johnston, 2000), such as the New Madrid seismic zone in North America. Even in Fennoscandia, the correlation of modeled current stress conditions with active seismicity is not simple, suggesting that other factors, such as unfavorable fault orientations and stress heterogeneity remaining from pre-glaciation conditions and earlier large triggered events (e.g., Gregersen et al., 2021), may influence the stress field (e.g., Wu, 1999). Nonetheless, the clear evidence of glacially triggered earthquakes since the Last Glacial Maximum (e.g., Steffen, Olesen, & Sutinen, 2021) suggests that the ongoing shrinkage of many modern ice sheets owing to recent global warming could impact earthquake occurrences in these regions (e.g., Hampel, Hetzel, & Maniatis, 2010; Pagli & Sigmundsson, 2008; Sauber, Rollins, Freymueller, & Ruppert, 2021).

The filling and desiccation of large lakes associated with glacial climate cycles also represent substantial surface loads and lead to deformation and redistribution of stress in the subsurface. Lithospheric rebound due to the regression of Lake Bonneville in Utah, together with deglaciation and unloading of the nearby Rocky Mountains, may have nearly doubled fault-slip rates and earthquake occurrences on normal faults in the eastern Basin and Range Province, in the early Holocene (Friedrich, Wernicke, Niemi, Bennett, & Davis, 2003; Hampel, Hetzel, & Maniatis, 2010; Hetzel & Hampel, 2005). Lake filling cycles in the Salton Trough basin in southern California during the last few thousand years have also been correlated with seismicity in an underlying fault-stepover zone and earthquake cycles on the nearby San Andreas fault, pointing to a direct linkage of earthquake hazard and climate-induced stresses (Brothers, Kilb, Luttrell, Driscoll, & Kent, 2011; Hill, Weingarten, Rockwell, & Fialko, 2023). Hill, Weingarten, Rockwell, and Fialko (2023) proposed that the past six major events on the southern San Andreas Fault took place when the water level of Lake Cahuilla was high, which increased Coulomb stress on the fault by several hundred kiloPascal and stressing rate by a factor of two.

In addition to the rapid retreat of the continental ice sheets and shrinking of large lakes in the early Holocene, associated rapid sediment transport may also produce substantial changes in surface loads that may trigger seismicity. For example, Calais, Freed, Van Arnsdale, and Stein (2010) proposed that $M > 7$ earthquakes in the intracontinental New Madrid seismic zone, which repeatedly occurred since the last ice age, resulted from unloading stresses associated with the accelerated fluvial erosion in the Mississippi River valley in the late Pleistocene, rather than from the direct or indirect far-field effects of the glacial unloading (Grollimund & Zoback, 2001; Pollitz, Kellogg, & Bürgmann, 2001). Similarly, Stein, Cloetingh, Sleep, and Wortel (1989) and Gradmann, Olesen, Keiding, and Maystrenko (2018) concluded that rapid coastal erosion and sediment transport during Pleistocene glaciations was a dominant source of stress changes in northern Norway.

As the massive ice sheets melted away at the end of the Pleistocene, sea-level rose globally by ~120 m at rates of 10–20 mm/yr, thus adding a substantial surface load in the ocean basins and along coastal margins (e.g., Lambeck & Chappell, 2001). The consequent elastic flexure produced shear stress changes in near-coastal regions reaching ~1 MPa (Brothers, Luttrell, & Chaytor, 2013; Luttrell & Sandwell, 2010). Luttrell and Sandwell (2010) used viscoelastic stress-modeling to find that nearshore (within a few hundred km) transform faults, such as the San Andreas, North Anatolian and Alpine fault systems, experienced failure-encouraging stress increments (Fig. 14.7), thus temporarily increasing their slip rate and decreasing earthquake recurrence intervals. However, they found that given the long timespans over which these stress changes occur, the stressing rates due to this process are modest compared to the rapid tectonic loading rates on the major strands of the plate-boundary faults. On the other hand, earthquake cycles on secondary, slow-slipping faults may have been more substantially perturbed. On the North Anatolian fault zone, where loading-induced stressing rates were orders of magnitude higher due to the very sudden filling of the Black Sea, it is plausible that the high stressing rates led to seismic failure of much of the fault zone within a relatively short time. This possibly led to the observed

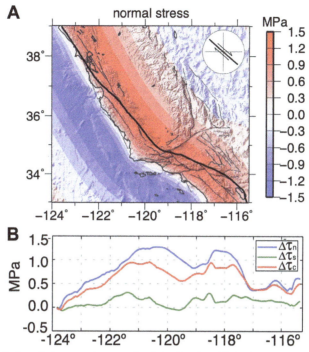

FIG. 14.7 Stress perturbation on San Andreas fault system due to sea-level rise since Last Glacial Maximum. (A) Normal-stress change (positive values represent slip-encouraging tension) on N50°W striking, vertical strike-slip faults following 120 m of sea-level rise. (B) Normal (blue), shear (green), and Coulomb stress (red) on the actual San Andreas fault plane whose surface trace is shown as bold black in (A). *(Modified from Luttrell, K., & Sandwell, D. (2010). Ocean loading effects on stress at near shore plate boundary fault systems. Journal of Geophysical Research, 115, B08411. https://doi.org/10.1029/2009JB006541.)*

synchronized timing of large earthquakes on the North Anatolian fault zone (Luttrell & Sandwell, 2010). On the Cascadia subduction zone, the shallow and coupled offshore megathrust was exposed to increased compression and thus slip is discouraged during sea-level rise, whereas slip on the creeping downdip portion of the fault beneath the continent is accelerated (Luttrell & Sandwell, 2010). The associated stress changes on the subduction thrust are relatively small, but they could possibly affect the depth extent of ruptures during times of high versus low sea level (Luttrell & Sandwell, 2010). Brothers, Luttrell, and Chaytor (2013) proposed that abundant slope failures observed in the sedimentary records of passive continental margins around 15–8 ka were caused by the seismic reactivation of suitably oriented and near-critically stressed near-coastal normal faults. Depending on the local tectonic environment, fault geometries and rate of stress associated with ice-age sea-level change, earthquake cycles on nearshore faults may be substantially perturbed, especially on slow-moving faults for which the climatically induced stress changes are proportionally higher.

3.2 Consequences of recent climate change

Since the 1800s, human activities have been the main driver of climate change (IPCC, 2021). As modern Earth's climate warms at unprecedented rates, it promotes faster melting of the polar ice caps (Hugonnet et al., 2021; The IMBIE team, 2018, 2020), thereby raising sea levels (e.g., Frederikse et al., 2020). It also accelerates global and regional trends in precipitation and water runoff and enhances regional floods and droughts (e.g., Dai, 2016; Gudmundsson, Leonard, Do, Westra, & Seneviratne, 2019). Moreover, groundwater, which serves as a water-shortage buffer, is subject to major fluctuations and often rapid depletion in some regions (e.g., Green et al., 2011). Particularly, groundwater extraction in coastal regions can cause subsidence that adds to relative sea-level rise and increases the risk of floods (e.g., Shirzaei et al., 2021; Tay et al., 2022; Wu, Wei, & D'Hondt, 2022). Accelerating changes in the global water distribution and rising temperatures, occurring over timescales ranging from months to centuries, cause the Earth to move through (visco-)elastic, poroelastic, and/or thermoelastic deformation, inducing stresses that may modulate tectonic seismicity and possibly reactivate formerly stable faults.

In polar regions, glaciers and ice sheets are undergoing massive ice loss on timescales similar to the seismic cycle, with thinning of meters per year in regions of ablation (Mouginot et al., 2019; Shepherd et al., 2012). Glacier surges, during which ice flow velocities transiently increase by orders of magnitude, can abruptly modify the distribution of ice load through catastrophic glacier collapse (Kääb et al., 2018). In addition, several regions such as Patagonia (Dietrich et al., 2010), Alaska (Larsen, Motyka, Freymueller, Echelmeyer, & Ivins, 2005), the Alps (Barletta et al., 2006), and Greenland (Adhikari et al., 2021), have been impacted by rapid GIA as a result of glacial unloading following the end of the Little Ice Age, a period of regional cooling from the early 14th century to the mid-19th century during which glaciers and ice sheets expanded. Present-day ice mass loss induces an instantaneous elastic response of the solid Earth, while past ice loss contributes a prolonged viscoelastic response. These two responses may act simultaneously and produce among today's fastest surface uplift rates on Earth exceeding 30 mm/yr and are also associated with millimeter/year horizontal deformation (Adhikari et al., 2021; Hu & Freymueller, 2019). This combined glacial unloading effect can influence seismicity in volcanic (e.g., Sigmundsson et al., 2010) and tectonic dip-slip and strike-slip settings (Rollins, Freymueller, & Sauber, 2021; Sauber, Plafker, Molnia, & Bryant, 2000; Sauber, Rollins, Freymueller, & Ruppert, 2021). Rollins, Freymueller, and Sauber (2021) illustrated the viscoelastic stress changes associated with a scenario of 1 m/yr glacial unloading over 200 years, alongside a vertical strike-slip fault. Rapid GIA acts to unclamp the fault, promoting fault failure, and the Coulomb stress (Fig. 14.4) amplitude reaches 0.5 MPa with an asymmetrical distribution caused by the unloading location with respect to the fault (Fig. 14.8A).

A particularly interesting region to investigate the impact of present-day ice retreat, thinning surges, and GIA following the Little Ice Age is Southern Alaska, a rapidly deforming and seismically active setting with thrust and strike-slip faults. In the Bering glacier region, Sauber and Molnia (2004) found an increase/decrease in the seismicity rate associated with ice thinning/thickening over the 1995–2000 period, modulated by glacier surges. Similarly, Sauber and Ruppert (2008) suggested that the rapid ice wastage between 2002 and 2006 in the Icy Bay may have affected seismicity rates in the area. At longer timescales, Rollins, Freymueller, and Sauber (2021) showed that rapid GIA following the thinning of the Bay Icefield since 1770, and elsewhere in southeast Alaska since 1900, likely promoted Coulomb stress at the location of the 1958 Mw 7.8 Fairweather Fault earthquake and of at least 23 of 30 known Mw ≥5.0 earthquakes in southeast Alaska (Fig. 14.8B).

In other polar regions, where tectonic deformation is occurring at much lower rates, correlations between recent ice melting and seismic activity are more difficult to establish. In Greenland, earthquakes mainly occur along continental margins, which coincide with the ice sheet margin and where most of the recent ice melting occurs. While Olivieri and Spada (2015) could not find a clear correlation between seismicity and recent glacial unloading, they interestingly suggest

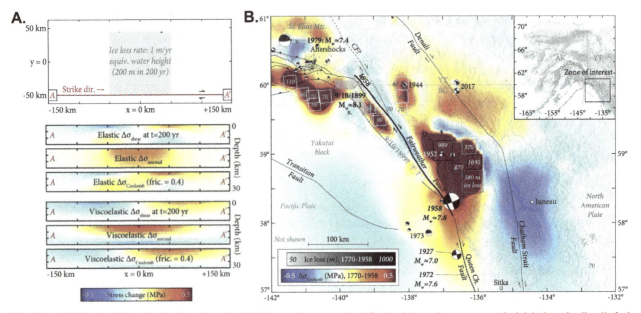

FIG. 14.8 (A) Elastic and viscoelastic effect of cumulative 200-year shear, normal, and Coulomb stress changes on a vertical right-lateral strike-slip fault caused by a constant ice unloading rate of 1 m/yr equivalent water height (gray) for a fault located at the edge of the 100 km by 100 km zone of unloading. Stresses are positive for unclamping. (B) Coulomb stress changes as of 1958 at 10 km depth induced by rapid GIA since 1770 in Southern Alaska. Ice mass loss in meters of equivalent water height is shown by gray squares. Mw ≥5.0 earthquakes occur preferably in regions where rapid-GIA driven Coulomb stress promotes failure. *(Modified after Rollins, C., Freymueller, J. T., & Sauber, J. M. (2021). Stress promotion of the 1958 Mw ~ 7.8 Fairweather Fault earthquake and others in Southeast Alaska by glacial isostatic adjustment and inter-earthquake stress transfer. Journal of Geophysical Research: Solid Earth, 126(1), e2020JB020411.)*

that GIA in response to the melting of the late-Pleistocene ice sheet may have promoted regional moderate-size crustal earthquakes, whereas recent ice melting may be associated with local shallower and smaller events. However, the link between seismicity and recent ice melting may be more complex. Indeed, Lough, Wiens, and Nyblade (2018) reported 27 extensional shallow to mid-crustal intraplate earthquakes in East Antarctica that occurred in 2009, a region with no significant recent changes in ice mass (Martin-Español, Bamber, & Zammit-Mangion, 2017; Whitehouse, Bentley, Milne, King, & Thomas, 2012), and suggested the influence of preexisting tectonic weakness or subglacial hydrological processes on the seismic activity.

Recent climate change also impacts groundwater storage through an increase in frequency and intensity of extreme weather events, such as droughts, as well as anthropogenic activities (IPCC, 2022). Sustained aquifer depletion over years to centuries can cause the Earth's surface to subside by up to tens of centimeters per year and will also impact the regional stress field and potentially influence seismic activity. In California, decades of continual groundwater pumping have already resulted in tens of meters of water level drop and meters of irreversible land subsidence (Galloway, Jones, & Ingebritsen, 1999), at rates modulated by climate forcing with more groundwater used during dry periods related to the evolving El Nino/La Nina cycles (e.g., Faunt, Sneed, Traum, & Brandt, 2016). The Central Valley, which is surrounded by active faults and has experienced multiannual periods of drought, has been particularly intensively studied for changes in groundwater, surface deformation, and Coulomb stress on nearby faults. GNSS (Amos et al., 2014; Argus et al., 2017) and InSAR (Ojha, Werth, & Shirzaei, 2019; Smith et al., 2017) observations indicate subsidence rates ranging from a few millimeters/year to 0.3 m/yr, inducing Coulomb stress changes of a few Pascal to a few kiloPascal in the crust (Carlson, Shirzaei, Werth, Zhai, & Ojha, 2020; Johnson, Fu, & Bürgmann, 2017a, 2017b). Moreover, Lundgren, Liu, and Ali (2022) investigated viscoelastic effects accompanying groundwater unloading using a 150-year historical groundwater change record. They showed that the associated cumulative Coulomb stress change on the San Andreas Fault is of the order of 0.01–0.02 MPa, comparable to Coulomb stress increases triggering seismicity following large earthquakes (Stein, 1999). Kundu, Vissa, and Gahalaut (2015) and Kundu et al. (2019) argued that stress changes due to long-term groundwater extraction and hydrological unloading substantially contributed to the stress levels at the times of the 2015 M7.8 Gorkha earthquake and 2017 M7.3 Iran-Iraq border earthquake. While these studies show clear groundwater-loss-induced strain and stress changes, assessing the impact on seismicity or the timing of individual large earthquakes is not straightforward at these longer timescales.

Young, Kreemer, and Blewitt (2021) document vertical uplift and horizontal extension around Great Salt Lake during the 2012–16 drought, showing that in addition to the load reduction associated with lake-level drop, additional water storage loss must have occurred in surrounding aquifers. They also suggest that earthquakes within the region affected by Great Salt Lake and surrounding groundwater occur preferentially during dry multiyear periods and the earthquake rate appears anticorrelated with the lake elevation rate. In turn, earthquakes outside the load region exhibit no relationship.

Other examples of earthquake triggering by long-term groundwater loss include the 2011 Lorca earthquake in Spain (Gonzalez, Tiampo, Palano, Cannav, & Fernandez, 2012), earthquake swarms near the Dead Sea Fault (Wetzler et al., 2019), and the 2016 Petermann Ranges and other shallow intraplate earthquakes in central Australia (Wang et al., 2019). These earthquakes occurred at very shallow depths, in regions where Coulomb stress induced by long-term fluctuations in nearby groundwater promoted failure, providing insights into poroelastic processes that could modify seismic hazard in regions where large aquifer depletion occurs.

In contrast, temperature changes of a few degrees over the next century, as projected by climate models (IPCC, 2021), are expected to only induce negligible thermoelastic deformation and stress. These longer-term changes are indeed substantially smaller than annual temperature variations of a few tens of degrees that result in deformation of a few millimeters (Dong, Fang, Bock, Cheng, & Miyazaki, 2002; Xu et al., 2017) and stress of a few hundred Pascals (Fig. 14.9; Johnson, Fu, & Bürgmann, 2017b). Similarly, the changes in stressing rate on near-coastal faults due to the anticipated acceleration in sea-level rise by several millimeters/year (IPCC, 2021), are unlikely to lead to a notable rise in earthquake occurrences or hazards.

3.3 Seasonal hydrological and atmospheric loads

The large-scale seasonal water cycle and associated mass redistribution couple with the solid Earth and cause periodic loading strain and stress variations. The hydrologically induced loading stress change may modulate seismicity, if they are large enough in amplitude and consistent with the background tectonic stress in orientation. In the Nepal Himalaya, Bettinelli et al. (2008) reported that the seasonal summer monsoon produces ∼2–4 kPa loading stress change on the Main Himalayan Thrust zone, and the seismicity rate in winter is twice that in the summer at all magnitudes above the detection threshold ($M > 2.2$). In Japan, Heki (2003) showed that the snow load on the western flank of the backbone mountain ranges

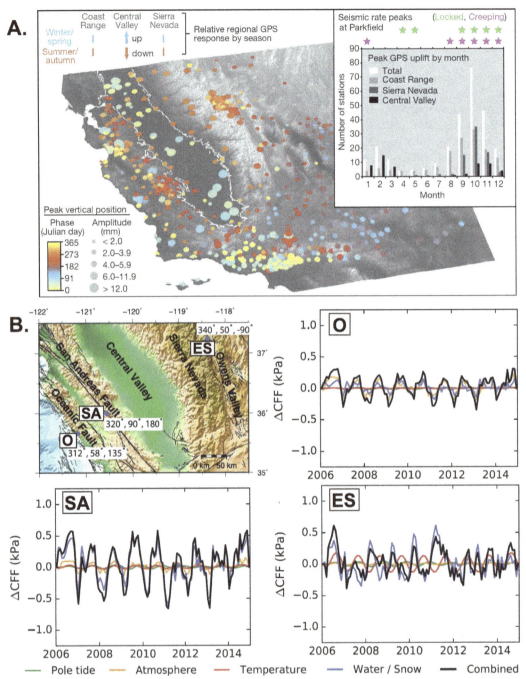

FIG. 14.9 (A) Phase and amplitude of peak GNSS vertical position in California. The histograms in the inset show the results of peak GNSS uplift in the Coast Range, Sierra Nevada, and Central Valley, respectively. The comparison with seismicity rate variation indicates that both the locked and creeping sections of the San Andreas Fault at Parkfield exhibit more events in dry later summer and autumn, corresponding to the fault unclamping caused by the unloading of seasonal terrestrial water mass. (B) Coulomb stress changes (with a friction coefficient $\mu = 0.4$) at three locations of three faults (O: Oceanic Fault; SA: central San Andreas Fault; and ES: eastern Sierra range front normal fault) from different sources of periodic stress. Strike, dip, and rake directions of the corresponding receiver fault are labeled at each location. ((A) Modified from Amos, C. B., Audet, P., Hammond, W. C., Bürgmann, R., Johanson, I. A., & Blewitt, G. (2014). Uplift and seismicity driven by groundwater depletion in Central California. Nature, 509(7501), 483–486. (B) Modified from Johnson, C. W., Fu, Y., & Bürgmann, R. (2017b). Stress models of the annual hydrospheric, atmospheric, thermal, and tidal loading cycles on California faults: Perturbation of background stress and changes in seismicity. Journal of Geophysical Research, 122, 10605–10625. https://doi.org/10.1002/2017JB014778.)

causes a few kiloPascal stress change, and the spring thaw decreases compressional stress and correlates with an increase in the number of $M \geq 7$ earthquakes. In California, Johnson, Fu, and Bürgmann (2017a, 2017b) modeled the seasonal loading strain and stress changes by annual snow and rainwater and showed that 1–5 kPa hydrologically induced loading stress changes are correlated with a ~10% increase in seismicity when the stress conditions favor slip on faults. This correlation also holds for damaging $M > 5.5$ earthquakes in catalogs reaching back to 1781 (Johnson, Fu, & Bürgmann, 2017a). Horizontal GNSS displacements have also been used to estimate seasonal strain variations and to discuss their potential association with seismicity in California (e.g., Kim, Bahadori, & Holt, 2021; Kreemer & Zaliapin, 2018). Kreemer and Zaliapin (2018) proposed that hydrologically induced seasonal strain may cause a larger stress release in an earthquake and fewer aftershocks. In the intracontinental New Madrid seismic zone, Craig, Chanard, and Calais (2017) identified annual and multiannual variations in microseismicity rates ($M \leq 2.3$) that coincide with stress variations driven by elastic hydrological loading. In southern Alaska, Johnson, Fu, and Bürgmann (2020) analyzed the shallow seismicity (<40 km) and found a seasonality with more events in winter, which correlates with ~10 kPa seasonal loading stress changes by surface snow and rainwater. Xue, Johnson, and Fu (2020) compared seasonal seismicity rate variations and water level changes of Lake Victoria and surrounding rift lakes in the western branch of the East African Rift Valley, and suggested the seasonal 0.2–0.8 kPa Coulomb stress changes due to lake water loading modulate seasonal seismicity rate variation. In western Taiwan, Hsu et al. (2021) reported a seasonal pattern with a higher seismicity rate from February to April and a lower rate from July to September, which is associated with annual water storage unloading.

In addition to hydrological loading by terrestrial water movement, atmosphere pressure loading (e.g., van Dam, Blewitt, & Heflin, 1994), solid Earth tides (e.g., Wahr, 1995), tidal and nontidal ocean loading (e.g., Agnew, 2015), and thermoelastic strain (e.g., Dong, Fang, Bock, Cheng, & Miyazaki, 2002) of the crust all produce elastic strain and stress on the surface and in the interior of the Earth. Gao, Silver, Linde, and Sacks (2000) proposed that seasonal changes in atmospheric pressure of ~2 kPa modulated seismicity in geothermal and volcanic regions of California over a 5-year period following the 1992 Landers earthquake. Johnson, Fu, and Bürgmann (2017b) modeled and compared seasonal stress changes on the faults in California from several sources, concluding that in most locations the hydrological water loading is the largest source of seasonal stress variation, followed by contributions from atmospheric and thermoelastic stress perturbations (Fig. 14.9B).

Besides analyzing the statistical correlation between environmentally induced stress changes and seismicity-rate variations, research was also carried out to investigate the possible triggering relationship with individual earthquakes. In California, Kraner, Holt, and Borsa (2018) used horizontal GNSS time series to calculate the strain and stress changes in the northern San Francisco Bay Area and found that the 2014 M 6.0 South Napa earthquake occurred during a period with peak seasonal Coulomb stress. Using a similar method, Kim, Holt, Bahadori, and Shen (2021) reported that the 2019 Ridgecrest earthquake sequence took place at a time with higher seasonal stress. Both studies suggested the possible earthquake triggering from non-tectonic seasonal loading in California (Johnson, Fu, & Bürgmann, 2017a). Hu, Xue, Bürgmann, and Fu (2021) modeled the stress perturbations by hydrological and industrial loads in the Salt Lake City region and assessed the possible correlation with the 2020 M5.7 Magna earthquake, but no definite link was found. Yao et al. (2022) reported that the 2019 M4.0 Ohio earthquake occurred when water level of Lake Erie was high. They calculated the elastic Coulomb stress change by lake-water loading, but a clear connection could not be found because of the uncertainty of the frictional coefficient in the region. It is generally difficult to confidently link modest seasonal stress changes with the occurrence of single large earthquakes, as it is not known when those events would have occurred without such perturbations.

Some studies also focused on poroelastic stress changes due to subsurface groundwater and aquifer pressure variations and their role in annual seismicity modulation. Kraft, Wassermann, and Schmedes (2006) identified a higher seismicity rate in summer coinciding with the increase of precipitation and groundwater level in the region of Mt. Hochstaufen, Germany. At the Long Valley Caldera, California, Montgomery-Brown, Shelly, and Hsieh (2019) showed the shallow seismicity rate significantly increased during the wet snowmelt season compared with the dry season and concluded that groundwater recharge and pore fluid pressure accelerated earthquake occurrences. Christiansen, Hurwitz, and Ingebritsen (2007) statistically analyzed earthquake records of the San Andreas Fault near Parkfield, California, identified an annual period in the creeping section and a semiannual period in the locked section of the fault, and suggested pore-pressure diffusion may play a role in the observed modulation. Seasonal seismicity at some volcanic centers of the western United States was also discovered and was inferred to be associated with pore-fluid pressure changes due to groundwater recharge (Christiansen, Hurwitz, Saar, Ingebritsen, & Hsieh, 2005; Saar & Manga, 2003).

3.4 "Earthquake weather"

Short-term weather events involve rapid spatiotemporal variations in atmospheric pressure, temperature, and precipitation. Atmospheric pressure changes produce relatively modest surface loads, but they can reach 10 kPa during the passage of great tropical cyclones (hurricanes and typhoons). Liu, Linde, and Sacks (2009) proposed that transient strain signals during the passage of 11 great typhoons in Taiwan reveal the occurrence of aseismic slip events on buried faults triggered by the associated low atmospheric-pressure transients. Hsu et al. (2015) considered strain from precipitation loading in addition to barometric pressure to find that much of the strain signal attributed to triggered slow-slip events by Liu, Linde, and Sacks (2009) can be explained by hydrological processes. However, some of the strain events associated with typhoons still appear to require a tectonic explanation (Hsu et al., 2015). Zhai et al. (2021) argued that the seismicity rate in an ongoing earthquake swarm in northeast Taiwan in August of 2009 was transiently subdued by the local reduction of atmospheric pressure by almost 6 kPa associated with the passage of typhoon Morakot's eye. In 2017, a burst of seismicity in the aftershock zone of the M5.7 Virginia earthquake in the eastern United States occurred just when atmospheric pressure dropped during the passage of Hurricane Irene 3–5 days following the mainshock, suggesting a causal relationship (Meng, Yang, & Peng, 2018). While only a few cases of earthquake or slow-slip triggering by short-term atmospheric pressure changes have been reported, the fact that transient stress changes in the crust from such events are comparable to those leading to seismicity rate changes due to other forcings suggests that this is likely a real phenomenon.

In addition to extreme storm events and associated atmospheric pressure changes, there are also daily temperature and atmospheric pressure cycles. Diurnal pressure variations are maximum near the equator, where the average amplitude reaches 300 Pa. Such atmospheric tides have been shown to modulate the motion of shallow slow-moving landslides (Schulz, McKenna, Kibler, & Biavati, 2009). However, daily modulation of seismicity has not been documented so far and is challenged by periods very close to those of the soli-lunar tides and the need to separate any subtle seismicity-rate changes from the effect of daily variations in seismic noise and thus network sensitivity.

Great cyclones also represent extreme precipitation and hydrological loading events, and the largest storms with m-level precipitation, such as Hurricane Harvey in Texas in 2017 (Milliner et al., 2018) and typhoon Hagibis in Japan in 2018 (Zhan, Heki, Arief, & Yoshida, 2021), led to multiday subsidence of the Earth's surface by up to ~20 mm, as measured by GNSS. Heki and Arief (2022) documented similar multiday subsidence events of 10–20 mm caused by loads in flooded areas during four large tropical rainstorms in Japan. Depending on the spatial extent of such load changes, storm-related stress changes at seismogenic depth may reach several kiloPascal; however, no earthquake triggering has been recognized in these cases. Typhoon Morakot in Taiwan produced up to 3 m of rainfall over the course of 5 days. The impact of the storm on the seismic network and noise levels made it impossible to resolve significant changes in seismicity during and in the days following the storm (Zhai et al., 2021). More comprehensive analysis is needed to properly assess the possible linkage of hydrological load transients of a few days and crustal seismicity.

In areas dominated by highly fractured rock in the uppermost crust, such as karst geology marked by caves and open fracture systems, individual rain storms appear to be capable of triggering earthquakes (e.g., Chmiel et al., 2022; Hainzl, Kraft, Wassermann, Igel, & Schmedes, 2006; Husen, Bachmann, & Giardini, 2007; Rigo, Béthoux, Masson, & Ritz, 2008). In these cases, the delay between peak rainfall and seismicity-rate increases can be just a few hours or days. Montgomery-Brown, Shelly, and Hsieh (2019) documented seismic swarm activity in the upper few kilometers of the crust near Long Valley Caldera in California, which follows stream-flow peaks with an average delay of 3–4 weeks (Fig. 14.10). Silverii, Montgomery-Brown, Borsa, and Barbour (2020) showed that this seismic activity is also correlated with horizontal displacement transients at nearby GNSS stations, which can be explained with a poroelastic model of a distributed pressure source. This suggests that pore pressure diffusion through fractured rock in the area drives both the shallow seismic swarms and the GNSS-measured deformation. These cases of quite rapid activation in shallow seismicity suggest effective triggering by rapidly increasing fluid pressure and extension in a highly permeable upper crust and/or fluid-pressure increases in saturated rocks due to near-surface hydrological loading (e.g., Chmiel et al., 2022; D'Agostino et al., 2018; Miller, 2008). Thus, the local geology and permeability structure of an area appear to be important factors that determine the susceptibility to weather-induced earthquake triggering by elastic surface loading, rapid pore-pressure diffusion, and/or poroelastic strain.

Great storms may leave a more enduring change in stress and seismicity through the enormous amount of erosion and sediment transport they cause. Steer et al. (2020) noted an increase in shallow (<15 km) earthquakes in the years following the August 2009 typhoon Morakot, which they attributed to crustal unloading due to rapid sediment export after the storm's passage. Zhai et al. (2021) also inferred a rise in background seismicity rate in the year following the event. However, Hsu et al. (2021) noted that these years of enhanced seismicity also correlate with a period of low water storage in south Taiwan, thus suggesting an example of multiyear hydrological load triggering that is unrelated to the typhoon.

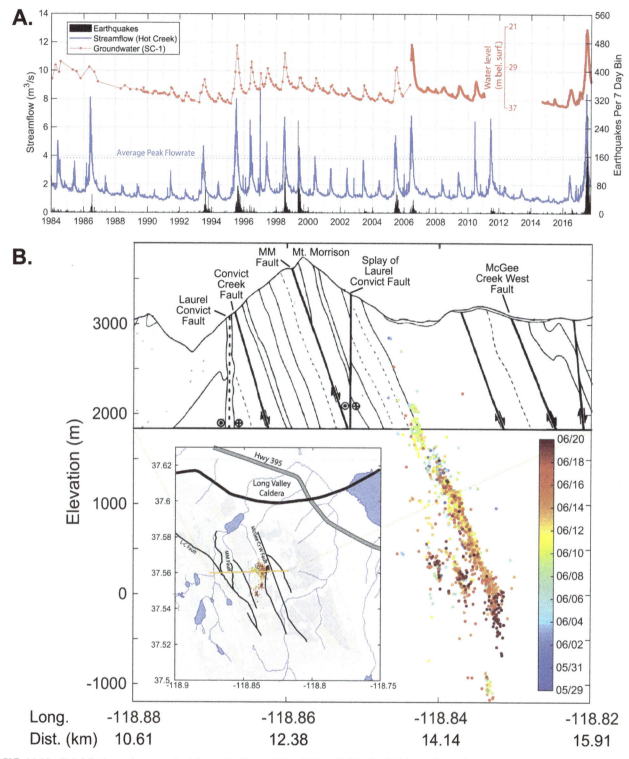

FIG. 14.10 Rainfall-triggered swarm seismicity south of Long Valley Caldera, California. (A) 35-year time series of streamflow in a nearby creek (blue line), local groundwater level (red line), and weekly earthquake counts (black bars) in the seismic swarm area. (B) 4 weeks of seismicity in June 2017, color-coded by time and superimposed on a simplified geologic cross section (orange line in inset map). The swarm propagated downward along the dipping bedrock bedding contacts and fractures. *(Modified from Montgomery-Brown, E. K., Shelly, D. R., & Hsieh, P. A. (2019). Snowmelt-triggered earthquake swarms at the margin of Long Valley Caldera, California. Geophysical Research Letters, 46, 3698–3705. https://doi.org/10.1029/2019GL082254.)*

As global weather becomes more energetic and extreme storm events more common (e.g., Otto, 2017), we can expect to see more cases of weather-triggered seismicity, especially in areas dominated by highly fractured and permeable rocks and regions exposed to extreme tropical storm systems and associated catastrophic rainfalls. Nonetheless, we don't expect "weather earthquakes" to become a significant source of natural hazard, given the relatively modest stress and pressure changes and earthquake magnitudes associated with the documented cases of triggering by individual weather events.

4 Lessons learned from climate-driven deformation and seismicity

4.1 Probing the Earth's constitutive properties using climate-driven deformation

Interactions between climate and solid Earth deformation include multiple forcings, such as heavy rain and cyclones (<days), annual and multiyear hydrologic loading cycles (years to decades), as well as unloading due to recent ice melting (years-decades), since the little ice age (~250 years) and since the last glacial maximum (~25,000 years). These interactions involve various physical processes acting over a range of spatial and temporal scales. The deformational response of the solid Earth to these forcings depends on the constitutive properties of the crust and mantle, that is its rheological structure.

Deformation caused by changes in surface loading, which dominates this interplay, has been a key observation to probe the Earth's mantle rheology (Fig. 14.11). Mantle viscosity has been inferred from observations related to GIA (e.g., Haskell, 1935; ; Lau et al., 2016; Mitrovica & Forte, 2004; Peltier & Andrews, 1976) while rheology of the upper mantle has been constrained using geodetic measurements following lake drainage (e.g., Austermann, Chen, Lau, Maloof, & Latychev, 2020; Bills, Adams, & Wesnousky, 2007). The time-dependent response of the Earth's mantle to stresses induced by surface loading has been shown to progress through instantaneous elastic strain, transient creep during which viscosity rapidly increases, to a steady-state viscous regime. This is supported by laboratory experiments investigating deformation of crystalline aggregates that have yielded constitutive laws allowing extrapolation to mantle conditions (e.g., Hansen, Zimmerman, & Kohlstedt, 2011; Hirth & Kohlstedt, 2003). However, the majority of climate-driven mantle rheology studies have focused on mantle elasticity and steady-state creep through observations at short and very long timescales, respectively. Therefore, the mantle is often described as a Maxwell body, characterized by a shear modulus for the elastic strain and a linear viscosity for the long-term strain rate.

However, over the past decades, emerging geodetic technologies and continuously growing spatiotemporal coverage have provided new opportunities to probe the solid Earth's rheology using its response to changes in surface loading. It is becoming increasingly clear that the Maxwell framework, with a single viscosity, is inadequate for capturing processes at intermediate timescales such as recent ice melting (years to decades; Nield et al., 2014) or since the Little Ice Age (~250 years). The inference of time- and stress-dependent viscosity is also supported by studies of postseismic relaxation of the mantle (e.g., Bürgmann & Dresen, 2008). Indeed, the viscoelastic relaxation of coseismic stresses in the lower crust and upper mantle with multiple relaxation times and/or power-law rheologies better explains the time series of postseismic deformation (e.g., Freed & Bürgmann, 2004; Pollitz, 2003; Trubienko, Garaud, & Fleitout, 2014). At the annual timescale, Chanard, Fleitout, Calais, Barbot, and Avouac (2018) demonstrated that Earth's response to seasonal surface loading, which is primarily driven by continental hydrology, can be used to constrain a lower bound on the globally averaged transient viscoelastic properties of the upper mantle using an approach relying on multiple geodetic techniques. Pollitz, Wech,

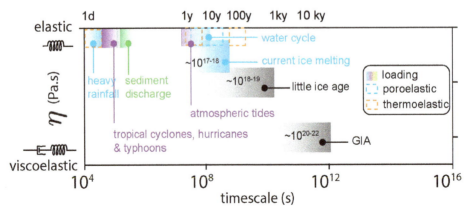

FIG. 14.11 Schematic illustration of climate-driven processes used to probe Earth's rheology.

Kao, and Bürgmann (2013) also explored phase delays in the deformational response to surface loading due to the viscoelastic Earth structure. These studies imply that long geodetic time series of hydrological load deformation could be used to obtain new rheological constraints over a wider range of periods. Reconciling observations at multiple timescales will require a fully consistent rheological model (Lau & Holtzman, 2019; Lau et al., 2021).

Variations in groundwater storage and associated poroelastic deformation can also be used to probe the mechanical properties of the Earth's shallow subsurface (e.g., Larochelle et al., 2022). Moreover, investigating the rheological processes behind irreversible aquifer compaction and depletion offers an opportunity to better understand the inner workings of groundwater systems, necessary to ensure sustainability of freshwater resources (Chaussard & Farr, 2019; Ojha, Shirzaei, Werth, Argus, & Farr, 2018; Smith et al., 2017).

While the Earth's response to periodic loading processes has been theoretically and numerically investigated, the role of climate-driven processes generating eigenstrain in the shallow subsurface, including fluctuations in groundwater pressure or near-surface temperature, in earthquake modulation or triggering remains poorly known. Stresses associated with seasonal eigenstrain and variations of mechanical properties at short wavelengths that are likely to occur at shallow depths can reach up to a few kiloPascals at a few kilometers depth (Ben-Zion & Allam, 2013), comparable to those induced by annual hydrological loading. Interestingly, these stresses may localize across fault damage zones where detailed seismic imaging shows strong variations in seismic velocities (up to 50%) relative to the surrounding host rock in the crust at 3–5 km depth (Allam, Ben-Zion, & Peng, 2014; Ben-Zion et al., 2003; Qin et al., 2018). Carefully conducted field studies combined with advances in modeling approaches are needed to better understand the role of poroelastic and thermoelastic processes and associated constitutive properties of the shallow crust in the Earth's response to climatic variations. The use of more advanced physical models of Earth's deformation and stress in response to climate and weather changes over a wide range of spatial and temporal scales will also improve our ability to link these processes to the occurrence of earthquakes.

4.2 Insights on frictional fault properties and state of stress in the Earth from periodic climate forcing

Just as the Earth's deformational response to external forcings reveals new insights about its constitutive properties, the occurrence of earthquakes or slow-slip events in response to these forcings can inform us about the environment and frictional mechanics of faults. Similar to solid Earth and ocean tides (Heaton, 1982; Vidale, Agnew, Johnston, & Oppenheimer, 1998), stresses resulting from climate forcing, of comparable amplitude, may not be sufficient to cause earthquakes. Faults may need to be critically stressed from long-term tectonic loading (Fig. 14.12A), as supported by laboratory experiments (Chanard et al., 2019; Noël, Pimienta, & Violay, 2019), and/or combined with favorable conditions including low-angle fault orientations and shallow depths (Cochran, Vidale, & Tanaka, 2004), resonant fault-frictional properties (Lowry, 2006; Perfettini, Schmittbuhl, Rice, & Cocco, 2001; Senapati, Kundu, & Jin, 2022), or high pore pressure in the fault zone (Bettinelli et al., 2008). Considering natural and laboratory observations, there is growing appreciation for the fact that faults' response to small climate-driven stress perturbations may indicate favorable frictional properties and conditions for rupture (Fig. 14.12A).

While Coulomb failure stress is often invoked to explain interactions between periodic loading and seismicity (e.g., Johnson, Fu, & Bürgmann, 2017a; Scholz, Tan, & Albino, 2019), some laboratory and natural observations suggest a time-dependent failure process (e.g., Beeler & Lockner, 2003). Models developed to explain these observations include rate-and-state spring-slider systems to simulate earthquake nucleation under periodic loading (Ader, Lapusta, Avouac, & Ampuero, 2014; Heimisson & Segall, 2018; Heimisson & Avouac, 2020) or Burridge-Knopoff models (e.g., Pétrélis, Chanard, Schubnel, & Hatano, 2021). More realistic descriptions such as 2-D faults (Ader, Lapusta, Avouac, & Ampuero, 2014; Perfettini, Schmittbuhl, & Cochard, 2003) or even a 3D interacting fault population in a rate-and-state framework under periodic loading (Dublanchet, 2022) confirm prior theoretical results with numerical simulations. These models suggest that seismicity rate correlates with stress amplitude when the loading period is short compared to the characteristic nucleation time but correlates with stressing rate for longer periods (Fig. 14.12B). Models also suggest that the highest correlation should occur at periodic loading near the characteristic nucleation time. Perfettini, Schmittbuhl, Rice, and Cocco (2001) showed a potential fault-slip resonance, invoking enhanced modulation of a spring-slider system with rate-weakening rheology subject to periodic loading near a critical period. The frequency-dependent modulation of earthquakes inferred from the experimental results and model predictions has been invoked to explain why there is seasonal modulation in areas that lack evidence of modulation by the higher-frequency tides, suggesting that the earthquake nucleation time is close to a year (e.g., Ader & Avouac, 2013; Beeler & Lockner, 2003; Johnson, Fu, & Bürgmann, 2017a). Furthermore, by coupling the amplitude and frequency of periodic loading to mechanical parameters such as stress drop and frictional fault properties, some of these models provide a mechanical framework to further our understanding of periodic seismic modulation.

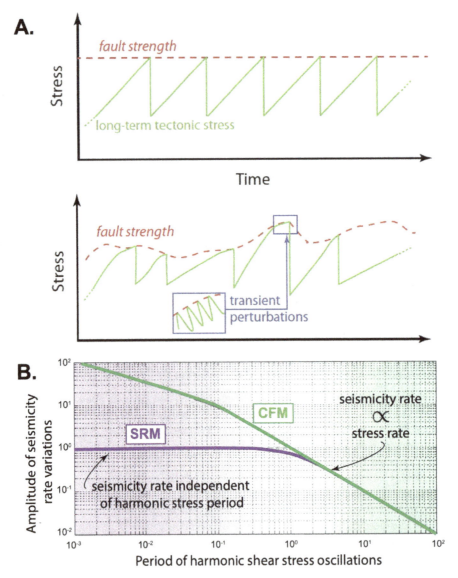

FIG. 14.12 (A) Schematic illustrating the state of stress and strength on a fault with a constant loading rate, stress drop, and strength from which perfectly periodic failure results (top) and with variable loading rate, stress drop, and strength in a more realistic scenario (bottom). In a natural system, a fault is stressed by external perturbations, including climatic forcings, in addition to long-term tectonic stress. Transient stress, even small, may initiate failure of a fault in a critical state of stress and detailed observations of triggered small earthquakes may provide information on the state of stress and frictional properties of faults. (B) Amplitude of seismicity-rate variations as a function of the shear stress oscillation period, according to the numerical (black) or asymptotic (gray) solutions for the Coulomb failure model (CFM) and spring-slider rate-and-state model (SRM). *(Figure modified from Ader T. J., Lapusta N., Avouac J.-P., Ampuero J.-P. (2014). Response of rate-and-state seismogenic faults to harmonic shear-stress perturbations. Geophysical Journal International, 198(1), 385–413. https://doi.org/10.1093/gji/ggu144.)*

5 Summary and future opportunities

In this chapter, we review the coupling effects of the solid Earth with climate and weather systems over a wide range of spatiotemporal scales. Large ice-sheet and lake-load changes, such as those associated with the Last Glacial Maximum and GIA, can trigger substantial earthquakes. More modest forcings due to climate processes at multiannual timescales may result in more subtle strain and seismicity responses. Annual hydrological loading and other seasonal forcings produce geodetically measurable, periodic deformation cycles and can significantly modulate seismicity on suitably oriented faults. In a few cases, even atmospheric pressure, hydrological surface loads, and/or subsurface fluid pressure transients associated with short-term weather events can affect earthquake occurrences. Most climate-driven stress and strain changes in the Earth are small and careful mechanical modeling and statistical analysis are required to discern their effect on earthquake occurrence. While there is solid evidence for climate-driven modulation of seismicity, the impact of ongoing climate change and sea-level rise caused by anthropogenic emissions on short- or long-term seismic hazards is probably modest.

Future studies are recommended to investigate the contributions from multiple processes, including atmospheric and hydrological surface loads as well as poroelastic and thermoelastic strain. These processes need to be carefully modeled to develop a comprehensive and quantitative understanding of the climate-driven stress tensor changes and associated solid Earth deformation and seismicity. We should seek evidence for, and causes of, variable seismicity modulation in different tectonic environments. We can also further improve our knowledge of fault-frictional properties based on observations of perturbed seismicity over a range of forcing periods. Finally, it may be possible to determine viscous rheological properties in the lower crust and upper mantle from geodetic observations of the spatiotemporal response of the Earth to periodic loads.

Acknowledgments

We thank everybody who provided figures, helpful discussions, and thoughtful comments during the preparation of this review, including Karen Lutrell, Emily Montgomery-Brown, Chris Rollins, Holger Steffen, Rebekka Steffen, Manoochehr Shirzaei, Patrick Wu, and Timothy Craig. Francesca Silverii and Corné Kreemer provided helpful review comments. We also acknowledge support from National Aeronautics and Space Administration's Earth Surface and Interior program and CNES's TOSCA program (HydroGeo project).

References

Aagaard, B., Knepley, M., & Williams, C. (2017). *PyLith v2. 2.1*. Davis, CA: Computational Infrastructure of Geodynamics.

Ader, T. J., & Avouac, J. P. (2013). Detecting periodicities and declustering in earthquake catalogs using the Schuster spectrum, application to Himalayan seismicity. *Earth and Planetary Science Letters, 377*, 97–105.

Ader, T. J., Lapusta, N., Avouac, J.-P., & Ampuero, J.-P. (2014). Response of rate-and-state seismogenic faults to harmonic shear-stress perturbations. *Geophysical Journal International, 198*(1), 385–413. https://doi.org/10.1093/gji/ggu144.

Adhikari, S., Ivins, E. R., & Larour, E. (2016). ISSM-SESAW v1.0: Mesh-based computation of gravitationally consistent sea-level and geodetic signatures caused by cryosphere and climate driven mass change. *Geoscientific Model Development, 9*(3), 1087–1109. https://doi.org/10.5194/gmd-9-1087-2016.

Adhikari, S., Milne, G. A., Caron, L., Khan, S. A., Kjeldsen, K. K., Nilsson, J., … Ivins, E. R. (2021). Decadal to centennial timescale mantle viscosity inferred from modern crustal uplift rates in Greenland. *Geophysical Research Letters, 48*(19), e2021GL094040.

Agnew, D. C. (1997). NLOADF: A program for computing ocean-tide loading. *Journal of Geophysical Research, 102*, 5109–5110.

Agnew, D. (2015). Earth tides. In *Treatise on geophysics* (pp. 151–178). Amsterdam: Elsevier. https://doi.org/10.1016/B978-0-444-53802-4.00058-0.

Allam, A. A., Ben-Zion, Y., & Peng, Z. (2014). Seismic imaging of a bimaterial interface along the Hayward fault, CA, with fault zone head waves and direct P arrivals. *Pure and Applied Geophysics, 171*(11), 2993–3011.

Altamimi, Z., Rebischung, P., Métivier, L., & Collilieux, X. (2016). ITRF2014: A new release of the international terrestrial reference frame modeling nonlinear station motions. *Journal of Geophysical Research: Solid Earth, 121*(8), 6109–6131.

Amelung, F., Galloway, D. L., Bell, J. W., Zebker, H. A., & Laczniak, R. J. (1999). Sensing the ups and downs of Las Vegas-InSAR reveals structural control of land subsidence and aquifer-system deformation. *Geology, 27*, 483–486.

Amos, C. B., Audet, P., Hammond, W. C., Bürgmann, R., Johanson, I. A., & Blewitt, G. (2014). Uplift and seismicity driven by groundwater depletion in Central California. *Nature, 509*(7501), 483–486.

Argus, D. F., Fu, Y., & Landerer, F. W. (2014). Seasonal variation in total water storage in California inferred from GPS observations of vertical land motion. *Geophysical Research Letters, 41*. https://doi.org/10.1002/2014GL059570.

Argus, D. F., Landerer, F. W., Wiese, D. N., Martens, H. R., Fu, Y., Famiglietti, J. S., … Watkins, M. M. (2017). Sustained water loss in California's mountain ranges during severe drought from 2012 to 2015 inferred from GPS. *Journal of Geophysical Research—Solid Earth, 122*, 10559–10585. https://doi.org/10.1002/2017JB014424.

Austermann, J., Chen, C. Y., Lau, H. C., Maloof, A. C., & Latychev, K. (2020). Constraints on mantle viscosity and Laurentide ice sheet evolution from pluvial paleolake shorelines in the western United States. *Earth and Planetary Science Letters, 532*, 116006.

Barbot, S., Moore, J. D., & Lambert, V. (2017). Displacement and stress associated with distributed anelastic deformation in a half-space. *Bulletin of the Seismological Society of America, 107*(2), 821–855.

Barletta, V. R., Ferrari, C., Diolaiuti, G., Carnielli, T., Sabadini, R., & Smiraglia, C. (2006). Glacier shrinkage and modeled uplift of the Alps. *Geophysical Research Letters, 33*(14), L14307. https://doi.org/10.1029/2006GL026490.

Beeler, N., & Lockner, D. (2003). Why earthquakes correlate weakly with the solid earth tides: Effects of periodic stress on the rate and probability of earthquake occurrence. *Journal of Geophysical Research, 108*(B8). https://doi.org/10.1029/2001JB001518.

Ben-Zion, Y., & Allam, A. (2013). Seasonal thermoelastic strain and postseismic effects in Parkfield borehole dilatometers. *Earth and Planetary Science Letters, 379*, 120–126.

Ben-Zion, Y., & Leary, P. (1986). Thermoelastic strain in a half- space covered by unconsolidated material. *Bulletin of the Seismological Society of America, 76*, 1447–1460.

Ben-Zion, Y., Peng, Z., Okaya, D., Seeber, L., Armbruster, J. G., Ozer, N., et al. (2003). A shallow fault-zone structure illuminated by trapped waves in the Karadere–Duzce branch of the North Anatolian Fault, western Turkey. *Geophysical Journal International*, *152*(3), 699–717. https://doi.org/10.1046/j.1365-246X.2003.01870.x.

Berger, J. (1975). A note on thermoelastic strains and tilts. *Journal of Geophysical Research*, *80*, 274–277.

Bettinelli, P., Avouac, J.-P., Flouzat, M., Bollinger, L., Ramillien, G., Rajaure, S., & Sapkota, S. (2008). Seasonal variations of seismicity and geodetic strain in the Himalaya induced by surface hydrology. *Earth and Planetary Science Letters*, *266*(3–4), 332–344. https://doi.org/10.1016/j.epsl.2007.11.021.

Bevis, M., Alsdorf, D., Kendrick, E., Fortes, L. P., Forsberg, B., Smalley, R., Jr., & Becker, J. (2005). Seasonal fluctuations in the mass of the Amazon River system and Earth's elastic response. *Geophysical Research Letters*, *32*, L16308. https://doi.org/10.1029/2005GL023491.

Bills, B. G., Adams, K. D., & Wesnousky, S. G. (2007). Viscosity structure of the crust and upper mantle in western Nevada from isostatic rebound patterns of Lake Lahontan shorelines. *Journal of Geophysical Research*, *112*(B6), B06405. https://doi.org/10.1029/2005JB003941.

Biot, M. A. (1956). Thermoelasticity and irreversible thermodynamics. *Journal of Applied Physics*, *27*(3), 240–253.

Blewitt, G., Hammond, W. C., & Kreemer, C. (2018). Harnessing the GPS data explosion for interdisciplinary science. *Eos*, *99*. https://doi.org/10.1029/2018EO104623.

Blewitt, G., Lavallée, D., Clarke, P., & Nurudinov, K. (2001). A new global model of Earth deformation: Seasonal cycle detected. *Science*, *294*, 2342–2345. https://doi.org/10.1126/science.1065328.

Bollinger, L., Perrier, F., Avouac, J.-P., Sapkota, S., Gautam, U., & Tiwari, D. R. (2007). Seasonal modulation of seismicity in the Himalaya of Nepal. *Geophysical Research Letters*, *34*(8), L08304. https://doi.org/10.1029/2006GL029192.

Borsa, A. A., Agnew, D. C., & Cayan, D. R. (2014). Ongoing drought-induced uplift in the western United States. *Science*, *345*. https://doi.org/10.1126/science.1260279.

Boussinesq, J. (1885). *Application des Potentiels a l'Etude de l'Equilibre et du Mouvement des Solides E 'lastiques* (p. 508). Paris: Gauthier-Villars.

Brandes, C., Steffen, H., Steffen, R., & Wu, P. (2015). Intraplate seismicity in northern Central Europe is induced by the last glaciation. *Geology*, *43*, 611–614. https://doi.org/10.1130/G36710.1.

Brothers, D., Kilb, D., Luttrell, K., Driscoll, N., & Kent, G. (2011). Loading of the San Andreas fault by flood-induced rupture of faults beneath the Salton Sea. *Nature Geoscience*, *4*(7), 486–492. https://doi.org/10.1038/ngeo1184.

Brothers, D. S., Luttrell, K. M., & Chaytor, J. D. (2013). Sea-level-induced seismicity and submarine landslide occurrence. *Geology*, *41*(9), 979–982. https://doi.org/10.1130/G34410.1.

Bürgmann, R., & Dresen, G. (2008). Rheology of the lower crust and upper mantle: Evidence from rock mechanics, geodesy, and field observations. *Annual Review of Earth and Planetary Sciences*, *36*, 531–567.

Byerlee, J. D. (1978). Friction of rocks. *Pure and Applied Geophysics*, *116*. https://doi.org/10.1007/BF00876528.

Calais, E., Freed, A. M., Van Arnsdale, R. B., & Stein, S. (2010). Triggering of New Madrid seismicity by late-Pleistocene erosion. *Nature*, *466*. https://doi.org/10.1038/nature09258.

Carlson, G., Shirzaei, M., Werth, S., Zhai, G., & Ojha, C. (2020). Seasonal and long-term groundwater unloading in the Central Valley modifies crustal stress. *Journal of Geophysical Research: Solid Earth*, *125*(1), e2019JB018490.

Caron, L., Ivins, E. R., Larour, E., Adhikari, S., Nilsson, J., & Blewitt, G. (2018). GIA model statistics for GRACE hydrology, cryosphere, and ocean science. *Geophysical Research Letters*, *45*, 2203–2212. https://doi.org/10.1002/2017GL076644.

Cathles, L. M. (1975). *Viscosity of the Earth's mantle*. Vol. *1362*. Princeton University Press.

Chanard, K., Avouac, J. P., Ramillien, G., & Genrich, J. (2014). Modeling deformation induced by seasonal variations of continental water in the Himalaya region: Sensitivity to Earth elastic structure. *Journal of Geophysical Research - Solid Earth*, *119*, 5097–5113. https://doi.org/10.1002/2013JB010451.

Chanard, K., Fleitout, L., Calais, E., Barbot, S., & Avouac, J.-P. (2018). Constraints on transient viscoelastic rheology of the asthenosphere from seasonal deformation. *Geophysical Research Letters*, *45*, 2328–2338. https://doi.org/10.1002/2017GL076451.

Chanard, K., Fleitout, L., Calais, E., Rebischung, P., & Avouac, J.-P. (2018). Toward a global horizontal and vertical elastic load deformation model derived from GRACE and GNSS station position time series. *Journal of Geophysical Research: Solid Earth*, *123*, 3225–3237. https://doi.org/10.1002/2017JB015245.

Chanard, K., Métois, M., Rebischung, P., & Avouac, J.-P. (2020). A warning against over-interpretation of seasonal signals measured by the global navigation satellite system. *Nature Communications*, *11*, 1375. https://doi.org/10.1038/s41467-020-15100-7.

Chanard, K., Nicolas, A., Hatano, T., Petrelis, F., Latour, S., Vinciguerra, S., & Schubnel, A. (2019). Sensitivity of acoustic emission triggering to small pore pressure cycling perturbations during brittle creep. *Geophysical Research Letters*, *46*(13), 7414–7423. https://doi.org/10.1029/2019GL082093.

Chaussard, E., & Farr, T. G. (2019). A new method for isolating elastic from inelastic deformation in aquifer systems: Application to the San Joaquin Valley, CA. *Geophysical Research Letters*, *46*(19), 10800–10809.

Chaussard, E., Wdowinski, S., Cabral-Cano, E., & Amelung, F. (2014). Land subsidence in Central Mexico detected by ALOS InSAR time-series. *Remote Sensing of Environment*, *140*, 94–106. https://doi.org/10.1016/j.rse.2013.08.038.

Chen, J., Cazenave, A., Dahle, C., Llovel, W., Panet, I., Pfeffer, J., & Moreira, L. (2022). Applications and challenges of GRACE and GRACE follow-on satellite gravimetry. *Surveys in Geophysics*, 1–41.

Chmiel, M., Godano, M., Piantini, M., Brigode, P., Gimbert, F., Bakker, M., ... Chapuis, M. (2022). Brief communication: Seismological analysis of flood dynamics and hydrologically triggered earthquake swarms associated with Storm Alex. *Natural Hazards and Earth System Sciences*, *22*(5), 1541–1558.

Christiansen, L. B., Hurwitz, S., & Ingebritsen, S. E. (2007). Annual modulation of seismicity along the San Andreas Fault near Parkfield, CA. *Geophysical Research Letters*, *34*(L04306). https://doi.org/10.1029/2006GL028634.

Christiansen, L. B., Hurwitz, S., Saar, M. O., Ingebritsen, S. E., & Hsieh, P. A. (2005). Seasonal seismicity at western United States volcanic centers. *Earth and Planetary Science Letters*, *240*, 307–321.

Cocco, M., & Rice, J. R. (2002). Pore pressure and poroelasticity effects in Coulomb stress analysis of earthquake interactions. *Journal of Geophysical Research*, *107*(B2). https://doi.org/10.1029/2000JB000138. ESE 2-1–ESE 2-17.

Cochran, E. S., Vidale, J. E., & Tanaka, S. (2004). Earth tides can trigger shallow thrust fault earthquakes. *Science*, *306*(5699), 1164–1166. https://doi.org/10.1126/science.1103961.

Compton, K., Bennett, R. A., & Hreinsdóttir, S. (2015). Climate-driven vertical acceleration of Icelandic crust measured by continuous GPS geodesy. *Geophysical Research Letters*, *42*, 743–750. https://doi.org/10.1002/2014GL062446.

Craig, T. J., Calais, E., Fleitout, L., Bollinger, L., & Scotti, O. (2016). Evidence for the release of long-term tectonic strain stored in continental interiors through intraplate earthquakes. *Geophysical Research Letters*, *43*, 6826–6836. https://doi.org/10.1002/2016GL069359.

Craig, T. J., Chanard, K., & Calais, E. (2017). Hydrologically-driven crustal stresses and seismicity in the New Madrid Seismic Zone. *Nature Communications*, *8*(1), 2143. https://doi.org/10.1038/s41467-017-01696-w.

D'Agostino, N., Silverii, F., Amoroso, O., Convertito, V., Fiorillo, F., Ventafridda, G., & Zollo, A. (2018). Crustal deformation and seismicity modulated by groundwater recharge of karst aquifers. *Geophysical Research Letters*, *45*, 12253–12262. https://doi.org/10.1029/2018GL079794.

Dai, A. (2016). Historical and future changes in streamflow and continental runoff: A review. In Q. Tang, & T. Oki (Eds.), *Geophysical monograph: Vol. 221. Terrestrial water cycle and climate change: Natural and human-induced impacts* (pp. 17–37). Hoboken: AGU, Wiley. https://doi.org/10.1002/9781118971772.

Darwin, G. H. (1882). XLVI. On variation of the vertical due to elasticity of the Earth's surface. *Philosophical Magazine Series*, *5*(14), 409–427. https://doi.org/10.1080/14786448208628439.

Davis, J. L., Elósegui, P., Mitrovica, J. X., & Tamisiea, M. E. (2004). Climate-driven deformation of the solid Earth from GRACE and GPS. *Geophysical Research Letters*, *31*, L24605. https://doi.org/10.1029/2004GL021435.

Dietrich, R., Ivins, E. R., Casassa, G., Lange, H., Wendt, J., & Fritsche, M. (2010). Rapid crustal uplift in Patagonia due to enhanced ice loss. *Earth and Planetary Science Letters*, *289*(1–2), 22–29.

Dong, D., Fang, P., Bock, Y., Cheng, M. K., & Miyazaki, S. I. (2002). Anatomy of apparent seasonal variations from GPS-derived site position time series. *Journal of Geophysical Research: Solid Earth*, *107*(B4), ETG-9.

Drake, N. F. (1912). Destructive earthquakes in China. *Bulletin of the Seismological Society of America*, *2*(1), 40–91.

Dublanchet, P. (2022). Seismicity modulation in a 3-D rate-and-state interacting fault population model. *Geophysical Journal International*, *229*(3), 1804–1823.

Dutilleul, P., Johnson, C. W., Bürgmann, R., Wan, Y., & Shen, Z. K. (2015). Multifrequential periodogram analysis of earthquake occurrence: An alternative approach to the Schuster spectrum, with two examples in Central California. *Journal of Geophysical Research: Solid Earth*, *120*(12), 8494–8515.

Dziewonski, A., & Anderson, D. L. (1981). Preliminary reference Earth model. *Physics of the Earth and Planetary Interiors*, *25*, 297–356.

Enzminger, T. L., Small, E. E., & Borsa, A. A. (2018). Accuracy of snow water equivalent estimated from GPS vertical displacements: A synthetic loading case study for western U.S. mountains. *Water Resources Research*, *54*, 581–599. https://doi.org/10.1002/2017WR021521.

Fallou, L., Marti, M., Dallo, I., & Corradini, M. (2022). How to fight earthquake misinformation: A communication guide. *Seismological Research Letters*, *93*(5), 2418–2422. https://doi.org/10.1785/0220220086.

Fang, M., Dong, D. N., & Hager, B. H. (2014). Displacements due to surface temperature variation on a uniform elastic sphere with its centre of mass stationary. *Geophysical Journal International*, *196*(1), 194–203. https://doi.org/10.1093/gji/ggt335.

Farrell, W. E. (1972). Deformation of the Earth by surface loads. *Reviews of Geophysics and Space Physics*, *10*, 761–797. https://doi.org/10.1029/RG010i003p00761.

Faunt, C. C., Sneed, M., Traum, J., & Brandt, J. T. (2016). Water availability and land subsidence in the Central Valley, California, USA. *Hydrogeology Journal*, *24*(3), 675–684.

Frederikse, T., Landerer, F., Caron, L., Adhikari, S., Parkes, D., Humphrey, V. W., … Wu, Y. H. (2020). The causes of sea-level rise since 1900. *Nature*, *584*(7821), 393–397.

Freed, A. M., & Bürgmann, R. (2004). Evidence of power law flow in the Mojave desert mantle. *Nature*, *430*, 548–551. https://doi.org/10.1038/nature02784.

Friedrich, A. M., Wernicke, B. P., Niemi, N. A., Bennett, R. A., & Davis, J. L. (2003). Comparison of geodetic and geologic data from the Wasatch region, Utah, and implications for the spectral character of Earth deformation at periods of 10 to 10 million years. *Journal of Geophysical Research*, *108*(B4), 2199.

Fu, Y., Argus, D. F., & Landerer, F. W. (2015). GPS as an independent measurement to estimate terrestrial water storage variations in Washington and Oregon. *Journal of Geophysical Research—Solid Earth*, *120*, 552–566. https://doi.org/10.1002/2014JB011415.

Fu, Y., & Freymueller, J. T. (2012). Seasonal and long-term vertical deformation in the Nepal Himalaya constrained by GPS and GRACE measurements. *Journal of Geophysical Research*, *117*, B03407. https://doi.org/10.1029/2011JB008925.

Galloway, D. L., Jones, D. R., & Ingebritsen, S. E. (Eds.). (1999). *Vol. 1182. Land subsidence in the United States* US Geological Survey.

Gao, S. S., Silver, P. G., Linde, A. T., & Sacks, I. S. (2000). Annual modulation of triggered seismicity following the 1992 Landers earthquake in California. *Nature*, *406*(6795), 500–504. https://doi.org/10.1038/35020045.

Gauer, L. M., Chanard, K., & Fleitout, L. (2023). Data-driven gap filling and spatio-temporal filtering of the GRACE and GRACE-FO records. *Journal of Geophysical Research: Solid Earth*, e2022JB025561.

Gilbert, G. K. (1890). *Lake Bonneville. Vol. 1*. USGS Monograph. https://doi.org/10.3133/m1.

Gonzalez, P., Tiampo, K., Palano, M., Cannav, F., & Fernandez, J. (2012). The 2011 Lorca earthquake slip distribution controlled by groundwater crustal unloading. *Nature Geoscience, 5*, 821–825.

Gradmann, S., Olesen, O., Keiding, M., & Maystrenko, Y. (2018). The regional 3D stress field of Nordland, northern Norway—Insights from numerical modelling. In O. Olesen, et al. (Eds.), *Neotectonics in Nordland—Implications for petroleum exploration (NEONOR2)* (pp. 215–240). NGU Report 2018.010.

Grapenthin, R., Sigmundsson, F., Geirsson, H., Árnadóttir, T., & Pinel, V. (2006). Icelandic rhythmics: Annual modulation of land elevation and plate spreading by snow load. *Geophysical Research Letters, 33*, L24305. https://doi.org/10.1029/2006GL028081.

Green, T. R., Taniguchi, M., Kooi, H., Gurdak, J. J., Allen, D. M., Hiscock, K. M., ... Aureli, A. (2011). Beneath the surface of global change: Impacts of climate change on groundwater. *Journal of Hydrology, 405*(3–4), 532–560.

Gregersen, S., LIndholm, C., Korja, A., Lund, B., Uski, M., Oinonen, K., ... Kieding, M. (2021). Seismicity and sources of stress in Fennoscandia. In H. Steffen, O. Olesen, & R. Sutinen (Eds.), *Glacially-triggered faulting* (pp. 177–197). Cambridge, UK: Cambridge University Press. https://doi.org/10.1017/9781108779906.014.

Grollimund, B., & Zoback, M. D. (2001). Did deglaciation trigger New Madrid seismicity? *Geology, 29*, 175–178.

Gudmundsson, L., Leonard, M., Do, H. X., Westra, S., & Seneviratne, S. I. (2019). Observed trends in global indicators of mean and extreme streamflow. *Geophysical Research Letters, 46*(2), 756–766.

Hainzl, S., Kraft, T., Wassermann, J., Igel, H., & Schmedes, E. (2006). Evidence for rainfall triggered earthquake activity. *Geophysical Research Letters, 33*, L19303.

Hampel, A., Hetzel, R., & Maniatis, G. (2010). Response of faults to climate-driven changes in ice and water volumes on Earth's surface. *Philosophical Transactions of the Royal Society A: Mathematical, Physical and Engineering Sciences, 368*(1919), 2501–2517.

Hampel, A., Hetzel, R., Maniatis, G., & Karow, T. (2009). Three-dimensional numerical modeling of slip rate variations on normal and thrust fault arrays during ice cap growth and melting. *Journal of Geophysical Research, 114*, B08406. https://doi.org/10.1029/2008JB006113.

Hansen, L. N., Zimmerman, M. E., & Kohlstedt, D. L. (2011). Grain boundary sliding in San Carlos olivine: Flow law parameters and crystallographic-preferred orientation. *Journal of Geophysical Research: Solid Earth, 116*(B8).

Haskell, N. A. (1935). The motion of a fluid under a surface load. *Physics, 6*, 256.

Heaton, T. H. (1982). Tidal triggering of earthquakes. *Bulletin of the Seismological Society of America, 72*(6A), 2181–2200. https://doi.org/10.1785/BSSA07206A2181.

Heimisson, E. R., & Avouac, J.-P. (2020). Analytical prediction of seismicity rate due to tides and other oscillating stresses. *Geophysical Research Letters, 47*(23), e2020GL090827. https://doi.org/10.1029/2020GL090827.

Heimisson, E. R., & Segall, P. (2018). Constitutive law for earthquake production based on rate-and-state friction: Dieterich 1994 revisited. *Journal of Geophysical Research, 123*(5).

Heki, K. (2003). Snow load and seasonal variation of earthquake occurrence in Japan. *Earth and Planetary Science Letters, 207*(1–4), 159–164. https://doi.org/10.1016/S0012-821X(02)01148-2.

Heki, K. (2004). Dense GPS array as a new sensor of seasonal changes of surface loads. In R. S. J. Sparks, & C. J. Hawkesworth (Eds.), *Geophys. Monogr. Ser.: Vol. 150. The state of the planet: Frontiers and challenges in geophysics* (pp. 177–196). Washington, DC: AGU. https://doi.org/10.1029/150GM15.

Heki, K., & Arief, S. (2022). Crustal response to heavy rains in Southwest Japan 2017–2020. *Earth and Planetary Science Letters, 578*, 117325.

Hersbach, H., Bell, B., Berrisford, P., Hirahara, S., Horányi, A., Muñoz-Sabater, J., ... Thépaut, J. N. (2020). The ERA5 global reanalysis. *Quarterly Journal of the Royal Meteorological Society, 146*(730), 1999–2049.

Hetzel, R., & Hampel, A. (2005). Slip rate variations on normal faults during glacial-interglacial changes in surface loads. *Nature, 435*, 81–84.

Hill, R. G., Weingarten, M., Rockwell, T. K., & Fialko, Y. (2023). Major southern San Andreas earthquakes modulated by lake-filling events. *Nature, 618*(7966), 761–766. https://doi.org/10.1038/s41586-023-06058-9.

Hirth, G., & Kohlstedt, D. (2003). Rheology of the upper mantle and the mantle wedge: A view from the experimentalists. *Geophysical Monograph-American Geophysical Union, 138*, 83–106.

Hsu, Y. J., Chang, Y. S., Liu, C. C., Lee, H. M., Linde, A. T., Sacks, S. I., ... Chen, Y. G. (2015). Revisiting borehole strain, typhoons, and slow earthquakes using quantitative estimates of precipitation-induced strain changes. *Journal of Geophysical Research: Solid Earth, 120*, 4556–4571. https://doi.org/10.1002/2014JB011807.

Hsu, Y. J., Kao, H., Bürgmann, R., Lee, Y. T., Huang, H. H., Hsu, Y. F., ... Zhuang, J. (2021). Synchronized and asynchronous modulation of seismicity by hydrological loading: A case study in Taiwan. *Science Advances, 7*(16), eabf7282. https://doi.org/10.1126/sciadv.abf7282.

Hu, X., & Bürgmann, R. (2020). Aquifer deformation and active faulting in Salt Lake Valley, Utah, USA. *Earth and Planetary Science Letters, 547*, 116471. https://doi.org/10.1016/J.EPSL.2020.116471.

Hu, Y., & Freymueller, J. T. (2019). Geodetic observations of time-variable glacial isostatic adjustment in Southeast Alaska and its implications for Earth rheology. *Journal of Geophysical Research: Solid Earth, 124*(9), 9870–9889.

Hu, X., Xue, L., Bürgmann, R., & Fu, Y. (2021). Stress perturbations from hydrological and industrial loads and seismicity in the Salt Lake City region. *Journal of Geophysical Research: Solid Earth, 126*, e2021JB022362. https://doi.org/10.1029/2021JB022362.

Hugonnet, R., McNabb, R., Berthier, E., Menounos, B., Nuth, C., Girod, L., ... Kääb, A. (2021). Accelerated global glacier mass loss in the early twenty-first century. *Nature, 592*(7856), 726–731.

Husen, S., Bachmann, C., & Giardini, D. (2007). Locally triggered seismicity in the central Swiss Alps following the large rainfall event of August 2005. *Geophysical Journal International, 171*(3), 1126–1134.

IPCC. (2021). In V. Masson-Delmotte, P. Zhai, A. Pirani, S. L. Connors, C. Pean, S. Berger, ... B. Zhou (Eds.), *Climate change 2021: The physical science basis. Contribution of working Group I to the sixth assessment report of the intergovernmental panel on climate change*. Cambridge, UK/New York, NY: Cambridge University Press. https://doi.org/10.1017/9781009157896.

IPCC. (2022). In H.-O. Pörtner, D. C. Roberts, M. Tignor, E. S. Poloczanska, K. Mintenbeck, A. Alegría, ... B. Rama (Eds.), *Climate change 2022: Impacts, adaptation, and vulnerability. Contribution of working group II to the sixth assessment report of the intergovernmental panel on climate change*. Cambridge, UK/New York, NY: Cambridge University Press (3056 pp.) https://doi.org/10.1017/9781009325844.

Ivins, E. R., James, T. S., & Klemann, V. (2003). Glacial isostatic stress shadowing by the Antarctic ice sheet. *Journal of Geophysical Research: Solid Earth, 108*(B12).

Johnson, C. W., Fu, Y., & Bürgmann, R. (2017a). Seasonal water storage, stress modulation, and California seismicity. *Science, 356*, 1161–1164. https://doi.org/10.1126/science.aak9547.

Johnson, C. W., Fu, Y., & Bürgmann, R. (2017b). Stress models of the annual hydrospheric, atmospheric, thermal, and tidal loading cycles on California faults: Perturbation of background stress and changes in seismicity. *Journal of Geophysical Research, 122*, 10605–10625. https://doi.org/10.1002/2017JB014778.

Johnson, C. W., Fu, Y., & Bürgmann, R. (2020). Hydrospheric modulation of stress and seismicity on shallow faults in southern Alaska. *Earth and Planetary Science Letters, 530*, 115904. https://doi.org/10.1016/j.epsl.2019.115904.

Johnston, A. C. (1987). Suppression of earthquakes by large continental ice sheets. *Nature, 330*(6147), 467–469.

Kääb, A., Leinss, S., Gilbert, A., Bühler, Y., Gascoin, S., Evans, S. G., ... Yao, T. (2018). Massive collapse of two glaciers in western Tibet in 2016 after surge-like instability. *Nature Geoscience, 11*(2), 114–120.

Kierulf, H. P., Steffen, H., Barletta, V. R., Lidberg, M., Johansson, J., Kristiansen, O., & Tarasov, L. (2021). A GNSS velocity field for geophysical applications in Fennoscandia. *Journal of Geodynamics, 146*, 101845.

Kim, J., Bahadori, A., & Holt, W. E. (2021). Crustal strain patterns associated with normal, drought, and heavy precipitation years in California. *Journal of Geophysical Research: Solid Earth, 126*, e2020JB019560. https://doi.org/10.1029/2020JB019560.

Kim, J., Holt, W. E., Bahadori, A., & Shen, W. (2021). Repeating nontectonic seasonal stress changes and a possible triggering mechanism of the 2019 Ridgecrest earthquake sequence in California. *Journal of Geophysical Research: Solid Earth, 126*, e2021JB022188. https://doi.org/10.1029/2021JB022188.

Knowles, L. A., Bennett, R. A., & Harig, C. (2020). Vertical displacements of the Amazon Basin from GRACE and GPS. *Journal of Geophysical Research: Solid Earth, 125*(2). https://doi.org/10.1029/2019JB018105.

Kraft, J., Wassermann, E., & Schmedes, H. I. (2006). Meteorological triggering of earthquake swarms at Mt. Hochstaufen, SE-Germany. *Tectonophysics*, 0040-1951. *424*(3–4), 245–258. https://doi.org/10.1016/j.tecto.2006.03.044.

Kraner, M. L., Holt, W. E., & Borsa, A. A. (2018). Seasonal nontectonic loading inferred from cGPS as a potential trigger for the M6.0 South Napa earthquake. *Journal of Geophysical Research: Solid Earth, 123*(6), 5300–5322. https://doi.org/10.1029/2017JB015420.

Kreemer, C., & Zaliapin, I. (2018). Spatio-temporal correlation between seasonal variations in seismicity and horizontal dilatational strain in California. *Geophysical Research Letters, 45*, 9559–9568. https://doi.org/10.1029/2018GL079536.

Kundu, B., Vissa, N. K., & Gahalaut, V. K. (2015). Influence of anthropogenic groundwater unloading in indo-Gangetic plains on the 25 April 2015 Mw 7.8 Gorkha, Nepal earthquake. *Geophysical Research Letters, 42*. https://doi.org/10.1002/2015GL066616.

Kundu, B., Vissa, N. K., Gahalaut, K., Gahalaut, V. K., Panda, D., & Malik, K. (2019). Influence of anthropogenic groundwater pumping on the 2017 November 12 M7.3 Iran-Iraq border earthquake. *Geophysical Journal International, 218*(2), 833–839. https://doi.org/10.1093/gji/ggz195.

Lambeck, K., & Chappell, J. (2001). Sea level change through the last glacial cycle. *Science, 292*(5517), 679–686.

Lambeck, K., & Purcell, A. (2003). *Glacial rebound and crustal stress in Finland*. Oliluoto, Finland: Tech. Rep.

Landerer, F. W., Flechtner, F. M., Save, H., Webb, F. H., Bandikova, T., Bertiger, W. I., ... Yuan, D. N. (2020). Extending the global mass change data record: GRACE Follow-On instrument and science data performance. *Geophysical Research Letters, 47*(12), e2020GL088306.

Larochelle, S., Chanard, K., Fleitout, L., Fortin, J., Gualandi, A., Longuevergne, L., ... Avouac, J. P. (2022). Understanding the geodetic signature of large aquifer systems: Example of the Ozark Plateaus in Central United States. *Journal of Geophysical Research: Solid Earth, 127*(3), e2021JB023097.

Larsen, C. F., Motyka, R. J., Freymueller, J. T., Echelmeyer, K. A., & Ivins, E. R. (2005). Rapid viscoelastic uplift in Southeast Alaska caused by post-Little Ice ge glacial retreat. *Earth and Planetary Science Letters, 237*(3–4), 548–560. https://doi.org/10.1016/j.epsl.2005.06.032.

Lau, H. C., Austermann, J., Holtzman, B. K., Havlin, C., Lloyd, A. J., Book, C., & Hopper, E. (2021). Frequency dependent mantle viscoelasticity via the complex viscosity: Cases from Antarctica. *Journal of Geophysical Research: Solid Earth, 126*(11), e2021JB022622.

Lau, H. C., & Holtzman, B. K. (2019). "Measures of dissipation in viscoelastic media" extended: Toward continuous characterization across very broad geophysical time scales. *Geophysical Research Letters, 46*(16), 9544–9553.

Lau, H. C., Mitrovica, J. X., Austermann, J., Crawford, O., Al-Attar, D., & Latychev, K. (2016). Inferences of mantle viscosity based on ice age data sets: Radial structure. *Journal of Geophysical Research: Solid Earth, 121*(10), 6991–7012.

Li, W., van Dam, T., Li, Z., & Shen, Y. (2016). Annual variation detected by GPS, GRACE and loading models. *Studia Geophysica et Geodaetica, 60*, 608–621.

Liu, C., Linde, A. T., & Sacks, I. S. (2009). Slow earthquakes triggered by typhoons. *Nature, 459*(7248), 833–836.

Longman, I. (1962). A Green's function for determining the deformation of the earth under surface mass loads. 1. Theory. *Journal of Geophysical Research, 67*(2), 845–850.

Longman, I. (1963). A Green's function for determining the deformation of the earth under surface mass loads. 2. Computations and numerical results. *Journal of Geophysical Research, 68*(2), 485–489.

Lough, A. C., Wiens, D. A., & Nyblade, A. (2018). Reactivation of ancient Antarctic rift zones by intraplate seismicity. *Nature Geoscience, 11*(7), 515–519.

Lowry, A. R. (2006). Resonant slow fault slip in subduction zones forced by climatic load stress. *Nature, 442*(7104), 802–805. https://doi.org/10.1038/nature05055.

Lund, B., Schimdt, P., & Hieronymus, C. (2009). *Stress evolution and fault stability during the Weichselian glacial cycle*. Stockholm, Sweden: Swedish Nucl. Fuel and Waste Manage. Co (SKB).

Lundgren, P., Liu, Z., & Ali, S. T. (2022). San Andreas fault stress change due to groundwater withdrawal in California's Central Valley, 1860–2010. *Geophysical Research Letters, 49*(3), e2021GL095975.

Luttrell, K., & Sandwell, D. (2010). Ocean loading effects on stress at near shore plate boundary fault systems. *Journal of Geophysical Research, 115*, B08411. https://doi.org/10.1029/2009JB006541.

Martens, H. R., Rivera, L., & Simons, M. (2019). LoadDef: A Python-based toolkit to model elastic deformation caused by surface mass loading on spherically symmetric bodies. *Earth and Space Science, 6*(2), 311–323. https://doi.org/10.1029/2018EA000462.

Martin-Español, A., Bamber, J. L., & Zammit-Mangion, A. (2017). Constraining the mass balance of East Antarctica. *Geophysical Research Letters, 44*, 4168–4175. https://doi.org/10.1002/2017GL072937.

Melini, D., Gegout, P., Spada, G., & King, M. (2014). *REAR—A regional ElAstic rebound calculator. User manual for version 1.0*. Available at: http://hpc.rm.ingv.it/rear.

Meng, X., Yang, H., & Peng, Z. (2018). Foreshocks, b value map, and aftershock triggering for the 2011 Mw 5.7 Virginia earthquake. *Journal of Geophysical Research: Solid Earth, 123*, 5082–5098. https://doi.org/10.1029/2017JB015136.

Michel, A., & Boy, J. P. (2022). Viscoelastic Love numbers and long-period geophysical effects. *Geophysical Journal International, 228*(2), 1191–1212.

Miller, S. A. (2008). Note on rain-triggered earthquakes and their dependence on karst geology. *Geophysical Journal International, 173*(1), 334–338. https://doi.org/10.1111/j.1365-246X.2008.03735.x.

Milliner, C., Materna, K., Bürgmann, R., Fu, Y., Moore, A. W., Bekaert, D., … Argus, D. F. (2018). Tracking the weight of Hurricane Harvey's stormwater using GPS data. *Science Advances, 4*, eaau2477. https://doi.org/10.1126/sciadv.aau2477.

Milne, G. A., Davis, J. L., Mitrovica, J. X., Scherneck, H. G., Johansson, J. M., Vermeer, M., & Koivula, H. (2001). Space-geodetic constraints on glacial isostatic adjustment in Fennoscandia. *Science, 291*(5512), 2381–2385.

Missiakoulis, S. (2008). Aristotle and earthquake data: A historical note. *International Statistical Review, 76*(1), 130–133.

Mitrovica, J. X., & Forte, A. M. (2004). A new inference of mantle viscosity based upon joint inversion of convection and glacial isostatic adjustment data. *Earth and Planetary Science Letters, 225*(1–2), 177–189.

Mogi, K. (1958). Relations between the eruptions of various volcanoes and the deformations of the ground surfaces around them. *Bulletin. Earthquake Research Institute, University of Tokyo, 36*, 99–134.

Montgomery-Brown, E. K., Shelly, D. R., & Hsieh, P. A. (2019). Snowmelt-triggered earthquake swarms at the margin of Long Valley Caldera, California. *Geophysical Research Letters, 46*, 3698–3705. https://doi.org/10.1029/2019GL082254.

Mouginot, J., Rignot, E., Bjørk, A. A., Van den Broeke, M., Millan, R., Morlighem, M., … Wood, M. (2019). Forty-six years of Greenland ice sheet mass balance from 1972 to 2018. *Proceedings of the National Academy of Sciences, 116*(19), 9239–9244.

Munier, R., Adams, J., Brandes, C., Brooks, G., Dehls, J., Gibbons, S. J., … Tassis, G. (2020). *International database of glacially induced faults*. PANGAEA.

Nield, G. A., Barletta, V. R., Bordoni, A., King, M. A., Whitehouse, P. L., Clarke, P. J., … Berthier, E. (2014). Rapid bedrock uplift in the Antarctic Peninsula explained by viscoelastic response to recent ice unloading. *Earth and Planetary Science Letters, 397*, 32–41.

Noël, C., Pimienta, L., & Violay, M. (2019). Time-dependent deformations of sandstone during pore fluid pressure oscillations: Implications for natural and induced seismicity. *Journal of Geophysical Research: Solid Earth, 124*(1), 801–821.

Ojha, C., Shirzaei, M., Werth, S., Argus, D. F., & Farr, T. G. (2018). Sustained groundwater loss in California's Central Valley exacerbated by intense drought periods. *Water Resources Research, 54*. https://doi.org/10.1029/2017WR022250.

Ojha, C., Werth, S., & Shirzaei, M. (2019). Groundwater loss and aquifer system compaction in San Joaquin Valley during 2012–2015 drought. *Journal of Geophysical Research: Solid Earth, 124*(3), 3127–3143.

Olivieri, M., & Spada, G. (2015). Ice melting and earthquake suppression in Greenland. *Polar Science, 9*(1), 94–106.

Otto, F. E. L. (2017). Attribution of weather and climate events. *Annual Review of Environment and Resources, 42*(1), 627–646. https://doi.org/10.1146/annurev-environ-102016-060847.

Ouellette, K. J., de Linage, C., & Famiglietti, J. S. (2013). Estimating snow water equivalent from GPS vertical site-position observations in the western United States. *Water Resources Research, 49*, 2508–2518. https://doi.org/10.1002/wrcr.20173.

Pagli, C., & Sigmundsson, F. (2008). Will present day glacier retreat increase volcanic activity? Stress induced by recent glacier retreat and its effect on magmatism at the Vatnajokull ice cap, Iceland. *Geophysical Research Letters, 35*, L09304. https://doi.org/10.1029/2008GL033510.

Peltier, W. R., & Andrews, J. T. (1976). Glacial-isostatic adjustment. I. The forward problem. *Geophysical Journal International, 46*(3), 605–646.

Perfettini, H., Schmittbuhl, J., & Cochard, A. (2003). Shear and normal load perturbations on a two-dimensional continuous fault. 2. Dynamic triggering. *Journal of Geophysical Research, 108*(B9). https://doi.org/10.1029/2002JB001805.

Perfettini, H., Schmittbuhl, J., Rice, J. R., & Cocco, M. (2001). Frictional response induced by time-dependent fluctuations of the normal loading. *Journal of Geophysical Research, 106*(B7), 13455–13472. https://doi.org/10.1029/2000JB900366.

Pétrélis, F., Chanard, K., Schubnel, A., & Hatano, T. (2021). Earthquake sensitivity to tides and seasons: Theoretical studies. *Journal of Statistical Mechanics: Theory and Experiment, 2021*(2), 023404.

Pisarska-Jamroży, M., Belzyt, S., Börner, A., Hoffmann, G., Hüneke, H., Kenzler, M., … van Loon, A. J. (2018). Evidence from seismites for glacio-isostatically induced crustal faulting in front of an advancing land-ice mass (Rügen Island, SW Baltic Sea). *Tectonophysics, 745*, 338–348. https://doi.org/10.1016/j.tecto.2018.08.004.

Pollitz, F. F. (1996). Coseismic deformation from earthquake faulting on a spherical Earth. *Geophysical Journal International, 125*(1), 1–14. https://doi.org/10.1111/j.1365-246X.1996.tb06530.x.

Pollitz, F. F. (2003). Transient rheology of the uppermost mantle beneath the Mojave Desert, California. *Earth and Planetary Science Letters, 215*(1–2), 89–104.

Pollitz, F., Kellogg, L., & Bürgmann, R. (2001). Sinking mafic body in a reactivated lower crust: A mechanism for stress concentration in the New Madrid seismic zone. *Bulletin of the Seismological Society of America, 91*, 1882–1897.

Pollitz, F. F., Wech, A. G., Kao, H., & Bürgmann, R. (2013). Annual modulation of non-volcanic tremor in northern Cascadia. *Journal of Geophysical Research, 118*. https://doi.org/10.1002/jgrb.50181.

Prawirodirdjo, L., Ben-Zion, Y., & Bock, Y. (2006). Observation and modeling of thermoelastic strain in Southern California integrated GPS network daily position time series. *Journal of Geophysical Research, 111*, B02408. https://doi.org/10.1029/2005JB003716.

Qin, L., Ben-Zion, Y., Qiu, H., Share, P. E., Ross, Z. E., & Vernon, F. L. (2018). Internal structure of the San Jacinto fault zone in the trifurcation area southeast of Anza, California, from data of dense seismic arrays. *Geophysical Journal International, 213*(1), 98–114.

Reasenberg, P. (1985). Second-order moment of central California seismicity, 1969–1982. *Journal of Geophysical Research: Solid Earth, 90*(B7), 5479–5495.

Rice, J. R., & Cleary, M. P. (1976). Some basic stress diffusion solutions for fluid-saturated elastic porous media with compressible constituents. *Reviews of Geophysics and Space Physics, 14*, 227–241.

Rigo, A., Béthoux, N., Masson, F., & Ritz, J. F. (2008). Seismicity rate and wave-velocity variations as consequences of rainfall: The case of the catastrophic storm of September 2002 in the Nîmes Fault region (Gard, France). *Geophysical Journal International, 173*(2), 473–482.

Rodell, M., Houser, P. R., Jambor, U. E. A., Gottschalck, J., Mitchell, K., Meng, C. J., ... Toll, D. (2004). The global land data assimilation system. *Bulletin of the American Meteorological Society, 85*(3), 381–394.

Rollins, C., Freymueller, J. T., & Sauber, J. M. (2021). Stress promotion of the 1958 Mw ~ 7.8 Fairweather Fault earthquake and others in Southeast Alaska by glacial isostatic adjustment and inter-earthquake stress transfer. *Journal of Geophysical Research: Solid Earth, 126*(1), e2020JB020411.

Saar, M. O., & Manga, M. (2003). Seismicity induced by seasonal groundwater recharge at Mt. Hood, Oregon. *Earth and Planetary Science Letters, 214*, 605–618. https://doi.org/10.1016/S0012-821X(03)00418-7.

Sauber, J. M., & Molnia, B. F. (2004). Glacier ice mass fluctuations and fault instability in tectonically active southern Alaska. *Global and Planetary Change, 42*(1–4), 279–293.

Sauber, J., Plafker, G., Molnia, B. F., & Bryant, M. A. (2000). Crustal deformation associated with glacial fluctuations in the eastern Chugach Mountains, Alaska. *Journal of Geophysical Research: Solid Earth, 105*(B4), 8055–8807.

Sauber, J., Rollins, C., Freymueller, J. T., & Ruppert, N. A. (2021). Glacially induced faulting in Alaska. In H. Steffen, O. Olesen, & R. Sutinen (Eds.), *Glacially-triggered faulting* (pp. 353–365). Cambridge: Cambridge University Press. https://doi.org/10.1017/9781108779906.026.

Sauber, J., & Ruppert, N. A. (2008). Rapid ice mass loss: Does it have an influence on earthquake occurrence in southern Alaska. In *Active tectonics and seismic potential of Alaska* (pp. 369–384). Geophysical Monograph Series.

Sayles, R. W. (1913). Earthquakes and rainfall. *Bulletin of the Seismological Society of America, 3*(2), 5.

Scholz, C. H., Tan, Y. J., & Albino, F. (2019). The mechanism of tidal triggering of earthquakes at mid-ocean ridges. *Nature Communications, 10*(1), 1–7. https://doi.org/10.1038/s41467-019-10605-2.

Schulz, W. H., McKenna, J. P., Kibler, J. D., & Biavati, G. (2009). Relations between hydrology and velocity of a continuously moving landslide—Evidence of pore-pressure feedback regulating landslide motion? *Landslides, 6*(3), 181–190.

Schuster, A. (1897). On lunar and solar periodicities of earthquakes. *Proceedings of the Royal Society of London, 61*(369–377), 455–465.

Schuster, A. (1898). On the investigation of hidden periodicities with application to a supposed 26 day period of meteorological phenomena. *Terrestrial Magnetism, 3*(1), 13–41.

Sella, G. F., Stein, S., Dixon, T. H., Craymer, M., James, T. S., Mazzotti, S., & Dokka, R. K. (2007). Observation of glacial isostatic adjustment in "stable" North America with GPS. *Geophysical Research Letters, 34*, 2.

Senapati, B., Kundu, B., & Jin, S. (2022). Seismicity modulation by external stress perturbations in plate boundary vs. stable plate interior. *Geoscience Frontiers, 13*(3), 101352. https://doi.org/10.1016/j.gsf.2022.101352.

Shepherd, A., Ivins, E. R., Barletta, V. R., Bentley, M. J., Bettadpur, S., Briggs, K. H., ... Zwally, H. J. (2012). A reconciled estimate of ice-sheet mass balance. *Science, 338*(6111), 1183–1189.

Shirzaei, M., Freymueller, J., Törnqvist, T. E., Galloway, D. L., Dura, T., & Minderhoud, P. S. (2021). Measuring, modelling and projecting coastal land subsidence. *Nature Reviews Earth & Environment, 2*(1), 40–58.

Sigmundsson, F., Pinel, V., Lund, B., Albino, F., Pagli, C., Geirsson, H., & Sturkell, E. (2010). Climate effects on volcanism: Influence on magmatic systems of loading and unloading from ice mass variations, with examples from Iceland. *Philosophical Transactions of the Royal Society A: Mathematical, Physical and Engineering Sciences, 368*(1919), 2519–2534.

Silverii, F., Montgomery-Brown, E. K., Borsa, A. A., & Barbour, A. J. (2020). Hydrologically induced deformation in Long Valley caldera and adjacent Sierra Nevada. *Journal of Geophysical Research: Solid Earth, 125*, e2020JB019495. https://doi.org/10.1029/2020JB019495.

Smith, R. G., Knight, R., Chen, J., Reeves, J. A., Zebker, H. A., Farr, T., & Liu, Z. (2017). Estimating the permanent loss of groundwater storage in the southern San Joaquin Valley, California. *Water Resources Research, 53*(3), 2133–2148.

Steer, P., Jeandet, L., Cubas, N., Marc, O., Meunier, P., Simoes, M., ... Shyu, J. B. H. (2020). Earthquake statistics changed by typhoon-driven erosion. *Scientific Reports, 10*(1), 1–11. https://doi.org/10.1038/s41598-020-67865-y.

Steer, P., Simoes, M., Cattin, R., & Shyu, J. B. H. (2014). Erosion influences the seismicity of active thrust faults. *Nature Communications, 5*. https://doi.org/10.1038/ncomms6564.

Steffen, H., Olesen, O., & Sutinen, R. (2021). Glacially triggered faulting: A historical overview and recent developments. In H. Steffen, O. Olesen, & R. Sutinen (Eds.), *Glacially-triggered faulting* (pp. 3–19). Cambridge: Cambridge University Press. https://doi.org/10.1017/9781108779906.003.

Steffen, R., & Steffen, H. (2021). Reactivation of non-optimally orientated faults due to glacially induced stresses. *Tectonics*, *40*(11), e2021TC006853.

Steffen, R., Steffen, H., Weiss, R., Lecavalier, B. S., Milne, G. A., Woodroffe, S. A., & Bennike, O. (2020). Early Holocene Greenland-ice mass loss likely triggered earthquakes and tsunami. *Earth and Planetary Science Letters*, *546*, 116443. https://doi.org/10.1016/j.epsl.2020.116443.

Steffen, R., Wu, P., Steffen, H., & Eaton, D. W. (2014). The effect of earth rheology and ice-sheet size on fault-slip and magnitude of postglacial earthquakes. *Earth and Planetary Science Letters*, *388*, 71–80. https://doi.org/10.1016/j.epsl.2013.11.058.

Stein, R. S. (1999). The role of stress transfer in earthquake occurrence. *Nature*, *402*, 605–609.

Stein, S., Cloetingh, S., Sleep, N., & Wortel, R. (1989). Passive margin earthquakes, stresses and rheology. In S. Gregersen, & P. Basham (Eds.), *Earthquakes at North-Atlantic passive margins: Neotectonics and postglacial rebound, NATO ASI series C 266* (pp. 231–259). Dordrecht: Springer.

Štěpančíková, P., Rockwell, T. K., Stemberk, J., Rhodes, E. J., Hartvich, F., Luttrell, K., ... Hók, J. (2022). Acceleration of Late Pleistocene activity of a Central European fault driven by ice loading. *Earth and Planetary Science Letters*, *591*, 117596. https://doi.org/10.1016/j.epsl.2022.117596.

Taber, S. (1914). Seismic activity in the Atlantic coastal plain near Charleston, South Carolina. *Bulletin of the Seismological Society of America*, *4*(3), 108–160.

Tamisiea, M., Mitrovica, J., & Davis, J. (2007). GRACE gravity data constrain ancient ice geometries and continental dynamics over Laurentia. *Science*, *316*(5826), 881–883.

Tapley, B. D., Bettadpur, S., Ries, J. C., Thompson, P. F., & Watkins, M. M. (2004). GRACE measurements of mass variability in the Earth system. *Science*, *305*(5683), 503–505.

Tay, C., Lindsey, E. O., Chin, S. T., McCaughey, J. W., Bekaert, D., Nguyen, M., ... Hill, E. M. (2022). Sea-level rise from land subsidence in major coastal cities. *Nature Sustainability*, 1–9.

The IMBIE team. (2018). Mass balance of the Antarctic Ice Sheet from 1992 to 2017. *Nature*, *558*, 219–222. https://doi.org/10.1038/s41586-018-0179-y.

The IMBIE Team. (2020). Mass balance of the Greenland Ice Sheet from 1992 to 2018. *Nature*, *579*, 233–239. https://doi.org/10.1038/s41586-019-1855-2.

Trubienko, O., Garaud, J. D., & Fleitout, L. (2014). Models of postseismic deformation after megaearthquakes: The role of various rheological and geometrical parameters of the subduction zone. *Solid Earth Discussions*, *6*(1), 427–466.

Tsai, V. C. (2011). A model for seasonal changes in GPS positions and seismic wave speeds due to thermoelastic and hydrologic variations. *Journal of Geophysical Research*, *116*, B04404. https://doi.org/10.1029/2010JB008156.

Ueda, T., & Kato, A. (2019). Seasonal variations in crustal seismicity in San-in district, southwest Japan. *Geophysical Research Letters*, *46*, 3172–3179. https://doi.org/10.1029/2018GL081789.

Vachon, R., Schmidt, P., Lund, B., Plaza-Faverola, A., Patton, H., & Hubbard, A. (2022). Glacially induced stress across the Arctic from the Eemian interglacial to the present—Implications for faulting and methane seepage. *Journal of Geophysical Research: Solid Earth*, *127*(7), e2022JB024272.

van Dam, T., Whitehouse, P., & Liu, L. (2024). Mass loss of cryosphere. In Y. Aoki, & C. Kreemer (Eds.), *GNSS monitoring of the terrestrial environment: Earthquakes, volcanoes, and climate change* (pp. 215–242). Elsevier.

van Dam, T. M., Blewitt, G., & Heflin, M. (1994). Detection of atmospheric pressure loading using the global positioning system. *Journal of Geophysical Research*, *99*, 23939–23950.

van Dam, T. M., & Wahr, J. (1987). Displacements of the Earth's surface due to atmospheric loading: Effects on gravity and baseline measurements. *Journal of Geophysical Research*, *92*, 1281–1286.

van Dam, T., Wahr, J., & Lavallée, D. (2007). A comparison of annual vertical crustal displacements from GPS and gravity recovery and climate experiment (GRACE) over Europe. *Journal of Geophysical Research*, *112*, B03404. https://doi.org/10.1029/2006JB004335.

van Dam, T. M., Wahr, J. M., Milly, P. C. D., Shmakin, A. B., Blewitt, G., Lavallee, D., & Larson, K. M. (2001). Crustal displacements due to continental water loading. *Geophysical Research Letters*, *28*, 651–654.

Vidale, J. E., Agnew, D. C., Johnston, M. J., & Oppenheimer, D. H. (1998). Absence of earthquake correlation with earth tides: An indication of high preseismic fault stress rate. *Journal of Geophysical Research*, *103*(B10), 24567–24572. https://doi.org/10.1029/98JB0059.

Wahr, J. (1995). *Earth tides, global earth physics, A handbook of physical constants* (pp. 40–46). Washington, DC: AGU. Reference Shelf.

Wahr, J., Khan, S. A., van Dam, T., Liu, L., van Angelen, J. H., van den Broeke, M. R., & Meertens, C. M. (2013). The use of GPS horizontals for loading studies, with applications to Northern California and Southeast Greenland. *Journal of Geophysical Research—Solid Earth*, *118*, 1795–1806. https://doi.org/10.1002/jgrb.50104.

Wang, R., & Kümpel, H. J. (2003). Poroelasticity: Efficient modeling of strongly coupled, slow deformation processes in a multilayered half-space. *Geophysics*, *68*(2), 705–717. https://doi.org/10.1190/1.1567241.

Wang, H., Xiang, L., Jia, L., Jiang, L., Wang, Z., Hu, B., & Gao, P. (2012). Load love numbers and Green's functions for elastic earth models PREM, iasp91, ak135, and modified models with refined crustal structure from crust 2.0. *Computers & Geosciences*, *49*, 190–199.

Wang, S., Xu, W., Xu, C., Yin, Z., Bürgmann, R., Liu, L., & Jiang, G. (2019). Changes in groundwater level possibly encourage shallow earthquakes in Central Australia: The 2016 Petermann ranges earthquake. *Geophysical Research Letters*, *46*, 3189–3198. https://doi.org/10.1029/2018GL080510.

Wetzler, N., Shalev, E., Göbel, T., Amelung, F., Kurzon, I., Lyakhovsky, V., & Brodsky, E. E. (2019). Earthquake swarms triggered by groundwater extraction near the Dead Sea fault. *Geophysical Research Letters*, *46*(14), 8056–8063.

White, A. M., Gardner, W. P., Borsa, A. A., Argus, D. F., & Martens, H. R. (2022). A review of GNSS/GPS in Hydrogeodesy: Hydrologic loading applications and their implications for water resource research. *Water Resources Research*, *58*(7), e2022WR032078. https://doi.org/10.1029/2022WR032078.

Whitehouse, P. L., Bentley, M. J., Milne, G. A., King, M., & Thomas, I. D. (2012). A new glacial isostatic adjustment model for Antarctica: Calibrated and tested using observations of relatives sea level change and present day uplift rates. *Geophysical Journal International, 190*, 1464–1482.

Wu, P. (1999). Modelling postglacial sea levels with power-law rheology and a realistic ice model in the absence of ambient tectonic stress. *Geophysical Journal International, 139*, 691–702.

Wu, P., & Johnston, P. (2000). Can deglaciation trigger earthquakes in N. America? *Geophysical Research Letters, 27*, 1323–1326. https://doi.org/10.1029/1999GL011070.

Wu, P., Steffen, R., Steffen, H., & Lund, B. (2021). Glacial isostatic adjustment models for earthquake triggering. In H. Steffen, O. Olesen, & R. Sutinen (Eds.), *Glacially-triggered faulting* (pp. 383–401). Cambridge: Cambridge University Press. https://doi.org/10.1017/9781108779906.029.

Wu, P. C., Wei, M., & D'Hondt, S. (2022). Subsidence in coastal cities throughout the world observed by InSAR. *Geophysical Research Letters, 49*(7), e2022GL098477.

Wunsch, C., Heimbach, P., Ponte, R. M., Fukumori, I., & ECCO-GODAE Consortium Members. (2009). The global general circulation of the ocean estimated by the ECCO-Consortium. *Oceanography, 22*(2), 88–103.

Xu, X., Dong, D., Fang, M., Zhou, Y., Wei, N., & Zhou, F. (2017). Contributions of thermoelastic deformation to seasonal variations in GPS station position. *GPS Solutions, 21*(3), 1265–1274. https://doi.org/10.1007/s10291-017-0609-6.

Xue, L., Johnson, C. W., Fu, Y., & Bürgmann, R. (2020). Seasonal seismicity in the Western branch of the East African rift system. *Geophysical Research Letters*. https://doi.org/10.1029/2019GL085882.

Yao, D., Huang, Y., Xue, L., Fu, Y., Gronewold, A., & Fox, J. L. (2022). Seismicity around southern Lake Erie during 2013–2020 in relation to lake water level. *Seismological Research Letters*. https://doi.org/10.1785/0220210343.

Young, Z. M., Kreemer, C., & Blewitt, G. (2021). GPS constraints on drought-induced groundwater loss around great salt Lake, Utah, with implications for seismicity modulation. *Journal of Geophysical Research, 126*(10), e2021JB022020. https://doi.org/10.1029/2021JB022020.

Zaliapin, I., & Ben-Zion, Y. (2020). Earthquake declustering using the nearest-neighbor approach in space-time-magnitude domain. *Journal of Geophysical Research: Solid Earth, 125*(4), e2018JB017120.

Zhai, Q., Peng, Z., Chuang, L. Y., Wu, Y. M., Hsu, Y. J., & Wdowinski, S. (2021). Investigating the impacts of a wet typhoon on microseismicity: A case study of the 2009 typhoon Morakot in Taiwan based on a template matching catalog. *Journal of Geophysical Research: Solid Earth, 126*(12), e2021JB023026.

Zhan, W., Heki, K., Arief, S., & Yoshida, M. (2021). Topographic amplification of crustal subsidence by the rainwater load of the 2019 typhoon Hagibis in Japan. *Journal of Geophysical Research: Solid Earth, 126*(6), e2021JB021845.

Zhuang, J., Ogata, Y., & Vere-Jones, D. (2002). Stochastic declustering of space-time earthquake occurrences. *Journal of the American Statistical Association, 97*, 369–382.

Chapter 15

Influence of climate change on magmatic processes: What does geodesy and modeling of geodetic data tell us?

Freysteinn Sigmundsson[a], Michelle Parks[b], Halldór Geirsson[a], Fabien Albino[c], Peter Schmidt[d], Siqi Li[a,*], Finnur Pálsson[a], Benedikt G. Ófeigsson[b], Vincent Drouin[b], Guðfinna Aðalgeirsdóttir[a], Eyjólfur Magnússon[a], Andy Hooper[e], Sigrún Hreinsdóttir[f], John Maclennan[g], Erik Sturkell[h], and Elisa Trasatti[i]

[a]Institute of Earth Sciences, Science Institute, University of Iceland, Reykjavík, Iceland, [b]Icelandic Meteorological Office, Reykjavík, Iceland, [c]Université Grenoble Alpes, Université Savoie Mont Blanc, CNRS, IRD, Université Gustave Eiffel, ISTerre, Grenoble, France, [d]University of Uppsala, Uppsala, Sweden, [e]COMET, University of Leeds, Leeds, United Kingdom, [f]GNS Science, Lower Hutt, New Zealand, [g]University of Cambridge, Cambridge, United Kingdom, [h]University of Gothenburg, Gothenburg, Sweden, [i]Istituto Nazionale di Geofisica e Vulcanologia, Rome, Italy

1 Introduction

Glacier mass loss due to anthropogenic climate change is now occurring in all glacierized regions in the world (e.g., Aðalgeirsdóttir et al., 2020; Björnsson et al., 2013; Fox-Kemper et al., 2021; Hugonnet et al., 2021; IPCC, 2021; Zemp et al., 2019). The ice mass loss affects not only volcanoes covered by retreating glaciers but also volcanoes up to considerable distances away from retreating glaciers. A 1 m reduction of ice, with a density of $910 \, kg/m^3$, corresponds to a normal stress change on the surface of the Earth of $(910 \, kg/m^3) \times (9.8 \, m/s^2) \times (1 \, m) = 9 \, kPa$. This pressure change effect decays with depth in the crust and mantle, dependent on the width of the surface load involved. Pressure change also occurs outside the extent of the unloading involved. At depth where magma is generated and stored, the effect may be on the order $\sim 1 \, kPa/yr$. If ice retreat continues over decades and begins to correspond to load change of tens of meters or more over considerable areas, then influence on subsurface magmatic processes may be anticipated. We here evaluate how GNSS measurements contribute to the improved understanding of the effect of ice unloading. We use Iceland as a laboratory as glaciers cover $\sim 10\%$ of Iceland and have generally been retreating since ~ 1890 (Aðalgeirsdóttir et al., 2020; Björnsson et al., 2013).

There is considerable evidence for variations in loads on the surface of the Earth affecting magmatic activity. On a global scale, periodicities in glaciations, according to Milankovitch cycles, have also been identified in tephra records and phase shifts between the glacial and volcanic records suggest a peak in volcanism during deglaciation. However, these records are noisy and the correlation coefficients identified are generally too low for precise constraints on the relative timing (Kutterolf et al., 2019). Global sea-level change has also been suggested to influence magmatic activity (Crowley et al., 2015; McGuire et al., 1997). On a more local scale, an anticorrelation has been found between volcanism and glaciations in California for the last 800,0000 years, indicating volcanism has been modulated by changes in climate (Glazner et al., 1999). A further study of the correlation between timing of volcanic activity and the growth and retreat of glaciers in California found a statistically significant cross-correlation between changes in eruption frequency and the first derivative of the glacial time series, implying that the temporal pattern of volcanism is influenced by the rate of change in ice volume (Jellinek et al., 2004). Variations in recent surface load variation on Earth have also been suggested to modulate magmatic activity, such as the observation that the 2011–2016 uplift (mapped with GNSS geodesy) of the Long Valley caldera during a period of drought was much faster than drought-induced rebound of surrounding terrain, suggesting that the removed load may have caused decompression and magmatic inflation in the Long Valley Caldera magmatic system (Hammond et al., 2019).

---

*Now at Institute of Geophysics, China Earthquake Administration, China.

One of the strongest pieces of evidence for glacial retreat influencing magmatic processes comes from studies of environmental conditions and eruption activity at the end of the last glaciation in Iceland. During the Weichselian glaciation (~70–12 kyr BP), Iceland was covered by a >300-km wide ice sheet that was depressing the surface and causing downward flexure (Sigmundsson, 1991). By the end of the Pleistocene, widespread deglaciation reversed this process causing the Earth's surface to rebound in response to ice loss. As a result, a major increase in volcanic activity occurred in Iceland (e.g., Slater et al., 1998), with an estimated 30–50-fold increase in eruption rates and change in the chemical signature of extruded lavas (e.g., Maclennan et al., 2002; Sigvaldason et al., 1992). Jull and McKenzie (1996) could fit this observation with a model of the effect of deglaciation on magma generation.

There are over 30 active central volcanoes in Iceland, half of which lie beneath glaciers (Fig. 15.1). Volcanism results from the interaction of the plate spreading and a hotspot (hotspot-ridge interaction). In Iceland, the N-American and Eurasian plates diverge at ~19 mm/yr. Excessive mantle upwelling occurs in a mantle plume beneath Iceland. The interaction of a spreading ridge and upwelling of anomalously hot material within the mantle has formed Iceland. As Icelandic glaciers are retreating fast (Aðalgeirsdóttir et al., 2020; Belart et al., 2020), it may thus only be a matter of time when deglaciation again affects eruptive activity here—with the potential for more frequent or larger eruptions. Furthermore, modulation of the crustal stress in response to the changing surface load may affect seismicity with implications for changes in the long-term seismic hazard. This needs to be understood in the overall context of volcanism and glaciations that are responsible for shaping Iceland (e.g., Sigmundsson et al., 2020).

Present-day retreat of the glaciers in Iceland is causing the Earth's surface to rebound in response to the unloading of the ice, by the process of glacial isostatic adjustment (GIA), on timescales ranging from seasonal to centurial. GIA is here taken to include both the viscoelastic response of the Earth, as well as the instantaneous elastic response to load changes. GIA in Iceland is well observed by Global Navigation and Satellite System (GNSS) and Interferometric Synthetic Aperture Radar (InSAR) geodetic measurements; the central part of Iceland is uplifting at rates of >30 mm/yr as a result of ongoing GIA (e.g., Árnadóttir et al., 2009; Compton et al., 2015; Drouin & Sigmundsson, 2019).

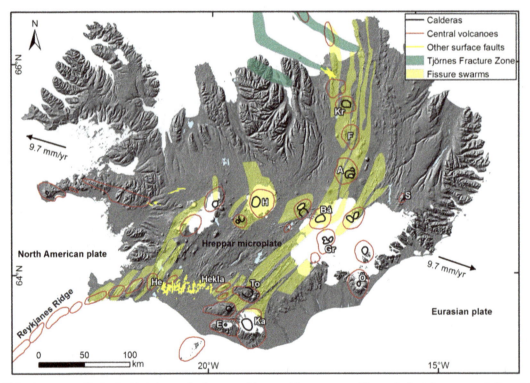

FIG. 15.1 Volcanic systems of Iceland, with their central volcanoes, calderas, and fissure swarms. Names of selected systems are indicated: Kr = Krafla, F = Fremri Námar, A = Askja, S = Snæfell, Ö = Öræfajökull, Bá = Bárðarbunga, Gr = Grímsvötn, E = Eyjafjallajökull, Ka = Katla, He = Hengill, To = Torfajökull, H = Hofsjökull, and Hekla. Also shown are active faults and fractures in the South Iceland Seismic zone *(yellow)*, and the three main strands of the Tjörnes Fracture Zone (TFZ) offshore Northern Iceland *(green)*. Glaciers in *white*. *(Reproduced from Sigmundsson, F., Einarsson, P., Hjartardóttir, Á. R., Drouin, V., Jónsdóttir, K., Árnadóttir, T., Geirsson, H., Hreinsdóttir, S., Li, S., & Ofeigsson, B. G., 2020. Geodynamics of Iceland and the signatures of plate spreading.* Journal of Volcanology and Geothermal Research, 391, *106436.)*

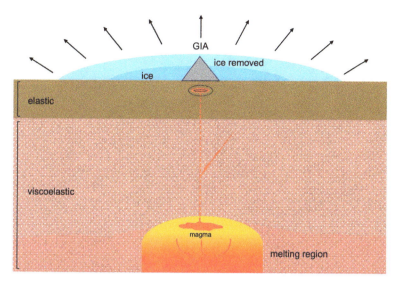

FIG. 15.2 A schematic view of the processes involved at subglacial volcanoes as a result of deglaciation, indicating the melting region, and magma accumulation in viscoelastic and elastic surroundings. Magma generation due to pressure release under a retreating ice cap may increase, new pathways of magma may form, and the stability of magmatic bodies at shallow depth may decrease or increase, depending on their geometry and rheology.

The glacier mass loss modifies the stress field in the crust and upper mantle and may alter volcanic activity in three ways (Fig. 15.2) (Sigmundsson et al., 2010, 2013): (i) increases melt generation at depth due to decompression (Jull & McKenzie, 1996; Pagli & Sigmundsson, 2008; Schmidt et al., 2013), (ii) influences magma migration (e.g., Hooper et al., 2011), and (iii) affects the stability of magma bodies—bringing a volcano either closer to, or further from failure (Albino et al., 2010). In the latter case, larger volumes of melt need to accumulate before an eruption is triggered, which can increase the hazards of an eruption.

A recent review of these and other effects of climate change on volcanoes is given by Aubry et al. (2022). Here, we consider the role of geodetic measurements, in particular GNSS observations, for providing an estimate of the effects on magma plumbing systems.

2 Ongoing glacier load changes at volcanoes

The ongoing subsurface climate change effect on volcanoes is due to load changes taking place at retreating ice caps. To estimate the effects, an estimate of load changes is needed. A global study by Hugonnet et al. (2021) spanning 2000–2019 has revealed accelerated retreat of glaciers in many regions of the Earth, and that in general thinning rates of glaciers outside ice sheet peripheries doubled over the study period. However, there are contrasting glacier fluctuations in different parts of the world. Areas of major glacier retreat in volcanic regions include Iceland, Alaska, and the Southern Andes (Fig. 15.3).

In Iceland, global warming due to anthropogenic climate change has resulted in the mass loss of the glaciers since ~1890. The loss of 540 ± 130 gigatonnes (Gt) or $16 \pm 4\%$ of the 1890 volume has been estimated for all glaciers in Iceland for the period ~1890–2019 (Fig. 15.3) and half of this mass loss occurred in 1995–2019 (Aðalgeirsdóttir et al., 2020). The largest ice cap, Vatnajökull, has lost most ice. Projections for the other main ice caps of Iceland, Hofsjökull and Langjökull indicate that they will likely disappear within the next 150–200 years (Aðalgeirsdóttir et al., 2006; Björnsson & Pálsson, 2008). Simulations for Vatnajökull indicate that it may lose up to 80%–90% of its volume by 2300, depending on climate scenarios (Schmidt et al., 2019).

The rate of reduction in surface loading due to climate change in Iceland has varied with time in the period since 1890, including a time period when the retreat temporarily halted (Fig. 15.3). Temperature records, glacier mass balance observations (both in situ and geodetic), GNSS observations of ice elevation change, and modeling indicate significant glacier fluctuations in the last century, with increased mass loss in the 1930–1950s, near zero or positive balance in 1960–1980s and increased mass loss again after mid-1990s, as well as large interannual variability (e.g., Aðalgeirsdóttir et al., 2020; Belart et al., 2020; Björnsson et al., 2013).

The assumption of constant deglaciation rates for the whole 20th century has been found to yield a reasonable fit to the observed long-term, nation-wide, GIA signal in Iceland in earlier studies (e.g., Árnadóttir et al., 2009). However, studies of more recent GIA deformation with increasing spatial and temporal resolution display notable variations in GIA deformation rates (e.g., Compton et al., 2015), and indicate the assumption of a constant rate of ice mass reduction is insufficient

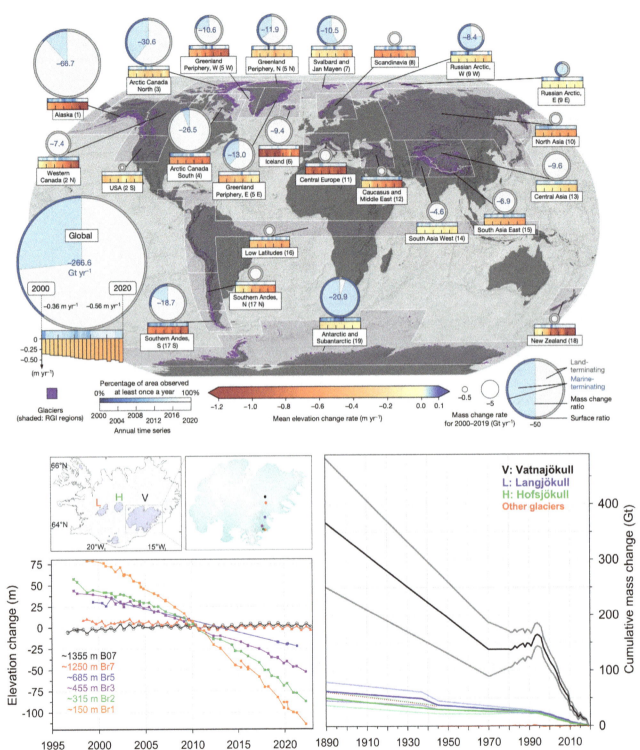

FIG. 15.3 Glacier retreat, globally and in Iceland. *Upper*: worldwide regional glacier mass changes and their temporal evolution from 2000 to 2019 as estimated by Hugonnet et al. (2021). Regional and global mass change rates with time series of mean surface elevation change rates for glaciers *(indigo)* of 19 regions (white-delimited indigo polygons; region numbers indicated in parentheses), shown on top of a world hillshade. Regions 2, 5, 9, and 17 are further divided (N, S, E, and W indicating north, south, east, and west, respectively) to illustrate contrasting temporal patterns. Mass change rates are represented by the area of the disk delimiting the inside wedge, which separates the mass change contribution of land-terminating *(light gray)* and marine-terminating *(light blue)* glaciers. Mass change rates larger than 4 Gt/yr are printed in blue inside the disk (in units of Gt/yr). The outside ring discerns between land *(gray)* and marine-terminating *(blue)* glacier area. Annual time series of mean elevation change (in m/yr) and regional data coverage are displayed on time friezes at the bottom of the disks. *Lower left*: surface elevation change at mass balance survey sites on Breiðamerkurjökull outlet of Vatnajökull ice cap as measured with biannual GNSS observations by the Institute of Earth Sciences, University of Iceland. Elevation in 2010 is used as reference for all sites. Iceland map shows glaciers (main ice caps: V = Vatnajökull, L = Langjökull, H = Hofsjökull). *Colored dots* on Vatnajökull show location of the GNSS observation sites. *Lower right*: cumulative mass change (Gt) of three main ice caps in Iceland and the sum of other glaciers in Iceland as inferred by Aðalgeirsdóttir et al. (2020). *(Upper panel: Reproduced with permission from Hugonnet, R., McNabb, R., Berthier, E., Menounos, B., Nuth, C., Girod, L. et al., 2021. Accelerated global glacier mass loss in the early twenty-first century.* Nature, *592(7856), 726–731. Springer Nature.)*

and in need of updating (e.g., Drouin & Sigmundsson, 2019). It is important to better understand these variations and analyze further if the immediate elastic response to load changes is larger than previously considered.

3 Uplift and deformation of volcanoes due to climate change and magma movements

Extensive continuous GNSS observations (cGNSS) in Iceland and InSAR observations have revealed the ongoing rise of Iceland (Fig. 15.4). Vertical velocities have been interpreted in a series of GIA models. However, an updated model considering the newest set of observations is though needed. A few examples of the vertical component of deformation at continuous GNSS (cGNSS) stations are shown in Fig. 15.5. A station at Jökulheimar (JOKU) has uplifted about 70 cm in the last three decades. The station is within the volcanic zone of Iceland, but far from central volcanoes. The inferred uplift is due to GIA. The initial observed uplift rate in the 1990s was significantly lower than for the later period.

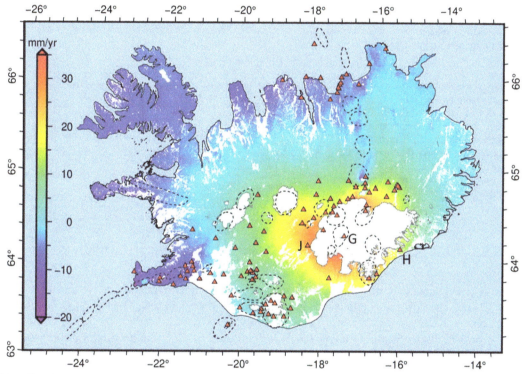

FIG. 15.4 Map of location of continuous GNSS stations (cGNSS) *(red triangles)* and InSAR map of uplift. Estimated average vertical velocities induced by GIA between summer 2015 and summer 2020 in Iceland, derived from Sentinel-1 interferometry, using the procedure described by Drouin and Sigmundsson (2019). Outlines of central volcanoes indicated. Letters indicate cGNSS stations mentioned in text. H=HOFN, J=JOKU, G=GFUM.

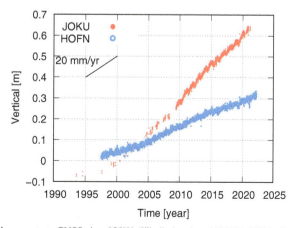

FIG. 15.5 Time series of vertical displacement at cGNSS sites JOKU (Jökulheimar) and HOFN (Höfn). See Fig. 15.4 for location of stations.

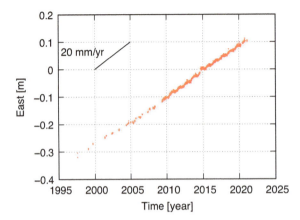

FIG. 15.6 Time series of the difference in east component of displacement of cGNSS stations JOKU and HOFN. See Fig. 15.4 for location of stations. See Fig. 15.5 for vertical displacement of the sites.

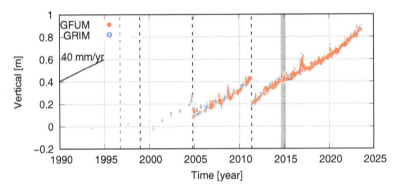

FIG. 15.7 Time series of vertical displacement at cGNSS site GFUM, and nearby campaign site GRIM, at Grímsvötn volcano. See Fig. 15.4 for location of GFUM. Vertical hatch lines represent eruptions at Grímsvötn (and in 1996 at the nearby Gjálp eruption site). The shaded time period in late 2014 and 2015 marks the timing of the Bárðarbunga eruption. The uplift is a combination of GIA and local volcano signals relating to pressure changes in the magmatic system.

The temporal pattern can though not be described as a steady acceleration until the present, rather there are some fluctuations in the uplift rate. The slow initial rate occurs in a time period when there was a pause in the ice loss of the nearby Vatnajökull ice cap.

Although the vertical displacement of station JOKU is dominated by GIA, the horizontal displacement is not. The station is within the plate boundary deformation zone at the boundary of the North American and Eurasian plates. The difference in the east component of the displacement at this site and the Höfn (HOFN) site in SE-Iceland shows mostly the effect of plate spreading (Fig. 15.6). When interpreting regional deformation fields, it is thus evident one has to consider both GIA and plate movements.

An example of a station where volcano deformation is superimposed on GIA and plate movement is shown in Fig. 15.7, at the Grímsvötn volcano. Such superposition of displacement components originating from different sources provides challenges for interpretation.

4 Modeling the effects

4.1 Decomposing observed deformation fields to infer GIA and volcano processes

The observed ground displacement field in a volcanic region influenced by glacial retreat is influenced by at least three processes that can influence to a varying degree: (i) GIA, (ii) plate movements, and (iii) magmatic processes. If the respective velocities at a GNSS site are v_{GIA}, v_{PS}, and $v_{volcano}$, then the observed velocity, $v_{observed}$, can be expressed as.

$$v_{observed} = v_{GIA} + v_{PS} + v_{volcano} + v_{noise} \qquad (15.1)$$

where the last term represents the noise. Here, $v_{volcano}$ can include, for example, the deformation caused by magma movements and viscoelastic response due to previous volcanic processes. The velocity due to plate movements, v_{PS}, may be constant over a study area if it is an intraplate volcano but varies across a study area if it is within a plate boundary deformation zone. v_{noise} is the noise in the observation data as well as unmodeled deformation. In order to evaluate volcano deformation, there is a need to decompose observed deformation into these components. Therefore, as GIA may play an increasing role as glacial retreat continues, then climate change has an influence on how to model volcano processes from GNSS data. Most GIA models are of global nature and do not consider well contributions from retreating ice caps covering volcanoes. Iceland is an exception. Due to low asthenospheric viscosity, the main GIA signal is attributed to present-day retreat of the Icelandic ice caps, which has been mapped. However, experience shows that GIA signals may rapidly change with time, and one has to consider the possibility that observed deformation in a particular time window is not well described by a previously existing GIA model. A preferred approach is to evaluate a GIA model which is most up-to-date for the time period considered, but this may not be feasible. Another approach, less favorable, is to scale a previously existing GIA model to reproduce better observations (e.g., Geirsson et al., 2012). Long-time series of GNSS observations are then particularly valuable for the decomposition of observed deformation fields. An example of this procedure is shown in Fig. 15.8, from a study by Li et al. (2021). In this study, GIA was corrected using a scaled model from Auriac (2014) and the plate spreading correction uses a model by Drouin and Sigmundsson (2019), before analyzing the local volcano deformation.

When modeling the effects of load changes on magmatic systems, there is the need to initially map the crustal deformation field in the area. Once data have been analyzed and time series of change are available, one should consider how plate movements contribute to these time series. Then there is the need to consider a GIA model that in essence captures the role of load changes in producing ground displacement, over a wide area surrounding the volcanic region of interest, so the associated subsurface stress changes can be realistically modeled. Thereafter, one can proceed to model local volcano deformation sources. The selection of appropriate rheology to include is important to derive realistic information on what is happening in volcano interiors; the assumption of elastic behavior may only be appropriate when modeling deformation on short timescales. The shape, depth, and size of magma bodies in volcano roots will have an effect on how ice retreat influences magmatic systems. Determining these is extra challenging in volcanic regions affected by GIA. If GIA signals are not accounted for, this can lead to issues for derived volcano models, for example, on the determination of volume change related to magma inflow. When a new period of inflation begins at a volcano undergoing GIA, it is important to have an estimate of the GIA signal, so the signals due to magmatic processes at depth can be properly determined.

4.2 Influence on magma generation

Modeling has demonstrated that both the increase in magma volume and the chemical signature of the lavas can be explained by decompressional melting, due to the stress changes in the mantle resulting from the reduction in surface loads (Eksinchol et al., 2019; Jull & McKenzie, 1996; Maclennan et al., 2002), but uncertainties remain (Cooper et al., 2020).

To evaluate the effect of pressure release on magma generation, one has to consider how mantle melting takes place under volcanoes in general. A large volume of magma within the Earth is generated by decompression melting, when hot mantle material upwells. The upwelling leads to pressure decrease, and if the mantle is warm enough then magma is generated. For each volcanic region on Earth, there is the need to couple a model of a pressure change inferred from a GIA model with a mantle melting model. The total melting rate in a mantle undergoing both upwelling as well as GIA response can be written as the material derivative of F, the melt fraction by weight:

$$\frac{DF}{Dt} = \left(\frac{\partial F}{\partial P}\right)_S \left(\frac{\partial P}{\partial t} + \overline{\mathbf{V}} \cdot \nabla P\right) \tag{15.2}$$

Here P is the pressure, T is the temperature, $\partial P/\partial t$ is the in situ pressure change due to GIA decompression, and the last term is the pressure change due to upwelling, with $\mathbf{V}$ being the velocity vector of the solid matrix. $(\partial F/\partial P)_S$ is the partial derivative of the degree of melting with respect to pressure at a constant entropy. This equation needs to be evaluated within a melting regime in the mantle (Fig. 15.9). Schmidt et al. (2013) expanded on the work of Pagli and Sigmundsson (2008) and Árnadóttir et al. (2009), by combining the subsurface stress field from a revised three-dimensional (3D) GIA model with a model of mantle melting in the Icelandic mantle. Their study showed that glacially induced pressure changes in the mantle increase melt production rates by 100%–135%, or an additional 0.21–0.23 km^3 of magma per year beneath Iceland.

There is, however, a large uncertainty in if, how and when this magma reaches the surface, because it has to travel from the melting region, spanning the depth from the base of the lithosphere and potentially down to 250 km, all the way up through the lithosphere. The bulk of volatile-free melting under Iceland takes place between about 120 km depth and the base of the crust (Matthews et al., 2016), and the arrival of the generated melt to the surface will be delayed. Studies of Icelandic basalts have indicated melt ascent velocities ranging from 30 m/yr to more than 1 km/yr

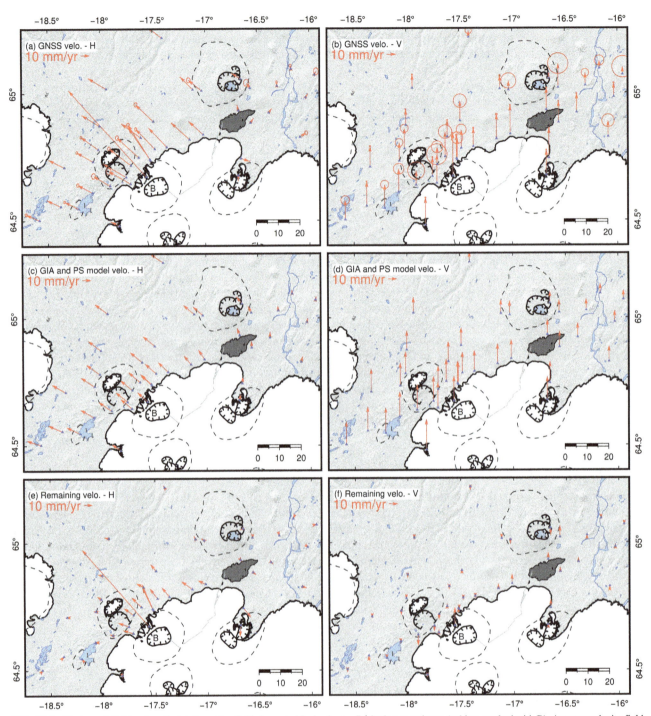

FIG. 15.8 Horizontal *(left)* and vertical deformation fields in central Iceland, near Bárðarbunga volcano (caldera marked with B). Average velocity field from GNSS measurements covering the period 2015–2018, following an eruption and dike injection in 2014–2015. Panels (A) and (B) show inferred average velocity relative to a stable Eurasian plate, with ellipses at the end of the arrows indicating 95% confidence intervals. Panel (C) shows the model horizontal velocity field of combined contributions of scaled GIA model by Auriac (2014) and plate spreading, and (D) is the average vertical GIA velocity. Panels (E) and (F) show residual velocities after subtracting velocities shown in the middle panels from observed velocities that can be used for modeling magmatic processes. Data from Li et al. (2021), who interpreted the data with a series of models, including a model considering the viscoelastic response from both magma withdrawal and associated caldera collapse.

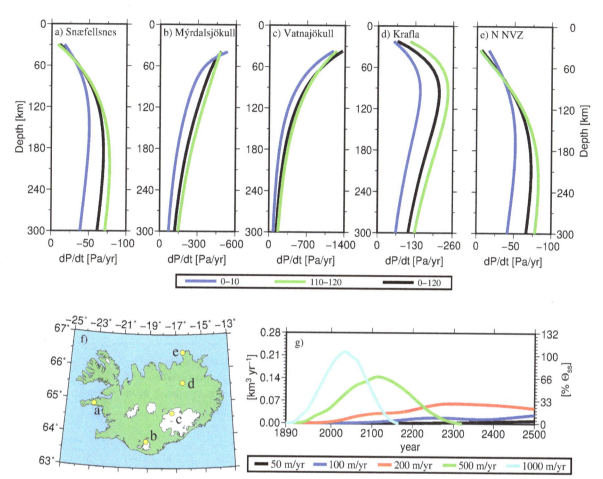

FIG. 15.9 (A)–(E) Predicted glacially induced pressure changes (dP/dt) in the mantle as a function of depth and time at selected locations across Iceland. (A)–(E) The dP/dt as a function of depth averaged over the time periods (model years) 0–10 (blue), 110–120 *(green)*, and 0–120 *(black)*. (F) Map showing the locations of the depth profiles (A)–(E). (G) Increase in melt supply rate, ΔMSR, at the base of the elastic lithosphere due to deglaciation of Iceland between 1890 and 2010, assuming a mantle potential temperature of 1500°C, a bulk water content of 125 ppm, and melt ascent velocities of 50 m/yr *(black)*, 100 m/yr *(blue)*, 200 m/yr *(red)*, 500 m/yr *(green)*, and 1000 m/yr *(cyan)*. Left vertical scale gives the ΔMSR in units of km³/yr; the scale on the right, [% Θ_{ss}], gives it as a percentage of 0.21 km³/yr, the estimated steady state melt production rate under Iceland. Note that all curves are based on a time-varying decompression rate between 1890 and 2010, rather than the mean over the period. *(Reproduced from Schmidt, P., Lund, B., Hieronymus, C., Maclennan, J., Árnadóttir, T., & Pagli, C., 2013. Effects of present-day deglaciation in Iceland on mantle melt production rates. Journal of Geophysical Research: Solid Earth, 118(7), 3366–3379.)*

(e.g., Eksinchol et al., 2019; Maclennan et al., 2002), where the upper range implies additional magma generated by ongoing GIA would constitute an increasingly significant contribution to melt supply of magmatic systems in Iceland at present (Fig. 15.9). However, alternate views have been presented, suggesting a timescale of hundreds of years between perturbations in climate and volcanic activity (Swindles et al., 2018).

4.3 Influence on magma emplacement

In addition to increased magma generation, stress changes associated with ice retreat can also alter the capacity for storing magma within the crust. A magma intrusion in 2007–2008 at Upptyppingar to the north of the Vatnajökull ice cap, aligned almost perpendicular to the zone of plate spreading, and was inferred to be under the influence of ice unloading (Hooper et al., 2011). Using numerical modeling, Hooper et al. (2011) estimated the direction of maximum extensional stress acting on the dike. They also modeled the effect of the ice load decrease since 1890 on the stress field. They concluded that the direction of maximum extensional stress had been rotated by ice mass change from the direction expected from plate spreading alone, leading to enhanced capture of magma within the crust.

While the effect of ice retreat for this off-rift dike was to increase capacity for magma storage, Hooper et al. (2011) also calculated the effect it would have for dikes intruded perpendicular to the direction of plate spreading, such as the latter part of the dike that propagated from Bárðarbunga in 2014. They found that for dikes with this orientation, the capacity for storage is decreased. As dikes intruded in the rift zones are usually oriented in this manner, the overall effect of ice retreat on the capacity for magma storage in the crust is likely to be a decrease, leading to an increase in erupted volumes.

4.4 Influence of ice retreat on the stability of shallow magma bodies

Numerical models, e.g., using the finite element method, are needed to quantify the stress changes caused by the modulation of ice cap loading and to evaluate its effect on the stability of the shallow magma reservoirs (Albino et al., 2010). The aim of such models is to evaluate how external processes modify host rock stresses and consequent effects on the stability of magma bodies, by enhancing or preventing the initiation of an intrusion emanating from a magma body. The model results will also depend on the failure criterion used.

Initially, one can consider an axisymmetrical geometry: the magma body in the crust is modeled as an ellipsoidal cavity with a given overpressure (ΔP_m) and the glacier changes are modeled as an unloading disk applied at the surface (ΔP_s). One thus needs an estimate of the glacial retreat, as well as an indication of the shape, size, and depth of the magma body (Fig. 15.10). To undertake modeling of such processes, there is the need to furthermore assume a rheology for the model domain (the host rock), as well as a failure criteria. The simplest models assume the host rock medium is elastic and homogeneous. Numerical models of this type show that a surface unloading causes: (i) a decrease of the overpressure within a magma body, with the amplitude depending on the shape of the reservoir and the compressibility of the magma and (ii) a change (increase or decrease) of the failure pressure (ΔP_f), the pressure required to initiate an intrusion; the sign of the change depends on the shape of the reservoir as well as the spatial distribution of the unload.

A model as described above has been applied to the subglacial Katla volcano, located under the Mýrdalsjökull ice cap in S-Iceland. The most recent confirmed eruption at Katla breaking the ice cover, with explosive activity and tephra fall, occurred in 1918. An ice unloading model for the area, spanning a 5-years period between 1999 and 2004, has both a seasonal surface snow load change between summer and winter with a maximum amplitude of up to 6 m at the center of the

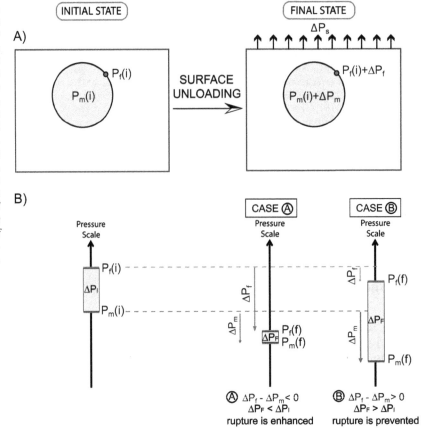

FIG. 15.10 (A) Evolution of magma pressure, P_m, and threshold pressure for failure, P_f, during an unloading event. (B) Evolution of the ability of the system to erupt. ΔP_I and ΔP_F represent, respectively, the difference between magma pressure and threshold pressure for failure before and after an unloading event. The final state depends on the initial state and the change in P_m and P_f: $\Delta P_F = \Delta P_I + (\Delta P_f - \Delta P_m)$. In the case $\Delta P_F < \Delta P_I$ (case A), rupture likelihood is enhanced and may occur or not depending on the initial state. In the case $\Delta P_F > \Delta P_I$ (case B) no eruption will occur. *(Modified from Sigmundsson, F., Pinel, V., Lund, B., Albino, F., Pagli, C., Geirsson, H., & Sturkell, E. (2010). Climate effects on volcanism: Influence on magmatic systems of loading and unloading from ice mass variations with examples from Iceland, special issue on climate forcing of geological and geomorphological hazards. Philosophical Transactions of the Royal Society A, 368, 1–16.)*

Mýrdalsjökull ice cap superposed on a long-term thinning of −4m along the edges of the ice cap during the period 1999–2004 (Pinel et al., 2007). These surface pressure changes correspond to stress changes of 40–60 kPa which is the same order of magnitude as static stress changes observed for earthquake triggering (King et al., 1994).

In addition to the ice loading model, a model for the magmatic system is required. For a shallow magma body under Katla, Albino et al. (2010) model a radially symmetric molten magma body of oblate sill-shaped geometry (half-width = 2.5 km and half-height = 0.5 km) with the center located at 3 km below the surface based on the results of two-dimensional (2D) seismic surveys (Gudmundsson et al., 1994). Using model set up like this, Albino et al. (2010) found that the seasonal unloading at Mýrdalsjökull between winter and summer induces a decrease in magma pressure ΔP_m of 30 kPa (considering a magma bulk modulus of 10 GPa) and a decrease in failure pressure ΔP_f of 45 kPa. Because $|\Delta P_m| < |\Delta P_f|$, the magma body moves closer to failure conditions by an amount of 15 kPa during unloading periods (spring–summer). If the failure criteria for the onset of magma movements relates, e.g., to tensile strength of 1–10 MPa (Haimson & Rummel, 1982), then such modulation of stress can provide an explanation for preferred eruptions of Katla during the months from May to October when the seasonal load is reducing (from the load maximum in early summer to load minimum in late fall), as observed in historical records (Eliasson et al., 2006). Thus, provided that the volcano is already very close to failure conditions, the stress modulation resulting from seasonal unloading may become significant in terms of eruption triggering.

For the long-term thinning of the Mýrdalsjökull ice cap, the models of Albino et al. (2010) show a decrease in magma pressure ΔP_m of 2.4 kPa/yr and a decrease in failure pressure $\Delta P_f < 1$ kPa. In that scenario in which $|\Delta P_m| > |\Delta P_f|$, the reservoir moves away from failure by an amount of 1.4 kPa/yr. Thus, for this particular example, the influence of long-term thinning appears limited. However, the host rock medium does not behave purely elastic beneath Iceland due to the presence of the hot spot and the associated intense volcanic activity. If a Maxwell viscoelastic layer below 10 km depth is assumed (Fig. 15.2), the results are very different. Modeling presented by Sigmundsson et al. (2013) shows that in such a case the magma body is moving away from failure by an amount of 8.3 kPa/yr which corresponds to about ∼1 MPa over 120 years (Sigmundsson et al., 2013). This result shows that the viscoelastic response has a large influence on the long-term evolution of volcanic systems even for shallow reservoirs located in the brittle upper crust. Although the conditions to initiate an intrusion will be more difficult through time, it implies that more magma can accumulate in the reservoir prior to an eruption resulting potentially in large erupted volumes.

5 Discussion

The knowledge of the state of stress surrounding magmatic systems and more importantly its evolution is key information to better understanding the influence of climate change on subsurface magmatic processes and triggering mechanism of an eruption. Commonly, an increase in magma pressure is the driving mechanism to initiate and propagate magma intrusions toward the surface from a shallow magma body. Magma intrusions initiate when magma overpressure exceeds a mechanical threshold. The conditions of failure of magma bodies depend on the state of stress surrounding the reservoir and the tensile strength of the host rocks. As a result, external processes that modify the host rock stresses potentially modify the stability of magma reservoirs by enhancing or preventing the initiation of an intrusion. In Iceland, many of the active volcanoes are covered by glaciers, therefore, the interactions between glaciers and volcanism are frequent and they have been extensively studied (e.g., Hooper et al., 2011; Pagli & Sigmundsson, 2008; Sigmundsson et al., 2010). Lessons learned in Iceland can provide guidance for evaluation of the effects of climate change on magmatic processes in other parts of the world.

6 Conclusions

GNSS observations are critical to measure GIA signals in volcanic regions and for the development of GIA models. Based on GIA models, stress and pressure changes in the crust and the mantle resulting from changes in ice loading due to climate change can be estimated. These form the basis for evaluation of increased melt generation due to climate change, as well as for estimates of eventual changes in conditions of subsurface magma emplacement and triggering effects on magmatic systems. Uncertainties in the estimates are large, and a detailed model of a magmatic system needs to be incorporated when evaluating the effects. Stress changes within and surrounding a shallow magma body, as a result of ice retreat due to climate change, may bring a magma body either closer or further from failure depending on the compressibility of the magma and the shape, size, and location of it.

Acknowledgment

This work was supported by the ISVOLC project funded by the Icelandic Science Fund (project number 239615-051) and the University of Iceland Research Fund.

References

Aðalgeirsdóttir, G., Jóhannesson, T., Björnsson, H., Pálsson, F., & Sigurðsson, O. (2006). The response of Hofsjökull and southern Vatnajökull, Iceland, to climate change. *Journal of Geophysical Research, 111*, F03001. https://doi.org/10.1029/2005JF000388.

Aðalgeirsdóttir, G., Magnússon, E., Pálsson, F., Thorsteinsson, T., Belart, J., Jóhannesson, T., et al. (2020). Glacier changes in Iceland from ~1890 to 2019. *Frontiers in Earth Science, 8*, 520.

Albino, F., Pinel, V., & Sigmundsson, F. (2010). Influence of surface load variations on eruption likelihood: Application to two Icelandic subglacial volcanoes. *Geophysical Journal International, 181*, 1510–1524. https://doi.org/10.1111/j.1365-246X.2010.04603.x.

Árnadóttir, T., Lund, B., Jiang, W., Geirsson, H., Björnsson, H., Einarsson, P., et al. (2009). Glacial rebound and plate spreading: Results from the first countrywide GPS observations in Iceland. *Geophysical Journal International, 177*(2), 691–716.

Aubry, T. J., Farquharson, J. I., Rowell, C. R., Watt, S. F. L., Pinel, V., Beckett, F., et al. (2022). Impact of climate change on volcanic processes: Current understanding and future challenges. *Bulletin of Volcanology, 84*, 58.

Auriac, A. (2014). *Solid earth response to ice retreat and glacial surges in Iceland inferred from satellite radar interferometry and finite element modelling*. Ph.D. thesis Faculty of Earth Sciences, University of Iceland.

Belart, J., Magnússon, E., Berthier, E., Gunnlaugsson, Á.Þ., Pálsson, F., Aðalgeirsdóttir, G., et al. (2020). Mass balance of 14 Icelandic glaciers, 1945–2017: Spatial variations and links with climate. *Frontiers in Earth Science, 8*, 163.

Björnsson, H., & Pálsson, F. (2008). Icelandic glaciers. *Jökull, 58*, 365–386.

Björnsson, H., Palsson, F., Gudmundsson, S., Magnusson, E., Adalgeirsdottir, G., Johanneson, T., et al. (2013). Contribution of Icelandic ice caps to sea level rise: Trends and variability since the little ice age. *Geophysical Research Letters, 40*, 1546–1550. https://doi.org/10.1002/grl.50278.

Compton, K., Bennett, R. A., & Hreinsdóttir, S. (2015). Climate-driven vertical acceleration of Icelandic crust measured by continuous GPS geodesy. *Geophysical Research Letters, 42*(3), 743–750.

Cooper, C. L., Savov, I. P., Patton, H., Hubbard, A., Ivanovic, R. F., Carrivick, J. L., et al. (2020). Is there a climatic control on Icelandic volcanism? *Quaternary Science Advances, 1*, 100004.

Crowley, J. W., Katz, R. F., Huybers, P., Langmuir, C. H., & Park, S.-H. (2015). Glacial cycles drive variations in the production of oceanic crust. *Science, 347*, 1237–1240.

Drouin, V., & Sigmundsson, F. (2019). Countrywide observations of plate spreading and glacial isostatic adjustment in Iceland inferred by sentinel-1 radar interferometry, 2015–2018. *Geophysical Research Letters, 46*, 8046–8055. https://doi.org/10.1029/2019GL082629.

Eksinchol, I., Rudge, J. F., & Maclennan, J. (2019). Rate of melt ascent beneath Iceland from the magmatic response to deglaciation. *Geochemistry, Geophysics, Geosystems*. https://doi.org/10.1029/2019GC008222.

Eliasson, J., Larsen, G., Tumi Gudmundsson, M., & Sigmundsson, F. (2006). Probabilistic model for eruptions and associated flood events in the Katla caldera, Iceland. *Computational Geosciences, 10*, 179–200.

Fox-Kemper, B., Hewitt, H. T., Xiao, C., Aðalgeirsdóttir, G., et al. (2021). Ocean, cryosphere and sea level change. In V. P. Masson-Delmotte, et al. (Eds.), *Climate change 2021: The physical science basis. Contribution of working group I to the sixth assessment report of the intergovernmental panel on climate change* (pp. 1211–1362). Cambridge, United Kingdom and New York, NY, USA: Cambridge University Press. https://doi.org/10.1017/9781009157896.011.

Geirsson, H., LaFemina, P., Arnadottir, T., Sturkell, E., Sigmundsson, F., Travis, M., et al. (2012). Volcano deformation at active plate boundaries: Deep magma accumulation at Hekla volcano and plate boundary deformation in South Iceland. *Journal of Geophysical Research: Solid Earth, 117*(B11).

Glazner, A. F., Manley, C. R., Marron, J. S., & Rojstaczer, S. (1999). Fire or ice: Anticorrelation of volcanism and glaciation in California over the past 800,000 years. *Geophysical Research Letters, 26*, 1759–1762.

Gudmundsson, O., Brandsdóttir, B., Menke, W., & Sigvaldason, G. E. (1994). The crustal magma chamber of the Katla volcano in South Iceland revealed by 2-D seismic undershooting. *Geophysical Journal International, 119*(1), 277–296.

Haimson, B., & Rummel, F. (1982). Hydrofracturing stress measurements in the Iceland research drilling project drill hole at Reydarfjordur, Iceland. *The Journal of Geophysical Research, 87*, 6631–6649.

Hammond, W. C., Kreemer, C., Zaliapin, I., & Blewitt, G. (2019). Drought-triggered magmatic inflation, crustal strain, and seismicity near the Long Valley Caldera, Central Walker Lane. *Journal of Geophysical Research: Solid Earth, 124*, 6072–6091. https://doi.org/10.1029/2019JB017354.

Hooper, A., Ófeigsson, B., Sigmundsson, F., Lund, B., Einarsson, P., Geirsson, H., et al. (2011). Increased capture of magma in the crust promoted by ice-cap retreat in Iceland. *Nature Geoscience, 4*(11), 783–786.

Hugonnet, R., McNabb, R., Berthier, E., Menounos, B., Nuth, C., Girod, L., et al. (2021). Accelerated global glacier mass loss in the early twenty-first century. *Nature, 592*(7856), 726–731.

IPCC. (2021). Summary for policymakers. In V. Masson-Delmotte, P. Zhai, A. Pirani, et al. (Eds.), *Climate change 2021: The physical science basis. Contribution of working group I to the sixth assessment report of the intergovernmental panel on climate change* Cambridge University Press.

Jellinek, A. M., Manga, M., & Saar, M. O. (2004). Did melting glaciers cause volcanic eruptions in eastern California? Probing the mechanics of dike formation. *Journal of Geophysical Research, 109*, B09206. https://doi.org/10.1029/2004JB002978.

Jull, M., & McKenzie, D. (1996). The effect of deglaciation on mantle melting beneath Iceland. *Journal of Geophysical Research: Solid Earth, 101*(B10), 21815–21828.

King, G. C., Stein, R. S., & Lin, J. (1994). Static stress changes and the triggering of earthquakes. *Bulletin of the Seismological Society of America, 84*(3), 935–953.

Kutterolf, S., Schindlbeck, J. C., Jegen, M., Freundt, A., & Straub, S. M. (2019). Milankovitch frequencies in tephra records at volcanic arcs: The relation of kyr-scale cyclic variations in volcanism to global climate changes. *Quaternary Science Reviews, 204*, 1–16.

Li, S., Sigmundsson, F., Drouin, V., Parks, M. M., Ofeigsson, B. G., Jónsdóttir, K., et al. (2021). Ground deformation after a caldera collapse: Contributions of magma inflow and viscoelastic response to the 2015–2018 deformation field around Bárðarbunga, Iceland. *Journal of Geophysical Research - Solid Earth, 126*, e2020JB020157. https://doi.org/10.1029/2020JB020157.

Maclennan, J., Jull, M., McKenzie, D., Slater, L., & Grönvold, K. (2002). The link between volcanism and deglaciation in Iceland. *Geochemistry, Geophysics, Geosystems, 3*(11), 1–25.

Matthews, S., Shorttle, O., & Maclennan, J. (2016). The temperature of the Icelandic mantle from olivine-spinel aluminum exchange thermometry. *Geochemistry, Geophysics, Geosystems, 17*(11), 4725–4752.

McGuire, W., Howarth, R., Firth, C., et al. (1997). Correlation between rate of sea-level change and frequency of explosive volcanism in the Mediterranean. *Nature, 389*, 473–476.

Pagli, C., & Sigmundsson, F. (2008). Will present day glacier retreat increase volcanic activity? Stress induced by recent glacier retreat and its effect on magmatism at the Vatnajökull ice cap, Iceland. *Geophysical Research Letters, 35*(9).

Pinel, V., Sigmundsson, F., Sturkell, E., Geirsson, H., Einarsson, P., Gudmundsson, M. T., et al. (2007). Discriminating volcano deformation due to magma movements and variable surface loads: Application to Katla subglacial volcano, Iceland. *Geophysical Journal International, 169*(1), 325–338.

Schmidt, L. S., Aðalgeirsdóttir, G., Pálsson, F., Langen, P. L., Guðmundsson, S., & Björnsson, H. (2019). Dynamic simulations of Vatnajökull ice cap from 1980 to 2300. *Journal of Glaciology, 66*(255), 97–112.

Schmidt, P., Lund, B., Hieronymus, C., Maclennan, J., Árnadóttir, T., & Pagli, C. (2013). Effects of present-day deglaciation in Iceland on mantle melt production rates. *Journal of Geophysical Research: Solid Earth, 118*(7), 3366–3379.

Sigmundsson, F. (1991). Post-glacial rebound and asthenosphere viscosity in Iceland. *Geophysical Research Letters, 18*(6), 1131–1134.

Sigmundsson, F., Albino, F., Schmidt, P., Lund, B., Pinel, V., Hooper, A., et al. (2013). Multiple effects of ice load changes and associated stress change on magmatic systems. In B. McGuire, & M. A. Maslin (Eds.), *Climate forcing of geological hazards* (pp. 108–123). Wiley-Blackwell.

Sigmundsson, F., Einarsson, P., Hjartardóttir, Á. R., Drouin, V., Jónsdóttir, K., Árnadóttir, T., et al. (2020). Geodynamics of Iceland and the signatures of plate spreading. *Journal of Volcanology and Geothermal Research, 391*, 106436.

Sigmundsson, F., Pinel, V., Lund, B., Albino, F., Pagli, C., Geirsson, H., et al. (2010). Climate effects on volcanism: Influence on magmatic systems of loading and unloading from ice mass variations with examples from Iceland, special issue on climate forcing of geological and geomorphological hazards. *Philosophical Transactions of the Royal Society A, 368*, 1–16.

Sigvaldason, G. E., Annertz, K., & Nilsson, M. (1992). Effect of glacier loading/deloading on volcanism: Postglacial volcanic production rate of the Dyngjufjöll area, Central Iceland. *Bulletin of Volcanology, 54*, 385–392.

Slater, L., Jull, M., McKenzie, D., & Gronvöld, K. (1998). Deglaciation effects on mantle melting under Iceland: Results from the northern volcanic zone. *Earth and Planetary Science Letters, 164*(1), 151–164.

Swindles, G. T., Watson, E. J., Savov, I. P., Lawson, I. T., Schmidt, A., Hooper, A., et al. (2018). Climatic control on Icelandic volcanic activity during the mid-Holocene. *Geology, 46*(1), 47–50.

Zemp, M., Huss, M., Thibert, E., Eckert, N., McNabb, R., Huber, J., et al. (2019). Global glacier mass changes and their contributions to sea-level rise from 1961 to 2016. *Nature, 568*(7752), 382.

Index

Note: Page numbers followed by *f* indicate figures and *t* indicate tables.

A

Accelerometers, 72–73
Acoustic wave (AW), 163–165, 163–164f
ActiveMQ exchange, 122–123
Afterslip, 99–101
Airborne Topographic Mapper (ATM) data, 219
Akaike Bayesian information criterion (ABIC), 71
Åknes landslide, Norway, 248, 248f
Alaska subduction zone, 89t
    deep slow slip events, 85–88, 86f
    shallow SSEs, 90
Almenningar landslide, Iceland, 248–249, 249f
Alpine fault system, 266–267
Altimetry, 215
Amplitude metrics
    challenges, 210, 210f
    signal-to-noise ratio (SNR), 207
    surface reflectivity (SR), 208
Anatolian fault system, 266–267
ANSS/Comcat catalog, 51, 51f
Antarctica, 220, 228
    G/GFO mass change observations, 215
    glacial isostatic adjustment (GIA) models, 229–230
    McMurdo Research Station, 215
    Southern Antarctic Peninsula, 217–218
    surface mass balance (SMB) model, 218–219
Antenna reference point (ARP), 17
Arctic (Thule Airbase), 215
Arctic oscillation (AO), 197, 198f
Asama volcanic eruption, 161, 168, 171–172
Atmosphere sounding, 26
    GNSS radio occultation (GNSS-RO)
        advantages of, 32
        data processing, 31
        physical principle of, 31, 31f
    ground-based ionosphere sounding
        GNSS ionosphere delay, 28–29, 29f
        slant ionosphere delays and their spatiotemporal variations, 30
        vertical ionosphere delays and global ionosphere map, 29–30
    ground-based troposphere sounding
        GNSS troposphere tomography, 27–28
        GNSS tropospheric delay, 26–27
        water vapor content, 27
Atmospheric pressure changes, 272
AutoRegressive Moving Average Model ARMA(1,1) model, 198, 200

B

Bam earthquake, 101–102
Bárðarbunga volcano, 148, 292f, 294f
Baseline Inferences for Fennoscandian Rebound Observations, Sea-level, and Tectonics (BIFROST) project, 224
Bayesian approach, 71–72
Bayesian framework, 119
Bayesian information criterion (BIC), 200
BEFORES algorithm, 118–119, 122
BeiDou Navigation Satellite System (BDS), 14, 20–21, 21f, 163
Blackman-Tukey method, 169, 170f
Block models, 42
Bootstrap resampling analysis, 114
Burridge-Knopoff models, 275

C

Calbuco (Chile) eruptions, 169–171, 170f
Cà Lita landslide, Italy, 250–251, 251f
Campi Flegrei caldera, 142–143, 143f
Carrier-phase-based relative positioning, 22–24
Carrier-phase measurement, 12
Cascadia subduction zone, 83–84, 89t, 266–267
    deep SSEs and interseismic coupling, 85–90, 86f
    shallow SSEs, 90
Center of mass (CoM), 15–16, 16f
Central limit theorem, 55–56
Centroid moment tensor (CMT), 119, 121f
Chi-Chi earthquake, 90–91
Clausius-Clapeyron equation, 197–198
Climate change and GNSS
    earthquakes and volcanism, 6–7
    glacier motions, measurements of, 6
    long- and short-term responses, 5
Climate-driven deformation and seismicity, 276–277
    changing climate and weather-driven deformation and seismicity
        earthquake weather, 272–274, 273f
        ice age cycles, 264–267, 265f, 267f
        recent climate change, consequences of, 267–269, 270f
        seasonal hydrological and atmospheric loads, 269–271
    earthquake triggering and modulation, 263, 264f
    fault-frictional properties and stress from climate forcing, 275, 276f
    measuring climate-driven deformation, 258–261, 258f, 260f, 262f
    probing Earth's constitutive properties
        mantle rheology, 274, 274f
        shallow subsurface, 275
        surface loading, changes in, 274–275
    and stress
        absolute ambient stress, 257
        modeling, 261–263, 262f
        surface loading and poro- and thermoelastic eigenstrain, 257, 258f
        surface processes and associated stress variations, 257, 259f
        tectonic stressing rates, 257
        tectonics and crustal deformation, 257
Climate feedback effect, 189
Coherency metrics
    challenges, 210, 210f
    continuous model functions, 209
    entropy, 208
    power ratio (PR), 209
    statistical algorithms, 209
    thresholding, 209
    trailing edge slope, 209
Continuous GNSS (cGNSS), 85–88, 291–292, 291f
Continuously operating reference stations (CORS), 18, 246
Convolutional neural networks (CNN), 111
Corner moment, 44, 54
Coseismic-based methods, for EEW
    BEFORES algorithm, 118–119
    centroid moment tensor (CMT), GlobalCMT solution, 119, 121f
    estimated offsets and UNR offsets, variance reduction and azimuthal variation between, 118–119, 120f
    GlarmS, 119
    Green's functions, 119–122
    kinematic slip inversions, 119–122

Coseismic-based methods, for EEW *(Continued)*
  pre-event average window/single point, 118
  short-term average/long-term average (STA/LTA), 118–119
  slip models, 119, 121*f*
  static coseismic offsets, 118
Coseismic slip models, 77–78
COSMIC-2 program, 32
Costa Rica, 89*t*
  deep slow slip events, 88
  shallow slow slip events, 87*f*, 90–91
Cost function, 70–71
Coulomb failure criterion, 263
Coulomb failure model (CFM), 276*f*
Coulomb failure stress, 275
Coulomb stress change, 262*f*, 263, 265*f*, 267–271, 268*f*, 270*f*
Creep events
  in circum-Pacific region, 83, 84*f*
  episodic slow slip and creep events *(see* Episodic slow slip and creep events*)*
Creeping landslides. *See* Slow moving landslides
Cross-validation, 71
Crustal deformation, in plate boundary zones. *See* Transient deformation, in tectonic plate boundary zones
Crustal strain accumulation, in seismic potential evaluation
  characteristic earthquake model, 53
  data, 47–48, 47–48*f*
  geodetic potency *vs.* earthquake numbers approach, 47
    discrete plots, 47
    linear relationship, 46–47
  geodetic strain and moment rates, estimation of
    from strain rate to moment rate, 43
    from velocities to strain rates, 42–43
  global strain rate model (GSRM), 47, 51, 51*f*, 53
  magnitude-frequency relationship, 42
  Mediterranean-Asia area and western United States, 50–53, 51–53*f*
  probabilistic seismic hazard assessment (PSHA), 42
  results, 48–50, 49–50*f*, 49*t*
  seismic moment distribution, 43–45
    corner moment, 44
    cumulative seismic moment distribution, 54–58
    Gutenberg-Richter law, 43–44
    long-term regional/local moment, ratio of, 44–45
    long-term seismic moment, 44
    maximum likelihood estimation, 44
    Pareto index, 44
    tapered Pareto distribution, 44
    upper incomplete gamma function, 44
  seismic-to-geodetic moment ratio, 42
    approach, 46
    continental plate boundaries, 45
    definition, 45
  mid-oceanic ridges, analyses of, 45
  subduction zones, 45
  uncertainties, 45
Cryosphere, GNSS and
  elastic surface displacements
    geodynamic processes, in cryospheric regions, 220
    glacier dynamics, 219–220
    half-space loading models, 219
    theory, 217–219, 217–218*f*
  GNSS interferometric reflectometry (GNSS-IR), 231, 235
    ice mass balance, 233, 233*f*
    permafrost areas, freeze and thaw movements in, 233–234*f*, 234
    principles and methodology, 231–232, 232*f*
    snow depth, 232–233, 233*f*
  MIDAS-determined velocities, 216
  permanent GNSS stations, 215
  Polar Earth Observing Network (POLENET), 215
  residual uplift time series, from permanent stations, 216, 216*f*
  seasonal signals, 216
  transient displacements using SSA, 216, 216*f*
  viscoelastic response of Earth to cryospheric change *(see* Glacial isostatic adjustment (GIA), viscoelastic response of Earth to cryospheric change*)*
CubeSat, 7, 210
Cyclone Global Navigation Satellite System (CYGNSS), 7
  delay-Doppler maps (DDMs), 207, 207*f*
    amplitude metrics, 207–208
    coherency metrics, 208–210
  hurricane wind speed retrieval, 205, 210–212, 211–212*f*
  mission of, 205
  surface reflectivity, observations of, 205, 206*f*
Cyclones, 272, 274

D

Damping, 70
Deep-ocean Assessment and Reporting of Tsunamis (DART), 67, 111
Deep-seated gravitational slope deformation (DSGSD), 244
DeepShake, 111
Deep slow slip events, 85–90, 86*f*, 88*f*, 89*t*
Delaunay triangulation scheme, 113
Delay-Doppler maps (DDMs), 207, 207*f*
  amplitude metrics, 207
    signal-to-noise ratio (SNR), 207
    surface reflectivity (SR), 208
  coherency metrics, 208–210
Denali earthquake, in Alaska, 73, 163–164
Detrended SNR (dSNR), 184–185, 185*f*
Differential code biases (DCBs), 162
Differential code phase biases (DCPBs), 21–22
Differential positioning. *See* Relative positioning
Digital elevation maps (DEMs), 138
Direct acoustic waves from epicenters, 163, 163*f*, 165–166, 165–167*f*
Discrete Laplacian operator, 70
Discrete wave number approach, 73–74
Doppler Orbitography and Radiopositioning integrated by satellite (DORIS), 198–199, 201
Doppler sounding, 161
Double-difference analysis, 183
Durres, Albania, earthquake, 117
Dynamic atmosphere corrections (DAC), 186

E

Earth-centered, earth-fixed (ECEF) reference frame, 12, 16
Earthquake early warning (EEW), 123–124
  algorithm development, 122–123
  convolutional neural networks (CNN), 111
  coseismic-based methods
    BEFORES algorithm, 118–119
    centroid moment tensor (CMT), GlobalCMT solution, 119, 121*f*
    estimated offsets and UNR offsets, variance reduction and azimuthal variation between, 118–119, 120*f*
    GlarmS, 119
    Green's functions, 119–122
    kinematic slip inversions, 119–122
    pre-event average window/single point, 118
    short-term average/long-term average (STA/LTA), 118–119
    slip models, 119, 121*f*
    static coseismic offsets, 118
  DeepShake, 111
  design elements, generalized flowchart for, 111–112, 112*f*
  distributed acoustic sensing (DAS) fiber optic cables, 111–112, 112*f*
  inertial sensors, 112
  Japan Meteorological Agency (JMA) early warning system, 112
  Mexican Seismic Alert System (SASMEX), 111
  peak ground displacements (PGD) scaling
    Chilean megathrust events, 117
    coefficients, 114, 115*t*
    distance weighting, 115–116
    Durres, Albania, earthquake, 117
    exponential distance weighting, 115
    Greece, regional coefficients, 117
    ground motion prediction equation (GMPE), 114
    Iniskin earthquake, 117
    Maduo, China, earthquake, 117
    magnitude-dependent attenuation, 114
    mining-related earthquake, 117
    Napa event, regression analysis, 117
    Network of the Americas (NOTA), earthquakes in, 117
    Pedernales, Ecuador, earthquake, 117, 118*f*
    PGD dataset, log residual between, 115–116, 116*f*
    principle of, 114

Index

signal-to-noise ratio criterion, 115–116
single-magnitude estimate, methods for, 115
three-component displacement waveforms, 114
travel time mask approach, 114–115
weak determinism, 114–115
real-time GNSS, from historical perspective
accuracy and noise characteristics, 114
PPP with ambiguity resolution (PPP-AR), 113
precise point positioning (PPP), 112–113
relative positioning, 112–113
RTD software, 113
triangulated relative positioning method, 113
recurrent neural networks (RNN), 111
seismic algorithms, 112
ShakeAlert system, 111
Earthquakes, 78
afterslip, geodetic evidence for, 83, 85f
climate change, triggered by, 6–7
coseismic static displacement, observation of
aerial/satellite optical images, subpixel correlation of, 67
Deep-ocean Assessment and Reporting of Tsunamis (DART), 67
differential air-bone Lidar topographic data, 67
earthquake-induced vertical displacements, estimates of, 67
GNSS-A technique, 67
interferometric synthetic aperture radar (InSAR), 67
ocean bottom pressure gauges (OBPG), 67
crustal deformation, in plate boundary zones (see Transient deformation, in tectonic plate boundary zones)
definition, 42
down-dip slip rate, 42
dynamic displacement
accelerometers, 72–73
broadband seismometers, 72–73
GNSS and accelerometric data, joint use of, 73
high-rate GNSS data, 73
1 sample-per-second GNSS dynamic displacement, 72, 72f
sidereal filtering, 72
single-epoch measurements, for Mw 7.1 Hector Mine earthquake, 73
elastic strain rate, 42
forecasts, 42
"geodetic" fault slip rates, 42
geodetic potency vs. earthquake numbers approach, 47
discrete plots, 47
linear relationship, 46–47
geodetic strain and moment rates, estimation of
from strain rate to moment rate, 43
from velocities to strain rates, 42–43
glacially triggered earthquakes, 266
GNSS measurements, 2–3

GNSS static coseismic displacement, for imaging slip distribution
Akaike criterion and ABIC, 71
Bayesian approaches, 71–72
cost function, 70–71
covariance matrix, 71
cross-validation, 71
discrete Laplacian operator, 70
hyperparameter, 71
imaging fault coseismic slip, 68
L-curve, 71
likelihood function, 68
linear system, 70–71
megathrust earthquakes, slip for, 71
Metropolis algorithm, 71–72
Monte Carlo Markov chain (MCMC), 71–72
probability density function (pdf), 71–72
regularization constraints, 70–71
regularization weight, 71
slip inversion, forward model for, 68, 70f
smoothing constraints, 70
spatially variable regularization schemes, 71
static Green's function, 68
surface displacements, 68
3D coseismic displacements, examples of, 68, 69f
variance-covariance matrix, 68
high-rate GNSS, as small aperture seismic array, 76–77, 76f
ionospheric disturbance (see Ionospheric disturbance, by earthquakes and volcanic eruptions)
kinematic slip inversion, 65
Ecuador Mw 7.8 2016 earthquake, 74–76, 75f
Maule (Chile) Mw 8.8 2010 earthquake, 74–76, 75f
multi-time window approach, 73–74, 74f
onset times/rupture propagation velocity, 74
time and frequency approaches, 74
large earthquakes, 65
magnitude-frequency relationship, 42
probabilistic seismic hazard assessment (PSHA), 42
seismic moment distribution, 43–45, 54–58
seismic-to-geodetic moment ratio, 42
approach, 46
continental plate boundaries, 45
definition, 45
mid-oceanic ridges, analyses of, 45
subduction zones, 45
uncertainties, 45
seismic wavefield, 65
shear strain rate, 42
static displacements, from GNSS time series, 66–67, 66f
static offset/coseismic displacement, 65
teleseismic data for, 65
triggering and modulation, 263, 264f
weather, 257, 272–274, 273f

Earth rotation effect, 12, 16
East Atlantic (EA) index, 197
Ecuador, 89t
Mw 7.8 2016 earthquake, 72, 72f, 74–77
shallow slow slip events, 87f, 90–91
EEW. See Earthquake early warning (EEW)
Eikonal equation, 31
Elastic Earth model, 217
Elastic strain rate, 42
Elastic surface displacements
geodynamic processes, in cryospheric regions, 220
glacier dynamics, 219–220
half-space loading models, 219
theory, 217–219, 217–218f
Electronic distance measurement (EDM) surveying, 1
El Mayor-Cucapah earthquake, 95–96, 96f
El Nino/La Nina cycles, 269
El Niño-Southern Oscillation (ENSO), 197
El Yunque forest, Puerto Rico, 249–250, 250f
Entropy, 208
EPIC algorithm, 111, 122–123
Episodic slow slip and creep events
faults, slow slip events on
North Anatolian fault, in Turkey, 91
offshore faults flanking Mount Etna volcano, in Italy, 91
San Andreas faults, 91
shallow-dipping reverse faults, 91, 92f
Superstition Hills fault, 91, 92f
slow slip events and earthquakes, interplay between, 91–94, 93f
subduction zones
continuous GNSS (cGNSS) networks, 85–88
deep slow slip events, 85–90, 86f, 88f, 89t
shallow slow slip events, 85–88, 87f, 89t, 90–91
tectonic tremor, 83–84
ubiquity of slow slip events, 94
Episodic tremor and slip (ETS) episodes, 85–88
Error sources, GNSS, 15f
receiver-related errors
receiver antenna PCO and PCV, 17, 17f
receiver clock offset, 17
receiver hardware delay, 18
receiver observation noise, 18
reference frame misalignment, 18
satellite-related errors
Earth rotation effect (Sagnac effect), 16
phase wind-up effect, 16
relativistic effects, 16
satellite antenna PCO/PCV, 15–16, 16f
satellite clock offset, 15
satellite hardware delay, 16
satellite orbit error, 15
signal propagation-related errors
ionospheric delay, 17
multipath effects, 17
tropospheric delay, 17
tide displacement, 18
Etna volcano, 144–146, 150, 152

European Centre for Medium-Range Weather Forecasts (ECMWF) atmospheric reanalysis, 192
Extensometers, 252

F

Fairweather fault earthquake, 268
Fastlane software, 114
Fault friction coefficient, 263
Fennoscandia, 265–266, 265f
FinDer algorithm, 111, 122–123
Finite element method, 137–140, 296
First-order autoregressive (AR1) model, 198, 200
Flex power events, 208
Flood inundation mapping
  GNSS-reflectometry (GNSS-R) for, 205
    amplitude metrics, 207–208
    challenges, 210, 210f
    coherency metrics, 208–210
    delay-Doppler maps (DDMs), 207, 207f
    instruments, 205
    sensitivity of surface-reflected signals, reasons for, 206–207
    surface reflectivity, CYGNSS observations of, 205, 206f
  microwave sensors, 205
  MODIS/Landsat, 205
Flux gate approach, 219–220
Frequency division multiple access (FDMA) technique, 18
Fresnel reflection coefficient, 210–211
Fukutoku-Okanoba (Japan) eruptions, 169–171, 170–171f
Funu landslide, in Democratic Republic of the Congo, 243, 244f

G

Galileo, 1, 20–21, 21f, 163
Generalized central limit theorem, 55–56
"Geodetic" fault slip rates, 42
Geodetic phase-delay analysis, 181–182
Geodetic potency vs. earthquake numbers approach, 47
  discrete plots, 47
  linear relationship, 46–47
Geodetic strain and moment rates, estimation of
  from strain rate to moment rate, 43
  from velocities to strain rates, 42–43
Geomagnetic fields, 166
Geostationary satellites, 162–163
G-FAST algorithm, 114–115, 117–119, 122–123
GIPSY-OASIS II software, 113
Glacial isostatic adjustment (GIA), 180, 215, 219, 276–277, 292–293, 294f
  deformation and associated stress changes, 264–266
  deformation rates, 289–291
  and earthquake occurrence, relationship between, 266
  in Iceland, 288–291
  mantle viscosity, 274

viscoelastic response of Earth to cryospheric change
  Antarctica, 229–230
  deglaciated regions, 224–225
  GNSS measurements, 222–223, 223f
  Greenland, 230–231, 230f
  horizontal deformation, 221, 223
  ice-covered areas, viscoelastic deformation in, 226–229, 227f, 229f
  modeling, 220
  past cryospheric change, global-scale response to, 225–226
  problem, components of, 221–222
  spatial distribution of present-day GIA-related uplift and subsidence, estimate of, 220, 221f
  time-decaying process, 221
Glacially induced faults (GIFs), 265f
Glacially triggered earthquakes, 266
Glacier mass loss, 287
Glacier motions, measurements of, 6
GlarmS algorithm, 118–119, 122
Global ionospheric map (GIM) files, 162
Global mapping function (GMF), 190
Global navigation satellite system (GNSS), 1, 161, 288
  and climate change
    earthquakes and volcanism, 6–7
    glacier motions, measurements of, 6
    long- and short-term responses, 5
  for earthquake early warning (see Earthquake early warning (EEW))
  Earth's deformation measurements
    offshore, 3–5, 3f
    onshore, 2–3
  future of, 7
  GNSS radio occultation (GNSS-RO)
    advantages of, 32
    data processing, 31
    physical principle of, 31, 31f
  ground-based ionosphere sounding, 28–30
  ground-based troposphere sounding, 26–28
  ionospheric seismology/volcanology, GNSS-TEC technique (see Ionospheric seismology/volcanology)
  landslide monitoring, technique for (see Landslides)
  measurements (see Measurements, GNSS)
  meteorology
    applications, 189
    features, 189
    weather and climate change monitoring, role in (see Weather and climate change monitoring, GNSS meteorology in)
  positioning
    carrier-phase-based relative positioning (see Relative positioning)
    precise point positioning (PPP), 18–22, 19f
    real-time GNSS, 24–26
  for sea level measurements, 186
    coastal GNSS-IR installations, with single antennas, 183–186, 183–185f
    coastal GNSS-R installations, with two/more antennas, 181–183, 181–182f

    reflected GNSS signals, 180–181, 180f
    sensing sea level variability with GNSS, 185–186
    traditional tide gauge measurements, 179–180, 180f
  sites, 1
  vs. synthetic aperture radar (SAR), 1, 2t
  transient deformation, in tectonic plate boundary zones (see Transient deformation, in tectonic plate boundary zones)
  for tsunami early warning (see Tsunami early warning (TEW))
  unconventional use of, 5
Global navigation satellite system reflectometry (GNSS-R), 32, 231
  based on dual antennas, 34–35, 34f
  coastal GNSS-R installations, with two/more antennas, 181–183, 181–182f
  for flood inundation mapping, 205
    amplitude metrics, 207–208
    challenges, 210, 210f
    coherency metrics, 208–210
    delay-Doppler maps (DDMs), 207, 207f
    instruments, 205
    sensitivity of surface-reflected signals, reasons for, 206–207
    surface reflectivity, CYGNSS observations of, 205, 206f
  hurricane wind speed retrieval, 210–212, 211–212f
Global'naya NAvigatsionnaya Sputnikovaya Sistema (GLONASS), 1, 20–22, 21f
Global positioning system (GPS), 1, 20–21, 21f, 26, 161, 163
Global pressure and temperature 3 (GPT3), 190, 192
Global pressure and temperature (GPT) model, 190
Global strain rate model (GSRM), 47, 51, 51f
Global warming, 289
GNSS. See Global navigation satellite system (GNSS)
GNSS interferometric reflectometry (GNSS-IR), 32
  coastal GNSS-IR installations, with single antennas, 183–186, 183–185f
  for cryosphere, 231, 235
    ice mass balance, 233, 233f
    permafrost areas, freeze and thaw movements in, 233–234f, 234
    principles and methodology, 231–232, 232f
    snow depth, 232–233, 233f
  single antennas, based on, 32–34, 33f
GNSS radio occultation (GNSS-RO)
  advantages of, 32
  data processing, 31
  physical principle of, 31, 31f
Gorkha earthquake, 99–100, 269
GPS effective isotropic radiated power (GPS EIRP), 208, 210
GRACE Follow-On (GRACE-FO), 215, 259–261

Gravity Recovery and Climate Experiment
   (GRACE), 215, 259–261, 264–265
Great Salt Lake, 269
Greenland, 217, 220
   earthquakes, 268–269
   G/GFO mass change observations, 215
   glacial isostatic adjustment (GIA) models,
      230–231, 230f
   ice mass change, 224
   marine-terminating glaciers, dynamic ice loss
      from, 219
   permanent GNSS stations, 215
   surface mass balance (SMB) model, 218–219
Green's functions, 68, 73–74, 102, 119–122,
   217–219, 261–262
Grívstötn eruption, 148, 149f, 150, 152, 292f
Ground-based GNSS meteorology
   applications, 189
   features, 189
   weather and climate change monitoring, role
      in (see Weather and climate change
      monitoring, GNSS meteorology in)
Ground motion prediction equation (GMPE),
   114
Groundwater storage, climate change
   impact on, 269
GSRM. See Global strain rate model (GSRM)
Gutenberg-Richter earthquake frequency-
   magnitude distribution, 263
Gutenberg-Richter law, 43–44

H

Half-space loading models, 219
Helheim glacier
   annual dynamic ice loss, 219–220
   discharge, 216, 219–220
   mass loss of, 219
Hikurangi subduction zone, in New Zealand, 89t,
   93–94
   deep SSEs and interseismic coupling, 85–88,
      86f
   shallow SSEs and slow earthquakes, 85–88,
      87f, 90–91
Höfn (HOFN) site, in SE-Iceland
   east component of displacement, 292, 292f
   location of, 291–292, 291f
   vertical displacement, time series of, 291–292,
      291f
Hokkaido-toho-oki earthquake, 164, 164f, 167,
   167f, 169
Hurricane
   Harvey, 205, 212, 261, 272
   Michael, 212
   wind speed retrieval, GNSS-R for, 210–212,
      211–212f
Hurricane Weather Research and Forecasting
   (HWRF) model, 212

I

Ice age climate cycles, 264–267, 265f, 267f
ICE-5G model, 225
ICE-6G model, 225
Ice mass balance, 232–233, 233f

Ice mass loss, 267–268, 268f, 287
Ice melting, 258f, 268–269, 274–275
Ice unloading model, 296–297
Icy Bay, 268
IGW. See Internal gravity wave (IGW)
Illapel earthquake, 66–67, 66f, 101–102
Inclinometers, 252
Indian Regional Navigation Satellite System
   (IRNSS), 20–21, 21f
Inertial sensors, 112
Iniskin, Alaska, earthquake, 117, 123
InSAR. See Interferometric synthetic aperture
   radar (InSAR)
Integrated Global Radiosonde Archive Version 2
   (IGRA2), 192
Integrated Surface Database (ISD), 192
Integrated water vapor (IWV), 27, 189–193
   annual cycle of, 195–196, 196–197f
   diurnal cycle of, 194–195, 195f
   extreme weather events, 190–191, 193–194f
   interannual variations and associated climatic
      teleconnections, 197, 198f
   long-term trends and climate change,
      197–200, 199–200f
Interferometric synthetic aperture radar
   (InSAR), 67, 97–98, 192–193, 201, 215,
   243–244, 252, 261, 288
Inter-frequency biases (IFBs), 21–22
Internal gravity wave (IGW)
   propagating upward from focal area and
      tsunami, 163f, 164
   signatures, 164–165, 164f, 167
International GNSS Service (IGS), 162, 189
International Ocean Discovery Program
   (IODP), 90
International Terrestrial Reference Frame
   (ITRF), 18, 226
Ionospheric disturbance, by earthquakes and
   volcanic eruptions, 161
   acoustic wave (AW), 163–165, 163–164f
   Asama volcanic eruption, 161, 168
   dense GNSS array, 161, 168
   direct acoustic waves from epicenters, 163,
      163f
   Doppler sounding, 161
   Hokkaido-toho-oki earthquake, 164, 164f
   internal gravity wave (IGW), 163f, 164
      propagating upward from focal area and
         tsunami, 163f, 164
      signatures, 164–165, 164f, 167
   moment magnitudes (M_w), from disturbance
      amplitudes, 166–167, 167f
   Soufriére Hills eruption, 168
   TEC changes, 161–162, 168
   type 1 disturbance, 168–171, 168f,
      170–171f
   type 2 disturbance, 168–169, 168f, 171–174,
      172–173f
Ionospheric piercing points (IPP), 162
Ionospheric seismology/volcanology
   acoustic wave (AW), 163–165, 163–164f
   direct acoustic waves from epicenters, 163,
      163f, 165–166, 165–167f
   GNSS-TEC observations

   earthquakes and volcanic eruptions, finding
      signals related to, 162
   multi-GNSS, 163
   phase difference and TEC, 161–162
   from STEC to VTEC, 162
   internal gravity wave (IGW)
      propagating upward from focal area and
         tsunami, 163f, 164
      signatures, 164–165, 164f, 167
   ionospheric disturbance, by volcanic
      eruptions (see Ionospheric disturbance,
      by earthquakes and volcanic eruptions)
   moment magnitudes (M_w), from disturbance
      amplitudes, 166–167, 167f
Ionospheric Volcanic Power Index (IVPI), 168
Iquique earthquake, in Chile, 91, 93f
Iran-Iraq border earthquake, 269
Isobaric mapping functions (IMF), 190
ISSM-SESAW, 261–262
IWV. See Integrated water vapor (IWV)
Izmit, Turkey earthquake, 68, 69f

J

Jakobshavn Isbræ (JI) glacier, 219
   dynamic ice loss, 219–220
   mass loss, 219
   weekly discharge, 219–220
Japan Meteorological Agency (JMA), 111–112
Japan Trench, 87f, 89t, 90
Jökulheimar (JOKU) station, Iceland
   east component of displacement, 292, 292f
   location of, 291–292, 291f
   vertical displacement, time series of, 291–292,
      291f

K

Kaikoura, New Zealand, earthquake, 93–94, 93f,
   117, 123
Kangerlussuaq (KG) glacier, 219–220
Karst geology, 272
Katla volcano, 296–297
Kelud (Indonesia) eruptions, 169–171, 170f
Kuchinoerabu-jima volcano, 172
Kumamoto earthquake, 94

L

Landers earthquake, 101–102, 271
Landslides
   cascading hazards, example of, 243
   deformation rate of, 243
   Funu landslide, in Democratic Republic of the
      Congo, 243, 244f
   global navigation satellite system (GNSS)
      Åknes landslide, Norway, 248, 248f
      Almenningar landslide, Iceland, 248–249,
         249f
      Cà Lita landslide, Italy, 250–251, 251f
      El Yunque, Puerto Rico, 249–250, 250f
      equipment and data processing, 245–247,
         246–247f
      and other deformation methods, 243–244,
         252
      instantaneously failing landslides, 244

Landslides *(Continued)*
   landslide motion, triggering factors for, 243–245
   movement types, 245, 245f
   multiparametric geophysical instrumentation, 252
   slow moving landslides, 244
   terminology, 244
   and their secondary effects, 243–244
Laplace equation, 68
L'Aquila earthquake, 68, 69f, 73
Large-scale traveling ionospheric disturbances (LSTID), 162
Last glacial maximum (LGM), 220, 224–225, 265f, 266, 276–277
Laurentide Ice Sheet, 224–225
Law of large numbers, 55
Least-squares ambiguity decorrelation adjustment (LAMBDA) method, 20
Least-squares collocation method, 42–43
Leica Spider software, 250–251
Light-time equation, 11
Linear combination observables, 12–15
   geometry-free combination, 14–15
   ionosphere-free combination, 14
   wide-lane- and narrow-lane combinations, 13–14
Line-of-sight geometry, 166, 172, 174
Lithospheric rheology, 83
LoadDef models, 261–262
Lomb-Scargle periodogram (LSP), 184, 232
Long-term average (LTA), 118–119
Long Valley caldera magmatic system, 287
Lorca earthquake, in Spain, 269
Low-frequency earthquakes (LFEs), 83–84, 90

M

Machine learning techniques, 111
Maduo, China, earthquake, 117
Magma lens, 140–141, 141f
Magma plumbing, 289
Magmatic processes, climate change influence on, 289, 289f, 297
   central volcanoes, in Iceland, 288, 288f
   glacial isostatic adjustment (GIA), 288
   glacier load changes at volcanoes, 289–291, 290f
   global sea-level change, influence on, 287
   Long Valley caldera magmatic system, 287
   magma emplacement, 295–296
   magma generation, 288, 293–295, 295f
   pressure change effect, 287
   stability of shallow magma body, ice retreat influence on, 296–297, 296f
   surface load variation, 287
   Weichselian glaciation, Iceland, 288
Magna earthquake, 271
Mantle melting model, 293
Manual file transfer protocol (mftp), 247
Mass eruption rates (MER), 170–171
Maule earthquake, in Chile, 73–76, 75f, 101–102, 117
Maximum likelihood estimation, 44, 48
Mean sea level (MSL), 192

Measurements, GNSS
   error sources, 15f
      receiver-related errors, 17–18
      reference frame misalignment, 18
      satellite-related errors, 15–16
      signal propagation-related errors, 17
      tide displacement, 18
   observation equations
      carrier-phase measurement, 12
      linear combination observables, 12–15
      pseudorange measurements, 11–12
   offshore deformation measurements, 3–5, 3f
   onshore deformation measurements, 2–3
Medium-scale traveling ionospheric disturbances (MSTID), 162
Melbourne-Wübbena combination, 15, 20
Merapi (Indonesia) eruptions, 169–171, 170f
Metropolis algorithm, 71–72
Mexican hut wavelet, 164–165
Mexican Seismic Alert System (SASMEX), 111
Microwave sensors, 205
Middle America subduction zone, in Mexico, 85–88, 86f
Midgaard glacier, 216, 219
Minimum scalloping, 162
Mining-related earthquake, PGD studies, 117
Miyakejima eruption, 149, 150f
Modified SLM (MSLM), 29–30
Mogi model, 131–132
Monte Carlo Markov chain (MCMC), 71–72
Mount St Helens volcanic eruption, 149, 161
Multifrequency GNSS positioning, 20–22, 21f
Multifrequential periodogram, 263
Multipaths, 5
Multiple GNSS (multi-GNSS), 162–163, 189, 200
Multi-time window approach, 73–74, 74f
Mýrdalsjökull/Eyjafjallajökull volcanoes, 219
Mýrdalsjökull ice cap, in S-Iceland, 296–297
MyShake, 111–112

N

Nankai subduction zone, in Japan, 83–84, 89t
   deep SSEs, 88–90
   shallow SSEs, 85–88, 87f, 90, 94
Napa earthquake, 117
Navier's equation, 130, 132
Navigation with Indian Constellation (NavIC), 1, 163
New Madrid earthquake, 45
Niell mapping functions (NMF), 190
North Anatolian fault in Turkey, 50–51, 53, 91
North Atlantic Oscillation (NAO), 197, 216
Northern Oscillation Index (NOI), 197, 198f
Northridge earthquake, 161
Numerical models, 296
Numerical weather model (NWM), 190, 192

O

Observable-specific biases (OSBs), 18–20, 22
Observation space representation (OSR), 24
Ocean bottom pressure gauges (OBPG), 67
Okmok eruption, 150–151, 151f

Onsala Space Observatory, 179, 180f, 183
Optical imaging monitoring techniques, 252

P

Pacific Tsunami Warning Center, 113
Pamir-Hindu Kush region, 48
Papanoa earthquake, in Mexico, 91
Papkovich-Boussinesq displacement potentials, 132
Pareto distribution, 44, 55–56
Pareto index, 44, 55–56
Parkfield earthquake, 66–67, 101
Partial water vapor pressure, 148
PCOs. *See* Phase-center offsets (PCOs)
PCV. *See* Phase-center variation (PCV)
Peak ground displacements (PGD) scaling, for EEW
   Chilean megathrust events, 117
   coefficients, 114, 115t
   distance weighting, 115–116
   Durres, Albania, earthquake, 117
   exponential distance weighting, 115
   Greece, regional coefficients, 117
   ground motion prediction equation (GMPE), 114
   Iniskin earthquake, 117
   Maduo, China, earthquake, 117
   magnitude-dependent attenuation, 114
   mining-related earthquake, 117
   Napa event, regression analysis, 117
   Network of the Americas (NOTA), earthquakes in, 117
   Pedernales, Ecuador, earthquake, 117, 118f
   PGD dataset, log residual between, 115–116, 116f
   principle of, 114
   signal-to-noise ratio criterion, 115–116
   single-magnitude estimate, methods for, 115
   three-component displacement waveforms, 114
   travel time mask approach, 114–115
   weak determinism, 114–115
Peak-to-peak amplitude (PPA), 194
Pedernales, Ecuador, earthquake, 72, 72f, 117, 118f
Permafrost, 233–234f, 234
Permanent service for mean sea level (PSMSL), 179–180, 186
Peru-Chile Trench, 90
Phase-center offsets (PCOs)
   for receiver antenna, 17–19, 17f
   for satellite antenna, 15–16, 16f, 18–19
Phase-center variation (PCV)
   for receiver antenna, 17–19, 17f
   for satellite antenna, 15–16, 16f, 18–19
Phase wind-up effect, 12, 16
Pinatubo volcanic eruption, 161, 168–169
Plinian-type continuous eruptions, 168–170
PLUM algorithm, 111, 122–123
Point-source approximation, 133, 134f
Poisson rate, 48–49
Poisson's ratio, 101
Polar Earth Observing Network (POLENET), 215

Pore pressure, 261
　change in, 263
　diffusion, 271–272
　in fault zone, 275
Poroelastic deformation/rebound, 101–102
Poroelastic stress change model, 263
Postseismic deformation and GNSS, 83
　afterslip, 99–101
　availability of datasets, 96–97
　fundamental science and seismic hazard assessment, use in, 95
　Green's functions, 102
　interferometric synthetic aperture radar (InSAR), 98
　interseismic deformation, 102
　observations at great distances from earthquakes, 97–98, 97f
　poroelastic deformation/rebound, 101–102
　rheology-related questions, 95
　role in triggering earthquakes and aftershocks, 95
　signals within daily GNSS timeseries, 95–96, 96f
　variety of, 94, 95f
　viscoelastic relaxation, 94, 98–99, 100f, 102
Potsdam mapping factors (PMF), 190
Power-law (PL) model, 198, 200
Power ratio (PR), 209
Precipitable water vapor (PWV), 27, 191
Precise point positioning (PPP), 19f, 112–113
　continuously operating reference stations (CORS), 18
　ionosphere-free combination observable, 18–19
　multiconstellation and multifrequency GNSS positioning, 20–22, 21f
　PPP ambiguity resolution (PPP-AR), 20, 113
　real-time PPP, 25
　　convergences of, 25–26
　　service, flowchart of, 25, 26f
　vs. relative positioning, 24
　standard single-point positioning, 18
　static and kinematic PPP, 19–20
　undifferenced, uncombined PPP model, 19
Preliminary reference earth model (PREM), 217–218, 261–262
Probabilistic seismic hazard assessment (PSHA), 42
Probability density function (pdf), 71–72
Probability mass function (pmf), 48–49
Product distribution layer (PDL), 122–123
Propagator matrix method, 137
Pseudorange measurements, 11–12
P-wave energy analysis, 77
Pylith, 261–262

Q

Quasistatic equilibrium equation, 130
Quasi-zenith satellite system (QZSS), 1, 14, 20–21, 21f, 163

R

RACMO2.3p2 model, 217–218
Radio frequency interference (RFI), 210
Radiometers, 205
Radiosondes, 189, 192
Rainfall-triggered swarm seismicity, 272, 273f
Rayleigh surface wave (RW), 163–164, 163–164f, 169
Ray-tracing model, 28–29
Real-time GNSS
　IGS real-time products, 24, 25t
　positioning, from historical perspective
　　accuracy and noise characteristics, 114
　　PPP with ambiguity resolution (PPP-AR), 113
　　precise point positioning (PPP), 112–113
　　relative positioning, 112–113
　　RTD software, 113
　　triangulated relative positioning method, 113
　real-time PPP, 25
　　convergences of, 25–26
　　service, flowchart of, 25, 26f
Real-time kinematic (RTK) positioning, 23–24
Real-time service (RTS), 24
REAR, 261–262
Receiver-independent exchange format (RINEX) files, 162
Receiver-related errors
　receiver antenna PCO and PCV, 17, 17f
　receiver clock offset, 17
　receiver hardware delay, 18
　receiver observation noise, 18
Recurrent neural networks (RNN), 111
Redoubt eruption, 150–151, 151f
Regional Navigation Satellite System (RNSS), 1
Relative positioning
　basics of, 22–23, 23f
　vs. precise point positioning (PPP), 24
Relativistic effects, 16
Reverse fault earthquakes, 166–167
RHtestV4 software, 199–200
Ridgecrest, California, earthquake, 119, 271
Right-hand circular polarized (RHCP) antenna, 180–183
Root-mean-square difference (RMSD), 211, 211f
Root-mean-square (RMS) error, 183
RTD software, 113
RTKLIB software, 247, 249–251, 249f
Ryukyu Trench, 88, 89t

S

Saffir-Simpson hurricane scale (SSHS), 193, 193f
Sagnac effect. See Earth rotation effect
Sakurajima volcano, 145, 149–150, 152, 171, 172f
San Andreas fault, 51, 53, 91, 94, 266–267, 267f, 270f
SAR. See Synthetic aperture radar (SAR)
Sarychev Peak volcano, 169
SASMEX. See Mexican Seismic Alert System (SASMEX)
Satellite laser ranging (SLR), 2
Satellite-related errors
　Earth rotation effect (Sagnac effect), 16
　phase wind-up effect, 16
　relativistic effects, 16
　satellite antenna PCO/PCV, 15–16, 16f
　satellite clock offset, 15
　satellite hardware delay, 16
　satellite orbit error, 15
Schuster spectrum, 263
Schuster test, 263
Seafloor deformation, Tohoku-oki earthquake, 4, 4f
Sea level changes, 215
　climate change, effect on, 179
　GNSS, for sea level measurements, 186
　　coastal GNSS-IR installations, with single antennas, 183–186, 183–185f
　　coastal GNSS-R installations, with two/more antennas, 181–183, 181–182f
　　reflected GNSS signals, 180–181, 180f
　　sensing sea level variability with GNSS, 185–186
　　traditional tide gauge measurements, 179–180, 180f
　viscoelastic response to cryospheric change, 222, 225
Seasonal hydrological and atmospheric loads, 269–271
Sea surface height (SSH), 181
Seismic algorithms, 112
Seismic array analysis, 76, 76f
Seismic moment distribution, 43–45
　corner moment, 44
　cumulative seismic moment distribution
　　analytic results, for each regimes, 55–56
　　approximation equations, 56–58, 56–57f
　　existence of two regimes, 54–55, 54–55f
　Gutenberg-Richter law, 43–44
　long-term regional/local moment, ratio of, 44–45
　long-term seismic moment, 44
　maximum likelihood estimation, 44
　Pareto index, 44
　tapered Pareto distribution, 44
　upper incomplete gamma function, 44
Seismic-to-geodetic moment ratio, 42
　approach, 46
　continental plate boundaries, 45
　definition, 45
　mid-oceanic ridges, analyses of, 45
　subduction zones, 45
　uncertainties, 45
Seismogenic depth, 43
ShakeAlert system, 111, 122–123
Shallow afterslip, 99–101
Shallow slow slip events, 85–88, 87f, 89t, 90–91
Shear strain rate, 42
Shear stress, 136, 136f
Shinmoedake eruption, 148
Shin-Moe volcano, 172, 174
Short-term average (STA), 118–119
Sichuan earthquakes, 101
Sidereal filtering, 72
Signal propagation-related errors
　ionospheric delay, 17
　multipath effects, 17
　tropospheric delay, 17

Signal-to-noise ratio (SNR), 32–34, 150–151, 181, 183–184, 207, 209–210, 231–232
Silent earthquakes. See Slow slip events (SSEs)
Single-difference analysis, 183
Single layer model (SLM), 29–30
Singular spectrum analysis (SSA), 216, 219
Skempton coefficient, 263
Slant TEC (STEC), 29–30, 161–162
    time series, of GPS satellite 6 (G06), 164, 164f
    to vertical TEC (VTEC), 162
Slant total delay (STD), 190
Slant water vapor (SWV), 27–28
Slant wet delay (SWD), 27–28
Slow moving landslides, 244, 246, 246f
Slow slip events (SSEs), 77–78
    accommodation of plate motion, role in, 83
    in circum-Pacific region, 83, 84f
    definition, 83
    episodic slow slip events (see Episodic slow slip and creep events)
Slow-slipping faults, 266–267
Smoothing approach, 70
Snow depth, 232–233, 233f
Snow water equivalent (SWE), 232–233
SNR. See Signal-to-noise ratio (SNR)
Soufriére Hills eruption, 168
South Napa earthquake, 271
Spatial sampling, of GNSS-R observations, 210
Spherical elastic Earth model, 261–262
SPOTL, 261–262
Spring-slider rate-and-state model, 275, 276f
SSEs. See Slow slip events (SSEs)
Standard single-point positioning, 18
State space representation (SSR) errors, 24
STATIC1D, 261–262
Statistical algorithms, 209
Statistical analysis, 263
Storegga landslides, 243
Strain gauges, 252
Strike-slip earthquakes, 166–167
Strike-slip faults, 42
Subionospheric points (SIP), 162
Sumatra-Andaman earthquake, 163–165, 169
Sumatran Global Positioning System (GPS) Array (SuGAr), 100f
Superstition Hills fault, 91, 92f, 94
Surface-breaching afterslip, 101
Surface mass balance (SMB) model, 215, 217–219
Surface reflectivity (SR), 205, 206f, 208
Surveying, 1
SWD. See Slant wet delay (SWD)
SWV. See Slant water vapor (SWV)
Synthetic aperture radar (SAR), 1, 2t, 142–143, 152, 205

T

Tapered Pareto distribution, 44, 46–47, 54–57, 54f
Technology Demonstration Satellite-1 (TechDemoSat-1), 7
Tectonic tremor, 83–84, 88, 90
Tehuantepec, Mexico, earthquake, 119

Teleconnection, 197, 198f
TERRASAR-X, 252
Terrestrial laser scanning (TLS), 249–250, 252
Terrestrial surveying, 1
TEW. See Tsunami early warning (TEW)
Thomson-Haskell propagator matrices, 119
Three-dimensional (3D) GIA model, 293
Three Sisters volcano, 143, 144f, 152
Thresholding, 209
Tidal analysis, 185
Tiltmeters, 252
Time series analysis, 263
Tohoku-oki earthquake, in Japan, 67–68, 69f, 73, 77–78, 90–91, 112
    afterslip, 99–100
    ionospheric disturbance, 161, 164, 167, 167f, 169
    postseismic deformation, 100f, 101–102
    seafloor deformation, 4, 4f
    slow slip events, 91, 94
Topcon Tools software package, 249–250
Total atmospheric pressure, 148
Total electron content (TEC), 12, 28–30, 148
Total water storage (TWS) measurements, 6–7
Trailing edge slope, 209
Transcrustal magma reservoir, 140–141, 141f
Transient deformation, in tectonic plate boundary zones
    episodic slow slip and creep events (see Episodic slow slip and creep events)
    postseismic deformation and GNSS, 83
        afterslip, 99–101
        availability of datasets, 96–97
        fundamental science and seismic hazard assessment, use in, 95
        Green's functions, 102
        interferometric synthetic aperture radar (InSAR), 98
        interseismic deformation, 102
        observations at great distances from earthquakes, 97–98, 97f
        poroelastic deformation/rebound, 101–102
        rheology-related questions, 95
        role in triggering earthquakes and aftershocks, 95
        signals within daily GNSS timeseries, 95–96, 96f
        variety of, 94, 95f
        viscoelastic relaxation, 94, 98–99, 100f, 102
Travel time mask approach, 114–115
Triangulated relative positioning method, 113
Trigonometric model hypothesis, 263
Tropical cyclones, 211–212
Tropospheric delay, 189–191
Tsunami early warning (TEW), 166. See also Earthquake early warning (EEW)
    algorithm development, 122–123
    Deep-ocean Assessment and Reporting of Tsunamis (DART), 111
    design elements, generalized flowchart for, 111–112, 112f
    GNSS time series for, 113

Japan Meteorological Agency (JMA), 111–112
    kinematic slip inversion, 119–122
    local and global seismic instruments, combination of, 111
    Pacific Tsunami Warning Center, 113
Tsunamis
    internal gravity waves, 164
    landslides, 243
2-D faults, 275
Typhoon
    Hagibis, in Japan, 272
    Haishen 2020
        daily average IWV and accumulated precipitations, 193, 193f
        GNSS IWV and hourly accumulated precipitations, 193, 194f
        pathway of, 193, 194f
    Morakot, in Taiwan, 272

U

Unifrequential Schuster periodogram, 263
University of Nevada-Reno (UNR) offsets, 118–119, 120f
Unmanned aerial vehicle (UAV), 67
Unmanned surface vehicle (USV), 67
Urakawa-oki earthquake, 161

V

Vatnajökull ice cap, Iceland, 220, 289, 295
Vatnajökull volcano, 219
Vertical land motion (VLM), 179–180, 180f, 186
Vertical TEC (VTEC), 28–30, 162, 171–172, 174
Very long baseline interferometry (VLBI), 2, 198–199, 201
Vienna mapping function 1 (VMF1), 190
Viscoelasticity, effect of, 137–138, 137–138f
Viscoelastic relaxation, 43, 94, 98–99, 100f, 102
Viscoelastic stress-modeling, 266–267
Volcanic explosivity index (VEI), 168–169
Volcanism, 6–7, 287–288
Volcano
    central volcanoes, in Iceland, 288, 288f
    decomposing observed deformation fields, 292–293, 294f
    glacial isostatic adjustment (GIA) model, 292–293, 294f
    glacier load changes at, 289–291, 290f
    ice mass loss, 287
    ionospheric disturbance (see Ionospheric disturbance, by earthquakes and volcanic eruptions)
    observed velocity, 292–293
    subglacial volcanoes, processes involved at, 289, 289f
    uplift and deformation, due to climate change and magma movements, 291–292, 291–292f
Volcano deformation, 3
    analytical model
        closed cylinder, 134, 135f
        dike and sill, 133, 134f

ellipsoidal source, 132, 133f
open cylinder, 135–136, 135f
spherical source, 131–132, 131f
Cartesian coordinate system, 130
caveats
  magma compressibility, 140
  multiple pressure sources, effect of, 140
  transcrustal magma reservoir, 140–141, 141f
  vertical and horizontal displacements, 133, 134f, 140
coeruptive deformation, 146–148
conduit, shear on, 136, 136f
flat Earth approximation, 130
location, of volcanoes, 142, 142f
material complexities, 136–139
  topography, effect of, 138–139, 139f
  vertical and lateral heterogeneities, effect of, 137
  viscoelasticity, effect of, 137–138, 137–138f
Navier's equation, 130
observational studies
  continuing observations, 152
  dense observations, of GNSS sites, 153
  modeling, 153
posteruptive deformation, 142, 148, 149f
pressure source and free surface, boundary condition on, 130–131, 130f
quasistatic equilibrium equation, 130
volcanic plumes, GNSS observations
  atmospheric disturbance, 148–150, 150f
  GNSS signal decay, 150–152, 151f
  total electron content (TEC), 148
during volcanic unrest, 142
  horizontal magma transport, 145–146, 147f
  magma accumulation, 142–144, 143–144f
  vertical magma transport, 144–145
Von Neumann entropy, 208
Voxel-based discretization model, 28

W

Water storage loss, 269
Water vapor, 191–193
  Earth's water and energy cycle, role in, 189
  GNSS IWV, weather and climate change monitoring, 189–193
    annual cycle of, 195–196, 196–197f
    diurnal cycle of, 194–195, 195f
    extreme weather events, 190–191, 193–194f
    interannual variations and associated climatic teleconnections, 197, 198f
    long-term trends and climate change, 197–200, 199–200f
  greenhouse warming effect, 189
  retrieval, 191–192
  and temperature, positive climate feedback effect between, 189
Wavelet transformation, 162, 164–165, 164f
Weak determinism, 114–115
Weather and climate change monitoring, GNSS meteorology in, 200–201
  data and methods
    tropospheric delay, 190–191
    water vapor retrieval, 191–192
  integrated water vapor (IWV), 189–193
    annual cycle of, 195–196, 196–197f
    diurnal cycle of, 194–195, 195f
    extreme weather events, 190–191, 193–194f
    interannual variations and associated climatic teleconnections, 197, 198f
    long-term trends and climate change, 197–200, 199–200f
Weather Research and Forecasting (WRF) model, 192–193
Weichselian glaciation, Iceland, 288
Wenchuan (China) Mw 7.92008 earthquake, 73, 76, 76f
White Noise (WN) model, 198, 200

Y

Young's modulus, 137

Z

Zenith hydrostatic delay (ZHD), 190–191
Zenith total delay (ZTD), 27, 190, 192–193
Zenith wet delay (ZWD), 190–191

www.ingramcontent.com/pod-product-compliance
Lightning Source LLC
Chambersburg PA
CBHW080750090425
24824CB00036B/2220